ULTRAVIOLET AND X-RAY SPECTROSCOPY OF THE SOLAR ATMOSPHERE

The solar atmosphere above the photosphere, the Sun's surface layers, is heated up to megakelvin temperatures and raised to a high level of dynamic activity through processes involving a pervading magnetic field. This book is concerned with one of the most important means of understanding the solar atmosphere, its ultraviolet and soft X-ray emission. The ultraviolet and X-ray spectra of the Sun's atmosphere provide valuable information about its nature – the heat and density of its various parts, its dynamics, and its chemical composition.

After a brief introduction, the book describes the principles governing spectral line and continuous emission, together with how spectral studies lead to deductions about physical properties. Spacecraft instrumentation from *Skylab*, the *P78-1* mission *SolarMax*, *Yohkoh*, *SOHO*, *TRACE*, and *Hinode* are described. With introductions to atomic physics and the diagnostic techniques used by solar spectroscopists, a list of emission lines in the ultraviolet and soft X-ray region, and a glossary of terms, this book is an ideal reference for graduate students and researchers in astrophysics and solar physics.

KENNETH J. H. PHILLIPS is a Research Fellow in the Department of Space and Climate Physics at the Mullard Space Science Laboratory, University College London. Prior to this he was a Senior Research Associate at the NASA Goddard Space Flight Center, and was involved in various spacecraft projects at the Rutherford Appleton Laboratory.

URI FELDMAN is employed by Artep Inc. a company specialising in Space Science and Plasma Physics research. In 1969 he was appointed as first director of the Wise Astronomical Observatory, Tel-Aviv University, responsible for its construction and operation. Between 1972 and his retirement from the Naval Research Laboratory, Washington, in 2001, he was involved with *Skylab*, *P78-1*, *Yohkoh* and *SOHO* space missions. In 1989 he was awarded the NRL E. O. Hulburt Annual Science and Engineering Award.

ENRICO LANDI received his Physics degree and Astrophysics Ph.D. Degree at the University of Florence, Italy, and has worked at the University of Florence, the Max-Planck Institut Für Aereonomie, Germany, and the Naval Research Laboratory, Washington, USA, where he is currently working. His interests cover X-ray, EUV and UV spectroscopy, atomic physics and solar physics.

Cambridge Astrophysics Series

Titles available in the series

7. Spectroscopy of Astrophysical Plasmas
edited by A. Dalgarno, D. Layzer

10. Quasar Astronomy
by Daniel W. Weedman

18. Plasma Loops in the Solar Corona
by R. J. Bray, L. E. Cram, C. Durrant, R. E. Loughhead

19. Beams and Jets in Astrophysics
edited by P. A. Hughes

22. Gamma-ray Astronomy 2nd Edition
by P. V. Ramana Murthy, A. W. Wolfendale

24. Solar and Stellar Activity Cycles
by Peter R. Wilson

25. 3k: The Cosmic Microwave Background Radiation
by R. B. Patridge

26. X-ray Binaries
by Walter H. G. Lewin, Jan van Paradijs, Edward P. J. van den Heuvel

27. RR Lyrae Stars
by Horace A. Smith

28. Cataclysmic Variable Stars
by Brian Warner

29. The Magellanic Clouds
by Bengt E. Westerlund

32. Accretion Processes in Star Formation
by Lee Hartmann

33. The Origin and Evolution of Planetary Nebulae
by Sun Kwok

35. The Galaxies of the Local Group
by Sidney van den Bergh

36. Stellar Rotation
by Jean-Louis Tassoul

37. Extreme Ultraviolet Astronomy
by Martin A. Barstow, Jay B. Holberg

38. Pulsar Astronomy 3rd Edition
by Andrew G. Lyne, Francis Graham-Smith

39. Compact Stellar X-ray Sources
edited by Walter H. G. Lewin, Michiel van der Klis

40. Evolutionary Processes in Binary and Multiple Stars
by Peter Eggleton

41. The Physics of the Cosmic Microwave Background
by Pavel D. Naselsky, Dmitry I. Novikov, Igor D. Novikov

42. Molecular Collisions in the Interstellar Medium 2nd Edition
by David Flower

43. Classical Novae 2nd Edition
edited by M. F. Bode, A. Evans

44. Ultraviolet and X-ray Spectroscopy of the Solar Atmosphere
by Kenneth J. H. Phillips, Uri Feldman, Enrico Landi

ULTRAVIOLET AND X-RAY SPECTROSCOPY OF THE SOLAR ATMOSPHERE

KENNETH J. H. PHILLIPS
Mullard Space Science Laboratory

URI FELDMAN
Artep Inc.

ENRICO LANDI
US Naval Research Laboratory

CAMBRIDGE
UNIVERSITY PRESS

CAMBRIDGE UNIVERSITY PRESS
Cambridge, New York, Melbourne, Madrid, Cape Town, Singapore, São Paulo, Delhi

Cambridge University Press
The Edinburgh Building, Cambridge CB2 8RU, UK

Published in the United States of America by Cambridge University Press, New York

www.cambridge.org
Information on this title: www.cambridge.org/9780521841603

First published 2008

Printed in the United Kingdom at the University Press, Cambridge

A catalogue record for this publication is available from the British Library

Library of Congress Cataloging-in-Publication Data
Phillips, Kenneth J. H.
Ultraviolet and x-ray spectroscopy of the solar atmosphere / Kenneth J. H. Phillips, Uri Feldman, Enrico Landi.
p. cm. — (Cambridge astrophysics series)
Includes bibliographical references and index.
ISBN 978-0-521-84160-3
1. Spectrum, Solar. 2. Ultraviolet spectroscopy. 3. X-ray spectroscopy.
I. Feldman, U. II. Landi, Enrico. III. Title.
QC455.P55 2008
523.7—dc22

2008008434

ISBN 978-0-521-84160-3 hardback

Contents

Preface

Even those not engaged in solar physics will have noticed a huge increase in space observations of the solar atmosphere over the past few years. The last ten years especially have seen several notable space missions launched by NASA, the European Space Agency (ESA), the Japanese and Russian space agencies, and several other organizations, among which have been the *Yohkoh* and *RHESSI* X-ray spacecraft, *SOHO*, *TRACE*, and *CORONAS-F* which have on board high-resolution instruments working in the extreme ultraviolet spectrum, and most recently the *Hinode* and *STEREO* missions, both launched in late 2006, all of which are making spectacular observations in the visible, ultraviolet, and soft X-ray regions. Major contributions to our knowledge have also been made by rocket-borne instruments such as SERTS and EUNIS, working in the extreme ultraviolet.

The increase in our understanding of the solar atmosphere giving rise to this emission has been enormous as a direct result of studying the data from these instruments. We have built on the knowledge gained from previous large solar missions such as the *Skylab* mission and *Solar Maximum Mission* to develop models for the solar atmosphere and for phenomena such as flares and coronal mass ejections. However, to the dismay of some but the excitement of most, we are now presented with a picture of the solar atmosphere that is far more dynamic and complex than we ever expected from early spacecraft or ground-based telescopes. Consequently, it has really been the case that as fast as we solve some problems, others are created that will obviously need great ingenuity in finding satisfactory physical explanations. A case in point is the ever-elusive coronal heating problem, one that has been with us ever since the megakelvin temperatures of the solar corona were discovered, from optical spectroscopy, in the 1940s.

Much of our knowledge has come from studying images of phenomena, monochromatic or in broad-band ranges. But quantitative work, just as with studies of stars, begins with spectroscopic studies. Ultraviolet and X-ray spectra give us information about temperatures, densities, flow velocities, filling factors (indicating to what extent features are being spatially resolved), thermal structure, element abundances, ionization state of solar plasmas in the outer parts of the solar atmosphere: the chromosphere, the corona and the enigmatic transition region. They also allow insight into all phenomena in the solar atmosphere, e.g. flares, jets, and prominences.

Though there have been review articles and portions of textbooks devoted to solar spectroscopy in the ultraviolet and X-ray regions, no monograph has yet been written that specifically addresses this aspect. This book is meant to fill this gap in the literature. First, we review basic concepts in studies of the solar atmosphere, with descriptions of the properties of the various parts of the solar atmosphere and phenomena (Chapter 1). Then basic concepts in radiation from the solar atmosphere, from the photosphere outwards, are given, with applications specific to ultraviolet and X-ray spectroscopy (Chapter 2). Basic atomic theory, needed in understanding spectral line and continuum formation and interpretation, is given in Chapter 3, with descriptions of how lines and continua are formed (Chapter 4). We expect these two chapters to be most useful for students, postgraduates and others who wish to have a detailed understanding of atomic physics and how it relates to spectroscopy. How narrow-band images and spectral line ratios particularly are used in 'diagnosing' parts of the solar atmosphere, i.e. finding densities, temperatures, and other parameters, is dealt with in Chapter 5, including specific examples of lines that are suitable for diagnostic purposes (Chapter 6). A description of current and recent instrumentation is given in Chapter 7. Later chapters (8, 9, 10) review recent literature giving applications of spectroscopic techniques in the ultraviolet and X-ray bands to the quiet and active Sun and the particular case of solar flares. As the literature is now so extensive, this review has necessarily been subjective and has quite likely omitted work that others will consider important. We apologize to any who feel that their work is not justly treated in these chapters. This book was prepared as the first results from *Hinode* and *STEREO* were first being made available and we have made an attempt to include some of them here, but unfortunately not much more than a brief mention.

One of the most significant findings using ultraviolet and X-ray spectroscopy in recent years has been the discovery of departures of element abundances in the solar atmosphere from those in the solar photosphere, with a clear link to the first ionization potential (FIP) of these elements. This result first became clear in the study of solar wind and solar energetic particles, but spectroscopically determined abundances showed that this was also true for the solar corona and maybe other parts of the solar atmosphere. A more personal view on this topic is given in Chapter 11, with additional evidence that is not yet in the published literature.

Notes on units (following the still current usage in solar physics, we use the c.g.s. system), lists of emission lines in the solar ultraviolet and soft X-ray spectrum, and a glossary are also given.

We are grateful to the many colleagues who have helped us in the task of writing this book. We thank those who have allowed us to use figures and other results from their published works; their names are acknowledged in figure captions. Data from many spacecraft missions are freely available from web-based sources, and we have taken advantage of obtaining images and spectra accordingly. We particularly acknowledge teams of scientists who have made these data sources available to us and to the solar community generally. We are also indebted to colleagues who offered their time to read portions of the book manuscript, and who have helped enormously by making very helpful suggestions or correcting our misconceptions. They are: A. K. Bhatia, J. L. Culhane, J. M. Laming, M. Landini, J. C. Raymond. Acknowledgement is made to colleagues who have worked on the CHIANTI database and software package, which is a collaborative project involving the US Naval Research Laboratory, Rutherford Appleton Laboratory, Mullard Space Science Laboratory, the Universities of Florence (Italy) and Cambridge (UK), and George Mason University (USA). We are grateful to our institutes, Mullard Space Science Laboratory (University College London) and Naval Research Laboratory (the US Department of the Navy) for their support during the writing of this book.

1

The solar atmosphere

1.1 Introduction

The solar atmosphere may be broadly defined as that part of the Sun extending outwards from a level known as the photosphere where energy generated at the Sun's core begins to escape into space as radiation. Regions progressively deeper than this level are characterized by increasing densities where there are continual interactions between atoms or ions and radiation. None of this radiation escapes into space. Such regions constitute the solar interior. The photosphere can loosely be regarded as the Sun's surface, though more precisely it is a thin layer. The base of the photosphere depends on the wavelength of the radiation. A common but arbitrary definition is where the *optical depth* τ, i.e. the integral of the absorption coefficient over path length, is unity for radiation of wavelength 5000 Å (corresponding to the green part of the visible spectrum), written $\tau_{5000} = 1$. We will give further definitions of these concepts in Chapter 2. The temperature of the $\tau_{5000} = 1$ level is approximately 6400 K. Much of the absorption of visible and infrared radiation, which makes up the bulk of the Sun's radiant energy escaping from the photosphere, is due to the negative hydrogen ion H^-. The absorption of photons of radiation $h\nu$ occurs by reactions like $h\nu + H^- \rightarrow H + e^-$, i.e. the breakup of the H^- ion to form a hydrogen atom and a free electron. At sufficiently high densities, the reverse reaction ensures the replenishment of H^- ions. The rapid fall-off of total density proceeding outwards from the Sun leads to a depletion of H^- ions, and thus a decrease in absorption coefficient and optical depth. The consequence of this is that the Sun appears to have a very sharp edge at optical wavelengths.

Energy is generated in the Sun's core by nuclear reactions, in particular fusion of four protons to form ^{4}He nuclei. The energy is transferred to the rest of the solar interior by radiation out to $0.667R_\odot$ where $R_\odot = 696\,000$ km is the solar radius, i.e. radius of the photosphere, and by convection from $0.667R_\odot$ to the photosphere. By the First Law of Thermodynamics, the temperature continuously falls with distance from the energy-generating region at the Sun's core. This fall-off of temperature would be expected to continue in the solar atmosphere, starting from the base of the photosphere where $\tau_{5000} = 1$. In fact the temperature rises, eventually reaching extremely large values, giving rise to an atmosphere that emits radiation at extreme ultraviolet and X-ray wavelengths. The nature of this emission, and what can be learned from it, are the subject matter of this book.

1.2 Chromosphere, transition region, and corona

Model solar atmospheres are a guide to the way in which temperature varies with height and help to provide definitions of atmospheric regions. In the well-known VAL models (Vernazza *et al.* (1981)), an initial temperature vs. height plot is used to calculate using radiative transfer equations of the emergent spectrum which is then compared with observed spectra, and adjustments to the initial temperature distribution made. According to their Model C (average solar atmosphere), there is a fall-off of temperature from the $\tau_{5000} = 1$ level, reaching ~4400 K at a height of about 500 km above $\tau_{5000} = 1$, the temperature minimum level. Above the temperature minimum, the temperature rises to form a broad plateau at ~6000 K, over a height range of approximately 1000–2000 km, then rises sharply. To account for large radiation losses due to hydrogen Ly-α emission, the VAL models require a small plateau at ~20 000 K on the sharp temperature rise. According to the model by Gabriel (1976), this rise continues on to temperatures of 1.4×10^6 K or more. For many authors, the solar *chromosphere* is defined to be the region from the temperature minimum up to where the temperature is approximately 20 000 K. The region of the solar atmosphere where the temperature reaches ~10^6 K and where the densities are very low compared with the chromosphere is the solar *corona.* A further region having temperatures of ~10^5 K known as the *transition region* is in solar atmospheric models a thin layer separating the dense, cool chromosphere and tenuous, hot corona. Figure 1.1 shows (solid curve) the

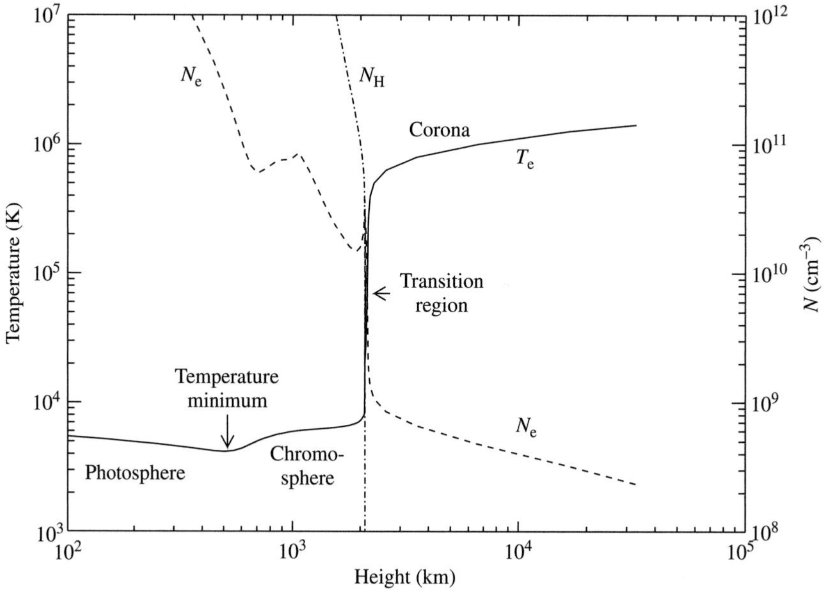

Fig. 1.1. The variation of temperature with height in the solar atmosphere (solid curve), based on one-dimensional model calculations of Vernazza *et al.* (1981), Fontenla *et al.* (1988), and Gabriel (1976). The chromospheric and transition region part of this model atmosphere is for the average quiet Sun. Based on these idealized representations of the solar atmosphere, the transition region is a thin layer with $T = 10^5$ K separating the chromosphere and corona. Variations in the number densities (cm^{-3}) of neutral hydrogen atoms (N_H) and electrons (N_e) (dot–dash and dash curves respectively) are also shown.

variation of temperature in the lower part of the solar atmosphere (heights up to 30 000 km above the $\tau_{5000} = 1$ level), together with variations in the number densities (cm^{-3}) of neutral hydrogen atoms (N_H) and free electrons (N_e), according to three quiet Sun models (Vernazza *et al.* (1981); Fontenla *et al.* (1990); Gabriel (1976)) which have been patched together over appropriate height ranges. Further details are given in Section 1.7 and Table 1.1.

The solar chromosphere is observable in visible wavelengths by forming images, known as spectroheliograms, in narrow wavelength bands around the cores of certain absorption or Fraunhofer lines that have their origins in the lower chromosphere. These include the hydrogen Balmer-α (or Hα) line at 6563 Å (red part of spectrum) and singly ionized calcium (Ca II) H and K lines at 3968 Å and 3933 Å (blue) respectively. The chromosphere also has a visible-wavelength emission-line spectrum, which is apparent at the beginning and end of the totality phase of total solar eclipses, when the Moon passes in front of the photosphere, covering up its intensely bright radiation. It is then clear that the chromosphere's extent is far greater than the 2000 km predicted by solar models and indeed its nature is far from being a uniform layer of the atmosphere. Rather, extending out from about 2000 km up to 10 000 km are transient pencil-shaped structures called spicules. In Hα spectroheliograms, they are dark features on the disk covering about 10% of the solar surface. Further, the white-light corona seen during total eclipses is observed to be highly non-uniform, consisting of streamers and loops that stretch out hundreds of thousands of kilometres above the limb. The white-light emission arises from the Thomson scattering of photospheric light by the fast-moving free electrons in the high-temperature corona. Observations, therefore, show that model atmospheres, while useful in describing the way in which temperature rises (and density decreases) with height, have limitations in that the high degree of structure in the chromosphere and corona is not properly reproduced. The nature of the transition region, predicted to be a thin layer separating the chromosphere and corona in models, may also be quite different in reality.

The bulk of the electromagnetic radiation from the chromosphere and corona has much higher photon energies than those of visible-wavelength radiation, which range from 1.6 eV to 3 eV. The characteristic radiation emitted by material with temperatures of between 10^5 K and a few MK has photon energies between 10 eV and a few 100 eV, corresponding to wavelengths between $\sim$1000 Å and less than $\sim$100 Å. The solar chromosphere and corona are therefore strong emitters of radiation with wavelengths between the ultraviolet and soft X-ray range.

The high temperatures of the chromosphere and corona, much exceeding the photospheric temperature, are plainly departures from those expected from physical considerations. A heating mechanism, due to some non-radiant energy source, is therefore required. This energy source, as well as the structured nature of the chromosphere and corona, is strongly correlated with the Sun's magnetic field, a fact that is readily observed: regions of the solar atmosphere hotter than their surroundings, as deduced from their ultraviolet and X-ray spectra, are highly correlated with regions in the photosphere associated with strong magnetic fields.

Below the photosphere, interaction between the atoms or ions and the radiation that they emit is complete, and thermodynamic equilibrium occurs. Towards the solar surface, some of this radiation begins to escape to space, and thus becomes visible to instruments on Earth. Analysis of the spectrum of this radiation shows that there is a local thermodynamic equilibrium (LTE), in which the emitted radiation is characterized by local values of temperature and density. Further out, there is an increasing departure from LTE, with the radiated energy from the photosphere interacting less and less with the surrounding material. Eventually, ions and electrons making up the hot corona, even though bathed in the strong photospheric radiation, are nearly unaffected by it.

The Sun is a second-generation star, formed by the coalescence of material left by first-generation stars in our Galaxy that underwent supernova explosions. This is reflected in the Sun's chemical composition, which is (by number of atoms) 90% hydrogen, nearly 10% helium, with trace amounts of heavier elements. In the chromosphere, with temperatures up to 10^4 K, the hydrogen is partially ionized, so that there are large numbers of free protons and electrons as well as neutral atoms of H and He. In the corona, the ionization of both hydrogen and helium is almost complete, so that the composition is free protons and electrons and He nuclei, with much smaller numbers of ions, including those of heavier elements. Thus, the corona is for many practical purposes a fully ionized plasma, having the associated properties that we discuss in Section 1.7.

1.3 The ultraviolet and X-ray spectrum of the solar atmosphere

The comparatively high temperatures and low densities of the solar atmosphere beyond the temperature minimum region result in a spectrum in the X-ray and ultraviolet ranges which at wavelengths less than ~1400 Å is made up of emission lines and continua. The lines and continua are due to specific ions or neutral atoms, which we will indicate by the appropriate element symbol and its degree of ionization. We use throughout this book a standard spectroscopic notation identifying the ionization degree with a roman numeral; thus C I indicates the spectrum of neutral carbon (read as 'the first spectrum of carbon'), C II indicates the spectrum of once-ionized carbon ions (C^+) ('the second spectrum of carbon'), and so on. The ions themselves are indicated by superscripted numbers showing the numbers of electrons removed, e.g. C^{+3} indicates three-times-ionized carbon, which emits the C IV spectrum.

Figure 1.2 (first panel) shows the ultraviolet disk spectrum of the quiet Sun in the range 800–1500 Å from the Solar Ultraviolet Measurements of the Emitted Radiation (SUMER) instrument on the *Solar and Heliospheric Observatory* (*SOHO*). On the logarithmic flux scale of this figure, the most striking feature is the presence of recombination edges in the continuum spectrum. In the range included, these are at 912 Å (the Lyman continuum edge, due to the recombination of hydrogen), 1101 Å (recombination to neutral carbon, C I), and 1197 Å (S I). At longer wavelengths, there are edges at 1527 Å (Si I), 1683 Å (Si I), 1700 Å (Fe I), and 1950 Å (Si I). Model atmosphere calculations, such as the Harvard Smithsonian Reference Atmosphere (Gingerich *et al.* (1971)) with subsequent corrections for line blanketing and absorption by molecules, give information about the location in the quiet Sun atmosphere where these continuum sources arise. Broadly, the continuum at wavelengths above the Si I edge at 1683 Å is emitted at photospheric layers, the region between the two Si I edges at 1527 Å and 1683 Å is emitted in the temperature minimum region, and the continuum shortward of 1527 Å is chromospheric in origin. Like the visible photospheric spectrum, spectroheliograms made in the ultraviolet continuum at $\lambda > 1683$ Å show limb darkening. Immediately below this wavelength, and continuing into the X-ray range, spectroheliograms in the continuum emission show limb brightening.

To the long-wavelength side of each recombination edge are members of line series that converge on each edge. Thus, members of the hydrogen Lyman line series converge on the Lyman continuum edge at 912 Å. The most prominent member of this series, and indeed the most intense (by two orders of magnitude) line of the entire solar ultraviolet spectrum, is the Ly-α line at 1215.67 Å. As this line can normally only be emitted in regions cool enough for

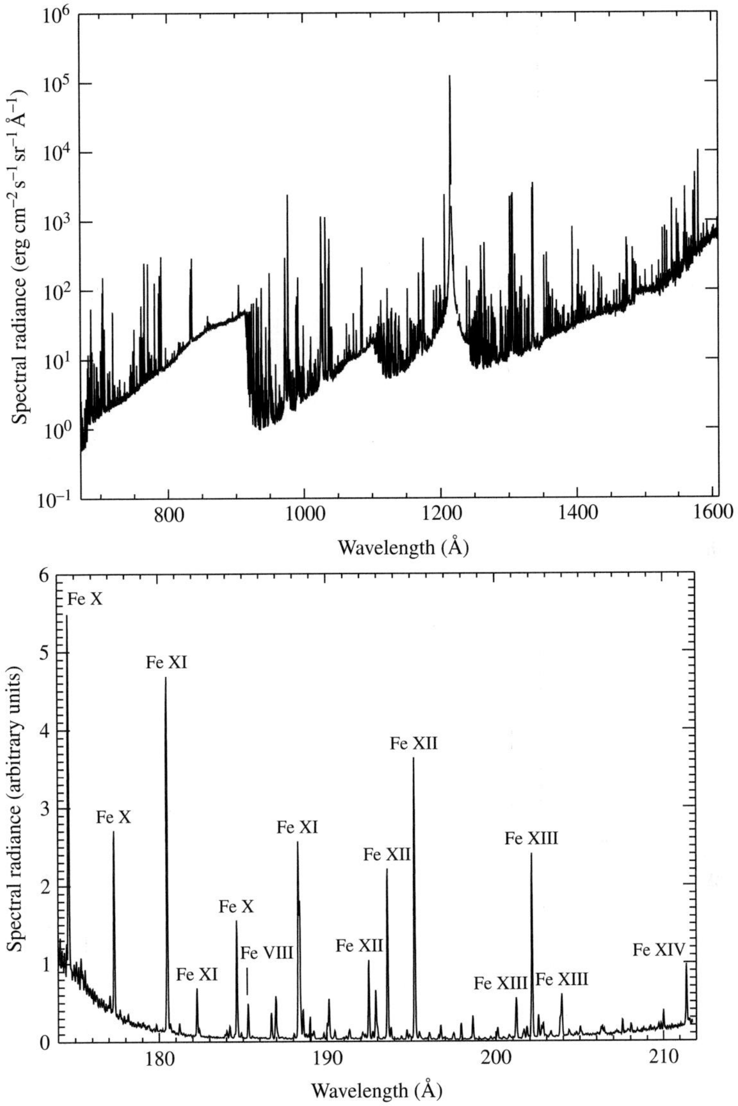

Fig. 1.2. Examples of solar ultraviolet, extreme ultraviolet, and X-ray spectra. (*First*) *SOHO* SUMER quiet Sun disk spectrum in the 675–1600 Å range, showing the Lyman lines (particularly Ly-α line at 1215.67 Å), the Lyman recombination edge at 911.8 Å and the C I recombination edge at 1101.2 Å. Courtesy *SOHO* SUMER team. (*Second*) Quiet Sun disk spectrum from the Extreme-ultraviolet Imaging Spectrometer (EIS) on the *Hinode* spacecraft, showing the diagnostically important lines of Fe ions between 174 Å and 212 Å, formed at coronal temperatures. Courtesy the *Hinode* EIS Team. (*Third*) Flare X-ray spectrum from the Flat Crystal Spectrometer on *Solar Maximum Mission*.

neutral hydrogen to exist, its emission has its origins almost entirely in the chromosphere, though Ly-α emission by recombination and resonant scattering has been observed in the high-temperature corona (Gabriel *et al.* (1971)). The high abundance of hydrogen results

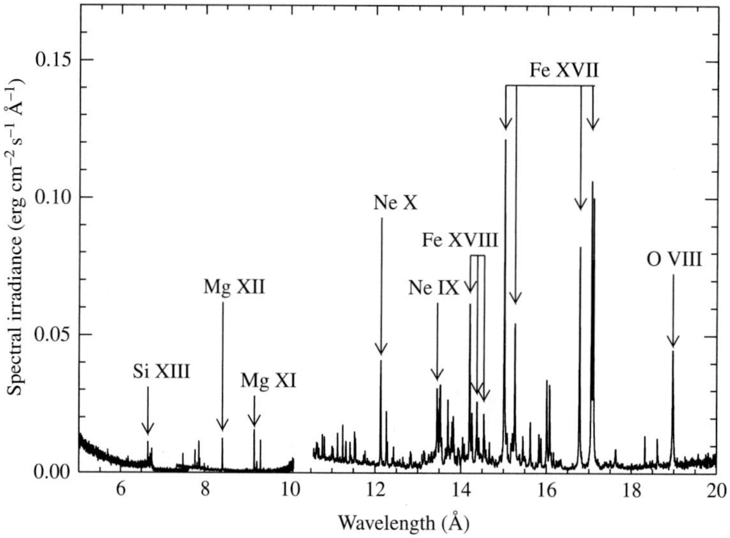

Fig. 1.2. *continued.*

in the line core being optically thick, and the line profile has a double reversal. There are two emission peaks separated by about 0.4 Å, with an absorption core that is formed high in the chromosphere, where the temperature is 40 000 K. The total width (FWHM) of the line is 0.7 Å, but the wings of the line, formed in the low chromosphere (temperature 6000 K), extend out to about 15 Å either side of the line core. The amount of the absorption dips in the line profile centre varies according to the solar feature emitting the Ly-α line – see Figure 1.3, based on measurements of Fontenla *et al.* (1988). Actually, the Ly-α line is in theory a doublet through the interaction of the spin of the single electron in hydrogen with the orbital angular momentum of the electron, but the separation of the two components (0.0054 Å) is far less than the width of the solar line profile, which is determined by Doppler (thermal velocity) and Stark (pressure) mechanisms as well as instrumental broadening (this is about 0.01 Å for the measurements shown in Figure 1.3). Other prominent lines in the Lyman series include Ly-β (1025.72 Å, also doubly reversed), Ly-γ (972.54 Å), and Ly-δ (949.74 Å). Helium emission lines are also prominent, those due to both neutral (He I) and ionized (He II) atoms. Since ionized helium is like hydrogen but with a nuclear charge of +2 instead of +1 atomic units, its spectrum consists of a sequence of Lyman-like spectral lines but with wavelengths that are a factor 4 smaller than those of H. The Ly-α line of He II is thus a prominent line at 303.79 Å, formed in the transition region. Neutral helium lines have a rather more complex sequence, with lines at 584.33 Å, 537.03 Å, 522.21 Å etc., all chromospherically formed.

Thousands of other emission lines occur in the 675–1600 Å quiet Sun spectrum (Figure 1.2, first panel), emitted in the chromosphere, transition region, and corona by ions of elements other than H or He. In the Appendix, we give a comprehensive list of these lines with wavelengths and identifications. The most intense emission lines are generally emitted by ions of abundant elements such as C, N, O, Ne, Mg, Si, S, Ar, Ca, and Fe. Broad groups of strong lines may be identified, such as the resonance lines of particular ions of different

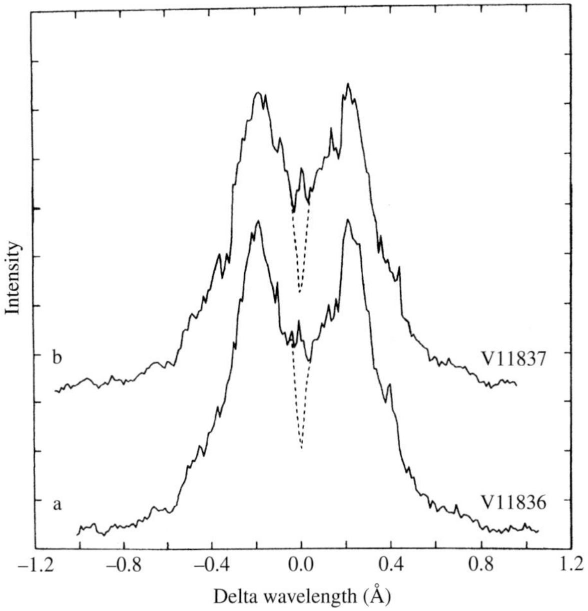

Fig. 1.3. Two examples of averaged quiet Sun Ly-α line profiles obtained by the Ultraviolet Spectrometer and Polarimeter on the *Solar Maximum Mission* spacecraft (the numbers on the plots are serial numbers of exposures from this instrument). The dashed curves are the raw data which are corrected (solid curve) for absorption by the geocorona. From Fontenla *et al.* (1988).

elements with common electron configurations. To define these spectral lines, we use a spectroscopic notation that describes both the atomic states and the transitions between them giving rise to the emission lines. We will use this notation here, but defer a full description to Chapter 3. Prominent among these lines are the resonance lines of Li-like ions, for which the transitions are $1s^2\,2s\,^2S_{1/2} - 1s^2\,2p\,^2P_{1/2,3/2}$. Because of the two possible orientations of the spin of the $2p$ electron in the excited state, the resonance lines are all doublets, with the lines having 2:1 flux ratios. The resonance lines in Li-like C (C IV) occur at 1548.20, 1550.77 Å, and are emitted at transition region temperatures (10^5 K). Similarly, the equivalent Li-like N (N V) lines are at 1238.82, 1242.80 Å. The Li-like O (O VI) lines (1031.91, 1037.61 Å) and Li-like Mg (Mg X) lines (609.80, 624.95 Å) have their origins in the corona (temperatures $\sim$1 MK). The separation of the doublet lines increases markedly with atomic number.

At shorter wavelengths the many prominent emission lines again include those originating in the chromosphere, transition region, and corona, up to temperatures of 2 MK. Figure 1.2 (second panel) shows a quiet Sun spectrum from the Extreme-ultraviolet Imaging Spectrometer (EIS) on the Japanese *Hinode* spacecraft, launched in 2006. The spectrum includes resonance lines of a range of Fe ions, from Fe VIII (formed at a little less than 1 MK) to Fe XIV ($\sim$2 MK). These strong lines arise from transitions involving electrons in the $n = 3$ shell. The Fe X line at 174.53 Å, emitted at 1 MK, and an Fe IX line at 171.07 Å (not shown in the figure), though always weak in Hinode EIS spectra because of the small effective area in this region, are important emission contributors to the so-called 171 Å filters of the *SOHO* Extreme-ultraviolet Imaging Telescope (EIT) and the *Transition Region And Coronal*

Explorer (*TRACE*) imaging instruments, which we will refer to frequently throughout this book. Nearby Fe XII lines, notably that at 195.12 Å, are emitted at a slightly higher temperature, 1.4 MK, and are the most important contributors to the EIT and *TRACE* 195 Å filters for the non-flaring Sun. In the quiet Sun spectrum shown in the figure, the principal lines of Fe X, Fe XI, and Fe XII are comparable to each other in flux, indicating an average temperature of 1.2 MK for the emitting region. At higher wavelengths, not included in this quiet Sun spectrum, is the strong Fe XV line at 284.16 Å, emitted at 3 MK, which is the only strong line within the range of the 284 Å EIT and *TRACE* filters.

At X-ray wavelengths, emission lines due to other highly ionized species exist, but are only strong when emitted by intense active regions on the Sun or more particularly solar flares. The ions emitting the lines are often those that have filled outer shells, such as the neon-like (electron configuration $1s^2\,2s^2\,2p^6$, i.e. $n = 2$ shell filled) or helium-like stages ($1s^2$, i.e. $n = 1$ shell filled). Thus, strong emission lines due to neon-like Fe (Fe XVII) occur at 15.01 Å, 16.78 Å, 17.05 Å, and 17.10 Å. Other strong lines include those due to helium-like C (C V) at 40.27 Å and helium-like O (O VII) at 21.60 Å. Such lines can be emitted by quiet coronal structures, with temperatures of about 1–2 MK, but become much more intense when emitted by active regions or flares, for which the temperatures are much higher still (Section 1.6). Figure 1.2 (third panel) shows the X-ray spectrum emitted by a hot flare plasma in its decaying phase observed with the *Solar Maximum Mission* Flat Crystal Spectrometer, a Bragg diffracting crystal spectrometer with several crystals covering a range extending from 1.4 Å to 22.4 Å. The Ly-α lines of H-like O (O VIII, 18.97 Å), Ne (Ne X, 12.13 Å), and Mg (Mg XII, 8.42 Å) are all prominent. Lines due to transitions between the $n = 2$ and $n = 3$ shells of Ne-like Fe (Fe XVII) are the most intense in this range, but there are also similar transitions in F-like Fe (Fe XVIII).

At the peak of strong flares, the hot plasma, with temperatures of 15–30 MK, is characterized by Fe ions in higher stages of ionization, up to the He-like (Fe XXV) or even H-like (Fe XXVI). Such ions need temperatures of ~20 MK and ~30 MK for their formation. The resonance line of Fe XXV, due to $1s^2 - 1s2p$ transitions, is at 1.85 Å, and is a prominent feature of X-ray spectra detected by crystal spectrometers on solar spacecraft missions dedicated to the study of flares. These include the US Air Force *P78-1* spacecraft, the NASA *Solar Maximum Mission* that flew in the 1980s, and the more recent *Reuven Ramaty High-Energy Solar Spectroscopic Imager* (*RHESSI*) with broad-band spectral resolution, having a range that extends downwards from ~4 Å (or in terms of equivalent photon energy units, energies above 3 keV). Throughout the entire X-ray range, and indeed throughout the EUV range, there is a continuum, emitted by free–free and free–bound processes. Relative to the lines, which become rather sparse in the higher-energy X-ray range, the continuum makes a larger contribution to the total emission. Also of increasing importance for X-ray line emission is the presence of *dielectronic* lines, formed by dielectronic recombination processes. An electron may recombine with an ion in a coronal plasma by a radiative process, in which the excess energy of the electron is removed from the atom by a photon of arbitrary energy, hence the formation of free–bound, or recombination, continuous emission. But resonance recombination processes also occur, in which an electron with a precisely defined energy recombines with an ion to produce a doubly excited ion, i.e. with two electrons instead of the normal one in excited shells. If one of these electrons de-excites, the resulting photon contributes to a dielectronic satellite line, so-called because the line generally accompanies a parent resonance line on the long-wavelength side. Such satellites are frequently an important way of diagnosing the flare plasma, giving information on its temperature in particular. This is discussed in Chapter 5.

1.4 Structure of the solar atmosphere

An important characteristic of the solar photosphere, which has consequences for the solar atmosphere, is convection. Convection occurs on large scales in the outer parts ($>0.667 R_\odot$) of the solar interior, where it dominates over radiation as the energy transfer process. Near the solar surface, a small-scale convection pattern consisting of cell structures or *granules* occurs, with typical sizes of ~100–1100 km and lifetimes of several minutes. Granules are visible in white-light images of the photosphere as slightly brighter areas with polygonal shapes that are characteristic of surface convection cells formed in the laboratory. Of more importance for our purposes are the *supergranules*, recognizable from the Doppler shifts of photospherically formed Fraunhofer lines. Material from the interior is convected upwards at the cell centre, then radially outwards and nearly horizontally with velocities of 0.4 km s^{-1} towards the cell boundaries, where it is convected downwards with velocities of about 0.1 km s^{-1}. Supergranules are thought to be pancake-like structures, approximately 30 000 km across and a few thousand km thick, and lasting from one to several days.

Spectroheliograms formed at the wavelengths of certain weak Fraunhofer lines, particularly the so-called G band (actually a band-head in the solar spectrum, at 4290–4310 Å, due to the CH molecule), show filigree-like strings of fine bright points, about 100 km across, in the dark lanes between granules. A pattern or network of bright points is recognizable, coincident with the supergranule boundaries. The bright points are associated with strong (~1000 G) magnetic field, probably in the form of magnetic flux tubes emerging from below the photosphere. The supergranular convective flow transports material and also concentrations of magnetic field towards the supergranule boundaries, so a network of magnetic field is formed, corresponding to the network of bright points. This field, with strengths of several hundred gauss, is detectable using the Zeeman effect, with magnetically sensitive absorption lines at visible wavelengths. The Michelson Doppler Imager (MDI) on *SOHO* routinely measures the line-of-sight component of the field, producing full Sun magnetograms such as those shown in Figure 1.4. The field may be positive (field lines directed towards the observer, or outwards from the Sun) or negative (field lines away from the observer, or towards the Sun), and exists all over the Sun's surface. (At the time of writing, the spectro-polarimeter part of the Solar Optical Telescope (SOT) on board the Japanese *Hinode* spacecraft is obtaining field strength and direction information, producing vector magnetic field maps.)

Magnetic field clumps at the photospheric level are associated with enhanced heating in the solar atmosphere, with consequent increased brightening at certain wavelengths in chromospheric and coronal features where the field is concentrated. The pattern of magnetic field concentrations in the quiet Sun results in a corresponding pattern of emission features formed in the chromosphere and transition region known as the quiet Sun chromospheric network. In visible wavelengths, the pattern can be seen in the Hα and ionized calcium (Ca II) H and K absorption lines. The cores of these lines are formed in the low chromosphere. In the Hα line, the network takes the form of groups of spicules or pencil-shaped features pointing out approximately radially from the photosphere. In the Ca II lines, the network (sometimes called the calcium network) consists of bright structures roughly delineating the

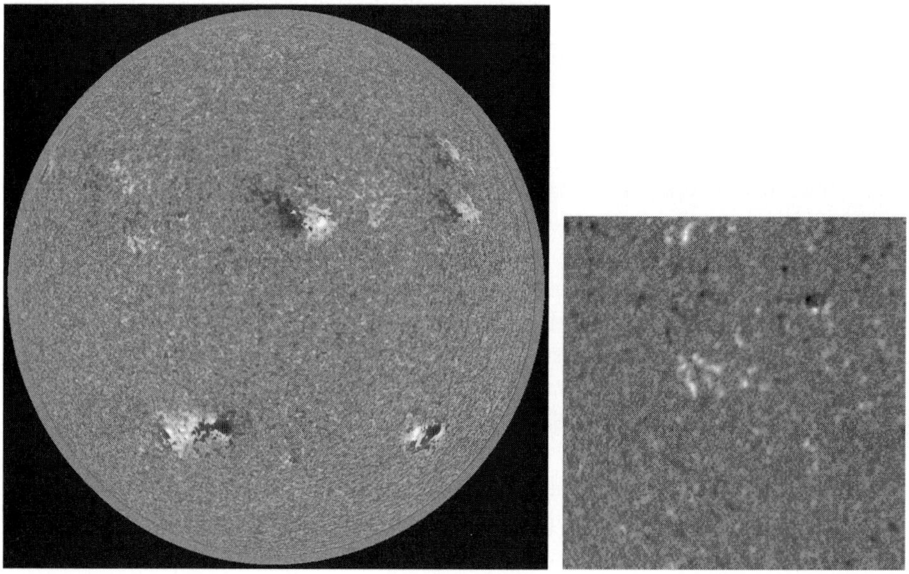

(a) (b)

Fig. 1.4. (*a*) Full Sun magnetogram from data obtained with the Michelson Doppler Imager on *SOHO* on 1997 September 11. Black areas indicate negative line-of-sight field concentrations; white areas positive line-of-sight field concentrations. A number of major active regions are evident, with concentrations of opposite field close to each other. (*b*) A portion of this magnetogram in the eastern part of the Sun, measuring 250 × 250 pixel^2, devoid of active regions but showing a mixture of positive and negative polarities. These are field concentrations at supergranule boundaries, and are present all over the Sun's surface away from active regions. The pixel size is 1 arcsecond, or 725 km. In both images (as in nearly all solar images in this book), north is at the top, east to the left. Courtesy the *SOHO* MDI Team.

supergranular cell boundaries, so appearing as a honeycomb structure over the whole Sun. It shows most clearly at the apparent centre of the Sun's disk as seen from Earth.

The network is especially prominent when viewed at the wavelengths of ultraviolet emission lines that are formed in the upper chromosphere (10^4 K) or at transition region temperatures ($\sim 10^5$ K), i.e. at greater altitudes above the photosphere than the Hα or Ca II H and K lines (see Figure 1.1). Thus, in the chromospherically formed emission lines of neutral helium (He I), e.g. the resonance line at 584 Å, the network over the entire Sun gives rise to a mottled appearance to the Sun, like the surface of an orange. This is illustrated by full Sun ultraviolet images from the SUMER instrument on the *SOHO* spacecraft (Figure 1.5). The He I resonance line is formed in the upper chromosphere, where the temperature is about 20000 K. The network has much the same nature at higher temperatures, 10^5 K, characteristic of the transition region.

With imaging spectrometers, such as the Coronal Diagnostic Spectrometer (CDS) on the *SOHO* spacecraft, the network can be traced at different temperatures forming images in extreme ultraviolet emission lines of ions having a large variety of characteristic temperatures. This is illustrated in Figure 1.6, where images in chromospheric lines (He I, He II), transition region lines (O III, O IV, O V, Ne VI), and coronal lines (Mg VIII, Mg IX, Mg X, Si VIII) are

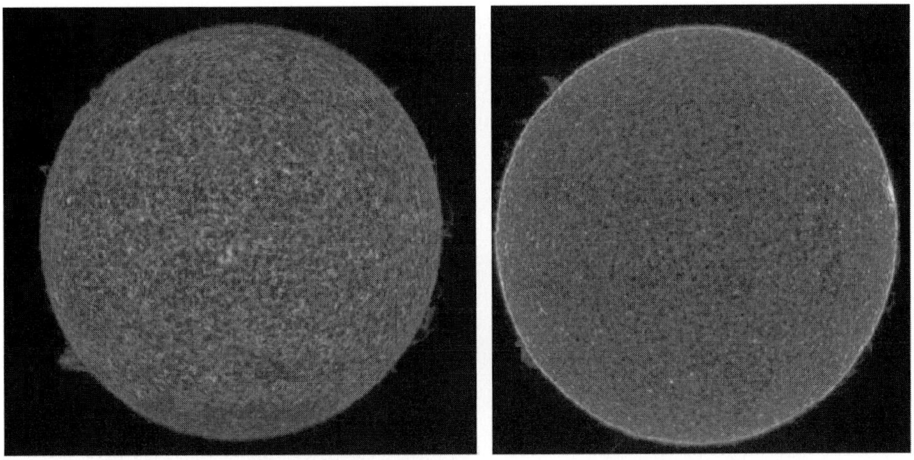

(a) (b)

Fig. 1.5. (*a*) Image of the Sun in the light of the chromospheric emission line due to He I at 584.33 Å, formed at 20 000 K, between 1996 March 2 and 4, near solar minimum. The image is actually a composite, made up of eight horizontal scans each in turn made up of 1600 exposures. The smallest picture element is 4 × 4 SUMER pixels (1 pixel = 1 arcsec, or 725 km). (*b*) C IV emission line at 1548 Å, formed at 10^5 K. Both images are composed of smaller images taken with the SUMER instrument between 1996 February 4 and 5. Courtesy the *SOHO* SUMER Team.

compared. The chromospheric and transition region images are made up of emission clumps of a few arcsec dimensions, the clumps forming the honeycomb structure of the network. In the coronal images, the network disappears, the main emission structures being bright but ill-defined structures. Note also that the helium line images have slightly less contrast than the transition region line images, the reason being (Gallagher *et al.* (1998)) that these lines have a very high optical depth owing to the large abundance of helium in the solar atmosphere – the line radiation suffers repeated absorption and re-emission before emerging from each clump of emission.

The nature of the network, its heating, and its progressively smaller contrasts at greater altitudes, have been modelled by Gabriel (1976) and others. Figure 1.7 illustrating Gabriel's (1976) model shows a cross-section through a supergranular cell from just beneath the photosphere to the corona. In this model, heating occurs along the field lines by magnetohydrodynamic (MHD) waves. The heating is greatest where the field lines are most concentrated, at the boundaries of the supergranules, in particular the network bright points. In the figure, isotherms indicate the temperature levels across the field concentrations, higher over the supergranule boundaries, but dipping over the region of the supergranule centres. The field lines, as they diverge outwards at greater altitudes, form a canopy. One can see from this model how the network at chromospheric and transition region temperatures (log $T = 4$ to 5, T in K) is correctly predicted to be well defined but increasingly fuzzy at higher, i.e. coronal (log $T = 6$ or more), temperatures.

In both the images shown in Figure 1.5, small protrusions are visible at the limb (e.g. the one at the south-east limb in the He I image): these are identifiable as *prominences*,

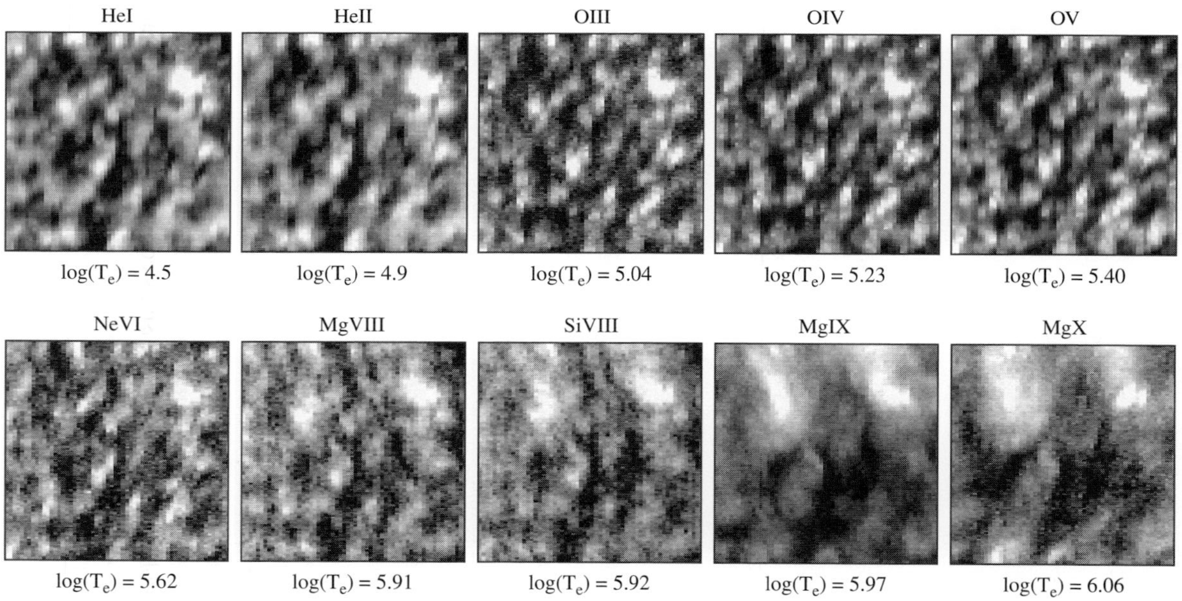

Fig. 1.6. Images from the *SOHO* Coronal Diagnostic Spectrometer (CDS) of the quiet Sun network, taken from the chromosphere (He I, He II lines), through the transition region (O III, O IV, O V, Ne VI) to the corona (Mg VIII, Si VIII, Mg IX, Mg X). These images were formed with the $4'' \times 240''$ arc slit of the CDS, with a scan mirror stepping 60 times in $4''$ arc steps with 50 s exposure times. From Gallagher *et al.* (1998).

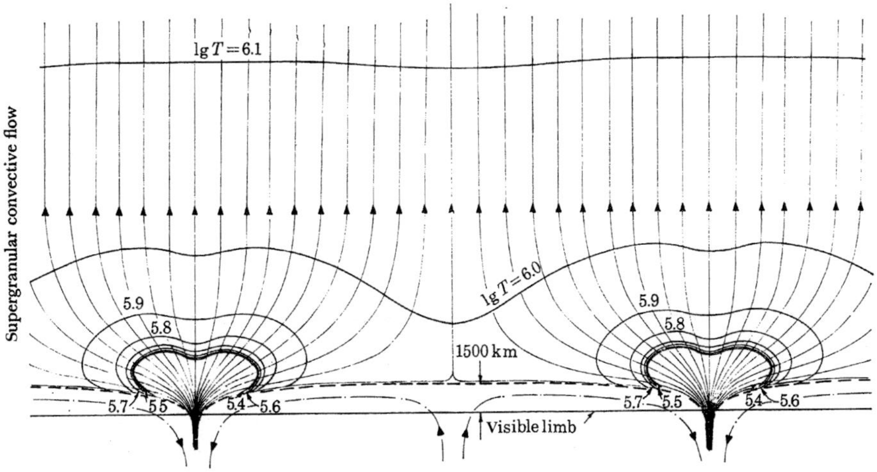

Fig. 1.7. Model of the chromosphere, transition region, and corona according to Gabriel (1976).

familiar as bright features in Hα spectroheliograms taken with ground-based telescopes. (The counterparts of prominences on the disk in Hα are known as filaments.) Quiescent prominences, like those in Figure 1.5, are those associated with quiet regions of the Sun, and generally extend up to ~50 000 km from the limb, with lengths up to several hundred thousand km. They are distributed over a wide latitude range, and may last for several solar rotations, i.e. several months.

The high-temperature corona is conspicuous at visible wavelengths during total eclipses, when the photosphere is obscured by the Moon, or with coronagraphs. Even then, only the portion of the corona off the limb is seen. With spacecraft instruments observing in the extreme ultraviolet and X-ray wavelengths, we are able to view the corona at any time that the instruments are operating and in addition we can see coronal structures on the solar disk. Figure 1.8 shows the solar corona on two occasions when the Sun was in an active state. The *SOHO* EIT image shows the corona in the 195 Å bandpass, dominated by the the Fe XII emission lines mentioned earlier. The image obtained with the *Yohkoh* SXT (thin aluminium filter, bandpass approximately 3–20 Å) shows the X-ray corona. In each image, the quiet Sun corona is seen to be made up of loop structures of varying sizes, the smallest being almost at the resolution limit of each instrument (down to 1 arcsec, or 725 km). At the poles there is often an extensive coronal *hole*, or region of decreased emission at extreme ultraviolet or X-ray wavelengths. Polar plumes, giving the Sun's corona a 'bar magnet' appearance at either pole, are visible during periods of low solar activity in white-light solar eclipse images as well as those in the extreme ultraviolet. Also within coronal holes, visible on the solar disk in X-ray images such as those from the *Yohkoh* SXT and ultraviolet images from EIT, are the so-called *bright points*.

The MDI instrument on *SOHO* and the SOT on *Hinode* are able to measure the solar magnetic field at the photospheric level. It is much more difficult to measure the magnetic field at higher levels, though some advances are being made using the Zeeman effect in magnetically sensitive coronal emission lines in the infrared. Meanwhile, it can be deduced that the

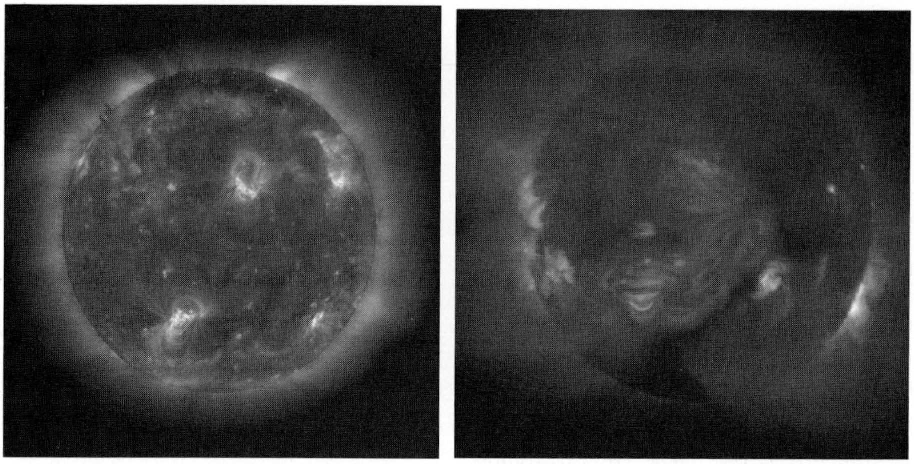

(a) (b)

Fig. 1.8. (*a*) Image of the Sun in the 195 Å passband of the EIT instrument on the *SOHO* spacecraft on 1997 September 11. The 195 Å passband includes the intense 195 Å Fe XII emission line formed at $T = 1.4$ MK, so this image shows coronal structures including the extensive loop arcades making up active regions and quiet Sun loops as well as loops and streamers off the limb. Note that the MDI magnetogram in Figure 1.4 corresponds to the date of this image. A coronal hole is evident at the Sun's north pole, within which are numerous bright points on the disk and, off the limb, super-radial polar plumes. A small prominence on the north-east limb appears as a fine dark structure. Courtesy the *SOHO* EIT Team. (*b*) Image of the Sun from the Soft X-ray Telescope on the *Yohkoh* spacecraft viewed through the thin Al filter (bandpass 3–20 Å) on 1992 May 16. In this image, the loop systems of active regions (e.g. that on the west limb, south of the equator), extensive loops and streamers at the limb, and a coronal hole at the south pole are visible. Courtesy the *Yohkoh* SXT Team.

loops and larger-scale arches in the corona are identifiable with magnetic flux tubes, the high-temperature coronal material being closely tied to the magnetic field lines. Most of the corona is in the form of these closed field structures, though coronal streamers, like their counterparts seen in white-light emission during eclipses, and the polar plumes in the coronal holes appear to have open field lines, or at least the field lines seem to connect to the heliosphere beyond the steady-state corona. These form the paths of the *solar wind*, a plasma consisting mostly of electrons and protons streaming outwards from the Sun. The so-called slow-speed solar wind, with velocities of about 400 km s^{-1}, originates in quiet regions, while the fast-speed solar wind, with velocities of about 800 km s^{-1}, originates in polar coronal holes.

1.5 Dynamics of the solar atmosphere

It has become increasingly evident over the past 20 years that what is generally known as the quiet solar atmosphere is in fact extremely dynamic, with flows, oscillations, transient brightenings, and explosive phenomena occurring throughout. This is independent of the solar activity in the form of active regions and flares (Section 1.6). Nearly all these phenomena are related to the Sun's magnetic field, which pervades the entire solar atmosphere above the magnetic canopy (Figure 1.7), and have significance for energy and mass supply. The dynamical nature has particular implications for the nature of the transition region.

Hα spicules have counterparts in ultraviolet images of the Sun such as those obtained by the SUMER instrument on *SOHO*, where they give rise to a ragged appearance of the limb. This is just visible in Figure 1.5(*a*), an image obtained in a He I chromospheric line. Upward velocities of ~100 km s^{-1} are observed, as well as an association with flaring X-ray bright points (Moore *et al.* (1977)). X-ray bright points like those in the EUV appear to be tiny loops, which are particularly prominent in regions of the Sun devoid of active regions, and are subject to flaring like larger active regions. Flares from them are correspondingly small, but as X-ray bright points are so numerous, they are possibly important for the energizing of the hot solar corona. Studies show that the thermal energy of the flaring plasma may be a significant fraction of the energy requirements of the corona.

Other examples of transient phenomena occurring in the quiet Sun have been observed in the ultraviolet range. The High Resolution Telescope Spectrometer (HRTS), flown on rockets and on the Space Shuttle-borne *Spacelab 2* mission, observed the C IV 1548, 1550 Å resonance doublet, finding many instances of broadening of the lines indicating directed flows in the emitting material. As the C IV lines are emitted at ~ 1.1×10^5 K, these transient phenomena are characteristic of transition region temperatures. They have downward and upward velocities of up to ~400 km s^{-1} and are explosive in nature (Brueckner & Bartoe (1983), Dere *et al.* (1989)). These transient phenomena have also been observed in other ultraviolet lines. Temporary brightenings in extreme ultraviolet transition region lines such as O V (e.g. at 629.7 Å) have been observed with the CDS instrument on *SOHO* (Harrison (1997)), and are known as blinkers. They appear to be associated mostly with the network, though they also occur in cell centres (Harra *et al.* (2000)).

A variety of oscillatory phenomena have been observed in the chromosphere, transition region, and corona, indicative of wave propagation, again with possible implications for the heating of the solar atmosphere and identification of a non-radiative energy term in the energy balance. Oscillations in the violet emission peaks of the Ca II H and K visible-wavelength lines have been observed for some years. Longer-period oscillations in network structures have also been found in ultraviolet lines observed by the SUMER instrument (Curdt & Heinzel (1998)). Here, the oscillations most likely arise through the buffeting of narrow flux tubes at chromospheric altitudes by solar granulation, which leads to the propagation of MHD waves, these in turn heating the network regions of the chromosphere.

An especially puzzling feature of the transition region ultraviolet line emission is the fact that, when compared with chromospheric lines that are presumed to be practically stationary, the transition region lines show a Doppler displacement to longer wavelengths (so-called red shift) indicating a downward velocity. This was first discovered in the data from the S082B slit spectrometer on the *Skylab* mission (Doschek *et al.* (1976a)) in lines ranging from Si IV (formed at ~80 000 K) to N V (1.8×10^5 K), with velocities of 7–9 km s^{-1}. The downward velocities that these red shifts imply appear to have a cos θ dependence, where θ is the angle from the disk centre (Peter & Judge (1999)). At higher temperatures, 6×10^5 K–1.5 MK, the red shifts become blue shifts, i.e. upward velocities. If the transition region line red shifts are interpreted as simple mass motions, a draining of all the coronal material in only a few hours is implied. This is clearly at odds with the observed persistence of the transition region and corona. Several explanations have been made for the red shifts in view of this difficulty, one being the production of pulses driven upwards at photospheric levels leading to heating of the chromosphere and transition region, but with a rebound of material going downwards to explain the red shift.

1.6 The active Sun

As described in Section 1.5, there are many dynamic events occurring all over the Sun even in its quiescent stages, so the Sun is never truly quiet. However, in addition to these phenomena, there are more specific and much more energetic processes that we may normally describe as solar activity. These include sunspots, active regions, flares, coronal mass ejections, and many eruptive events such as surges and the X-ray jets. A characteristic of much of this activity is the presence of a cycle that on average is 11.1 years long, operating at least since the mid seventeenth century, with sunspot numbers increasing from a minimum to the succeeding maximum in a mean of 4.8 years, and decreasing to the succeeding minimum in a mean of 6.2 years. Sunspot numbers, as defined by Wolf in the mid nineteenth century, are defined in terms of the number of groups (g) and total number of spots (f), $R = 10g + f$, since a property of spots is that they occur in groups which have varying degrees of complexity. Figure 1.9 shows monthly mean sunspot numbers over sunspot cycles 18 to 23, the numbering scheme starting with the cycle that reached its maximum in 1761. Cycles 18, 19, 21, and 22, covering the spacecraft era when many of the results given in this book were obtained, have been exceptionally active, with peak sunspot numbers of between 150 and 200, far greater than those in the eighteenth and nineteenth centuries when typical sunspot maximum peak values were between 50 and 100. At solar maximum, many large sunspot groups may occur on the Sun at any one time, but in a limited latitude range, roughly between 30° north and 30° south.

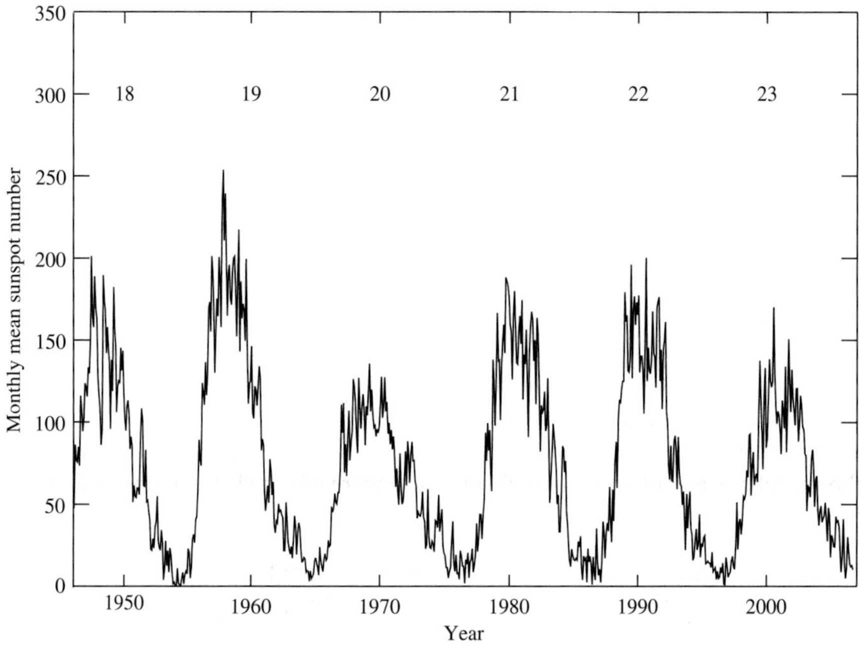

Fig. 1.9. Monthly mean sunspot numbers over the period 1946 to 2006, covering activity cycles 18 to 23. Courtesy NOAA.

At solar minimum, the Sun may have very few or even no spots at all for several days on its visible hemisphere. There is a dependence of average sunspot latitude on the phase of the sunspot cycle. The spots of a new cycle appear at latitudes $\sim$30° north and south, but later in the cycle the average latitude decreases, and may be within 5° north and south at the end of the cycle. Normally, there is an overlap of new-cycle and old-cycle spots for a short period near minimum. A plot of sunspot latitudes against time produces the well-known 'butterfly' diagram.

Sunspots are visible in white light as darker areas of the photosphere, where the temperature is about 4000 K. The interior of a well-developed spot, the *umbra*, is darker than the surrounding *penumbra*, which consists of nearly radial light filaments extending out to the photosphere. Bright patches known as *faculae* occur in the vicinity of spots, especially large groups, and are more prominent at the limb; their temperatures are a little higher, by 300 K, than the surrounding photosphere.

Sunspots form in the photosphere when magnetic flux ropes are buoyed up from the interior to the photosphere. In simple terms, the lower temperature in spots arises through an equilibrium of the gas and magnetic pressure inside the rope, $Nk_B T_{int} + B^2/8\pi$, and the gas pressure outside the rope, $Nk_B T_{ext}$, assuming a constant gas density N for the same level in the solar atmosphere (here, k_B is the Boltzmann constant). For $B \sim 4000$ G, as is typically measured by solar magnetographs, T_{int} is substantially less than T_{ext}. A complex sunspot group, such as those frequently occurring near the maxima of solar cycles, consists of a larger, leading spot (leading in the sense of solar rotation) and a smaller following spot or more usually group of spots.

Sunspots have a vertical structure or atmosphere extending into the chromosphere and beyond. In Hα spectroheliograms, there is a system of filaments making up a superpenumbra, while ultraviolet images show enhanced emission above spots and X-ray images show slightly depressed coronal emission and temperatures. A flow pattern, called the Evershed flow, exists in spot penumbrae, with material moving outwards near the photosphere forming a 'moat' with moving facular points. In ultraviolet wavelengths, the flow patterns are apparently inward within the spot umbra but outward in the penumbra. The lower temperatures near spot umbrae are sufficiently low for simple molecules to form, e.g. OH and H_2. Certain lines of the molecular hydrogen Lyman band, band-head at 1100 Å, are excited by photons from the intensely bright hydrogen Ly-α line at 1215.67 Å, as were observed in spectra from the HRTS instrument (Jordan *et al.* (1978)).

Active regions are regions of magnetic complexity in the solar atmosphere often associated with sunspots. They are born when magnetic field structures emerge from the interior, forming loops which extend into the corona, and which are detectable over a large wavelength range, from soft X-rays to decimetric radio waves. They are visible in Hα and calcium H and K spectroheliograms as bright patches or *plages*, situated in the chromosphere, which have a general correspondence with photospheric faculae seen in white-light emission. The chromospheric part of an active region is also apparent in the H Ly-α emission line and in the Mg II doublet absorption lines, known by analogy with the equivalent Ca II lines as the h and k doublet, at 2800 Å. In ultraviolet line and soft X-ray emission, formed in the corona, the coronal loop structure is obvious. Particularly striking examples are shown in Figure 1.8. The *SOHO* EIT image is taken through its 195 Å filter so includes strong Fe XII lines formed at $\sim$1.4 MK, while the soft X-ray emission that *Yohkoh* is sensitive to is formed at temperatures $\gtrsim$2 MK, so each image records the coronal counterpart of active regions.

Active regions evolve markedly over their lifetimes, which may be only a day or many months, depending on their size. In magnetograms, such as those from the *SOHO* MDI, the first indications are a bipolar region, with the leading polarity nearer the equator. The components rapidly separate, the inclination to lines of latitude decreases, and new magnetic flux may emerge between the two main parts. An active region birth in chromospherically formed lines occurs by a brightening of network structures, which become filled in by very bright plage. A strongly developed region may have sunspots within the first few days, with small flares occurring. In Hα images, low-lying loops called arch filament systems may develop, and at their footpoints are bright point-like features visible in the Hα line wings, known as Ellerman bombs or moustaches. Major active regions reach maximum development when the chromospheric plage is very bright, the sunspots and magnetic structure show great complexity, and there are many strong flares. Decay occurs when the region expands and fades, the sunspots disappear, and the flare occurrence diminishes; only the active region filament remains, slowly migrating polewards.

The most striking manifestation of solar activity is the solar *flare*, a sudden release of energy, up to 10^{32} erg, following rearrangement of magnetic fields in a complex active region. An impulsive phase, consisting of sub-second pulses of hard X-ray (wavelengths $\lesssim$ 0.6 Å or photon energies $\gtrsim$ 20 keV), ultraviolet and radio-wave emission, lasts for a minute or so, while a gradual phase may begin at or a little before the impulsive stage and peak a few minutes after it. The radiant energy from flares is often over all regions of the electromagnetic spectrum, from gamma-rays to km-wavelength radio emission. As much energy or more may appear as mass motions, e.g. in the form of cool, Hα-emitting material, while smaller amounts are released as non-thermal particles which are directly observed in space or whose presence is deduced from the hard X-ray and radio emission at the impulsive stage. The gradual phase is manifested by Hα and soft X-ray (wavelengths $\gtrsim$ 1 Å) emission. A common way of categorizing flare importance is through the amount of soft X-ray emission in the two spectral bands of the *Geostationary Operational Environmental Satellites* (*GOES*), geostationary Earth-orbiting satellites operated by the US National Oceanic and Atmospheric Administration that monitor solar X-ray emission. The spectral bands cover the wavelength ranges 0.5–4 Å and 1–8 Å (see Chapter 7, Section 7.5.1). Flares having peak fluxes in the 1–8 Å band of 10^{-6}, 10^{-5}, and 10^{-4} W m^{-2} are said to have X-ray importance C1, M1, and X1 respectively. (The *GOES* importance scale continues to smaller fluxes also, with 1–8 Å band fluxes of 10^{-7} and 10^{-8} W m^{-2} being called B1 and A1 levels.)

Flares in soft X-rays take the form of one or more bright coronal loops having a spectrum that initially includes emission lines of highly stripped ions such as He-like Fe, indicative of temperatures of $\sim$20–30 MK. Images in Hα often show bright ribbons that trace the footpoints of several such loops, separating with time as the loops expand. Flares on the limb show the loops and their time evolution particularly clearly, with the loops gradually climbing in altitude. In Hα images, loops (sometimes known as post-flare loops) form successively, apparently condensing out of the hot corona (Bruzek (1964), Phillips & Zirker (1977)). Loops in *TRACE* images may form impressive arcades – see Figure 1.10, an image of a large disk flare seen in the 195 Å *TRACE* filter which includes the Fe XII 195.12 Å emission line but also the much hotter lines of Fe XXIV (192.03 Å) and Ca XVII (192.82 Å): most of the loops in Figure 1.10 emit in Fe XII, but the arcade 'spine' structure is due to Fe XXIV and Ca XVII line emission. At the impulsive phase, the lines of highly excited ions such as He-like Fe, Ca, and S are both broadened and have Doppler-shifted short-component wavelengths. Many

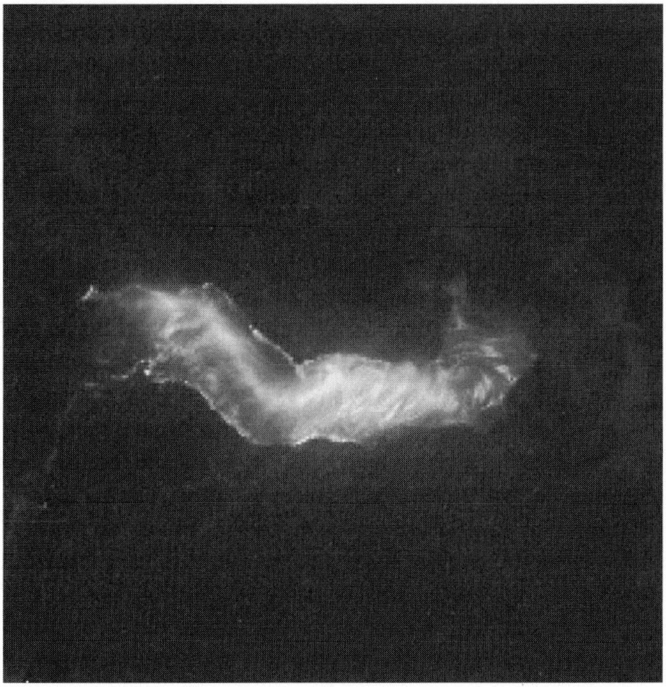

Fig. 1.10. The Bastille Day flare (on 2000 July 14) as imaged by *TRACE* in its 195 Å filter. In this disk flare, the arcade of flaring loops emit principally in the Fe XII line, but there is a 'spine' structure running along the tops of the loops, hot enough to produce Fe XXIV and Ca XVII lines within the range of the 195 Å filter. Courtesy the *TRACE* Team.

flare models are based on the idea that non-thermal electrons are accelerated in a magnetic reconnection event and are stopped in the chromosphere as they descend along flare loop legs, giving hard X-ray emission. Such models have become very popular over the years; the dumping of energy at the loop footpoints by the non-thermal electrons in these models gives rise to turbulent motions in the chromosphere (to account for the soft X-ray line broadening) and an evaporation, or convection of the now-heated plasma up the loop legs (thus explaining the soft X-ray line shift, interpreted as a Doppler shift of rising material). These ideas are explored in more detail in Chapter 10.

Often accompanying large flares, but sometimes independent of them, are the coronal mass ejections (CMEs). They were originally seen in white-light coronagraphs, sometimes not immediately recognized as ejections of mass but rather as transient phenomena. A CME is seen, most completely in spacecraft-borne coronagraphs, as a bubble-shaped ejection of coronal material, mass from 10^{14} g to 10^{16} g, with velocities ranging from only $\sim$100 km s^{-1} to more than 1000 km s^{-1}, or more than the Sun's escape velocity at the surface (618 km s^{-1}). With typical dimensions of several hundred thousand km, they are much larger than the whole magnetosphere of the Earth, so an Earth-directed CME has major implications for geomagnetic disturbances, known as magnetic storms. Following in the wake of a CME is often an erupting prominence. A marked feature of some CMEs is the eruption of a 'sigmoid' structure, an S-shaped feature seen in *Yohkoh* SXT images and in *TRACE* and

SOHO ultraviolet images. The sigmoid itself is thought to be the result of sheared dipole fields, often involving several active regions, in which the field lines greatly depart from a potential configuration. The excess energy in the field may be available to give rise to CMEs. Some of the properties of CMEs – densities, temperatures, mass, and kinetic energy – have been deduced from white-light emission (which like the white-light emission of the general corona is Thomson-scattered photospheric emission), radio emission (thermal free–free radiation), and ultraviolet spectral lines. The Ultraviolet Coronagraph Spectrometer (UVCS) on the *SOHO* mission (Kohl *et al.* (1995, 2006)) has used measurements of ultraviolet line fluxes, e.g. H Ly-α, the resonance doublet of O VI at 1031.9, 1037.6 Å, and other ions characteristic of the corona. The instrument observes the corona from $1.5R_\odot$ to $10R_\odot$, so by the time a CME reaches these altitudes it has already passed into its development stage. The CME core was long presumed to be the original erupting prominence because of its high density, but UVCS measurements show that the core not only has a chromospheric temperature ($\sim 10^4$ K) as expected, but rather a range of temperatures extending from 25 000 K to coronal values.

1.7 Physical conditions and energy balance

An empirical description or model of the solar atmosphere consists of a specification of physical conditions, in particular density, temperature, and pressure, as a function of location. Such a model for the average quiet Sun is illustrated in Table 1.1 and Figure 1.1, and shows the run of electron and neutral hydrogen atom number densities and electron temperature with height h above the τ_{5000} level. For heights $h < 1000$ km, this model is based on that of Fontenla *et al.* (1990) Model 1 (zero mechanical heating, radiative losses at height 270 km of 664 erg cm^{-2} s^{-1}); the minimum temperature (4170 K) is somewhat lower than values given earlier in this chapter. For heights in the range $1000 < h < 2110$ km, the model is that of Vernazza *et al.* (1981) Model C (average quiet Sun). At $h > 2140$ km, we have added values from Model A of Gabriel (1976), based on quiet Sun observational data in the ultraviolet. The heights in Gabriel's (1976) model, which have an arbitrary zero, give the region of rapid temperature rise at heights 450 km lower than in the models of Vernazza *et al.* (1981) and Fontenla *et al.* (1990), so the values in Table 1.1 are adjusted by this amount.

Like all solar atmospheric models, the model of Table 1.1 is idealized in that the transition region is portrayed as a narrow layer, only 200 km thick, separating the chromosphere from the much hotter corona. Also, it is based on average quiet Sun ultraviolet emission data and therefore ignores the network structure as well as the dynamic phenomena occurring in the chromosphere and transition region. Other models considered by Vernazza *et al.* (1981) and Fontenla *et al.* (1990) take account of fine structure.

Although the transition region is not an idealized thin layer, the model in Table 1.1 is useful for helping to understand the nature of parts of the solar atmosphere. Thus, we can see that the photosphere and chromosphere are largely made up of neutral hydrogen atoms (H), but for higher levels, where the temperature increases beyond about 10 000 K, the hydrogen atoms begin to be ionized, then higher still a nearly fully ionized gas results, made up of protons and electrons and some helium nuclei. The corona is nearly isothermal, with an electron temperature T_e between 1.0 MK and 1.4 MK, though variations occur for specific features like coronal holes, active regions, and bright points. The gas pressure is given by

$$p_{gas} = Nk_B T, \tag{1.1}$$

Table 1.1. *Atmospheric model for the quiet Sun: temperatures T, densities (neutral hydrogen N_H, electron N_e), total gas pressures P, as a function of height h above the $\tau_{5000} = 1$ level.*

h (km)	T (K)	log N_H (cm^{-3})	log N_e (cm^{-3})	log P (dyn cm^{-2})
0[a]	6 420	17.07	13.81	5.19
505	4 170	15.32	11.41	3.26
10 65[b]	6 040	13.32	10.93	1.41
2 016	7 360	10.95	10.23	−0.79
2 091	10 150	10.15	10.57	−0.97
2 100	65 890	7.43	9.96	−0.91
2 110	102 600	5.24	9.79	−0.88
2 150[c]	125 900	—	9.63	−0.65
2 292	501 200	—	9.03	−0.65
2 590	631 000	—	8.93	−0.65
3 531	794 300	—	8.82	−0.66
6 572	1000 000	—	8.68	−0.70
16 640	1259 000	—	8.51	−0.77
32 990	1413 000	—	8.37	−0.86

[a] Vernazza *et al.* (1981) (Model C, $h < 1000$ km),
[b] Fontenla *et al.* (1990) (Model 1, $1000 < h < 2110$ km), and
[c] Gabriel (1976) (Model A, $h > 2140$ km). See text for further details.

where k_B is the Boltzmann constant, N is the total particle number density, and T is the temperature. For the fully ionized quiet corona, the gas pressure is $(N_e + N_p + N_{He})k_B T_e$, where N_p is the number density of protons, i.e. fully ionized hydrogen atoms, and N_{He} the number density of He nuclei. With $N_e = N_p + 2N_{He}$ (the ionization of He results in two free electrons) and $N_{He} = 0.1N_p$, $N_e = N_p + 2N_{He} = 1.2N_p$. Thus, in the corona,

$$p_{gas} = 1.92 N_e k_B T_e. \tag{1.2}$$

Like all plasmas, the chromosphere and corona are strongly influenced by the Sun's magnetic field. A particle of mass m and charge q, moving with velocity $\mathbf{v}$, is subject to a force due to the magnetic field $\mathbf{B}$ equal to $\mathbf{F} = qm(\mathbf{v} \times \mathbf{B}/c)$, or with current density $\mathbf{j} = qm\mathbf{v}/c$,

$$\mathbf{F} = \mathbf{j} \times \mathbf{B}. \tag{1.3}$$

The magnetic or Lorentz force $\mathbf{F}$, acting perpendicularly to $\mathbf{j}$ and $\mathbf{B}$, constrains particles of mass m to move along the field lines in spirals, with gyro-radii equal to

$$a = mv_\perp c/qB. \tag{1.4}$$

For electrons, $q = -e$, for protons, $q = +e$, and for ions, $q = +Ze$, where Z is the charge of the ion in units of e. Since the electron mass m_e is very much less than the proton mass m_p, electrons move along field lines with relatively small gyro-radii compared with those of protons or of other ions.

At the Sun's surface, magnetic fields have strengths B of up to 4000 G in active regions and most particularly sunspots, but a few tens of G elsewhere. The magnetic field would be expected to spread out and diminish with height above the photosphere. Even so, magnetic field strengths B of ~10 G in the quiet corona would ensure that magnetic pressures p_{mag}, equal to $B^2/8\pi \approx 4$ dyn cm^{-2}, are an order of magnitude or more greater than coronal gas pressures (see Table 1.1). A useful parameter for distinguishing regions where the magnetic field has a controlling influence in the solar atmosphere is the plasma β, defined by

$$\beta = p_{\rm gas}/p_{\rm mag} = 8\pi N k_{\rm B} T_{\rm e}/B^2. \tag{1.5}$$

When $\beta < 1$, the magnetic field has a dominating influence. In the photosphere at $\tau_{5000} = 1$ ($h = 0$) where $T = 6400$ K, $B = 1000$ G for the magnetic elements making up the photospheric network and $\beta = 3$, so the magnetic elements are moved around by the convective motions of supergranules as the magnetic field is not dominant there. In fact they appear along the boundaries of the supergranular cells. In the upper chromosphere, where $T_{\rm e} \approx 10\,000$ K and $B \approx 100$ G, $\beta \approx 4 \times 10^{-4}$, and in the quiet solar corona, where $T_{\rm e} \approx 1.4$ MK and $B \approx 10$ G, $\beta \approx 0.1$. Thus, the magnetic field is the controlling influence in the solar atmosphere from the upper chromosphere outwards.

We may calculate typical values of the gyro-radii a in the solar atmosphere from Equation (1.4) with values typical of the solar atmosphere. The particle velocities in a completely thermalized gas or plasma have a Maxwell–Boltzmann distribution. The most probable thermal velocity $v_{\rm th}$ is related to the temperature T in such distributions by

$$v_{\rm th} = (2k_{\rm B}T_{\rm e}/m)^{1/2}. \tag{1.6}$$

For electrons ($m = m_{\rm e}$) in the photosphere (network elements), upper chromosphere, and inner corona, $v_{\rm th} = 440$, 550, and $\sim$6500 km s^{-1}. Equating these values of $v_{\rm th}$ to $v_\perp$ in Equation (1.4) leads to electron gyro-radii of 0.003 cm, 0.03 cm, and 4 cm respectively. These very small values mean that the electrons (as well as protons and ions) are tightly bound to the magnetic field lines, and so features in the quiet Sun atmosphere, in particular flux tubes and open streamers, are very well defined structures. This applies *a fortiori* to active regions and flares for which the magnetic field strengths are much larger.

The outer solar atmosphere, especially the corona, has the properties of a plasma including plasma oscillations, occurring when electrons are momentarily pulled apart from the positive ions which may be considered stationary. The plasma oscillation frequency is given by

$$\omega_{\rm p} = (4\pi N_{\rm e}e^2/m_{\rm e})^{1/2}. \tag{1.7}$$

The square-root dependence of $\omega_{\rm p}$ on $N_{\rm e}$ means that $\omega_{\rm p}$ decreases slowly through the solar atmosphere, from 6×10^9 s^{-1} (upper chromosphere) to 2×10^9 s^{-1} (inner corona). Electromagnetic waves travelling in the absence of a magnetic field in theory cannot propagate at less than the plasma frequency, a matter of importance in the radio wavelength range but not for X-ray or ultraviolet wavelengths. For the ultraviolet and X-ray range, electromagnetic waves have frequencies $\gtrsim 10^{15}$ s^{-1}. Another plasma quantity that is relevant for some purposes is the Debye screening length $h_{\rm D}$, given by $h_{\rm D} = 6.9(T_{\rm e}/N_{\rm e})^{1/2}$ cm. Outside of $h_{\rm D}$, the charge of a single electron or proton is shielded by particles of opposite charge. For typical densities and temperatures, $h_{\rm D}$ is 0.01 cm (upper chromosphere) to 0.2 cm (inner corona), not too different from electron gyro-radii.

Other wave phenomena relevant for solar dynamic events like flares include Alfvén and magnetohydrodynamic (MHD, or hydromagnetic) waves. Alfvén waves, a travelling oscillation of the magnetic field and ions of the plasma, have a velocity given by

$$v_{\rm A} = B/(4\pi N_{\rm ion} m_{\rm ion})^{1/2}, \tag{1.8}$$

where N_{ion} is the ion number density. Thus, for the quiet Sun inner corona (with $N_{\mathrm{ion}} \approx 10^9$ cm^{-3} and $B \approx 10$ G) v_{A} is ~700 km s^{-1}. This is little changed for hot plasmas in flare loops, for which $B \approx 100$ G and $N_{\mathrm{ion}} \approx 10^{11}$ cm^{-3}. Fast and slow mode MHD waves are hybrids of Alfvén and sound waves, with fast waves travelling at nearly the Alfvén speed, slow waves travelling at approximately the ion sound speed. Sound wave speeds for the solar atmosphere are approximately the most probable thermal speeds, which for protons in the quiet Sun corona are ~200 km s^{-1}. For flare plasmas having $T_{\mathrm{e}} \sim 20$ MK, they are ~800 km s^{-1}.

The corona's high temperature is maintained by a non-radiative heating mechanism which may be the dissipation of MHD waves or the occurrence of tiny flares, *nanoflares*, which produce heated plasma in localized regions that may disperse into larger volumes (Parker (1988)), or some other means. At any event, there is an energy balance in which the heating is counteracted by cooling. The chief cooling mechanisms for the corona or hot coronal plasmas such as those in active regions or flares are radiation and conduction, with minor contributions from mass flows. As coronal structures are either flux tubes or open-ended streamers, we can write the energy balance as

$$H(s) = R[T(s)] + \kappa \nabla T_{\mathrm{e}}, \tag{1.9}$$

where the length variable s is measured from the base of the corona, $R[T(s)]$ is the radiative power loss function, and the thermal conductivity is taken to be classical, i.e. that given by Spitzer (1962), in which energy is imparted from the hot coronal plasma down the loop or streamer legs to the relatively cool chromosphere or transition region. The classical thermal conductivity coefficient is

$$\kappa = \kappa_0 T_{\mathrm{e}}^{5/2}, \tag{1.10}$$

where κ is in erg cm^{-1} s^{-1}, with $\kappa_0 = 10^{-6}$ c.g.s. units. (A more accurate calculation is given by Braginskii (1965).) This expression for the conductivity is in the absence of a magnetic field. For a magnetized plasma like the corona, thermal conductivity is strongly reduced in directions transverse to the magnetic field **B**, while the classical expression is valid for conductivity parallel to **B**. Equation (1.10) is commonly applied to hot flare plasmas also (Chapter 10), though the applicability of a classical expression to such high-energy plasmas can be questioned: more probably, the conductivity is anomalous, i.e. the transfer of energy proceeds from hotter to cooler plasma by electron–wave interactions rather than electron–electron collisions as assumed in classical theory.

The radiation loss term $R(T)$ in Equation (1.9) is a function of temperature, as is illustrated in Figure 1.11. Both lines and continua contribute to $R(T_{\mathrm{e}})$, with particular ions predominating over small temperature ranges. Thus, H Ly-α is a strong radiator at ~20 000 K, as are ionized C, N, and O ions at transition region temperatures (~10^5 K). At coronal temperatures (1–1.4 MK), the extreme ultraviolet lines of Fe IX–Fe XII are important. At flare temperatures, the main radiators are lines of Fe XVII and Fe XVIII in the X-ray region at 10 MK, lines of highly ionized Fe including the X-ray lines of Fe XXV and dielectronic satellites at 20 MK, and continua at even higher temperatures.

We see that coronal structures must be maintained by a heating mechanism, defined by H in Equation (1.9). It is observed that active regions emerge as coronal magnetic loops in a range of complexity, as observed in X-rays (e.g. with the *Yohkoh* SXT instrument), or extreme ultraviolet (e.g. with *TRACE* or *SOHO* EIT), but in time these decay, with loops appearing to cool. The high-density structures as in newly emerged active region loops gradually decay

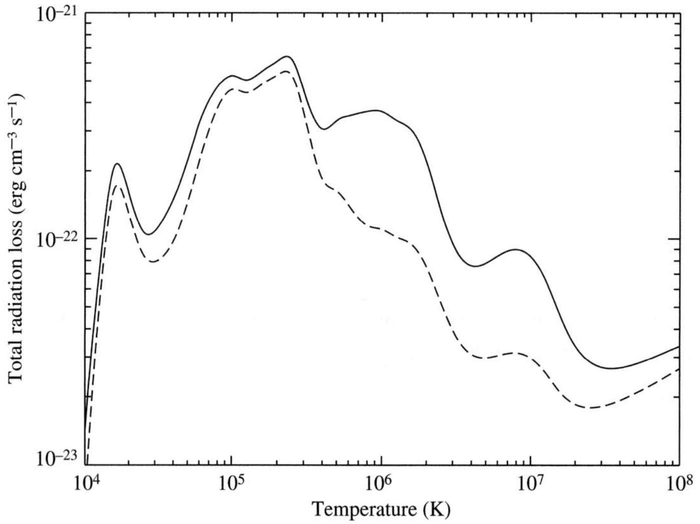

Fig. 1.11. Radiation loss for coronal material with photospheric (Grevesse & Sauval (1998)) and coronal abundances (Feldman & Laming (2000)) with unit emission measure (1 cm^{-3}). The curves were calculated with the CHIANTI (version 5.2) database and code (Landi *et al.* (2006a)).

by expansion, perhaps by a cross-field diffusion process. Thus, there is a gradual cooling and expansion of active region loop structures that most likely leads to the formation of much larger, cooler, and (in X-rays and extreme ultraviolet) fainter arches characterizing the general corona.

1.8 Ionization and excitation conditions

The solar upper atmosphere, particularly the corona, is a high-temperature region above the photosphere, but is practically unaffected by the photosphere's intensely bright radiation. The solar atmosphere from the upper chromosphere out to the corona is in an equilibrium, but one quite unlike LTE, which holds in the photosphere (Section 1.2). Rather, there is a balance between the number of ionizations per unit volume per unit time from an ionization stage of an element (e.g. X^{+m}, where $+m$ is the ion's charge) with numbers of recombinations per unit volume per unit time from the ionization stage X^{+m+1}. (We will deal with this further in Chapters 3 and 4.) This can be written

$$C_m N_e N_m = \alpha_{m+1} N_e N_{m+1}, \tag{1.11}$$

where N_m is the number density of X^{+m} ions, N_{m+1} that of X^{+m+1} ions. The quantity C_m is the total ionization rate coefficient, or number of ionizations from stage X^{+m} to X^{+m+1} per unit density and time (units cm^3 s^{-1}). Correspondingly, α_{m+1} is the rate coefficient of recombination from stage X^{+m+1} to X^{+m}. The number densities of ions in successive stages, N_m/N_{m+1}, is thus given by the ratio of the recombination to ionization rate coefficients, α_{m+1}/C_m.

For most coronal ions, the ionization potentials are generally several keV. Since there are practically no photons in the photospheric radiation field having these very high energies, radiative ionization is unimportant. Generally, only processes involving collisions with free electrons are significant in the ionizations since protons and heavier ions are too slow to be effective. Electron collisional ionization includes direct ionizations from the ion's ground state (which is the state most ions are in):

$$X^{+m} + e_1^- \rightarrow X^{+m+1} + e_1^- + e_2^-, \tag{1.12}$$

where the colliding free electron is e_1^- and the electron liberated from ion X^{+m} is e_2^-. Collisional ionization also includes autoionizations, which are two-step ionizations whereby a free electron first collides with an ion, leaving it in a doubly excited state above the ion's ionization limit, and the ion then stabilizes by releasing one of the doubly excited electrons.

There are two significant recombination processes in the solar atmosphere. The first is radiative recombination:

$$X^{+m+1} + e^- \rightarrow X^{+m} + h\nu, \tag{1.13}$$

where a previously free electron e^- is captured by the ion X^{+m+1} and its excess energy is removed by a photon represented by $h\nu$. The second process is dielectronic recombination, in which a free electron having a particular energy E is captured by ion X^{+m} to produce a doubly excited state, $X^{+m\star\star}$, i.e. ion X^{+m} with two electrons excited:

$$X^{+m+1} + e^- \rightarrow X^{+m\star\star} \tag{1.14}$$

followed by a stabilizing transition to produce a singly excited state in X^{+m}:

$$X^{+m\star\star} \rightarrow X^{+m\star} \tag{1.15}$$

followed by another stabilizing transition or transitions until ion X^{+m} reaches its ground state. In Equation (1.14), the capture of the incoming electron is radiationless, unlike Equation (1.13); recombination only occurs if the incoming electron has an energy E plus or minus a very small range ΔE. The energy range ΔE is related to the lifetime of the doubly excited state Δt by Heisenberg's Uncertainty Principle, $\Delta E \Delta t \approx \hbar$. As a doubly excited state is very unstable, Δt is often relatively small and ΔE correspondingly large, but still only a tiny fraction of E itself. Dielectronic recombination is therefore like a resonance process in nuclear reactions.

Calculation of the ion fractions of a large number of ions inevitably involves much atomic data referring to the ionization and recombination rate coefficients or cross-sections. In general, these data are based on calculations (e.g. dielectronic recombination rates are based on a formula due to Burgess (1965)), though laboratory measurements are now available for a limited number of ions. Continual improvements to the atomic data since the 1960s required adjustments to the resulting ion fractions, giving rise to changes in the interpretation of observed emission line data in terms of temperature of emitting regions. Figure 1.12 shows recent data for Fe ion fractions in the range 500 000 K–100 MK, which includes the temperatures of active regions and flare plasmas.

At very high temperatures, more than 100 MK, a significant number of electrons in the Maxwell–Boltzmann distribution move relativistically, i.e. v_e is an appreciable fraction of the speed of light c. This situation is not often encountered in the solar atmosphere, even for flares (where measured temperatures do not often exceed 30 MK), so relativistic corrections are generally neglected.

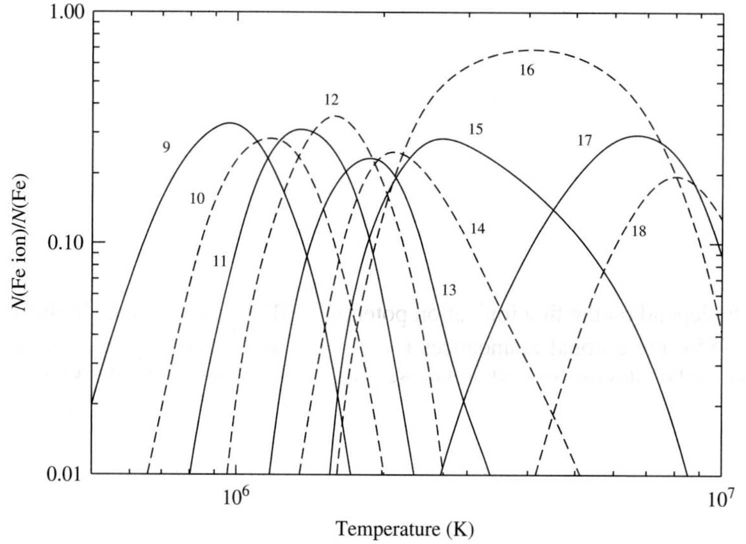

(a)

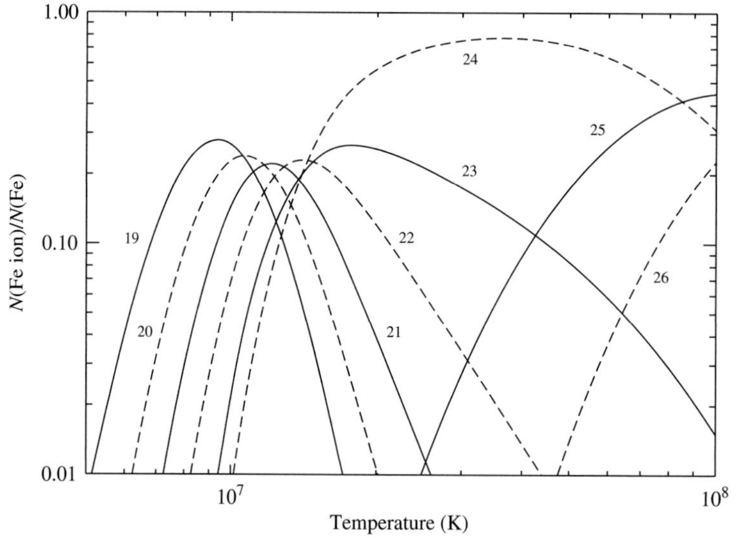

(b)

Fig. 1.12. Fe ion fractions in the range (*a*) 5×10^5 K $< T_e <$ 10 MK, including ions Fe^{+9} to Fe^{+18}; (*b*) 5 MK $< T_e <$ 100 MK, including ions Fe^{+19} to Fe^{+26} (fully stripped Fe ions). Ion stages are indicated by the numbers by the side of each curve (e.g. 9 means Fe^{+9}). The ion fractions are plotted as solid and dashed lines alternately. The plots are based on calculations of Mazzotta *et al.* (1998).

1.9 Chemical composition

The abundances of the elements in the Sun broadly reflect those of a second-generation star, i.e. one created from the material left over from first-generation stars that underwent supernova explosions in our Galaxy. Element abundances in the photosphere can be derived by analysis of Fraunhofer line strengths. Element abundances in the transition region and corona can be derived from emission line fluxes, generally those in the ultraviolet and X-ray spectral regions. *In situ* particle measurements from mass spectrometers in the slow solar wind also give element abundances. It has been found that the solar wind particle measurements and quiet Sun coronal abundances differ from the photospheric by amounts that appear to depend on the first ionization potential (FIP) of the element. In the discussion of Meyer (1985), the coronal abundances of elements with relatively low FIP, defined as $\lesssim$10 eV, are equal to photospheric abundances while those of high ($\gtrsim$10 eV) FIP are depleted by a factor of about 4. In the later discussion of Feldman and colleagues (Feldman & Laming (2000)), the low-FIP element coronal abundances are estimated to be enhanced by a factor 4, while high-FIP elements are photospheric: this has become known as the FIP effect. Other works (e.g. Fludra & Schmelz (1998)) are 'hybrid', i.e. the coronal abundances of low-FIP elements are enhanced, those of high-FIP elements are depleted. Further details including a discussion of the observational evidence are given in Chapter 11. Possible reasons for the dependence of coronal abundances on FIP include electric or magnetic fields operating on low-FIP elements in the chromosphere or lower layers of the solar atmosphere,'dragging out' the ions (but not the neutral atoms) into coronal loops including those of active regions or flares. They would not have any effect on high-FIP elements which are in a completely neutral state in the chromosphere.

1.10 Solar-type stars

The Sun is situated near the centre of the main sequence of nearby stars when plotted on the Hertzsprung–Russell diagram, i.e. a plot of luminosity vs. effective temperature (i.e. equivalent black-body temperature: see Figure 1.13). For stars with effective temperatures higher than the Sun's, the material is so hot that it consists mostly of ionized hydrogen and helium; practically all metals (i.e. elements heavier than helium) are also in a fully ionized state. Thus, the opacity of the material is low and energy is effectively transported by radiation. For stars with effective temperatures of about 6000 K, the sub-surface layers are cool enough for the metals to be only partly ionized, which increases the opacity considerably. For these stars, radiation ceases to be the most efficient energy transport and convection begins to be more important. This is the case for the Sun. For still cooler stars on the main sequence, e.g. the so-called K and M dwarf (dK, dM) stars, the entire stellar interior is convective.

The cool main sequence stars include those with coronae and signs of activity resembling those in the solar atmosphere. The reason is that the envelopes, through the action of convection and differential rotation, are sites of dynamos where magnetic fields are being constantly regenerated. The magnetic fields are convected to the stellar surfaces by the action of magnetic buoyancy, and as a result the stellar atmospheres are heated, like the Sun's, by non-radiative mechanisms. The amount of stellar activity depends on the speed of stellar rotation, which in turn depends on stellar age. The Sun's rotation speed at the equator, 2 km s^{-1}, is relatively modest, but those for some dK and dM stars (measured by the profiles of their Fraunhofer lines) are much greater, some tens of km s^{-1}. The combination of differential rotation (ω) and convection (α) gives rise to a so-called α–ω dynamo. The rapidly rotating dK and dM stars

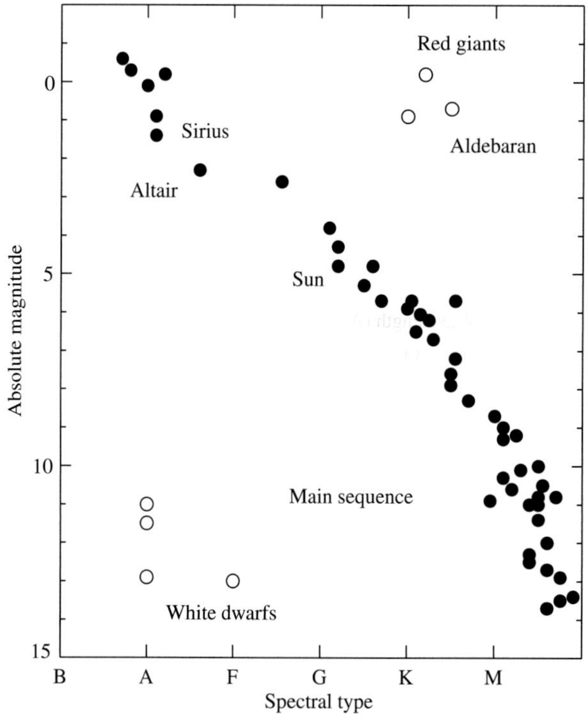

Fig. 1.13. Hertzsprung–Russell diagram for stars in the neighbourhood of the Sun. Spectral type (related to effective temperature) is plotted on the horizontal axis, absolute magnitude (related to stellar luminosity) along the vertical axis. The main sequence is the broad clustering of points running from upper left to lower right. Evolved stars on the red giant branch and white dwarf stars are also plotted. The Sun (marked) is situated near the centre of the main sequence.

(their large rotational speeds are a consequence of their relative youth) result in considerable stellar activity, including coronae that are strong X-ray emitters, detectable at the Earth, and extremely powerful flares, with visible-light increases that are greater than the total output of the stars in their non-flaring states. In visible light, their spectra are characterized by the Ca II H and K lines and members of the hydrogen Balmer sequence in emission: stars with the Hα line in emission have designations e after the spectral type (dKe, dMe).

The ultraviolet spectra of dKe and dMe stars in their non-flaring states include the same emission lines that are features of the quiet Sun spectrum, e.g. the well-known C IV lines at 1548, 1551 Å. As many of these very dim stars are close by (those further away would not be seen), their parallaxes are easily measurable, and so with precisely measured distances the surface fluxes of the line emission can be estimated. The line emission is found to be much larger than that of the quiet Sun, so confirming the much enhanced state of coronal and transition region activity in these stars. Their flares are also much more powerful than solar flares. Their X-ray spectra have been studied with the non-solar X-ray space missions *Chandra* and *XMM-Newton*. Again it has been found that the same lines from very highly ionized species in solar flares (e.g. the Fe XVII lines at 15–17 Å) exist in stellar flare spectra

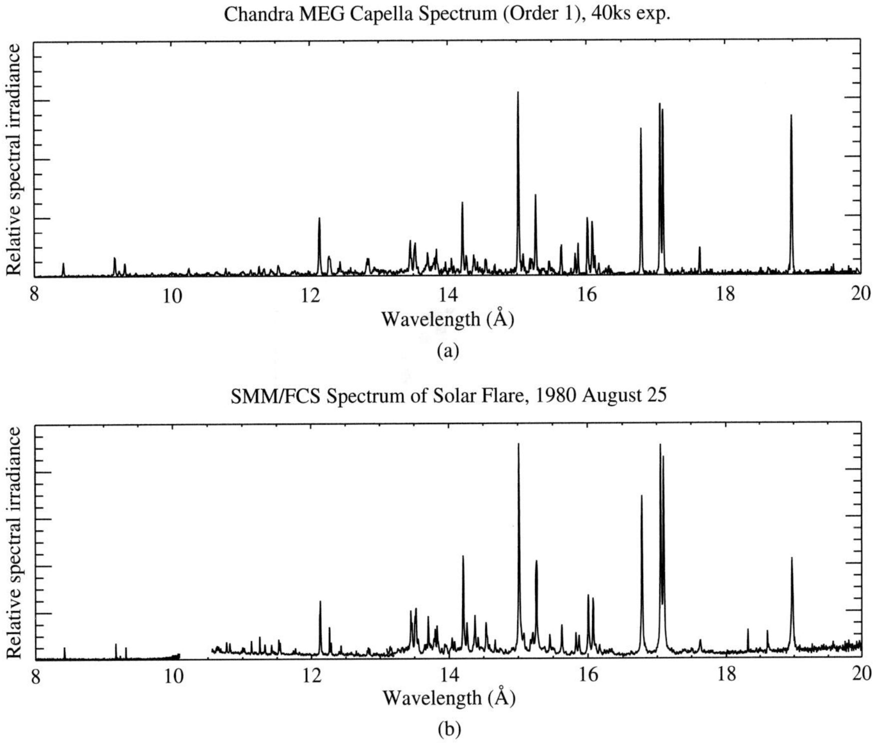

Fig. 1.14. (*a*) X-ray spectrum in the 8–20 Å range of Capella on 1999 September 24 obtained with the MEG spectrometer on *Chandra*. (*b*) Spectrum of the 1980 August 25 solar flare in the same range from the *SMM* Flat Crystal Spectrometer. For identifications of principal lines, see Figure 1.2 (third panel). From Phillips *et al.* (2001).

but with much increased flux. Figure 1.14 shows the striking similarity between the solar flare spectrum and the X-ray spectrum from Capella, a spectroscopic binary consisting of giant G stars (i.e. earlier spectral types than K and M flare stars).

Stellar binaries, in which there is interaction between two stars revolving close to each other, are often the sites of stellar activity in the ultraviolet and X-ray regions that is large even compared with dKe and dMe stars. A small class of such binaries, the RS CVn stars, consist of a main sequence F or G star, i.e. similar to the Sun, and an evolved K subgiant that is dimmer but has enhanced levels of chromospheric and coronal activity. The K star is thought to have evolved into a giant with some of its outer material gravitationally transferred to the G star. The result is that the coronae are extremely hot, with temperatures of tens of millions K, even in their quiescent state. Their X-ray spectra thus resemble those of solar flares. A particular feature of such binaries is the occurrence of flares that are much larger than the counterparts on dKe or dMe stars. Other binary systems producing X-ray coronae and flares include the contact binaries known as the W UMa stars.

A further category of flaring stars, with strong ultraviolet emission line spectra resembling those of the Sun, is the T Tauri stars, very youthful stars (ages $\sim 10^6$ years) not quite contracted on to the main sequence, still immersed in a surrounding gaseous cloud within which stars are still forming. The X-ray emission from very large flares occurring on young stellar objects within the Orion Nebula, recently observed with *Chandra* (Favata *et al.* (2005)), indicates that the flare loops are tens of millions of km long, stretching from the stellar surface to far greater distances than in solar flares. This has given rise to speculation that the Sun in its very early history was once like these young stellar objects, having enormous levels of coronal activity including extremely powerful flares.

2

Fundamentals of solar radiation

2.1 Wavelength ranges

Electromagnetic radiation emitted by the solar atmosphere is detected by Earth-based and spacecraft instruments, and through its *spectrum* of constituent frequencies (symbol ν, in Hz) or wavelengths (λ, in Å, nm, cm, etc.) we are able to deduce some of the properties of the emitting regions. We may divide the solar spectrum into regions according to wavelength ranges, although the exact boundaries are arbitrary and merely a matter of convenience. Visible light has wavelengths between approximately 3900 and 7500 Å; infrared radiation between 7500 Å and 100 μm; and radio radiation above about 100 μm. Short-wavelength radiation is known as near ultraviolet (NUV) for wavelengths between 2000 Å and 3900 Å; vacuum ultraviolet (VUV) between 1000 Å and 2000 Å; extreme ultraviolet (EUV) between 100 Å and 1000 Å, and soft X-rays between 1 Å and 100 Å. This short-wavelength radiation is emitted by the solar atmosphere extending from the chromosphere to the corona, and is particularly enhanced in active regions having some degree of magnetic complexity and in flares. For flares, a very hot plasma, with temperature T up to $\sim$30 MK, may be formed that lasts from a few minutes to several hours, and is a copious emitter of soft X-rays. All this radiation is emitted by particles that have a Maxwell–Boltzmann distribution. At the initial stages of flares, hard X-rays ($\lambda < 1$ Å) are emitted, with time scales of a few seconds, which are thought to arise from non-thermal electrons accelerated in the flare-initiating process as they are braked in the denser chromospheric material. Gamma-rays ($\lambda \ll 0.01$ Å) are emitted from exceptionally energetic flares. In this book, we will be concerned with the wavelength range 1–2000 Å, i.e. soft X-ray, extreme ultraviolet, and vacuum ultraviolet emission from the Sun's atmosphere.

For ultraviolet and X-ray radiation, it is sometimes more appropriate to use energies of the emitted *photons*, given by $h\nu$ where h is the Planck constant. As the energies of individual photons are extremely small, it is convenient to express them in electronvolts (eV), the energy acquired by an electron when accelerated through a potential of 1 volt, equal to 1.602×10^{-12} erg. In terms of photon energies in electronvolts, then, ultraviolet radiation spans the range 3.54 eV to 6.20 eV; vacuum ultraviolet 6.20 eV to 12.40 eV; extreme ultraviolet 12.40 eV to 124.0 eV; and soft X-rays 124.0 eV to 12.40 keV. Solar flare hard X-ray emission may have photon energies extending up to many hundreds of keV, and gamma-rays up to MeV energies.

2.2 Specific intensity and flux

We can define certain quantities which are commonly used when describing radiation in and from the solar atmosphere. (The following applies equally to stellar atmospheres.) See Figure 2.1 for illustration of the definitions that follow. First, we consider light with frequency in the range ν, $\nu + d\nu$ passing through a small area dA cm^2 into a solid angle $d\omega$ steradian (sr) over a time dt, with the light beam inclined at angle θ to the direction perpendicular to the area (z-axis in Figure 2.1). The *specific intensity* I_ν of radiation at a frequency ν Hz is defined by

$$dE = I_\nu \cos\theta \, dA \, dt \, d\nu \, d\omega, \tag{2.1}$$

where dE is the amount of energy (erg) crossing dA. It is thus the amount of energy passing through unit area (perpendicular to the beam of radiation) per unit time per unit frequency interval into unit solid angle. The c.g.s. units of I_ν are erg cm^{-2} s^{-1} sr^{-1} Hz^{-1}. If the source emits radiation equally in all directions, as for the energy source at the centre of the Sun, the radiation is said to be *isotropic*. Within the solar atmosphere, there is much more radiation coming from the direction of the Sun's centre than from the outside direction, i.e. the radiation is *anisotropic*.

Within the Sun, at a radial distance r, the *flux* at frequency ν, $\mathcal{F}_\nu$, is defined to be the net rate of energy flow across unit area, integrated over all solid angles 4π, i.e.

$$\mathcal{F}_\nu = \int_{4\pi} I_\nu(r) \cos\theta \, d\omega. \tag{2.2}$$

The c.g.s. units of $\mathcal{F}_\nu$ are erg cm^{-2} s^{-1} Hz^{-1}. We may define the elementary solid angle $d\omega$ in terms of polar coordinates, with angle θ measured from the z-axis (or co-latitude), and an azimuthal angle ϕ (see Figure 2.1):

$$d\omega = \sin\theta \, d\theta \, d\phi. \tag{2.3}$$

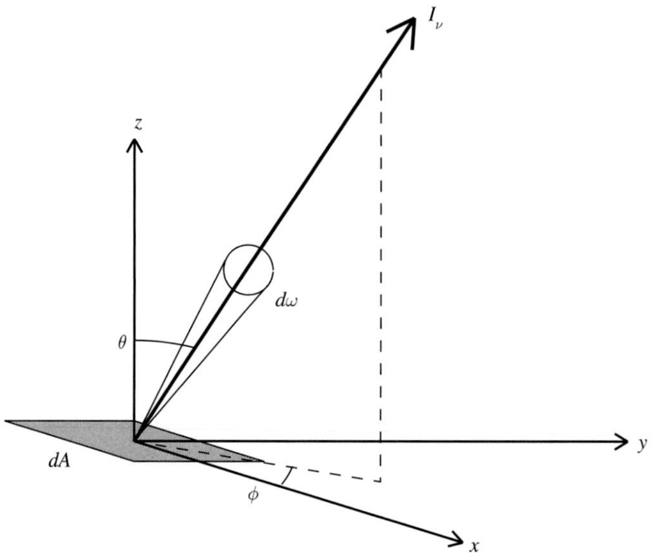

Fig. 2.1. Definition of specific intensity.

The flux in polar coordinates is then

$$\mathcal{F}_\nu = \int_0^{2\pi} d\phi \int_0^{\pi} I_\nu(r) \cos\theta \, \sin\theta \, d\theta. \tag{2.4}$$

Generally, within the Sun, I_ν is independent of ϕ. We may therefore rewrite Equation (2.4), with $\mu = \cos\theta$, as follows:

$$\mathcal{F}_\nu = 2\pi \int_{-1}^{+1} I_\nu(r)\, \mu \, d\mu \tag{2.5}$$

with $d\mu = -\sin\theta \, d\theta$.

In the solar atmosphere, since the radiation is anisotropic, we consider separately the outward (i.e. away from the Sun's centre) flux and inward (towards the Sun's centre) flux, i.e. split the integral of Equation (2.2) into $\mathcal{F}_+$ and $\mathcal{F}_-$ with θ measured from a direction towards the centre of the Sun or star ($\theta = 0$ or $\mu = 1$) to the outward direction ($\theta = \pi$ or $\mu = -1$), and ϕ going from 0 to 2π. Thus,

$$\mathcal{F}_+ = 2\pi \int_0^1 I_\nu(r)\, \mu \, d\mu \tag{2.6}$$

and

$$\mathcal{F}_- = 2\pi \int_0^{-1} I_\nu(r)\, \mu \, d\mu. \tag{2.7}$$

The net upward flux is $\mathcal{F} = \mathcal{F}_+ - \mathcal{F}_-$. At the solar surface, i.e. $r = R_\odot$, $\mathcal{F}_- = 0$ (no radiation coming from the outward direction), so the net flux is

$$\mathcal{F}_\nu(R_\odot) = 2\pi \int_0^1 I_\nu(R_\odot)\, \mu \, d\mu. \tag{2.8}$$

As the Sun can be considered symmetric and spherical, I_ν is only a function of the radial distance r and is independent of θ (or μ). Thus, $\mathcal{F}_\nu = \pi I_\nu(0)$. Because of this relation, some authors (see e.g. Böhm-Vitense (1989); Mihalas (1970)) prefer to define flux as $\mathcal{F}_\nu/\pi$. Note that, at the Sun's centre, I_ν is isotropic and $\mathcal{F}_\nu$ is therefore zero.

Also associated with specific intensity is the *mean specific intensity* J_ν,

$$J_\nu = \frac{1}{4\pi} \int_{4\pi} I_\nu \, d\omega = \frac{1}{2} \int_{-1}^{+1} I_\nu \, d\mu. \tag{2.9}$$

For isotropic radiation, $J_\nu = I_\nu$.

Although not used in this book, two other definitions are sometimes useful in other contexts. The *radiation density* is defined as

$$u_\nu = \frac{1}{c} \int_{4\pi} I_\nu \, d\omega = \frac{2\pi}{c} \int_{-1}^{+1} I_\nu \, d\mu = \frac{4\pi}{c} J_\nu, \tag{2.10}$$

where c is the velocity of light. *Radiation pressure* is defined by

$$P_\nu = \frac{1}{c} \int_{4\pi} I_\nu \, \mu^2 \, d\omega = \frac{2\pi}{c} \int_{-1}^{+1} I_\nu \, \mu^2 \, d\mu. \tag{2.11}$$

For isotropic I_ν, $P_\nu = 4\pi I_\nu / 3c$.

Note that throughout this section, we have defined I_ν and other quantities in terms of intervals of frequency ν (Hz), but we can equally well define them in terms of wavelength intervals. With $\nu\lambda = c$, the interval $d\nu$ is related to $d\lambda$ by $|d\nu| = (c/\lambda^2) d\lambda$.

2.3 Radiative transfer and the source function

In the Sun's photosphere, most of the radiation is in the visible and infrared parts of the spectrum and is optically thick. This means that an atom or ion may emit radiation which after a short path length is absorbed by an atom or ion of the same type, resulting in an excitation; the excited atom or ion may then de-excite, re-emitting a photon of similar energy but in general in a different direction. The solar atmosphere beyond the temperature minimum is characterized by increasing temperature and decreasing density. Some of the ultraviolet radiation from the lower, denser parts of the chromosphere may also be optically thick. For such situations, we must consider the issue of radiative transfer.

Consider a beam of radiation passing at angle θ through a slab of material which may add to or subtract from the radiation (Figure 2.2). The intensity of the beam $I_\nu(\theta)$ is increased by processes of *emission*, and decreased by processes of *absorption* (including scattering). Let j_ν be the emission coefficient at frequency ν, such that the amount of energy added to the beam dE_ν in a length element dl is given by

$$dE_\nu = j_\nu \, dA \, dt \, d\nu \, d\omega \, dl. \tag{2.12}$$

Similarly, the absorption coefficient κ_ν is defined as the amount of energy subtracted from the beam in a length element dl given by

$$dE_\nu = \kappa_\nu I_\nu \, dA \, dt \, d\nu \, d\omega \, dl. \tag{2.13}$$

Then the change of intensity over a length $dl = \sec\theta \, ds = ds/\mu$ of the material is

$$dI_\nu = [j_\nu - \kappa_\nu \, I_\nu(s)] \, ds/\mu. \tag{2.14}$$

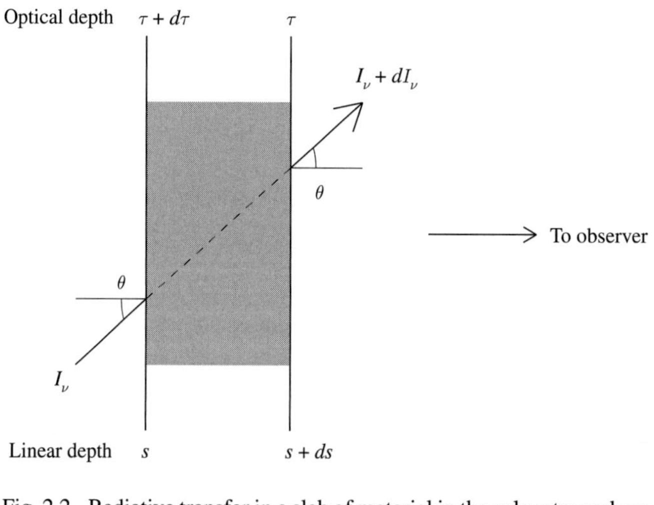

Fig. 2.2. Radiative transfer in a slab of material in the solar atmosphere. Distance s increases outwards from the Sun's surface, towards the observer, optical depth increases away from the observer. ds is the thickness of the slab, sec $\theta \, ds$ the path length of the beam, and $d\tau$ its optical depth.

Thus,

$$\mu \frac{dI_\nu}{ds} = j_\nu - \kappa_\nu I_\nu(s). \tag{2.15}$$

Defining optical depth by $d\tau_\nu = -\kappa_\nu \, ds/\mu$, we have the *equation of radiative transfer*,

$$\frac{dI_\nu}{d\tau_\nu} = I_\nu(\tau) - S_\nu(\tau), \tag{2.16}$$

where $S_\nu = j_\nu/\kappa_\nu$ is the *source function*.

Further details on radiative transfer may be found in standard texts such as those by Böhm-Vitense (1989) and Mihalas (1970).

2.4 Radiation from the solar atmosphere

2.4.1 Local thermodynamic equilibrium and effective temperature

A particular case of interest for the solar atmosphere is an ideal source known as a *black body*. A black body perfectly absorbs all radiation. In practice, one might imagine a black body to be the gas inside an isolated box with a tiny hole via which the radiation from the enclosure can be measured, but small enough that a negligible amount of radiation escapes. The black body is in a state of complete thermodynamic equilibrium, with no temperature differences inside the enclosure. The intensity of radiation integrated over all frequencies or wavelengths escaping from the hole is $I = (\sigma/\pi)T^4$ where T is the enclosure temperature in K and σ is the Stefan–Boltzmann constant. The flux $\mathcal{F}_\nu$ of black-body radiation, the so-called Planck function, is given by

$$\mathcal{F}_\nu = \pi B_\nu = \frac{2h\pi\nu^3/c^2}{\exp(h\nu/k_B T) - 1}, \tag{2.17}$$

where h is the Planck constant, k_B is the Boltzmann constant, and T is the black-body temperature in K.

Radiation passing through a black-body volume of gas suffers no change with time, i.e. there will be as much radiation absorbed as emitted, so by Equation (2.16) the intensity I_ν equals the source function S_ν which therefore equals B_ν. Thus, $j_\nu = \kappa_\nu B_\nu$. This is a particular case of Kirchhoff's law, namely, good absorbers are good emitters for the same temperature conditions.

In the photosphere, the radiation varies from place to place (in particular, it is decreasing towards the surface layers where most escapes into space) but sufficiently slowly that the emitting material and radiation are close to thermodynamic equilibrium. This is known as *local thermodynamic equilibrium* (LTE). In an LTE approximation, the photosphere's source function and specific intensity are given by $S_\nu = I_\nu = B_\nu$. The degree to which photospheric radiation approximates black-body radiation can be judged from Figure 2.3. This shows the amount of energy (erg) radiated by the quiet Sun per cm^2 per second per Å at one astronomical unit (1 AU = 1.496×10^{13} cm, i.e. the mean Earth–Sun distance), also known as the spectral irradiance. (This is defined further in Section 2.5.1.) Values for wavelengths greater than 2000 Å are from the Harvard–Smithsonian Reference Atmosphere (Gingerich *et al.* (1971)), those for wavelengths less than 1600 Å are from measurements with the SUMER instrument on *SOHO* (Curdt *et al.* (2001)). These curves are compared with the calculated black-body flux $\mathcal{F}_\nu$ for a temperature of 5778 K. The solar distribution is

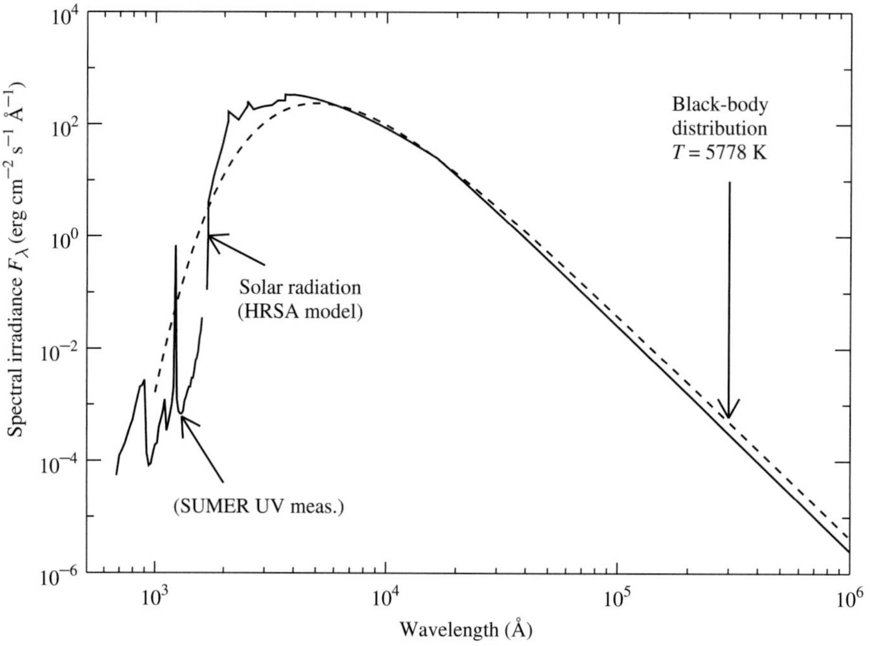

Fig. 2.3. Distribution of radiation, plotted as spectral irradiance, from a region of the Sun near Sun centre compared with the flux of a black body having $T = 5778$ K. The region from the ultraviolet (>2000 Å) to the infrared (2000 Å–10^6 Å) is from the Harvard–Smithsonian Reference Atmosphere (Gingerich *et al.* (1971)). For wavelengths less than 1610 Å, the plot is based on *SOHO* SUMER measurements for the quiet Sun (Curdt *et al.* (2001)).

crudely similar to the black-body distribution. More specifically, we may define a *brightness* temperature T_B of the photosphere to be such that $I_\nu = B_\nu(T_B)$. For visible wavelengths, $T_B \sim$6000 K, but has different values for other wavelength ranges (e.g. about 4400 K at far-infrared wavelengths), showing the departure of photospheric radiation from that of a black body.

The total amount of radiation at 1 AU over all frequencies has been accurately measured by spacecraft instruments in near-Earth orbit since 1978. Measurements of the total irradiance (i.e. spectral irradiance integrated over all frequencies) by the Active Cavity Radiometer Irradiance Monitor (ACRIM) instrument on *Solar Maximum Mission* (operational 1980–9) have a relative accuracy of better than 0.001%, and show decreases and increases of up to 0.3% occurring simultaneously with the passage of sunspots and faculae respectively. The mean value of the total solar irradiance is 1368 W m^{-2}, or 1.368×10^6 erg cm^{-2} s^{-1}. The total energy received from the Sun per second on a sphere, radius 1 AU, is therefore 3.85×10^{33} erg s^{-1}. This is equal to the solar *luminosity*. Dividing this number by the surface area of the solar photosphere, $4\pi R_\odot^2 = 6.09 \times 10^{22}$ cm^2, gives the power output per unit area of the Sun, 6.32×10^{10} erg cm^{-2}. Equating the solar irradiance to σT^4, the flux of a black body, gives $T = 5778$ K, which is known as the Sun's *effective temperature* $T_{\rm eff}$ (hence the black-body curve plotted in Figure 2.3 for this temperature). Note that this is a little lower than the solar photospheric temperature, around 6400 K.

There are, nonetheless, significant departures of the measured solar radiation distribution from the black-body distribution, particularly at ultraviolet and X-ray wavelengths, the range discussed in this book. One reason for this is the departure of the Sun's radiation from LTE. Non-local thermodynamic equilibrium (NLTE) applies in the solar atmosphere when the assumption of LTE breaks down, i.e. when radiation and the emitting and absorbing gas at a particular location in the solar atmosphere have significantly differing characteristic temperatures. This is relevant to the region where ultraviolet emission is formed as well as to the cores of strong Fraunhofer lines in the visible and near-ultraviolet spectrum.

2.4.2 Limb brightening and darkening: the Eddington–Barbier equation

Approximate solutions to the radiative transfer equation (Equation (2.16)) may be obtained under certain conditions. We write I_ν as a function of μ (i.e. $\cos\theta$) and τ_ν, $I_\nu(\mu, \tau_\nu)$. Multiplying Equation (2.16) by $e^{-\tau_\nu/\mu}$, we have

$$\frac{dI_\nu}{d(\tau_\nu/\mu)} e^{-\tau_\nu/\mu} = I_\nu\, e^{-\tau_\nu/\mu} - S_\nu\, e^{-\tau_\nu/\mu} \tag{2.18}$$

or

$$\frac{d(I_\nu\, e^{-\tau_\nu/\mu})}{d(\tau_\nu/\mu)} = -S_\nu\, e^{-\tau_\nu/\mu}, \tag{2.19}$$

which can be integrated over τ_ν/μ from 0 to ∞ to give the emergent intensity at $\tau_\nu/\mu = 0$ as

$$I_\nu(\mu, \tau_\nu = 0) = \int_0^\infty S_\nu(\tau_\nu)\, e^{-\tau_\nu/\mu}\, d(\tau_\nu/\mu). \tag{2.20}$$

If we approximate $S_\nu(\tau_\nu)$ by a linear function of τ_ν,

$$S_\nu(\tau_\nu) = A_\nu + B_\nu \tau_\nu, \tag{2.21}$$

we have

$$I_\nu(\theta, \tau_\nu = 0) = A_\nu + B_\nu \mu = S_\nu(\tau_\nu = \mu). \tag{2.22}$$

This equation tells us that for visible and near-ultraviolet light, the radiation from Sun centre ($\mu = 1$) has intensity $I_\nu = A_\nu + B_\nu$ but at the limb ($\mu = 0$) the intensity is less, $I_\nu = A_\nu$. This *limb darkening* depends on wavelength or ν, being much more pronounced for the near-ultraviolet continuum than for visible light. According to ground-based measurements of Mitchell (1981), the value of $I_\nu(0, 0)/I_\nu(1, 0)$ (i.e. limb intensity as a fraction of Sun centre intensity) is 0.20 at 5656 Å, 0.14 at 4500 Å, and 0.05 at 3000 Å. The limb darkening continues with decreasing wavelength into the ultraviolet until ~1600 Å. The reason is that the $\tau = 1$ level in the photosphere occurs when the temperature gradient dT/dr is negative, or (what is equivalent) the radiation emerges from higher, cooler layers as we approach the solar limb.

For wavelengths a little shorter than ~1600 Å, there is limb *brightening*, since for such wavelengths the radiation emerges from higher but hotter layers of the solar atmosphere, i.e. dT/dr is now positive.

For wavelengths in the extreme ultraviolet and X-ray regions, there is a limb brightening, not because of the positive gradient dT/dr but rather because the emission is optically thin and consequently a line of sight near the limb encounters a greater number of emitters of extreme ultraviolet radiation and X-rays. Lower down in the atmosphere, the temperatures are too low to emit significant amounts of such radiation. At the limb itself, there is a sudden upward jump of emission. Figure 2.4 (upper panel) shows this for an image in the 171 Å filter of the EIT instrument on *SOHO*. Figure 2.4 (lower panel) shows lines of sight (A, B) which intersect the Sun near the limb, including different amounts of coronal material emitting this radiation. Then at the limb (line of sight C) the amount of coronal material is doubled, with coronal material on the far side of the Sun becoming visible to an observer located in the

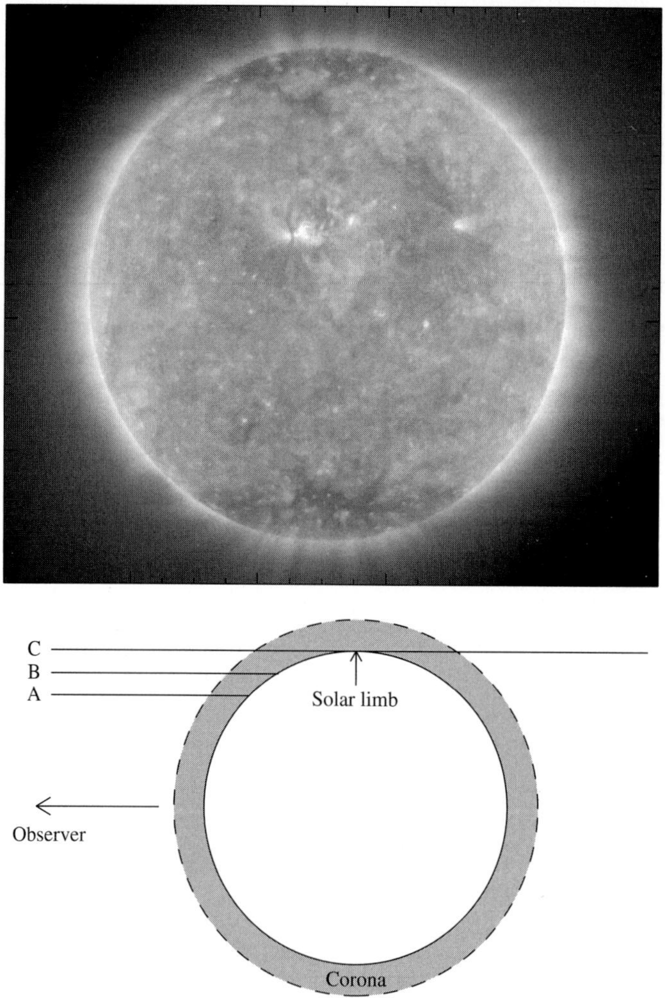

Fig. 2.4. (*Upper*) EIT 171 Å image on 2003 February 6 (13:00). Note how emission as a function of radial distance increases sharply at the limb. Courtesy the *SOHO* EIT Team. (*Lower*) Explanation of jump in limb emission: lines of sight A and B see progressively greater amounts of coronal material on the observer's side of the Sun, but the line of sight C sees in addition coronal material on the far side of the Sun.

direction indicated. Similar effects can be seen in EIT images made in the 195 Å and 284 Å images as well as in solar X-ray images.

We return to Equation (2.8) to find the flux at the solar surface. We insert Equation (2.22) into this equation to get

$$\mathcal{F}_\nu(\tau_\nu = 0) = 2\pi \int_0^1 (A_\nu + B_\nu\,\mu)\mu\, d\mu = \pi(A_\nu + \frac{2}{3}B_\nu) \tag{2.23}$$

leading to the *Eddington–Barbier* relation,

$$\mathcal{F}_\nu(\tau_\nu = 0) = \pi\, S_\nu(\tau_\nu = 2/3). \tag{2.24}$$

Thus, so long as Equation (2.21) holds, the surface flux is equal to π times the source function at optical depth $\tau_\nu = 2/3$. This gives us some intuition into understanding radiation from the solar atmosphere. Thus, radiation from the white-light continuum emitted at Sun centre emerges from the deepest photospheric layers, but for the continuum in the solar ultraviolet spectrum, we see higher up in the chromosphere. Features of the chromospheric network, therefore, that are not evident in white light become visible. Also for strong Fraunhofer lines, notably the hydrogen Hα and ionized calcium (Ca II) H and K lines, and the ionized magnesium (Mg II) h and k line doublet at 2800 Å, the $\tau_\nu = 2/3$ level is in the chromosphere, so spectroheliograms made in the cores of these lines reveal chromospheric structures such as spicules and the bright network.

If LTE holds, $\mathcal{F}_\nu(\tau_\nu = 0) = \pi \times B_\nu(\tau_\nu = 2/3)$. If we assume that the Sun's atmosphere is 'grey', i.e. has an absorption coefficient independent of ν, then

$$\mathcal{F}_\nu = \pi \times B_\nu[T(\tau = 2/3)], \tag{2.25}$$

so the distribution of radiation is like a black body's with a temperature equal to the surface that has $\tau = 2/3$. Integrating Equation (2.25) over all frequencies ν, we have $\mathcal{F} = \sigma T^4(\tau = 2/3)$. Now the definition of effective temperature is $\mathcal{F} = \sigma T_{\text{eff}}^4$, so $T_{\text{eff}}^4 = T^4(\tau = 2/3)$. Thus, the temperature at optical depth $\tau = 2/3$ is equal to the effective temperature. Although the solar atmosphere is not exactly grey, this result is not far wrong. A model atmospheric photospheric temperature at $\tau_{5000} = 1$ is 6420 K (Table 1.1); at $\tau_{5000} = 2/3$, nearly 100 km higher, the temperature is a little less, and is approximately equal to T_{eff}.

2.4.3 *Bright and dark features in solar images*

If Equation (2.19) is integrated over τ_ν from zero to a finite value τ'_ν/μ, we may examine the case of features such as prominences or spicules visible in solar ultraviolet and X-ray images. The feature is assumed to be viewed by a spacecraft with background radiation coming from the solar atmosphere beneath the feature. On an optical depth scale, optical depth τ'_ν/μ corresponds to the side of the feature nearest the Sun, optical depth 0 to the side of the feature towards the observer. We take $I_\nu(\mu, \tau'_\nu/\mu)$ to be the background radiation intensity, $I_\nu^{\text{B}}(\mu)$, and consider the feature to have a constant source function S_ν^{F}. Let the intensity of the feature be $I_\nu^{\text{F}}(\mu) = I_\nu^{\text{F}}(\mu, 0)$. Then we get

$$I_\nu^{\text{F}}(\mu) - I_\nu^{\text{B}}(\mu) = (S_\nu - I_\nu^{\text{B}})(1 - e^{-\tau'_\nu/\mu}). \tag{2.26}$$

The feature is bright in the image if $I_\nu^{\text{F}} > I_\nu^{\text{B}}$, but dark if $I_\nu^{\text{F}} < I_\nu^{\text{B}}$. The contrast becomes greater near the limb, i.e. as μ approaches zero, when $(1 - e^{-\tau'_\nu/\mu})$ approaches 1.

2.4.4 *The solar spectrum*

The solar spectrum from the visible to wavelengths of ~1600 Å is characterized by a continuum (an extension of the photospheric visible-light spectrum) with absorption lines. At smaller wavelengths, there is a spectrum consisting of emission lines superimposed on recombination continua (see Figure 1.2). For the photosphere to have an absorption line spectrum, the intensity of radiation emerging from deep layers of the Sun ($I_\nu^{\rm D}$) must be larger than the source function $S_\nu^{\rm L}$ for the region where the absorption lines are formed. Setting $I_\nu^{\rm D}$ equal to the Planck function $B_\nu^{\rm D}$ and the source function $S_\nu^{\rm L}$ to the Planck function $B_\nu^{\rm L}$, then $B_\nu^{\rm D} > B_\nu^{\rm L}$, i.e. the deeper layers must be hotter than those where the absorption lines are formed. This region can be identified with the photospheric layers up to the temperature minimum where there is a negative temperature gradient, $dT/dh < 0$ (see Figure 1.1).

For wavelengths less than ~1600 Å, the line-formation region of the solar atmosphere is where the temperature is rising with height, so the spectrum is the photospheric continuum with emission lines. The bulk of the emission at extreme ultraviolet and X-ray wavelengths is from the transition region and corona which are in general optically thin, so an emission-line spectrum without continuum is observed.

2.5 Radiant intensity and flux

2.5.1 *Definitions*

Spacecraft instruments, both imaging and spectral, make measurements of radiation coming from various parts of the solar atmosphere. The outputs of imaging instruments such as *TRACE* and EIT on *SOHO* are images of the Sun integrated over narrow wavelength bands that include emission lines and some continua, while spectrometers such as SUMER on *SOHO* (working in the ultraviolet) and *RHESSI* (working in the X-ray and gamma-ray range) measure spectra, i.e. the amount of radiation as a function of wavelength, or equivalent photon energy or frequency. Many spectrometers like SUMER can resolve spatial structures also, so can measure the amount of radiation from specific areas on the Sun, though some spectrometers, especially those in the X-ray range, measure the total amount of radiation from particular features such as solar flares. A familiar example of the latter is the ion chamber instruments on *GOES*.

To understand the units in which these measurements are, we first define some relevant quantities. First, *radiant flux* or *radiant power* is defined to be the amount of energy received from a source per unit time, and its c.g.s. unit is erg $\rm s^{-1}$. *Radiant intensity* is the amount of radiant power received from a source per unit solid angle, and has units erg $\rm s^{-1}\ sr^{-1}$. *Irradiance* is the amount of radiant flux or power received per unit area, and so has units of erg $\rm cm^{-2}\ s^{-1}$. *Radiance* is the amount of radiant intensity per unit area, with units erg $\rm cm^{-2}\ s^{-1}\ sr^{-1}$. These quantities are generally measured over particular wavelength bands, wide or narrow as use determines. Thus, the solar irradiance measured by instruments such as ACRIM on the *SMM* spacecraft was over very large bands so that the total amount of solar energy could be monitored.

Spectrometers measure the amount of radiant flux or radiant intensity in small wavelength bands. We define, following general usage in physics, *spectral irradiance* to be the radiant flux or radiant power per unit area per wavelength band, e.g. an angstrom (Å), or frequency band, e.g. a hertz ($\rm s^{-1}$). The c.g.s. unit of spectral irradiance is therefore erg $\rm cm^{-2}\ s^{-1}$ Å^{-1} or erg $\rm cm^{-2}\ s^{-1}\ Hz^{-1}$. *Spectral radiance* is the radiant intensity per wavelength or frequency

Table 2.1. *Radiation units defined.*

Quantity	Commonly called	c.g.s. unit	SI unit
Radiant flux or power		erg s^{-1}	W
Radiant intensity		erg s^{-1} sr^{-1}	W sr^{-1}
Irradiance	Flux	erg cm^{-2} s^{-1}	W m^{-2}
Radiance	Intensity	erg cm^{-2} s^{-1} sr^{-1}	W m^{-2} sr^{-1}
Spectral irradiance	Spectral flux	erg cm^{-2} s^{-1} $Å^{-1}$	W m^{-2} nm^{-1}
Spectral radiance	Spectral intensity	erg cm^{-2} s^{-1} sr^{-1} $Å^{-1}$	W m^{-2} sr^{-1} nm^{-1}

band, and its c.g.s. units are therefore erg cm^{-2} s^{-1} sr^{-1} $Å^{-1}$ or erg cm^{-2} s^{-1} sr^{-1} Hz^{-1}. Plotting the spectral radiance or irradiance against wavelength (e.g. in Å) or frequency (in Hz^{-1}) from a solar spectrometer thus gives the *spectrum* of the emitting source on the Sun. The spectrometer may be imaging, i.e. view an extended source such as part of the solar atmosphere, so obtaining solar irradiance. On the other hand, the instrument may have no imaging capabilities (e.g. whole-Sun instruments) and so measure solar radiance.

Frequently, the spectral intervals over which measurements are made are very narrow, and spectra are often plotted in terms of photons, taking account of the fact that the photons have different energies (equal to $h\nu = hc/\lambda$) for each part of the spectrum. Thus, the units of spectral irradiance may equivalently be photon cm^{-2} s^{-1} $Å^{-1}$ and those of spectral radiance photon cm^{-2} s^{-1} $Å^{-1}$ sr^{-1}.

Table 2.1 gives a summary of the c.g.s. and SI units of the radiation quantities defined above.

The above definitions, it must be said, are not often applied in plotting solar ultraviolet and X-ray spectra. The term flux is generally used instead of irradiance and intensity instead of radiance; spectral flux is often used instead of spectral irradiance, spectral intensity instead of spectral radiance. Table 2.1 will assist in interpretation of most solar physics literature where ultraviolet and X-ray spectra are given, including the discussion in Chapters 4, 5, and 6.

2.5.2 *Nature of the ultraviolet and X-ray solar spectrum*

As has been described already, the solar ultraviolet and X-ray spectrum has several components depending on the wavelength range. Here we anticipate a more detailed description of the ultraviolet and X-ray solar spectrum in Chapter 4.

At wavelengths above ~1600 Å and extending into the visible and infrared ranges, the photospheric continuum predominates, with absorption lines crossing this continuum. At wavelengths $\lesssim$ 1600 Å, emission lines predominate, with origins in the solar atmosphere above the photosphere and temperature minimum region. Continuous emission is also present, also originating in the solar atmosphere. The line emission may arise in the chromosphere, with temperatures of ~10 000 K, examples being the intense lines of hydrogen in the Lyman series between 1215.67 Å (Ly-α) and the series limit 911.75 Å. Others are emitted in the transition region or corona where the temperature is between 10^5 and a few 10^6 K, and still others are emitted in active regions or in flares with temperatures up to ~30 MK. Particular ion species are associated with each atmospheric region, so that the once-ionized Mg (Mg II) 2800 Å lines arise in the chromosphere but lines due to Li-like Mg

(Mg X) at 609.79, 624.95 Å are familiar lines in the extreme ultraviolet from the quiet corona. In X-ray flares, lines of He-like Mg (Mg XI) emit strong lines at 9.17–9.31 Å. The increasing degree of ionization of Mg atoms indicates the increasing temperature of the emitting region.

Spectral lines are not infinitely narrow in wavelength space but rather have intensity profiles extending over small wavelength intervals. Even in theory, these profiles are non-zero since the width of the upper level in energy units (ΔE) is related to the lifetime (Δt) of the upper level through Heisenberg's Uncertainty Principle, $\Delta E \Delta t \approx h/2\pi$. Much more significant line broadening occurs through the Doppler shifts of the emitting ions. If the ions have a Maxwell–Boltmann distribution, the line profile is said to be broadened by thermal Doppler broadening and has a Gaussian shape, with intensity distribution given by

$$I(\lambda) = I_0 \exp\left[- \frac{(\lambda - \lambda_0)^2}{\sigma^2} \right], \tag{2.27}$$

where I_0 is the peak intensity, λ_0 is the peak wavelength, and σ is the *line width*, defined as the interval $\sigma = \lambda_1 - \lambda_0$ where the intensity $I(\lambda_1)$ is $1/e$ times the peak intensity I_0. The full width of a Gaussian line profile at half the maximum intensity, usually abbreviated to FWHM, is given by

$$\mathrm{FWHM} = 2\sqrt{\ln 2}\, \sigma. \tag{2.28}$$

In terms of the emitting plasma temperature, the FWHM is

$$\mathrm{FWHM} = 2\sqrt{\ln 2}\, \frac{\lambda}{c} \left(\frac{2k_B T_{\mathrm{ion}}}{M} \right)^{1/2}, \tag{2.29}$$

where M is the ion mass and T_{ion} is the temperature (K) of the ions since it is the motion of the emitting ions that determines the line profile. If the ion and electron temperatures are equal, we may substitute T_e for T_{ion} in Equation (2.29). If there is significant microturbulence, then a term must be added to the right-hand side of Equation (2.29):

$$\mathrm{FWHM} = 2\sqrt{\ln 2}\, \frac{\lambda}{c} \left(\frac{2k_B T_{\mathrm{ion}}}{M} + \xi^2 \right)^{1/2}, \tag{2.30}$$

where ξ is the most probable velocity of the plasma turbulent motion.

Generally, for solar ultraviolet and X-ray line emission, thermal Doppler broadening and plasma turbulence are the most significant broadening mechanisms of solar origin, being much larger than that due to the finite lifetime of the upper level of the transition. The spectrometer observing the lines will itself broaden the line profile. This applies to instruments with diffracting gratings and with Bragg diffracting crystals. The instrumental profile will not necessarily be Gaussian or even symmetric about the line peak, though frequently, as in the case of Bragg crystals, the profile may be approximately *Lorentzian*:

$$I(\lambda) = \frac{I_0}{1 + (\lambda - \lambda_0)^2/d^2}, \tag{2.31}$$

where d is half of the full width of the Lorentzian profile. If the instrumental broadening has a Gaussian profile, with $\Delta\lambda_I$ the FWHM of the profile, the profile of a line with thermal Doppler broadening must be corrected by the addition in quadrature of $\Delta\lambda_I$ to FWHM in Equation (2.29).

In addition to line broadening, there may be line *shifts* because of bulk plasma motions. Sometimes spectral lines may have multiple components, for example a 'stationary' component (zero line shift) with components on the short- ('blue') or long-wavelength ('red')

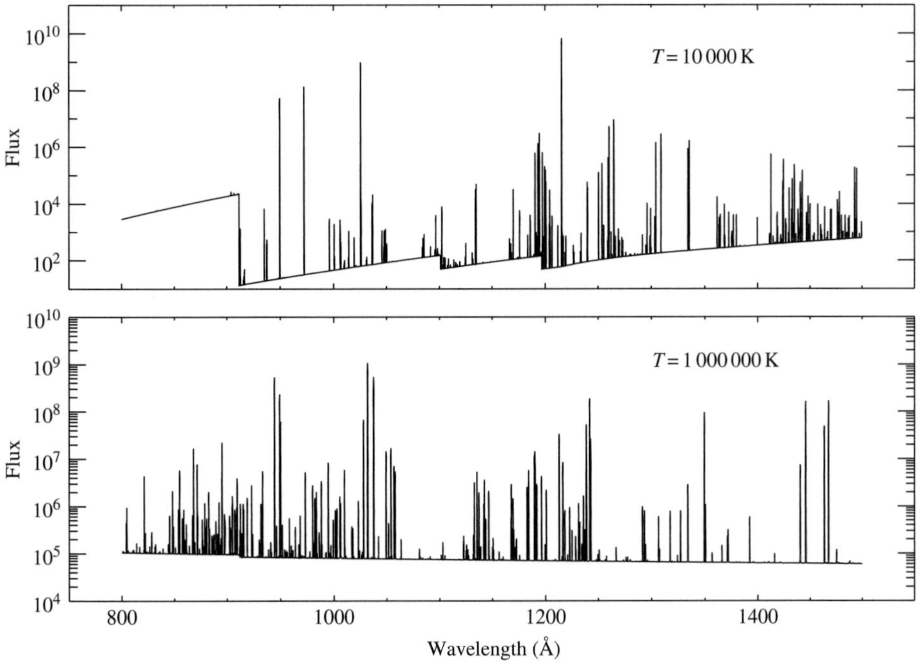

Fig. 2.5. CHIANTI spectra in the 800–1500 Å range for $T_e = 10\,000$ K (*top*) and $T_e = 1$ MK (*bottom*).

side of the stationary component, possibly blended with it. Note that for a star the emitting atmospheric structures are spatially unresolved, and the line profile may be broadened by stellar rotation. The line position will also be displaced if the star has significant Doppler motion relative to the Sun.

The main contributors of the continuum emission in the ultraviolet and X-ray regions are free–free and free–bound processes. In free–free emission or *bremsstrahlung*, a free electron moves in the Coulomb field of a target ion, with a loss of energy that appears as a photon of free–free emission. In free–bound emission or recombination radiation, a free electron is captured by an ion and the excess energy appears as free–bound emission. Two-photon emission has only a tiny contribution to the total continuum in solar spectra.

Atomic databases and codes enable ultraviolet and X-ray solar spectra to be predicted for particular conditions, e.g. temperature and density. As an example, we use the CHIANTI code (Dere *et al.* (1997), Young *et al.* (2003)) to predict the 800–1500 Å spectrum shown in Figure 1.2 (first panel). This spectrum, taken with the SUMER instrument on *SOHO*, was from a portion of the solar disk, and so both a relatively low chromospheric spectrum, with recombination continua of H, C, and S, and a coronal spectrum are apparent. This was synthesized in Figure 2.5 using the CHIANTI code for temperatures of 10 000 K and 1 MK to simulate the chromosphere and corona respectively. The recombination continua and H Lyman lines are evident in the chromospheric spectrum, while many of the other emission lines at coronal temperatures are evident in the coronal spectrum.

3

Fundamentals of atomic physics

3.1 Introduction

In this chapter, we will introduce the fundamental concepts and equations that describe atomic systems, and their interaction with external radiation. We will also describe the physical processes underlying the emission of spectral lines.

3.2 The one-electron atom: Bohr's theory

The first atom for which a theory was developed that could account for the properties of its spectrum was the H atom. One of the reasons is that the H spectrum is rather simple, and the H atom itself has the simplest structure among all elements, being composed of a nucleus of a single proton, and one bound electron.

The H spectrum consists of several series of lines whose wavelengths follow a regular pattern. When first observed, only one series at a time could be identified, since the wavelength ranges where each of these series was observed are located in distant regions of the spectrum and could not be observed by the same instrument. The peculiarity of the series of lines in the H spectrum was first expressed mathematically by Balmer, who found that the wavelengths λ of the lines in the series at around 5000 Å (the *Balmer series*) followed a simple formula, given by

$$\frac{1}{\lambda} = R\left(\frac{1}{2^2} - \frac{1}{n^2}\right). \tag{3.1}$$

The constant R, having the units of the inverse of a length, was measured and found to be 109 677.58 cm^{-1}, and n is a positive integer. Subsequent observations of the other series at different wavelengths allowed the validity of the Balmer formula to be extended to a more general form

$$\frac{1}{\lambda} = R\left(\frac{1}{n_1^2} - \frac{1}{n_2^2}\right) \qquad n_2 > n_1, \tag{3.2}$$

where the positive integer n_1 identifies the series (as in Table 3.1), and the positive integer n_2 identifies each of the lines within the series. As n_2 increases the separation between two consecutive members of the same series decreases, and when $n_2 \to \infty$ the wavelengths of the lines in the series converge towards a fixed value $\lambda = R/n_1^2$, called the series limit (i.e. for $n_1 = 1$ we have the Lyman limit, for $n_1 = 2$ the Balmer limit).

Table 3.1. *Lines of the two first series in the H spectrum.*

n_1	Series name	n_2	Line name	Wvl. (Å)
1	Lyman	2	Ly-α	1215.7
		3	Ly-β	1025.7
		4	Ly-γ	972.5
		$\vdots$	$\vdots$	
2	Balmer	2	Ba-α	6564.6
		3	Ba-β	4865.7
		4	Ba-γ	4341.7
		$\vdots$	$\vdots$	

Other important series are the Paschen series ($n_1 = 3$), the Brackett series ($n_1 = 4$), the Pfund series ($n_1 = 5$), and the Humphreys series ($n_1 = 6$).

A similar behaviour occurs in the spectrum of all ions with only one electron (i.e. He^+, Li^{2+}, Be^{3+}, etc.: the so-called *H-like* or *hydrogenic* ions). However, Equation (3.2) must be slightly modified to

$$\frac{1}{\lambda} = Z^2 R\left(\frac{1}{n_1^2} - \frac{1}{n_2^2}\right), \tag{3.3}$$

where Z is the atomic number of the ions and R has a slightly different value from the one measured for H. This means that the spectral lines of these ions will behave exactly as the H lines, but their wavelengths will be shorter than those of the corresponding series in the hydrogen spectrum.

The spectrum of hydrogen or of H-like ions was interpreted for the first time by Niels Bohr, who developed a theory that could account for the presence of series and predict their wavelengths. Bohr assumed that an H-like ion was composed of a nucleus with positive charge Ze and an electron of charge $-e$ bound in an orbit around the nucleus. Bohr further postulated that:

(1) The electron can only travel in a few orbits among the infinite possible orbits allowed by classical mechanics, characterized by a fixed angular momentum $l = n\hbar$, where n is an integer number and $\hbar$ is the Planck constant h divided by 2π.
(2) Although the electron is constantly accelerated as it travels along its orbit, it does not emit any radiation, even though this property violates Maxwell's theory, so that the energy in the orbit is constant.
(3) The electron emits radiation only when it jumps from one discrete orbit to another. Since the energy of each orbit is constant, and the energy of the system must be conserved, the energy of the emitted radiation is fixed, and the transition gives rise to a spectral line.

The wavelength λ of the spectral line is assumed to be related to the change in electron energy when the electron changes from an orbit 2 to an orbit 1, according to the relation

$$\Delta E = E_2 - E_1 = \frac{hc}{\lambda}, \tag{3.4}$$

where c is the speed of light. These assumptions allow us to calculate the energy of all the levels in H-like ions and to predict the wavelengths of their spectra. In fact, applying the first Bohr assumption to the equation of motion of the simplest orbit (the circular orbit), we have

$$F = ma \quad \Longrightarrow \quad \frac{Ze^2}{r^2} = \frac{mv^2}{r}, \tag{3.5}$$

where m is the mass of the electron, v the electron velocity, e the electron charge and r the electron–nucleus distance, with the nucleus assumed to be of infinite mass. The discrete values of the angular momentum l of the electron allow us to determine its velocity v:

$$|l| = n\frac{h}{2\pi} \quad \Longrightarrow \quad mvr = n\frac{h}{2\pi} \quad \Longrightarrow \quad v = n\frac{h}{2\pi mr}. \tag{3.6}$$

Combining Equations (3.5) and (3.6), we can determine the radius of the nth circular orbit as

$$r = \frac{n^2}{Z}a_0 \qquad \text{with} \qquad a_0 = \frac{h^2}{4\pi^2 me^2}. \tag{3.7}$$

The quantity a_0, having the dimensions of length, is called the *Bohr radius* and represents the radius of the first orbit ($n = 1$) of the H atom ($Z = 1$). It has a value of 0.529 Å. We can calculate the energy associated with each orbit of the electron as the sum of the kinetic and potential energies of the electron in the field of the nucleus. By assuming that at infinite distance from the nucleus ($r \to \infty$) the potential energy of the electron is zero, the energy of the nth orbit is

$$E = -\frac{Ze^2}{r} + \frac{1}{2}mv^2 = -\frac{Z^2}{n^2}\frac{2\pi^2 me^4}{h^2}. \tag{3.8}$$

Equation (3.8) has several important consequences. First, it shows that for a given H-like ion the energy of each orbit is determined by the quantum number n only. Second, energies of orbits are negative, that is, lower than the energy that the electron has at infinite distance from the nucleus. Third, the lowest energy level is the one for which $n = 1$. This state is called the *ground state*: since no level with lower energy exists, the atom in this state is stable, and no transition to lower levels can occur from the ground state. Fourth, applying the third Bohr assumption for the H-like ion, the wavelength λ of the photon emitted when the electron jumps from an orbit n_2 with higher energy to an orbit n_1 with lower energy can be obtained by combining Equations (3.4) and (3.8):

$$\frac{1}{\lambda} = Z^2\frac{2\pi^2 me^4}{ch^3}\left(\frac{1}{n_1^2} - \frac{1}{n_2^2}\right). \tag{3.9}$$

This equation is able to predict with great accuracy the wavelengths of the series of hydrogen and of H-like ions; from the comparison of Equation (3.9) with Equation (3.3), the constant R is determined as

$$R = \frac{2\pi^2 me^4}{ch^3}. \tag{3.10}$$

This quantity is called the Rydberg constant and has the units of an inverse length. Its value is 109 737.31 cm^{-1}. From Equation (3.9), we see that once n_1 is fixed, n_2 can have any integer value, provided it is greater than n_1. However, the larger n_2 is, the smaller is the decrease in the wavelength of the transition; for $n_2 \to \infty$, wavelengths converge to a limit:

$$\lim_{n_2\to\infty}\frac{1}{\lambda} = Z^2 R\frac{1}{n_1^2} = \frac{1}{\lambda_{\text{edge}}}. \tag{3.11}$$

Each series has its own limit. The energy of the series limit corresponds to the energy needed to bring an electron in the orbit n_1 to infinity, with zero final energy. Above this limit the electron is free from the nucleus and can assume any value of the energy; the process that brings a bound electron into a free state is called *ionization*. The energy necessary to bring a bound electron in the ground state n_1=1 of an atom or ion to infinity is called the *ionization potential*. For H, this energy is given by

$$I_{\rm H} = \frac{2\pi^2 m e^4}{h^2} = 13.6 \text{ eV}. \tag{3.12}$$

Transitions from these free states back to the level n_1, called *recombinations*, can have any wavelength shorter than $\lambda_{\rm edge}$ and form a continuum, called a *free–bound continuum*. The spectrum of a series of an element will therefore show an increasingly higher density of lines towards the series limit, followed by the free–bound continuum. Close to the recombination edge, the spectral lines are so close to each other that they are indistinguishable, so that they form a pseudo-continuum. An example of the spectrum of the Lyman series and the Lyman edge for neutral H, observed in the solar spectrum by the SUMER instrument on *SOHO*, is given in Figure 3.1.

The above treatment is correct if we assume that the mass of the nucleus is infinite. However, even though the mass M of the nucleus is much larger than that of the electron, it is not infinite. It is possible to show that both the electron and the nucleus orbit around the centre of mass of the system, but that the equations that govern the motion of the bound electron

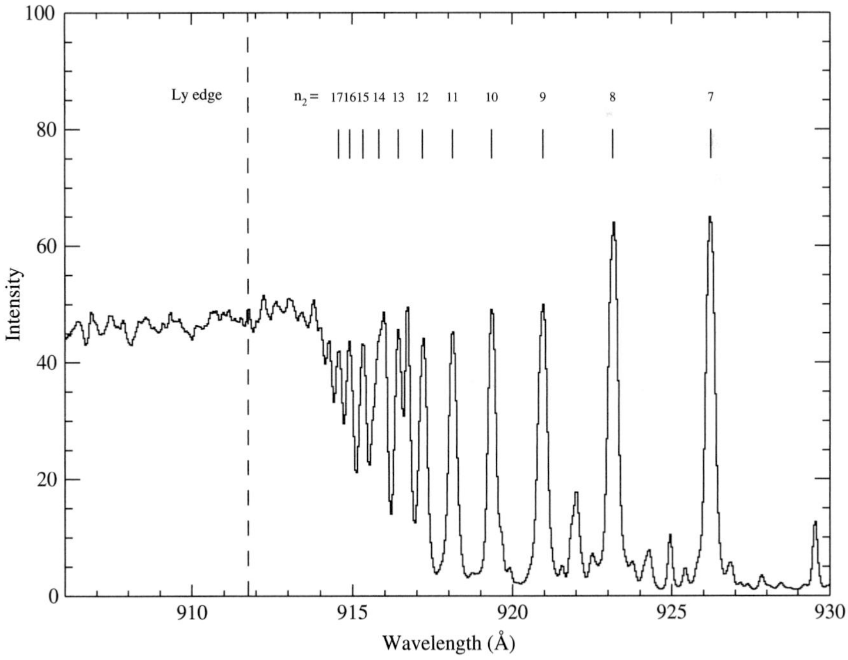

Fig. 3.1. Spectrum of the solar corona in the 906–930 Å wavelength range, observed by the SUMER instrument on board *SOHO*. The Lyman edge of H, and the Lyman lines with $7 \leq n_2 \leq 17$ are visible. Lines with $n_2 > 17$ are blended together in the pseudo-continuum between 914 Å and the Lyman series limit at 911.75 Å, indicated by the dashed line. Lines with $n_2 = 13$ and 14 are blended with lines from other elements. Courtesy *SOHO* SUMER team.

relative to the nucleus are the same as those given above, provided we replace the electron mass m with the *reduced mass*

$$\mu = \frac{mM}{m + M}. \tag{3.13}$$

Therefore, the energy of the orbits and the constant R for each H-like system will be

$$E = -\frac{Z^2}{n^2}\frac{2\pi^2\mu e^4}{h^2}, \qquad R = \frac{2\pi^2\mu e^4}{ch^3}. \tag{3.14}$$

This correction allows a more accurate prediction of the wavelengths of all the series. The value of the constant R is slightly different for each H-like system, and is equal to the Rydberg constant in the limit of infinite nuclear mass M.

Bohr's theory was very successful at predicting the spectra of H-like systems, but when applied to atomic systems with more than one electron it was unable to reproduce observed spectra. A more complete theory able to account for observed spectra of any atomic system makes use of quantum mechanics.

3.3 The one-electron atom: quantum theory

We will first introduce the quantum theory of atomic spectra by applying it to an H-like atomic system composed of a nucleus with positive charge Ze and one electron. We will make the assumptions that (i) the nucleus is point-like, and that (ii) this system is isolated, i.e. no external field or perturbation is applied to it.

In quantum mechanics, the status of a system is described by a function $\psi(\mathbf{x})$, called a *wavefunction*, such that the quantity $|\psi|^2 d\mathbf{x}$, where $d\mathbf{x}$ is an element in the space of the parameters $\mathbf{x}$, represents the probability of the system having the parameters in the element $d\mathbf{x}$. The evolution of the system described by the wavefunction ψ is determined by the Schrödinger equation

$$i\hbar\frac{\partial\psi}{\partial t} = H\psi, \tag{3.15}$$

where H is a linear Hermitian operator that corresponds to the classical Hamiltonian function, and is therefore called the *Hamiltonian* operator. In classical mechanics, the Hamiltonian function is equal to the total energy of the system when the coordinates are independent of time.

Since this atom is assumed to be isolated in space, its wavefunction can be expressed as the product of the wavefunction of the centre of mass of the atom and of the wavefunction of the relative motion of the electron and the nucleus. The former is a free-particle wavefunction, the latter describes the status of the electron in the atom and is the topic of the present discussion.

The wavefunction ψ of the bound electron can be determined by solving the Schrödinger equation. To start with, we consider the non-relativistic treatment of an isolated one-electron atom, so that the Hamiltonian H of this system can be expressed as

$$H = \frac{\mathbf{p}^2}{2\mu} + V(r), \qquad \mathbf{p} = \frac{\hbar}{i}\nabla, \qquad V(r) = -\frac{Ze^2}{r}, \tag{3.16}$$

where $\mathbf{p}$ is the momentum operator, $V(r)$ is the attractive potential of the nucleus, which depends on the distance r between the electron and the nucleus, and μ is the reduced mass.

Since the system is assumed to be unperturbed by external potentials at any time, we can use the stationary form of the Schrödinger equation:

$$H\psi = E\psi \qquad \Longrightarrow \qquad \left[-\frac{\hbar^2}{2\mu}\nabla^2 - \frac{Ze^2}{r}\right]\psi = E\psi, \tag{3.17}$$

where the Hamiltonian of the system includes the kinetic energy of the electron and the attractive potential of the nucleus $V(r)$, E is the electron energy, $\hbar = h/2\pi$ is the reduced Planck constant, ψ is the normalized wavefunction of the electron, and r is the electron–nucleus distance. A description of this system can be provided by a set of three mutually commuting operators, where with mutually commuting operators we indicate operators $\mathbf{T_1}$ and $\mathbf{T_2}$ which have the property

$$[\mathbf{T_1}, \mathbf{T_2}] = \mathbf{T_1 T_2} - \mathbf{T_2 T_1} = 0. \tag{3.18}$$

Since $V(r)$ is a central potential, a set of such operators includes the Hamiltonian H and the two angular momentum operators $\mathbf{L^2}$ and $\mathbf{L_z}$, where $\mathbf{L} = (\mathbf{r} \times \mathbf{p})$, $\mathbf{r}$ and $\mathbf{p}$ being the position and momentum operators, and $\mathbf{L_z}$ is the projection of $\mathbf{L}$ along the z-axis. The wavefunction of the system must therefore satisfy the relations

$$H\psi = E\psi, \tag{3.19}$$

$$\mathbf{L^2}\psi = \xi\psi, \tag{3.20}$$

$$\mathbf{L_z}\psi = m\psi, \tag{3.21}$$

where E, ξ and m are the eigenvalues of the three operators. To simplify the solution of Equation (3.17), we adopt the spherical coordinates

$$\begin{aligned} x &= r\ \sin\theta\ \cos\varphi, \\ y &= r\ \sin\theta\ \sin\varphi, \\ z &= r\ \cos\theta, \end{aligned} \tag{3.22}$$

where θ is the angle between $\mathbf{r}$ and the z axis ($0 \leq \theta \leq \pi$), and φ is the angle between the projection of $\mathbf{r}$ on the $x-y$ plane and the x axis ($0 \leq \varphi \leq 2\pi$), as shown in Figure 3.2.

In spherical coordinates, the $\mathbf{L^2}$ and $\mathbf{L_z}$ operators involve only angular coordinates:

$$\mathbf{L^2} = -\frac{1}{\sin\theta}\frac{\partial}{\partial\theta}\left(\sin\theta\frac{\partial}{\partial\theta}\right) - \frac{1}{\sin^2\theta}\frac{\partial^2}{\partial\varphi^2}, \tag{3.23}$$

$$\mathbf{L_z} = -i\frac{\partial}{\partial\varphi}, \tag{3.24}$$

while ∇^2 takes the form

$$\begin{aligned} \nabla^2 &= \frac{1}{r^2}\frac{\partial}{\partial r}\left(r^2\frac{\partial}{\partial r}\right) + \frac{1}{r^2\sin\theta}\frac{\partial}{\partial\theta}\left(\sin\theta\frac{\partial}{\partial\theta}\right) + \frac{1}{r^2\sin^2\theta}\frac{\partial^2}{\partial\varphi^2} \\ &= \frac{1}{r^2}\frac{\partial}{\partial r}\left(r^2\frac{\partial}{\partial r}\right) - \frac{\mathbf{L}^2}{r^2}. \end{aligned} \tag{3.25}$$

Because the radial coordinate r does not appear in Equations (3.23) and (3.24), and the $\mathbf{L_z}$ operator only involves the coordinate φ, the electron wavefunction can be expressed as the product of three functions with separate coordinates:

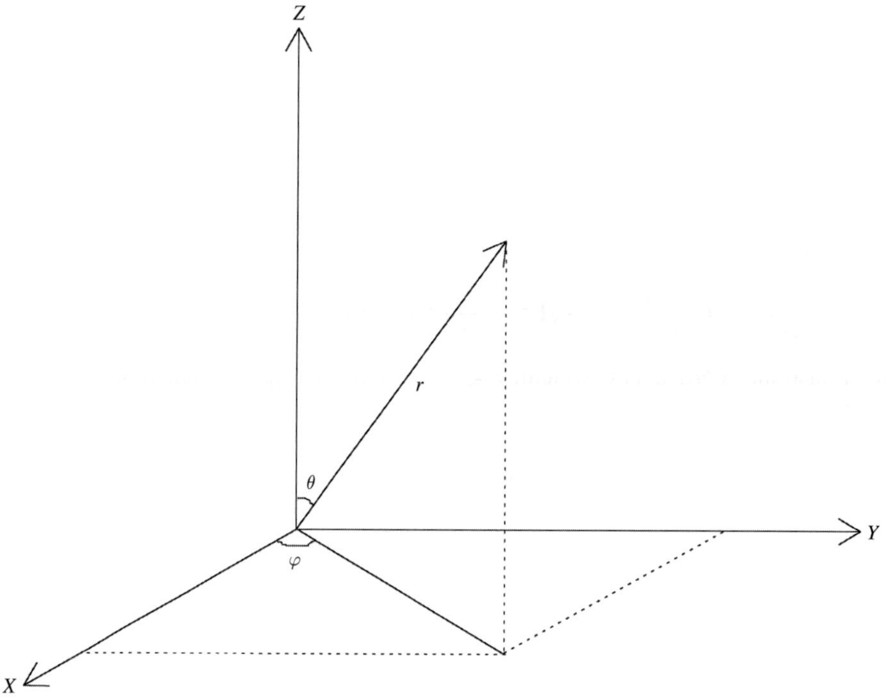

Fig. 3.2. Spherical coordinates reference system.

$$\psi(r, \theta, \varphi) = \frac{R(r)}{r} A(\theta) B(\varphi). \tag{3.26}$$

The solution of Equation (3.21) is straightforward:

$$\mathbf{L_z}\psi(r, \theta, \varphi) = m\psi(r, \theta, \varphi) \quad \Longrightarrow \quad B(\varphi) = e^{im\varphi}. \tag{3.27}$$

Since the electron wavefunction needs to be uniquely determined for any value of φ,

$$e^{im\varphi} = e^{im(\varphi+2\pi)} \quad \Longrightarrow \quad e^{2im\pi} = 1, \tag{3.28}$$

so that m must be an integer number. By applying the operator $\mathbf{L}^2$ to the function $\psi(r, \theta, \varphi)$ we can solve Equation (3.20) and determine $A(\theta)$:

$$\mathbf{L}^2\psi(r, \theta, \varphi) = \xi\psi(r, \theta, \varphi) \quad \Longrightarrow \quad A(\theta) = P_l^{|m|}(\cos\theta) \tag{3.29}$$

with

$$\xi = l(l+1), \qquad |m| \le l, \qquad l \ge 0. \tag{3.30}$$

Since m is an integer, l is also an integer, and the functions $P_l^{|m|}(\cos\theta)$ are the *associated Legendre polynomials*. Therefore, the function $Y(\theta, \varphi) = A(\theta)B(\varphi)$ will be characterized by the two integers m and l. The functions $Y_{lm}(\theta, \varphi)$, called *spherical harmonics*, are the eigenfunctions of both $\mathbf{L^2}$ and $\mathbf{L_z}$, and form a basis of orthogonal functions. When they are normalized in the solid angle, they can be expressed as

$$Y_{lm}(\theta, \varphi) = (-1)^{(m+|m|)/2} \sqrt{\frac{2l+1}{4\pi} \frac{(l-|m|)!}{(l+|m|)!}} P_l^{|m|}(\cos\theta) e^{im\varphi}. \tag{3.31}$$

Using Equation (3.25), the Schrödinger equation (Equation (3.17)) in spherical coordinates takes the form

$$\left[-\frac{\hbar^2}{2\mu r^2} \frac{\partial}{\partial r}\left(r^2 \frac{\partial}{\partial r} \right) + \frac{\hbar^2}{2\mu r^2} \mathbf{L^2} - \frac{Ze^2}{r} \right] \psi = E\psi. \tag{3.32}$$

By using Equations (3.26) and (3.29) with $\xi = l(l+1)$, the Schrödinger equation allows us to determine the radial function $R(r)$ by solving

$$-\frac{\hbar^2}{2\mu} \frac{\partial^2 R(r)}{\partial r^2} + V_l(r) R(r) = E\, R(r), \tag{3.33}$$

where

$$V_l(r) = -\frac{Ze^2}{r} + \frac{\hbar^2}{2\mu} \frac{l(l+1)}{r^2}. \tag{3.34}$$

The solution of Equation (3.33) must not diverge either for $r \to \infty$ or for $r = 0$, and be finite for every value of r. In order for solutions of Equation (3.33) to comply with these two requirements, it is necessary that the values of E be discrete, and related to l. By analogy to Bohr's theory, we can write the energy E as

$$E = -\frac{1}{n^2} \frac{Z^2 e^2}{2a_0}, \tag{3.35}$$

where a_0 is the Bohr radius. Also, in order to have finite solutions to Equation (3.33), the number n must be (i) a positive integer number, and (ii) related to l by

$$l \leq n - 1. \tag{3.36}$$

The number n is called the *principal quantum number*, l the *azimuthal quantum number*, and m the *magnetic quantum number*. The energy of the system depends solely on the principal quantum number n and is independent of the l and m quantum numbers. All these levels with the same energy are called *degenerate* levels.

It is important to note that the discrete nature of the solutions ψ to the Schrödinger equation, and of the quantum numbers n, l, and m, flows naturally as a consequence of Equation (3.17) and of the requirements that the solution be uniquely determined (Equation (3.28)) and be finite for any value of r (Equations (3.33), (3.35), and (3.36)). No further arbitrary assumptions such as those adopted by Bohr are now necessary.

It is also possible to demonstrate that the solutions of Equation (3.33) generate wavefunctions of the type in Equation (3.26) that satisfy the orthogonality condition

$$\int_0^\infty \int_0^\pi \int_{-\pi}^{+\pi} \psi^\star_{n'l'm'}(r, \theta, \varphi) \psi_{nlm}(r, \theta, \varphi) r^2 \sin\theta \, d\varphi \, d\theta \, dr = K_{nl} \delta_{n'n} \delta_{l'l} \delta_{m'm}, \tag{3.37}$$

where K_{nl} is a constant and the δs are Dirac δ functions. From now on, the wavefunctions $\psi_{nlm}(r, \theta, \varphi)$ (called *eigenfunctions*) will be indicated by their quantum numbers according to the notation

$$\psi_{nlm}(r, \theta, \varphi) = |n, l, m\rangle. \tag{3.38}$$

By normalizing the eigenfunctions $\psi_{nlm}(r, \theta, \varphi)$ to 1, Equation (3.37) can be rewritten with the notation

$$\langle n'l'm'|nlm\rangle = \delta_{n'n}\delta_{l'l}\delta_{m'm}. \tag{3.39}$$

If $|nlm\rangle$ and $|n'l'm'\rangle$ do not belong to a set of orthogonal wavefunctions, the expression $\langle n'l'm'|nlm\rangle$ indicates the component of the wavefunction $|nlm\rangle$ along the wavefunction $|n'l'm'\rangle$. Traditionally the values of the azimuthal quantum number l of a single electron are indicated by letters:

$l =$	0	1	2	3	4	5	6	7	8	9	...
	s	p	d	f	g	h	i	k	l	m	...

For example, an electron with $(n, l) = (5, 3)$ is called a $5f$ electron.

The above treatment of the one-electron atomic systems does not take into account an intrinsic property of the electrons, called *spin*. The electron spin consists of an angular momentum intrinsic to each electron, described by an operator indicated with **s**, whose eigenvalues are $s = \pm 1/2$. The electron spin was first introduced empirically in order to explain small discrepancies between the energy values predicted by the above model and measurements. At first, the angular momentum associated with the electron was interpreted as that due to the rotation of the electron, assumed to be a uniform spherical body, around its own axis; however, such interpretation was not satisfactory. Only a fully relativistic model, beyond the scope of this book, is able to predict the presence and the properties of the electron spin.

The electron spin could be neglected in the above model of the one-electron atom since its operator **s** commutes with the system's non-relativistic Hamiltonian in Equation (3.16), and with the $\mathbf{L}^2$ and $\mathbf{L_z}$ operators. However, the spin of the electron must be taken into account when relativistic corrections are introduced into the non-relativistic Hamiltonian, and when multi-electron atoms are considered. Usually the presence of the spin can be taken into account by multiplying the non-relativistic wavefunction in Equation (3.39) by a wavefunction depending on a quantum number m_s specifying the projection of the spin along the z axis.

Taking into account the presence of the spin, the total number of levels with the same energy is $2n^2$.

3.4 The multi-electron atom: quantum theory

3.4.1 Wavefunction symmetry

We will now deal with the case of a nucleus of charge Ze surrounded by two or more bound electrons. The presence of more than one electron in the system introduces the problem of *indistinguishable* particles. In classical mechanics, it is always possible, at least in principle, to determine the trajectory of each of two or more particles, so that even if they are identical under all aspects, they still can be distinguished at all times by following their motion. In quantum mechanics, the Heisenberg Uncertainty Principle prevents us from knowing the trajectory of each particle, so if two identical particles come close enough that their wavefunctions overlap, we lose the ability of distinguishing them. As a result, the probability describing a system with two or more identical particles must be the same if we exchange one or more particles. This means that, in the case of a system with two identical particles, if they are exchanged, the wavefunction ψ of the system must change only by a phase factor, because such a factor has no influence on the probability distribution, given by $|\psi|^2$. Therefore,

$$\psi(\mathbf{x}_2, \mathbf{x}_1) = e^{i\alpha}\psi(\mathbf{x}_1, \mathbf{x}_2), \tag{3.40}$$

where $e^{i\alpha}$ is the phase factor, and $\mathbf{x}_1$ and $\mathbf{x}_2$ are the position and spin coordinates of the two particles. By exchanging the role of the two particles again, we must obtain the original wavefunction, so that

$$e^{2i\alpha} = 1 \qquad \Longrightarrow \qquad \psi(\mathbf{x}_2, \mathbf{x}_1) = \pm\psi(\mathbf{x}_1, \mathbf{x}_2) \tag{3.41}$$

and therefore the wavefunction of a system including two identical particles must be either *symmetric* ($e^{i\alpha} = 1$) or *antisymmetric* ($e^{i\alpha} = -1$) to the exchange of the two particles. If we have more than two identical particles, the same rule applies to the wavefunction of the entire system: in fact, if two of those particles can be described by a symmetric (or antisymmetric) wavefunction, all the others must have the same property.

The *symmetry* of the total wavefunctions describing a system of identical particles depends on the spin of the particles themselves: integer spin corresponds to symmetric wavefunctions, half-integer spin corresponds to antisymmetric ones.

3.4.2 The Pauli exclusion principle

When we apply the symmetry properties to a system including two or more electrons, we find that it is impossible to have two electrons in the same state, i.e. characterized by the same quantum numbers. This property is called the *Pauli exclusion principle*. To understand this, let us consider a system with N electrons whose mutual interaction can be neglected. The electron spin is half integer, so the total wavefunction of the system must be antisymmetric, and it can be expressed by an antisymmetric combination of the product of the wavefunctions $\varphi_{n_i}(\mathbf{x}_i)$ of each electron i, where n_i represents the ensemble of quantum numbers that describe the status of the electron i, and $\mathbf{x}_i$ its position and spin coordinates. If we indicate by $\mathbf{P}$ the operator that exchanges the coordinates of two particles in the system, we can write

$$\psi = \sum (-1)^p \mathbf{P}\big[\varphi_{n_1}(\mathbf{x}_1)\varphi_{n_2}(\mathbf{x}_2)\ldots\varphi_{n_N}(\mathbf{x}_N)\big], \tag{3.42}$$

where the summation is carried out over all the possible permutations among the particles, and p is either 0 (even permutation) or 1 (odd permutation). Such a wavefunction can also be written in the form of a determinant (called a *Slater determinant*):

$$\psi = \det \begin{vmatrix} \varphi_{n_1}(\mathbf{x}_1) & \varphi_{n_1}(\mathbf{x}_2) & \cdots & \varphi_{n_1}(\mathbf{x}_N) \\ \varphi_{n_2}(\mathbf{x}_1) & \varphi_{n_2}(\mathbf{x}_2) & \cdots & \varphi_{n_2}(\mathbf{x}_N) \\ \vdots & \vdots & & \vdots \\ \varphi_{n_N}(\mathbf{x}_1) & \varphi_{n_N}(\mathbf{x}_2) & \cdots & \varphi_{n_N}(\mathbf{x}_N) \end{vmatrix}. \tag{3.43}$$

The wavefunction will be null if two (or more) columns of the determinant are identical, i.e. if the ensemble of quantum numbers n_i of two particles is the same, so that the states of the two particles are the same.

Another property of the symmetry of the wavefunction of a system is that it is constant with time. In fact, from the Schrödinger equation, the variation $d\psi$ of the wavefunction in the time interval dt is given by

$$d\psi = \frac{1}{i\hbar} H\psi \, dt, \tag{3.44}$$

so that the variation $d\psi$ has the same symmetry as that of ψ, since H is invariant with respect to particle exchange.

3.4.3 *Hamiltonian of a multi-electron atom: configurations and terms*

In the non-relativistic case, the Hamiltonian of an N-electron atom includes the same kinetic terms and the electron–nucleus interaction as the one-electron Hamiltonian, plus an additional term that describes the mutual interaction among the electrons:

$$\begin{aligned} H &= \sum_{i=1}^{N} \left(\frac{\mathbf{p}_\mathrm{i}^2}{2\mu_i} - \frac{Ze^2}{r_i} \right) + \sum_{i<j} \frac{e^2}{r_{ij}} \\ &= \sum_{i=1}^{N} \left(-\frac{\hbar^2}{2\mu_i} \nabla_i^2 - \frac{Ze^2}{r_i} \right) + \sum_{i<j} \frac{e^2}{r_{ij}}, \end{aligned} \tag{3.45}$$

where r_{ij} is the distance between the electrons i and j. In this Hamiltonian the relativistic effects are ignored. This Hamiltonian commutes with several operators, so that its wavefunctions can be characterized by their quantum numbers:

$$\begin{array}{ll} \mathbf{S} = \sum_i \mathbf{s}_i & \textit{Total spin angular momentum} \\ \mathbf{L} = \sum_i \mathbf{l}_i & \textit{Total orbital angular momentum} \\ \mathbf{J} = \mathbf{L} + \mathbf{S} & \textit{Total angular momentum} \\ \mathbf{P} & \textit{Parity} \end{array} \tag{3.46}$$

where $\mathbf{s}_i$ and $\mathbf{l}_i$ are the spin and orbital angular momenta operators of the ith electron, and $\mathbf{P}$ is the parity operator, whose eigenvalues can only be $p = \pm 1$: the levels with $p = +1$ are called 'even-parity' levels, those with $p = -1$ are called 'odd-parity' levels.

The definition of **S**, **L**, and **J** as given by Equations (3.46) is made possible by the fact that the Hamiltonian is a linear operator that commutes with the angular momenta of each electron; this is the case since relativistic corrections are neglected. Equations (3.46) define the so-called *L-S coupling*, or *Russell–Saunders* coupling. The quantum numbers L, S, J, and p of the operators defined by Equations (3.46) can then be used to describe the eigenfunctions of the non-relativistic Hamiltonian.

The eigenvalues L of the total orbital angular momentum **L** are usually indicated by letters, according to the notation

L:	0	1	2	3	4	5	6	7	8	9	...
	S	P	D	F	G	H	I	K	L	M	...

A level characterized by a given combination of quantum numbers L, S, J, and P is described by the notation

$$^{(2S+1)}L_J \tag{3.47}$$

for an even-parity level ($P = +1$) and

$$^{(2S+1)}L_J^{\mathrm{o}} \tag{3.48}$$

for an odd-parity level ($P = -1$). For example, the level with $S=1$, $L=2$, $J=3$ and $p = +1$ will be indicated with the notation $^3D_3$. The set of levels with the same S and L is called a *term*: if $S = 2$ and $L = 1$, the term is ^5P and includes the levels $^5P_1$, $^5P_2$ and $^5P_3$.

The solution of the time-independent Schrödinger equation $H\psi = E\psi$ with a Hamiltonian given by Equation (3.45) can only be obtained by approximate methods. One such method consists of expressing the Hamiltonian as the sum of two Hamiltonians:

$$H = H_0 + H_{\mathrm{I}} \tag{3.49}$$

with

$$H_0 = \sum_{i=1}^{N}\left(-\frac{\hbar^2}{2\mu_i}\nabla_i^2 + V_i(r_i)\right), \tag{3.50}$$

$$H_{\mathrm{I}} = \sum_{i=1}^{N}\left(-V_i(r_i) - \frac{Ze^2}{r_i}\right) + \sum_{i<j}\frac{e^2}{r_{ij}}, \tag{3.51}$$

where $V_i(r_i)$ is called the *central field* and approximates, for each electron, the potential given by both the nuclear attraction and the interactions with the other electrons. The basic assumption is in considering $H_{\mathrm{I}} \ll H_0$, so that H_{I} can be treated as a perturbation to H_0. The better the central field $V_i(r_i)$ is chosen, the more accurate the results are.

The Hamiltonian H_0 can be used to determine the wavefunctions of each individual electron by using the methods outlined in the previous section, and the potential $V_i(r_i)$ for the determination of the radial component of the wavefunction; the total wavefunction can be built as an antisymmetric linear combination of wavefunctions of the type

$$\psi = \varphi_{(nlm_lm_s)_1}\varphi_{(nlm_lm_s)_2}\cdots\varphi_{(nlm_lm_s)_N}, \tag{3.52}$$

where $\varphi_{(nlm_lm_s)_i}$ is the wavefunction of the one-electron atom given by Equation (3.26). The coefficients of the antisymmetric linear combination can be obtained by applying H_I as a perturbation to the system, and diagonalizing it.

The sets of quantum numbers $(nlm_lm_s)_i$ of each electron must be different from those of all the other electrons to satisfy the Pauli exclusion principle. The energy values corresponding to the final eigenfunctions of the Hamiltonian H_0+H_I will be given by the sum of the energies of each electron in the atom. These energies will depend on the adopted central field $V_i(r_i)$. Usually, these energy values depend on both the quantum numbers $(nl)_i$ of the ith electron, that is, the degeneracy in l of the energies in the one-electron atom is now removed. Since each of these energy values depends on the two quantum numbers $(nl)_i$ of each electron, the final energy of the level depends on the N sets of $(nl)_i$ numbers, but not on m_l and m_s. The ensemble of the N sets of quantum numbers $(nl)_i$ for a given atom is called a *configuration*, while a pair of quantum numbers (nl) is called a *shell*. Within the same configuration there are many individual levels, given by all possible combinations of the wavefunctions of single electrons allowed by the Pauli principle; their wavefunctions can be characterized by the total L, S, J numbers and therefore indicated by terms $^{(2S+1)}L$ and levels $^{(2S+1)}L_J$.

It is possible to demonstrate that the parity of all the levels within the same configuration is the same, and it is given by

$$P = (-1)^{\sum_i l_i}. \tag{3.53}$$

A configuration is usually indicated by the sets of shells (nl) and by the number of electrons in the same shell, having different m_l and m_s. For example, the configuration of an atom that has two electrons in the $(1, 0)$ shell, one electron in the $(2, 0)$ shell and three electrons in the $(2, 1)$ shell will be indicated with the notation

$$1s^2 2s^1 2p^3 \qquad \text{or} \qquad 1s^2 2s 2p^3. \tag{3.54}$$

Sometimes closed shells (see below) are unspecified, so that the above configuration can also be written as $2s2p^3$.

The ground configuration of an atom or ion with N electrons is defined as the one where all the electrons have the minimum possible energy. However, according to the Pauli exclusion principle, electrons cannot be characterized by the same set of quantum numbers, so that the ground configuration of the ion is not $1s^N$, but is the configuration where the shells corresponding to the lowest energies are filled. For example, if $N = 8$ (oxygen), the ground configuration is

$$1s^2 2s^2 2p^4$$

since the $1s$ and $2s$ allow for the presence of two electrons each, and the $2p$ shell, which can accommodate six electrons, is partially filled by the remaining four electrons. In this case, the $1s$ and $2s$ shells are called *closed shells*, since they are full, while the $2p$ shell is called an *open shell*, since it is only partially filled.

We indicate the wavefunctions of the system with the notation

$$\psi = |\gamma L S M_L M_S\rangle \tag{3.55}$$

where γ indicates the ensemble of the quantum numbers of all individual electrons, and M_L and M_S the eigenvalues of the $\mathbf{L_z}$ and $\mathbf{S_z}$ operators. In general, we also indicate with $\langle\alpha|\beta\rangle$ the integral

$$\langle\alpha|\beta\rangle = \int \psi_\alpha^\star \psi_\beta \, d^3r, \tag{3.56}$$

where $\psi_\alpha^\star$ is the complex conjugate of the wavefunction corresponding to $|\alpha\rangle$, ψ_β is the wavefunction corresponding to $|\beta\rangle$, and d^3r indicates integration over all coordinates. If $|\alpha\rangle$ and $|\beta\rangle$ are orthogonal, then $\langle\alpha|\beta\rangle = \delta_{\alpha\beta}$. If $|\beta\rangle$ is a linear combination of a suite of orthonormal wavefunctions $|\alpha_i\rangle$, such as $|\beta\rangle = \sum_i c_i|\alpha_i\rangle$, then $\langle\alpha_k|\beta\rangle = c_k$ provides the value of the coefficient c_k.

The matrix element of an operator $\mathbf{O}$ between levels $|\alpha\rangle$ and $|\beta\rangle$ will be indicated by

$$\langle\alpha|\mathbf{O}|\beta\rangle = \int \psi_\alpha^\star \mathbf{O} \psi_\beta \, d^3r, \tag{3.57}$$

where again d^3r indicates integration over all coordinates.

3.4.4 Relativistic corrections

The considerations in the previous section have been made in the non-relativistic limit. Relativistic corrections, however, need to be taken into account in order to obtain energy levels and system wavefunctions of sufficient accuracy for astrophysical applications. A complete discussion of a relativistic model for the multi-electron atom is beyond the scope of this book; here we will limit ourselves to the treatment of the case where relativistic effects are sufficiently small that they may be considered as a perturbation to the non-relativistic Hamiltonian. Such approximation holds for all elements up to Fe. In this approximation, the Hamiltonian of the multi-electron atom can be written as the sum of two terms:

$$H_{\rm tot} = H_{\rm nr} + H_{\rm rel}, \tag{3.58}$$

where $H_{\rm nr}$ is the non-relativistic Hamiltonian given by Equation (3.45), while $H_{\rm rel}$ includes the relativistic corrections, and can be expressed as the sum of many terms. These terms can be grouped into two main classes: *fine-structure* operators and *non-fine-structure* operators. The difference between the two lies in their commutation properties with the operators used to describe the non-relativistic Hamiltonian. Non-fine-structure operators commute with $\mathbf{S^2}$, $\mathbf{S_z}$, $\mathbf{L^2}$, and $\mathbf{L_z}$, so that they can be directly applied to the wavefunctions obtained from the non-relativistic Hamiltonian; their introduction provides additional contributions to the energy of each level of the system, but will not remove their degeneracy, nor will it change the level wavefunction.

By contrast, fine-structure operators do not commute with $\mathbf{S^2}$, $\mathbf{S_z}$, $\mathbf{L^2}$, and $\mathbf{L_z}$, so they will require a change in the set of mutually commuting operators and in the wavefunctions that describe the system. Their contributions to the energy of a level will depend on the quantum numbers of the additional operators, so that they will partially remove the degeneracy of the energy levels. A list of the most important relativistic corrections is given in Table 3.2.

In the rest of this chapter, we will consider only the most important of the relativistic corrections, namely the *spin–orbit interaction,* in order to introduce the new set of mutually commuting operators necessary to include the relativistic corrections.

Since $H_{\rm rel}$ is a perturbation of $H_{\rm nr}$, the problem is restricted to the calculation of the eigenfunctions and eigenvalues of the $H_{\rm rel}$ Hamiltonian; the latter will be added as corrections to the level energies of the non-relativistic case, while the former will be determined starting from the eigenfunctions of $H_{\rm nr}$.

The spin–orbit interaction can be written in the form

$$H_{\rm rel} = \sum_{i=1}^{N} \xi(r_i)\, \mathbf{l_i} \cdot \mathbf{s_i}, \tag{3.59}$$

where

$$\xi(r_i) = \frac{\hbar^2}{2m^2c^2} \frac{1}{r_i} \frac{\partial V(r_i)}{\partial r_i}. \tag{3.60}$$

For the individual electron i, it is possible to show that

$$\langle \alpha LSM'_L M'_S | \xi(r_i)\, \mathbf{l_i} \cdot \mathbf{s_i} | \alpha LSM_L M_S \rangle = \zeta_i(\alpha LS) \langle \alpha LSM'_L M'_S | \mathbf{L} \cdot \mathbf{S} | \alpha LSM_L M_S \rangle \tag{3.61}$$

Table 3.2. *Main relativistic corrections to the multi-electron atom.*

Non-fine-structure	Fine-structure
Mass variation	Spin–orbit
Darwin term	Mutual spin–orbit
Spin–spin contact	Spin–spin
Two-body orbit	Spin–other orbit
Orbit–orbit	

where

$$\zeta_i(\alpha LS) = \frac{\langle \alpha LM_L|\xi(r_i)\,\mathbf{L}\cdot\mathbf{l_i}|\alpha LM_L\rangle\langle SM_S|\mathbf{S}\cdot\mathbf{s_i}|SM_S\rangle}{L(L+1)S(S+1)}. \tag{3.62}$$

The Hamiltonian H_{rel} commutes with the total angular momentum $\mathbf{J}$, but not with $\mathbf{L}$ or $\mathbf{S}$, so that we need to adopt a new set of three mutually commuting operators: the total Hamiltonian H_{tot}, $\mathbf{J^2}$, and $\mathbf{J_z}$, where $\mathbf{J_z}$ is the projection of $\mathbf{J}$ along the z axis. Since H_{rel} is assumed to be a perturbation to H_{nr}, the system's wavefunctions can be built as linear combinations of the non-relativistic wavefunctions. We can indicate the new wavefunctions of the system by using L, S, and the two new quantum numbers J and M of the two new angular momenta, as $|\alpha LSJM\rangle$. These wavefunctions are obtained as a linear combination of the wavefunctions $|\alpha LSM_LM_S\rangle$ by the equation

$$|\alpha LSJM\rangle = \sum_{M_L,M_S} |\alpha LSM_LM_S\rangle\langle\alpha LSM_LM_S|\alpha LSJM\rangle, \tag{3.63}$$

where the coefficients of the linear combination $\langle\alpha LSM_LM_S|\alpha LSJM\rangle$ are called *Clebsch–Gordan coefficients*. Since $\mathbf{L}$ and $\mathbf{S}$ are mutually commuting operators, the operator $\mathbf{L}\cdot\mathbf{S}$ can be written as a function of the total angular momentum $\mathbf{J}$ by the formula

$$\mathbf{L}\cdot\mathbf{S} = \frac{1}{2}\left(\mathbf{J}^2 - \mathbf{L}^2 - \mathbf{S}^2\right). \tag{3.64}$$

By inserting Equation (3.64) into Equation (3.59) and using the new wavefunctions $|\alpha LSJM\rangle$, we obtain

$$H_{\text{rel}}|\alpha LSJM\rangle = \left[\frac{1}{2}\left(\mathbf{J}^2 - \mathbf{L}^2 - \mathbf{S}^2\right)\zeta(\alpha LS)\right]|\alpha LSJM\rangle \tag{3.65}$$

with $\zeta(\alpha LS) = \sum_i \zeta_i(\alpha LS)$. The eigenvalues of the H_{rel} Hamiltonian are therefore given by

$$E_{\text{rel}} = \frac{J(J+1) - L(L+1) - S(S+1)}{2}\zeta(\alpha LS) \tag{3.66}$$

and need to be added to the non-relativistic energies, so that the final energies of the $|\alpha LSJM\rangle$ states of the multi-electron atom depend not only on L and S, but also on J. Therefore, the inclusion of the spin–orbit interaction in the system's Hamiltonian has three main consequences:

(1) the system's wavefunctions need to be changed;
(2) the system is characterized by new quantum numbers;
(3) the energies of each level in each configuration and term are changed by amounts depending on J.

The separation of the energy levels within the same term, but with different J numbers, partially removes the degeneracy of the terms, and generates the so-called *fine structure* in the atom's spectrum of energies. The energies of levels with quantum numbers J and $J-1$ are now separated by

$$\Delta E = E_J - E_{J-1} = J\zeta(\alpha LS), \tag{3.67}$$

a property that is called the *Landé interval rule*. The coefficient $\zeta(\alpha LS)$ can be either positive or negative, according to the number of electrons in the atom.

It is important to note that, according to Equation (3.63), each wavefunction is a linear combination of non-relativistic wavefunctions belonging to different configurations. The wavefunctions are called *configuration interaction* wavefunctions. The largest contributions to the wavefunctions of a level $|\alpha\rangle$ from other configurations come from the wavefunctions of levels with energy close to the energy of level $|\alpha\rangle$.

3.4.5 Atomic levels above the ionization threshold

The excitation of one electron from an open shell generates a series of configurations with increasingly higher energies. For example, in boron (5 electrons) the ground configuration is

$$1s^2 2s^2 2p$$

so that the excitation of one electron of the open $2p$ shell to higher shells gives rise to a series of configurations

$$1s^2 2s^2 nl$$

with increasing energies until, when $n \to \infty$, the atom reaches the ionization threshold, beyond which the electron in the outer shell can break free. However, the other shells can also give rise to configuration series, such as, for example,

$$1s^2 2s 2p^2 \quad \to \quad 1s^2 2s 2pnl.$$

The lowest-energy configurations of such series have energies below the ionization threshold, but after a certain maximum shell nl the remaining members of such series have energies *above* the ionization threshold of the atom and are found in the continuum. It is also possible to excite an electron from a closed inner shell, and generate series of the kind

$$1s2s^2 2pnl,\ 1s2s2p^2 nl \ldots$$

Given the higher energy required to excite such an electron, all the configurations in such inner-shell series lie above the ionization threshold.

The effects of configuration interaction are particularly strong for levels with energy higher than the ionization limit, since there are continuum levels whose energies coincide with that of the bound levels. Therefore, the wavefunctions of such discrete levels become mixed with those of the continuum, whose contributions need to be accounted for.

The wavefunctions of the levels of the atom A above the ionization potential can be built by using a combination of a set of orthonormal wavefunctions φ_n, describing purely discrete atomic levels, and a set of orthonormal wavefunctions ϕ_E, describing a system composed of a free electron and the ion A^+. If H is the Hamiltonian of the system, let us assume that

$$\langle \varphi_m | \, H \, | \varphi_n \rangle = E_n \delta_{mn}, \tag{3.68}$$

$$\langle \phi_{E''} | \, H \, | \phi_{E'} \rangle = E' \delta\left(E'' - E'\right), \tag{3.69}$$

$$\langle \phi_{E'} | \, H \, | \varphi_n \rangle = V_{E',n}, \tag{3.70}$$

where the magnitude of $V_{E',n}$ indicates the strength of the interaction between the discrete level $|\varphi_n\rangle$ and the continuum state $|\phi_{E'}\rangle$. Equations (3.68) and (3.69) state that the two sets of $|\varphi_n\rangle$ and $|\phi_{E'}\rangle$ are wavefunctions of the Hamiltonian H in the case where there is no interaction between the continuum and the bound states, while Equation (3.70) indicates the presence of interactions between the bound levels and the system composed of the free electron and A^+. We need to determine the system's wavefunctions Ψ_E that account for such interaction. Let us consider for the sake of simplicity the case where we have only one discrete state $|\varphi_n\rangle$: such wavefunctions can be built as

$$\Psi_E = a\varphi_n + \int b_{E'} \phi_{E'} \, dE', \tag{3.71}$$

where E is the energy of the system. The coefficients a and $b_{E'}$ need to be calculated by solving the equations

$$\begin{aligned} E_n a + \int b_{E'} V_{E',n} \, dE' &= Ea, \\ V_{E',n} a + E' b_{E'} &= E b_{E'}, \end{aligned} \tag{3.72}$$

where E_n is the energy of the state $|\varphi\rangle$ and E is the energy of the system. The solution of this system is beyond the scope of this book. It was first described by Fano (1961). What is important to note is that now the wavefunctions Ψ_E of the bound state with energy E higher than the first ionization threshold include contributions from wavefunctions that describe the system given by the atom A^+ and a free electron. This means that the atom A, when excited to a level i above the first ionization threshold, has a finite, non-negligible probability of falling in a state described by a continuum wavefunction $|\phi_{E'}\rangle$, and therefore spontaneously ionizing, according to

$$A_i \rightarrow A_j^+ + \mathrm{e}(E_\mathrm{e}), \tag{3.73}$$

where j is a bound level of the ion A^+ and $\mathrm{e}(E_\mathrm{e})$ is the free electron with kinetic energy $E_\mathrm{e} = E_i - E_j$. This process is called *autoionization*. The probability for the system to be found in the bound level $|\varphi\rangle$ is given by

$$|a(E)|^2 = \frac{1}{|V_{E',n}|^2 \left(\pi^2 + z^2(E)\right)} \tag{3.74}$$

and it depends on the coefficient $V_{E',n}$. The function $z(E)$ is related to the phase shift of the asymptotic wave function due to the configuration interaction of $|\phi_{E'}\rangle$ with the state $|\varphi\rangle$. Equation (3.74) shows that the effects of configuration interaction between continuum states and the discrete bound state dilutes the latter in a band of states whose profile is given by a resonance curve with half-width given by $\pi |V_E|^2$. This means that if the atomic system is excited into the bound state $|\varphi\rangle$ at any given time, it will autoionize after a mean life

$$\tau = \frac{\hbar}{2\pi |V_{E',n}|^2}. \tag{3.75}$$

The *autoionization rate* A_a (also called *Auger rate*) from the bound state $|\varphi\rangle$ above the ionization threshold and the continuum state $\phi_{E'}$ is then defined as

$$A_a = \frac{2\pi}{\hbar} |\langle\varphi_n| \; H \; |\phi_{E'}\rangle|^2 = \frac{2\pi}{\hbar} \; |V_{E',n}|^2 \tag{3.76}$$

and each of the operators included in H contributes to it. The contributions of the electrostatic operator and the non-relativistic operators depend on the quantum numbers of the states φ_n and $\phi_{E'}$. It is possible to show that these contributions are non-zero only if some relations between the eigenvalues L, S, J, and P of the initial and final states, called *selection rules*, are fulfilled. The Coulomb interaction is the most important for autoionization.

The above treatment assumed the presence of a single bound state above the ionization threshold, but it can be extended to the case of many such states. Such a treatment can be found in Fano (1961).

3.5 Isoelectronic sequences

The considerations of the previous sections show that the overall structure of the energy levels of an atom or ion in a non-relativistic approximation is largely determined by the number of bound electrons. In fact, the mutual interaction of the bound electrons is crucial for determining the wavefunction and the energy levels; the parity of the wavefunctions, as well as the eigenvalues L and S, depend on the number of electrons. The relativistic corrections, if the approximation $H_{\rm rel} \ll H_{\rm nr}$ holds, are too weak to reach the importance of the electrostatic interactions between bound electrons; moreover, the latter do not depend on the nuclear charge. Therefore, it is reasonable to expect that the structure of the atomic levels be approximately the same for all the ions with the same number of bound electrons. The absolute energy of these levels will of course be much different, given the higher value of the nuclear charge Z in heavier elements. The energy separation between levels also increases as Z increases. However, the overall hierarchy and quantum numbers of energy levels will be of the same type in almost all cases.[1] Consequently, the transitions between atomic levels will be the same for all elements. Since the wavelengths of the spectral lines are calculated from the differences in the level energies, the spectra of all the ions with the same number of electrons will also be very similar.

Because of these properties, a group of ions that have the same number of electrons is called an *isoelectronic sequence*. Each sequence takes its name from the only neutral atom belonging to it: e.g. the atoms with eight bound electrons left belong to the *oxygen isoelectronic sequence*, since neutral oxygen has eight electrons, and they are called *O-like ions*.

An example of the spectral properties of the ions along an isoelectronic sequence is shown in Figure 3.3, where the same multiplet $2s^22p\,^2P - 2s2p^2\,^2P$ from four different ions is displayed for the B-like ions C^{+1} (C II), O^{+3} (O IV), Ne^{+5} (Ne VI) and Si^{+9} (Si X). Even though the ions have different Z, the multiplet shown in Figure 3.3 is basically identical

[1] There are a few exceptions, such as the level crossing within the triplet ^3P_J in the lowest energy configuration O-like systems at around $Z = 24$.

Table 3.3. *Wavelengths of a few transitions along the isoelectronic sequences of H, Li and Be.*

Trans.	Wavelength (Å)						
	Hydrogen isoelectronic sequence $1s\,^2S - 2p\,^2P$						
	H I	He II	C VI	Ne X	Si XIV	Ca XX	Fe XXVI
1/2–3/2	1215.67	303.78	33.73	12.13	6.180	3.019	1.778
1/2–1/2	1215.68	303.79	33.74	12.14	6.186	3.024	1.783
	Lithium isoelectronic sequence $2s\,^2S - 2p\,^2P$						
	N V	O VI	Ne VIII	Mg X	Si XII	Ca XVIII	Fe XXIV
1/2–3/2	1238.82	1031.91	770.41	609.79	499.41	302.19	192.03
1/2–1/2	1242.81	1037.62	780.33	624.94	520.67	344.76	255.11
	Beryllium isoelectronic sequence $2s^2\,^1S_0 - 2s2p\,^1P_1$						
	N IV	O V	Ne VII	Mg IX	Si XI	Ca XVII	Fe XXIII
	765.15	629.73	465.22	368.07	303.33	192.82	132.91

Note how as the atomic number Z increases, the wavelengths decrease, but the transitions still are of the same type.

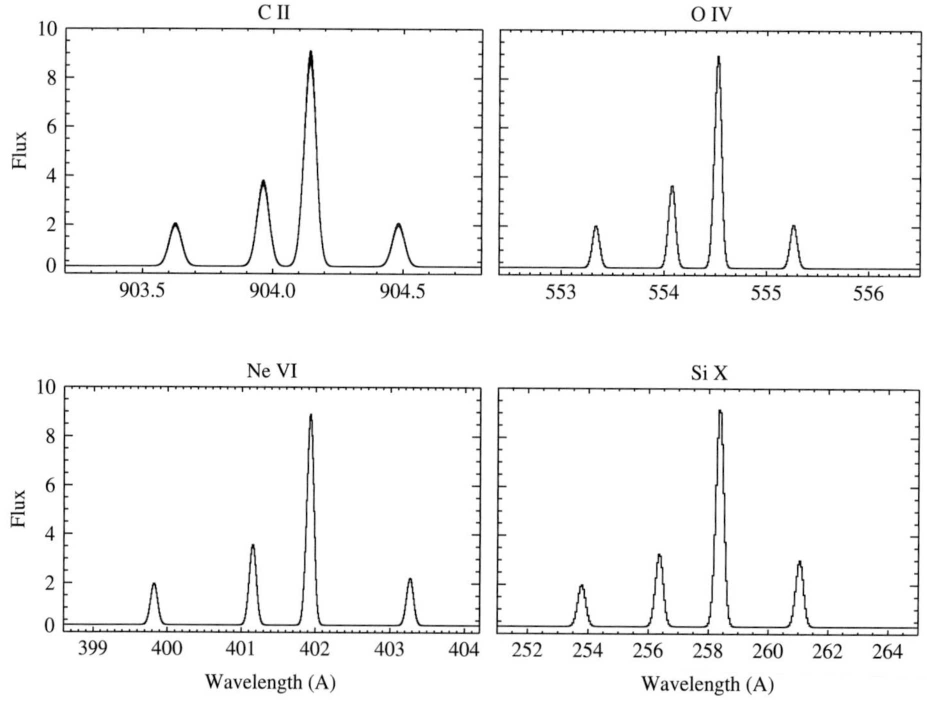

Fig. 3.3. Spectrum of the B-like $2s^22p\,^2P$ – $2s2p^2\,^2P$ along the boron isoelectronic sequence. The intensity is in arbitrary units.

(intensities included) in all ions. The only differences are in the wavelengths, which decrease as Z increases.

A list of the most intense transitions of a few ions from the hydrogen, lithium and beryllium isoelectronic sequences is given in Table 3.3. In each ion, the transitions are always of the same type, but their wavelengths are very different. Also, the relative intensities between lines within each ion in Table 3.3 are very similar throughout the sequences.

3.6 Interaction between atoms and electromagnetic fields

3.6.1 Hamiltonian of the electromagnetic radiation

In order to describe the interaction of the electromagnetic field with an atomic system, we will first introduce the quantum operators that describe the electromagnetic field. Let us consider a cubic cavity with linear dimension L, so that its volume is $V = L^3$. In such a cavity, the quantum operator corresponding to the electromagnetic field can be expressed as the sum of many harmonic oscillator operators, each corresponding to a *mode* of oscillation j of the field:

$$H_{\rm rad} = \sum_j \hbar\omega_j \left(\mathbf{a_j^\dagger a_j} + \frac{1}{2} \right), \tag{3.77}$$

where the operators $\mathbf{a_j}$ and $\mathbf{a_j^\dagger}$ are used to represent the vector potential $\mathbf{A}(\mathbf{r})$ of the field as

$$\mathbf{A}(\mathbf{r}) = \sum_j c\sqrt{\frac{2\pi\hbar}{\omega_j V}} \left[\mathbf{a_j}\mathbf{e}_j e^{i\mathbf{k}_j\cdot\mathbf{r}} + \mathbf{a_j^\dagger}\mathbf{e}_j^\star e^{-i\mathbf{k}_j\cdot\mathbf{r}} \right] \tag{3.78}$$

where $\mathbf{e}_j$ and $\mathbf{e}_j^\star$ are two unit vectors, called *polarization vectors*, perpendicular to each other and to the direction of $\mathbf{k}_j$. The vector $\mathbf{k}_j$ satisfies the condition

$$\mathbf{k}_j = \left(n_x\frac{2\pi}{L}, n_y\frac{2\pi}{L}, n_z\frac{2\pi}{L} \right), \tag{3.79}$$

where n_x, n_y and n_z are integers. Equation (3.78) can be written in the form given above since we adopt the gauge where the vector potential $\mathbf{A}$ satisfies the equation

$$\mathrm{div}\mathbf{A} = 0. \tag{3.80}$$

The advantage of describing $H_{\rm rad}$ according to Equation (3.77) is that the electromagnetic field is decomposed into the sum of independent harmonic oscillators j, or *modes*, each characterized by its own $\mathbf{k}_j$, angular frequency ω_j, and polarization vectors $\mathbf{e}_j$ and $\mathbf{e}_j^\star$. Each harmonic oscillator j is also characterized by the *occupation number* n_j indicating the number of identical oscillators in the mode j in the field. The occupation number can be interpreted as the number of individual radiation field quanta present in the cavity. These quanta are called *photons*.

Since the oscillator modes are independent of each other, the wavefunction of the total electromagnetic field is given by the product of the wavefunction of each mode, which can be characterized by the occupation number n_j of the mode:

$$|n_1, n_2, \ldots, n_j, \ldots\rangle = |n_1\rangle|n_2\rangle\ldots|n_j\rangle\ldots = |(n_l)\rangle, \tag{3.81}$$

where (n_l) indicates the ensemble of occupation numbers of all modes $(n_1, n_2, ..., n_j, ...)$, and with $|n_j\rangle$ the wavefunction of each mode, orthogonal to the wavefunction of all other modes. The operators $\mathbf{a_j}$ and $\mathbf{a_j^\dagger}$, when applied to the wavefunction of the field, have the following properties:

$$\mathbf{a_j}|n_1, n_2, \dots, n_j, ...\rangle = \sqrt{n_j}|n_1, n_2, ..., n_{j-1}, ...\rangle, \tag{3.82}$$

$$\mathbf{a_j^\dagger}|n_1, n_2, \dots, n_j, ...\rangle = \sqrt{n_j + 1}|n_1, n_2, ..., n_{j+1}, ...\rangle. \tag{3.83}$$

The operators $\mathbf{a_j}$ and $\mathbf{a_j^\dagger}$ are called *annihilation* and *creation operators*. If we interpret the occupation number n_j as the number of photons, the operators $\mathbf{a_j}$ and $\mathbf{a_j^\dagger}$ act on the electromagnetic field in order to destroy one photon or to create it, respectively.

The eigenvalues of the Hamiltonian $H_{\rm rad}$ can be calculated using Equations (3.82) and (3.83), so that the energy of the field will be given by

$$H_{\rm rad}|(n_l)\rangle = E|(n_l)\rangle \quad \Longrightarrow \quad E = \sum_j \hbar\omega_j\left(n_j + \frac{1}{2}\right). \tag{3.84}$$

This energy can be interpreted as the sum of the energies $\hbar\omega_j$ of each photon in each mode j.

When the number of photons in all modes is zero, Equation (3.84) still provides a non-zero energy. This zero-point energy can be very important in a few areas of physics outside the scope of this book, but here it will be ignored.

3.6.2 *Hamiltonian of the atom–field interaction*

In order to express the Hamiltonian for the interaction between the electromagnetic field and an atomic system, the operator $\mathbf{p}$ of each electron needs to be replaced by a new canonical operator given by

$$\mathbf{p}' = \mathbf{p} - \frac{e}{c}\mathbf{A}(\mathbf{r}) \tag{3.85}$$

while E needs to be replaced by $E + e\phi$, where ϕ is a scalar potential. Since we use the gauge where $\phi = 0$ and div$\mathbf{A} = 0$, it is possible to show that the operators $\mathbf{p}$ and $\mathbf{A}$ commute, i.e.

$$\mathbf{pA} = \mathbf{Ap}. \tag{3.86}$$

The Hamiltonian of the atomic system in the electromagnetic field can be written as

$$H = \sum_i \left(\frac{\left(\mathbf{p_i} - \frac{e}{c}\mathbf{A}(\mathbf{r_i})\right)^2}{2\mu} - \frac{Ze^2}{r_i}\right) + \sum_{i<j} \frac{e^2}{r_{ij}}, \tag{3.87}$$

where μ is the reduced mass, equal for all electrons. Using the commutation property in Equation (3.86), H can be reduced to the more familiar form

$$\begin{aligned} H &= H_0 + H_1 \\ &= \sum_i \left(\frac{\mathbf{p_i}^2}{2\mu} - \frac{Ze^2}{r_i}\right) + \sum_{i<j}\frac{e^2}{r_{ij}} \\ &\quad - \sum_i \frac{e}{\mu c}\mathbf{p_i}\mathbf{A}(\mathbf{r}_i) + \sum_i \frac{e^2}{2\mu c^2}\mathbf{A}^2(\mathbf{r}_i), \end{aligned} \tag{3.88}$$

where the Hamiltonian H_0 includes the first two terms of Equation (3.88) and is the Hamiltonian of an isolated atomic system with atomic number Z as described in the previous sections, while H_1 includes the last two terms of Equation (3.88), and can be considered the interaction Hamiltonian.

The ratio between the quadratic and the linear terms in $\mathbf{A}$ in Equation (3.88) is

$$R \approx \frac{e^2 A^2}{2\mu c^2} \times \frac{\mu c}{eAp} \propto \frac{v}{c} \frac{eA}{\mu v^2}. \tag{3.89}$$

In the case where the electromagnetic field is not very intense, $R \ll 1$ and the quadratic term can be neglected. This condition is always satisfied in the radiation fields typical of the solar atmosphere, so that the interaction Hamiltonian H_1 is simply given by the expression

$$H_1 = -\sum_i \frac{e}{\mu c} \mathbf{p_i} \mathbf{A}(\mathbf{r}_i). \tag{3.90}$$

3.6.3 The statistical equilibrium equations

The Hamiltonian for the system composed of the atomic system and the electromagnetic field can then be expressed as

$$H = H_0 + H_1 + H_{\mathrm{rad}}. \tag{3.91}$$

Our aim is to determine the wavefunction of the system and its temporal behaviour. The first consideration is that the ratio between H_1 and the kinetic term of the atomic Hamiltonian H_0 is given by

$$R' \approx \frac{eAp}{\mu c} \times \frac{2\mu}{p^2} \propto \frac{v}{c} \frac{eA}{\mu v^2}, \tag{3.92}$$

and it is of the same order as the ratio between the quadratic and linear terms given in Equation (3.89). This means that in solar radiation fields the Hamiltonian H_1 can always be considered a perturbation to the Hamiltonians $H_0 + H_{\mathrm{rad}}$, describing the field and the atomic systems as if they were not interacting.

The temporal behaviour of the wavefunctions of the system can be studied by applying the Schrödinger equation to the unperturbed wavefunctions of the electromagnetic field and of the atomic system. This is equivalent to approximating the field–atom interaction to an effect that is suddenly applied to an unperturbed system composed of an atom and a field previously not interacting with each other.

If we indicate the atomic wavefunctions with $|u_n\rangle$ and the radiation field wavefunctions with $|(n_l)\rangle$, where (n_l) is the ensemble of the occupation numbers of all the oscillation modes in the field, the unperturbed wavefunction of the system can be written as

$$|\alpha\rangle = |u_n\rangle|(n_l)\rangle. \tag{3.93}$$

The time-dependent wavefunction of the system, $|\psi(t)\rangle$, can be expressed as a time-dependent linear combination of the states $|\alpha\rangle$:

$$|\psi(t)\rangle = \sum_\alpha c_\alpha(t)|\alpha\rangle \exp\left(-i\frac{E_\alpha}{\hbar}t\right) \tag{3.94}$$

and must satisfy the Schrödinger equation

$$i\hbar\frac{\partial}{\partial t}\,|\psi(t)\rangle = (H_0 + H_{\rm rad} + H_1)\,|\psi(t)\rangle. \tag{3.95}$$

The probability of the system being in the state $|\beta\rangle$ is given by the square of the coefficient $c_\beta(t)$. If we consider long time intervals it is possible to show that this probability is given by

$$|c_\beta(t)|^2 = |c_\beta(0)|^2 + \frac{2\pi}{\hbar}t\sum_\alpha |c_\alpha(0)|^2|H_1^{\alpha\beta}|^2\delta\big(E_\beta - E_\alpha\big)$$
$$- \frac{2\pi}{\hbar}t\sum_\alpha |c_\beta(0)|^2|H_1^{\beta\alpha}|^2\delta\big(E_\beta - E_\alpha\big). \tag{3.96}$$

Equation (3.96) describes the temporal evolution of the probability of finding the system composed of the atom and the radiation field in the state $|\beta\rangle$. The quantities

$$P_{\beta\alpha} = |H_1^{\beta\alpha}|^2\delta\big(E_\beta - E\alpha\big), \qquad H_1^{\beta\alpha} = \langle\alpha|\,H_1\,|\beta\rangle \tag{3.97}$$

can be interpreted as the probability of having the system make a transition to and from the states $|\alpha\rangle$ and $|\beta\rangle$. The time derivative of Equation (3.96) yields the temporal variation of the probability of finding the system in the state $|\beta\rangle$, which can then be expressed as

$$\frac{d}{dt}|c_\beta(t)|^2 = \frac{2\pi}{\hbar}\left(\sum_\alpha |c_\alpha(0)|^2 P_{\alpha\beta} - \sum_\alpha |c_\beta(0)|^2 P_{\beta\alpha}\right). \tag{3.98}$$

Note that the right-hand side of Equation (3.98) does not depend on time. Equation (3.98) is the *equation for the statistical equilibrium* of the system, and expresses the effect of the field–atom interaction on the total wavefunction. The probability of finding the total system in the state $|\beta\rangle$ is increased by transitions from other states $|\alpha\rangle$ to $|\beta\rangle$ (first term on the right-hand side of Equation (3.98)) and decreased by transitions from $|\beta\rangle$ to other states $|\alpha\rangle$. The probability that such a transition occurs is given by Equation (3.97) and depends on $H_1^{\beta\alpha}$ and $H_1^{\alpha\beta}$.

3.6.4 The dipole approximation

The calculation of the transition probability $P_{\beta\alpha}$ can be carried out by explicitly considering the operators of H_1. In order to calculate $H_1^{\beta\alpha}$, some approximations are necessary. The most common one is the *dipole approximation*, which consists of assuming that $\mathbf{k}_j, \mathbf{r}_i \ll 1$ so that in Equation (3.78) $\exp(i\mathbf{k}_j{\cdot}\mathbf{r}) \approx 1$. In this case, indicating $\mathbf{p} = \sum_i \mathbf{p_i}$, the interaction Hamiltonian is simplified to

$$H_1 = -\frac{e}{\mu}\sum_j\sqrt{\frac{2\pi\hbar}{\omega_j V}}\big(\mathbf{a_j}\,\mathbf{p}\cdot\mathbf{e}_j + \mathbf{a_j^\dagger}\,\mathbf{p}\cdot\mathbf{e}_j^\star\big). \tag{3.99}$$

The advantage of this approximation is that the calculation of $H_1^{\beta\alpha}$ can be carried out by separating $H_1^{\beta\alpha}$ as the product of two terms, one involving the operator and wavefunctions of the radiation field, the other involving the operator and wavefunctions of the atomic system. If we assume that the initial state is $|\beta\rangle = |(n_l)\rangle|u_n\rangle$ and the final state is $|\alpha\rangle = |(n_l')\rangle|u_m\rangle$, we have

$$H_1^{\beta\alpha} = -\frac{e}{\mu}\sum_j \sqrt{\frac{2\pi\hbar}{\omega_j V}}\Big(\langle(n_l)|\, \mathbf{a_j}\, |(n_l')\rangle \langle u_n|\, \mathbf{p}\, |u_m\rangle \cdot \mathbf{e}_j$$

$$+ \langle(n_l)|\, \mathbf{a_j^\dagger}\, |(n_l')\rangle \langle u_n|\, \mathbf{p}\, |u_m\rangle \cdot \mathbf{e}_j^\star \Big). \tag{3.100}$$

Equation (3.100) shows that the field–atom interaction has effects on *both* the field *and* the atom. We will now deal with the atomic and radiation factors separately.

From the commutation properties of H_0 and $\mathbf{r} = \sum_i \mathbf{r}_i$, we obtain

$$\langle u_n|\, \mathbf{p}\, |u_m\rangle = 2\pi i \tfrac{\mu}{e} \nu_{mn} \mathbf{d}_{mn} \tag{3.101}$$

with

$$\nu_{mn} = \frac{E_n - E_m}{h}, \qquad \mathbf{d}_{mn} = \langle u_n|\, e\mathbf{r}\, |u_m\rangle. \tag{3.102}$$

The operator $e\mathbf{r}$ is called the *electric dipole operator* and can be thought of as the electric dipole of the ensemble of electrons relative to the nucleus.

The calculation of the radiation field terms of Equation (3.100) can be easily carried out knowing the properties of the operators $\mathbf{a_j}$ and $\mathbf{a_j^\dagger}$ (Equations (3.82) and (3.83)), and that the wavefunctions $|n_j\rangle$ of each mode are orthogonal to those of all the other modes, so that $\langle n_j|n_i\rangle = \delta_{ji}$. Using these properties, the only two final states $|(n_l')\rangle$ that provide non-zero contributions to $H_1^{\beta\alpha}$ for each mode j are

$$\langle n_l|\, \mathbf{a_j}\, |(n_l + 1_j)\rangle = \sqrt{n_j + 1}, \tag{3.103}$$

$$\langle n_l|\, \mathbf{a_j^\dagger}\, |(n_l - 1_j)\rangle = \sqrt{n_j}, \tag{3.104}$$

where the notations $|(n_l \pm 1_j)\rangle$ indicate the wavefunctions of the radiation field in the state identical to the state $|(n_l)\rangle$ except for the jth mode, whose occupation number has one photon more $(+1_j)$ or less (-1_j) than the $|(n_l)\rangle$ state. The energies of the initial and final state of the radiation field therefore differ by the energy of a single photon.

The transition probability $P_{\beta\alpha}$ is also proportional to $\delta\big(E_\beta - E_\alpha\big)$. Since the energies of the states of the total system are given by the sum of the energies of the atom and of the radiation field, the change in the system energy can have only two possible values:

$$\delta\big(E_{n,(n_l)} - E_{m,(n_l - 1_j)}\big) = \delta\big(E_n - E_m + h\nu_j\big) = \frac{1}{h}\delta\big(\nu_{nm} + \nu_j\big), \tag{3.105}$$

$$\delta\big(E_{n,(n_l)} - E_{m,(n_l + 1_j)}\big) = \delta\big(E_n - E_m - h\nu_j\big) = \frac{1}{h}\delta\big(\nu_{nm} - \nu_j\big). \tag{3.106}$$

In order to have a non-zero $H_1^{\beta\alpha}$ and therefore a non-zero transition probability $P_{\beta\alpha}$, $\delta\big(E_\beta - E_\alpha\big) = 1$, so that E_β must equal E_α. Since the energy $h\nu_j$ of the photon is always positive, Equations (3.105) and (3.106) then require that $\Delta E = E_n - E_m$ be negative or positive, respectively. Also, since the atomic level energies have fixed values, only the mode j whose energy is equal to the energy difference between the two levels will be able to interact with the atom, while all the other modes of the field will not contribute to $H_1^{\beta\alpha}$. If we assume that the state $|\beta\rangle = |n\rangle|(n_l)\rangle$ is the initial state of the interaction and the state $|\alpha\rangle = |m\rangle|(n_l - 1_j)\rangle$ is the final state, Equation (3.105) requires that the final energy E_m of the atom be *higher* than the initial energy E_n; at the same time, the occupation number of the mode j has been decreased by one photon: this is interpreted by saying that the atom–field interaction has caused the atom to be excited from one level n with lower energy to another level m with higher energy, and a photon of energy $h\nu_j$ equal to the difference of energy $E_n - E_m$ between the initial and final states to be destroyed, or *absorbed*. The case described by Equation (3.106) is the opposite: the atom has been de-excited from a state n with higher energy to a state m of lower energy, and while doing so, a photon of the mode j whose energy is equal to the energy lost by the atom has been created, or *emitted*, by the atom.

3.6.5 The Einstein coefficients

Equation (3.98) can be calculated explicitly using the above results and taking a few assumptions:

(1) Since we are interested in studying the behaviour of the atomic system irrespective of the state of the radiation field, we sum Equation (3.98) over all possible sets (n_l) in order to obtain the total probability p_n of finding the atomic system in the state n.
(2) We assume that no correlation exists among the modes j of the field, and between these modes and the atomic system; this assumption is well verified in the solar atmosphere.
(3) We assume that the radiation is not polarized, so that $|\mathbf{d}_{mn} \cdot \mathbf{e}_j|^2 = |\mathbf{d}_{mn} \cdot \mathbf{e}_j^\star|^2 = \frac{1}{3}|\mathbf{d}_{mn}|^2$.

The summation over the modes j is carried out as an integration over all the frequencies and the solid angle of the number of modes dN present in the volume V, solid angle $d\Omega$ and frequency interval $d\nu$. The number of modes dN with frequency between ν and $\nu + d\nu$ in the cubic cavity of volume V introduced in Section 3.6.1 is given by (Sakurai (1967), p.40)

$$dN = \frac{2V}{c^3}\nu^2 \, d\nu \, d\Omega, \tag{3.107}$$

where the factor 2 comes from the two possible directions of polarization.

Using these assumptions, and defining the mean specific intensity J_ν averaged over the solid angle as

$$J_\nu = \frac{1}{4\pi}\int_{4\pi} \frac{2h\nu^3}{c^2}\bar{n} \, d\Omega, \tag{3.108}$$

where $\bar{n}$ is the average number of photons at the frequency ν. Equation (3.98) can be expressed as

$$\begin{aligned}\frac{dp_n}{dt} = &\underset{(P_1)}{\sum_{m<n} p_m B_{mn} J_{\nu_{mn}}} + \underset{(P_2)}{\sum_{m>n} p_m A_{mn}} + \underset{(P_3)}{\sum_{m>n} p_m B_{mn} J_{\nu_{mn}}} \\ &- \underset{(D_1)}{\sum_{m>n} p_n B_{nm} J_{\nu_{mn}}} - \underset{(D_2)}{\sum_{m<n} p_n A_{nm}} - \underset{(D_3)}{\sum_{m<n} p_n B_{nm} J_{\nu_{mn}}},\end{aligned} \tag{3.109}$$

where p_n and p_m are the total probabilities of finding the atomic system in the states n and m, respectively, and

$$A_{mn} = \frac{64\pi^4\nu_{mn}^3}{3hc^3}|\mathbf{d}_{mn}|^2, \tag{3.110}$$

$$B_{mn} = \frac{32\pi^4}{3h^2c}|\mathbf{d}_{mn}|^2 = \frac{c^2}{2h\nu_{mn}^3}A_{mn}, \tag{3.111}$$

$$B_{nm} = B_{mn} \tag{3.112}$$

are the *Einstein coefficients*. The units of A_{mn} are s^{-1}, while those of B_{mn} and B_{nm} are erg^{-1} cm^2 s^{-1}. Note that sometimes Equation (3.109) is developed using the energy density $u_\nu = (4\pi/c)J_\nu$ in place of the mean specific intensity J_ν, so that the relation between the absorption and stimulated Einstein coefficients is changed to

$$B_{mn} = \frac{c^3}{8\pi h \nu_{mn}^3} A_{mn} \tag{3.113}$$

and u_ν should replace J_ν in Equation (3.109). The processes of population and de-population of the level n are illustrated schematically in Figure 3.4, where the meaning of P_1, P_2, P_3, D_1, D_2, and D_3 is illustrated. The Einstein coefficients and the average intensity $J_{\nu_{mn}}$ regulate the transitions to and from the atomic level n. These transitions can be of two kinds: emission of photons with frequency ν_{mn} (regulated by A_{mn} and B_{mn}) and absorption of photons with the same frequency (regulated by B_{nm}). However, while the process of absorption depends on the mean specific intensity $J_{\nu_{mn}}$ of the field, the process of emission can take place both with *and without* the presence of a radiation field. The former process, regulated by B_{mn}, is called *induced emission*, while the latter process, regulated by A_{mn}, is called *spontaneous emission*. The spontaneous emission process takes place any time the atom or ion is found in an excited level n, independent of the process that excited it. The inverse of the sum of all the A_{mn} values that involve transitions from the level n represents the *total spontaneous decay rate* of the level n, and its inverse

$$\tau = \frac{1}{\sum_m A_{mn}} \tag{3.114}$$

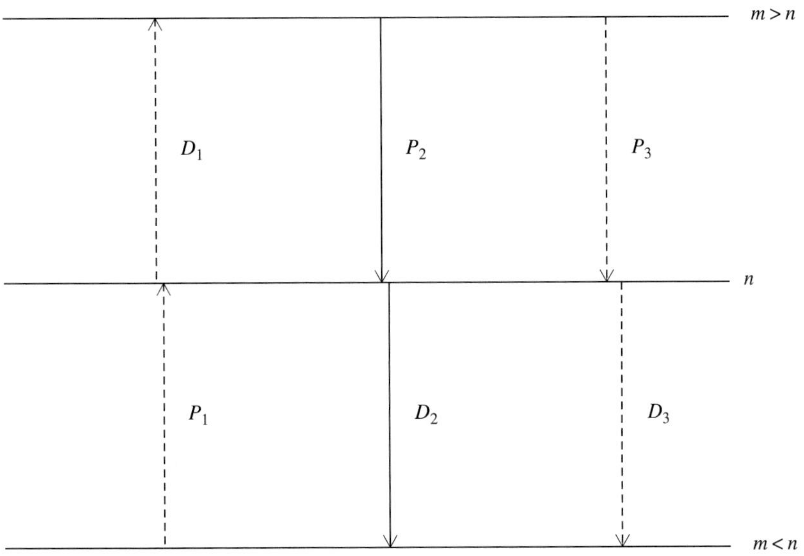

Fig. 3.4. Radiative processes that contribute to populate (P) and de-populate (D) level n, from Equation (3.109). *Full line*: spontaneous processes. *Dashed line*: stimulated processes. D_1 and P_1: absorption of radiation; D_2 and P_2: spontaneous emission; D_3 and P_3: induced emission.

is called *lifetime* of the level n, and is measured in seconds. Levels that can decay via the dipole transitions we have dealt with so far have very short lifetimes (i.e. 10^{-8} s or less). Levels with long lifetimes (i.e. longer than 10^{-4} s) are called *metastable* levels and have great importance for plasma diagnostics (see Chapter 5).

It is important to note that when the levels n and m are degenerate, the above arguments can be repeated, obtaining slightly different relations among the Einstein coefficients:

$$g_m B_{mn} = g_n B_{nm}, \tag{3.115}$$

$$g_n B_{nm} = \frac{c^2}{2h\nu_{mn}^3} g_m A_{mn}. \tag{3.116}$$

Also, even if the Einstein coefficients have been obtained assuming statistical equilibrium, their properties do not depend on the particular status of the atomic system or of the radiation field, so that they can be applied in circumstances where the system is not in thermal equilibrium.

3.6.6 The selection rules

The Einstein coefficients depend on the value of the dipole element $|\mathbf{d_{mn}}|^2$. When $|\mathbf{d_{mn}}|^2$ is zero, the transition cannot take place, and it is called a *forbidden* transition. If it is non-zero, the transition is called an *allowed* transition. The value of the dipole element $|\mathbf{d_{mn}}|^2$ depends on both the initial and final levels $|n\rangle$ and $|m\rangle$. The explicit expression of $|\mathbf{d_{mn}}|^2$ is

$$\mathbf{d}_{mn} = e \int \psi_m^\star(\mathbf{r})\, \mathbf{r}\, \psi_n(\mathbf{r})\, d^3r, \tag{3.117}$$

where d^3r indicates the integration over all the spatial coordinates, $\psi_m(\mathbf{r})$ and $\psi_n(\mathbf{r})$ are the wavefunctions of the states $|m\rangle$ and $|n\rangle$, respectively, and $\psi_m^\star$ denotes the complex conjugate of the wavefunction ψ_m.

An explicit solution of Equation (3.117) must be calculated for each case; here we limit ourselves to discussion of the conditions that need to be fulfilled so that the right-hand side of Equation (3.117) is non-zero and the transition is allowed. These conditions are called *selection rules*.

First, the operator $\mathbf{r}$ has odd parity, so that in order to have a non-zero integral the two wavefunctions must have opposite parity. This selection rule is valid for any angular momentum coupling scheme. In the LS-coupling scheme, the parity of an atomic state is given by Equation (3.53) and is the same for all the levels of the same configuration. So in LS-coupling allowed transitions can take place only between configurations of different parity. Moreover, the definition of the dipole operator $\mathbf{d}_{mn}$ involves a summation over individual terms $\langle u_n|\, \mathbf{r}_i\, |u_m\rangle$, where $\mathbf{r}_i$ is the position operator of each electron. However, the wavefunctions $|u_n\rangle$ and $|u_m\rangle$ are linear combinations of one-electron orbitals o_i, so that each element $\langle u_n|\, \mathbf{r}_i\, |u_m\rangle$ is given by a combination of integrals of the type

$$R_i = \int o_1^\star o_2^\star \ldots o_N^\star \mathbf{r}_i o_1 o_2 \ldots o_N\, d^3\mathbf{r}, \tag{3.118}$$

where N is the number of electrons in the atom. Each of these integrals will be non-zero only if all the orbitals are the same except the ith one. The only way of having all the terms in the sum providing $\mathbf{d}_{mn}$ is that all the orbitals in the wavefunction of the atom are the same except for one electron. This means that in LS-coupling allowed transitions can take place only between configurations that differ in the quantum numbers of only one electron.

In LS-coupling, Equation (3.117) yields non-zero dipole elements only if the states are two terms whose total angular momenta have eigenvalues that obey the following selection rules:

$$\Delta L = L - L' = 0, \pm 1, \qquad L = 0 \not\rightarrow L' = 0, \qquad \Delta S = S - S' = 0. \quad (3.119)$$

For example, the transition ${}^3P - {}^3D$ is allowed, while the transitions ${}^3F - {}^3P$ $(\Delta L > 1)$, ${}^1P - {}^3D$ $(\Delta S > 0)$ and ${}^3S - {}^3S$ $(L = 0 \rightarrow L' = 0)$ are forbidden.

These selection rules are not valid when the LS-coupling approximation breaks down, since then the wavefunctions of the atom can be built from the wavefunctions of different configurations, which will provide non-zero contributions to the dipole operator even for transitions which are forbidden in LS-coupling.

When the LS-coupling approximation no longer holds, the wavefunctions $|m\rangle$ and $|n\rangle$ can be characterized by the total angular momentum $\mathbf{J}$ and by the eigenvalues related to it. The selection rules for transitions between states described by such wavefunctions become

$$\Delta J = J - J' = 0, \pm 1, \qquad J = 0 \not\rightarrow J' = 0, \quad (3.120)$$

so that, for example, the transitions ${}^3P_1 - {}^3P_2$ and ${}^3P_2 - {}^3D_3$ are allowed, while the transitions ${}^3P_0 - {}^3P_0$ $(J = 0 \rightarrow J' = 0)$, ${}^3D_3 - {}^3P_0$ $(\Delta J > 1)$ are forbidden. In such a coupling scheme, transitions between terms with different total S can take place, and are called *intercombination* transitions.

3.6.7 Magnetic dipole and electric multipole transitions

When a transition is forbidden and the dipole element is zero, a radiative transition might still be possible. Such cases, neglected by the discussion of the previous sections, are due to the breakdown of the approximations we have made so far: the dipole approximation, and the neglect of relativistic corrections in the atomic Hamiltonian H_0.

The dipole approximation consists of neglecting all the terms in the expansion of the exponential $\exp\left(\pm i\mathbf{k}_j \cdot \mathbf{r}_i\right)$ in Equation (3.78) except the first one (see also Equation (3.99)). However, when the transition is forbidden, the higher-order terms of the expansion may provide non-zero transition probabilities, which can be very important in determining the lifetime of a level. The first of these higher-order terms $(\pm i\mathbf{k}_j \cdot \mathbf{r}_i)$ provides *electric quadrupole* transitions, the second $((\pm i\mathbf{k}_j \cdot \mathbf{r}_i)^2)$ provides *octupole transitions*, and so on. For these terms transition probabilities equivalent to those regulated by dipole transitions can be determined, and the equivalent of the Einstein coefficients can be calculated. The values of these coefficients are usually much smaller than those of the true Einstein coefficients, and the importance of these higher-order transitions decreases as the order of the $\pm i\mathbf{k}_j \cdot \mathbf{r}_i$ term increases. As for electric dipole transitions, selection rules can be found to determine which electric multipole transitions are possible. The selection rules for electric quadrupole transitions are given in Table 3.4.

Relativistic corrections when introduced in the atomic Hamiltonian allow the definition, in a similar way to the electric dipole operator, of a *magnetic dipole operator*, which allows the probability of some forbidden transitions to be non-zero. The equivalent of the Einstein coefficients for such transitions are much smaller than those for allowed transitions, but are of comparable size to those of electric quadrupole transitions. These transitions, called *magnetic*

Table 3.4. *Summary of selection rules in the LS-coupling scheme.*

	E_1	E_2	M_1
$\Delta S = S - S'$	0	0	0
$\Delta L = L - L'$	$0, \pm 1$	$0, \pm 1, \pm 2$	0
$\Delta J = J - J'$	$0, \pm 1$	$0, \pm 1, \pm 2$	± 1
Additional		$\mathrm{L+L'} \geq 2$	
		$\mathrm{J+J'} \geq 2$	

Transitions with $L = 0 \rightarrow L' = 0$ and $J = 0 \rightarrow J' = 0$ are also forbidden. E_1: electric dipole; E_2: electric quadrupole; M_1: magnetic dipole.

dipole transitions, also obey selection rules that determine which couples of levels can have non-zero transition probabilities. These selection rules are also given in Table 3.4. Magnetic multipole transition probabilities can be defined in a similar way to the electric multipole transitions.

One interesting feature of the selection rules is that, no matter which order of the electric or magnetic multipole is considered, the transition $^1S_0 \rightarrow {}^1S_0$ always has zero probability, and it is strictly forbidden. However, radiative decays are still possible in such transitions, but involve the emission of two photons instead of one, and give rise to the so-called *two-photon continuum*. Such a continuum can be observed in astrophysical plasmas under some conditions, but usually not in solar plasmas, so it will not be further discussed in this book. More information can be found in Goldman and Drake (1981) and references therein.

A detailed treatment of magnetic and electric multipole transitions is beyond the scope of this book and can be found in Cowan (1981).

3.7 Atomic codes

Given the importance of atomic data for the analysis of astrophysical X-ray, EUV and UV spectra, many computer codes have been developed that allow the calculation of energy levels, Einstein coefficients, and autoionization rates by solving the Schrödinger equation for multi-electron atoms. These computer codes were developed over the last forty or so years by many different groups, who adopted very different approaches and approximations.

It is well beyond the scope of this book to provide an overview of all the codes available in the literature, and we limit ourselves to listing a few of the most widely used ones:

(1) Superstructure – Eissner *et al.* (1974)
(2) CIV3 – Hibbert (1975)
(3) Cowan's code – Cowan (1981)
(4) HULLAC – Bar-Shalom *et al.* (1988)
(5) GRASP – Dyall *et al.* (1989)
(6) MCHF – Froese Fischer (2000)
(7) FAC – Gu (2003)

More details can be found in the review on atomic codes by Bautista (2000) and references therein.

4

Mechanisms of formation of the solar spectrum

4.1 Introduction

In the present chapter we will describe the mechanisms that generate the continuum and line radiation in the solar spectrum in the UV, EUV, and soft X-ray ranges. We will first describe the processes that can excite an atom or ion from an atomic level to another level with higher energy, the decay from which generates a photon, as described in Chapter 3. We will then discuss the processes of ionization or recombination of atoms and ions. We will define the contribution function of an optically thin spectral line and its dependence on key parameters of the plasma itself. We will also describe the processes that generate the free–free (or bremsstrahlung) and free–bound continuum radiation.

For simplicity, throughout the remainder of this chapter we will refer to a neutral atom or an ion simply as an *ion*, regardless of its state of ionization.

4.2 Collisional excitation and de-excitation of atomic levels

One of the main assumptions that we will make in this chapter is that ionization and recombination processes can be decoupled from excitation and de-excitation processes, so that they can be treated separately. This approximation is usually accurate in the solar atmosphere, since the time scales for ionization and recombination are usually longer than those for excitation and de-excitation. However, as we will see in Section 4.9, ionization and recombination effects on excited levels may be non-negligible in several atomic systems, but they can be taken into account using an approximate method that allows us to retain the separation between ionization and recombination on the one hand, and excitation and de-excitation on the other.

There are several physical processes that can cause a transition from one level to another in an ion in a plasma. In the solar atmosphere the most important processes that contribute to such transitions are collisional processes between ions and free electrons. These can be of two kinds: inelastic collisions or dielectronic recombination. Inelastic collisions are the most important of the two processes, since they are involved in the formation of nearly all the spectral lines below 2000 Å; dielectronic recombinations are important for the energy levels above the ionization potential of an ion and are responsible for the emission of dielectronic satellite lines in the X-ray wavelength range.

Other processes, such as proton–ion collisions, photoexcitation, radiation scattering, and ionization and recombination into excited levels, can also cause transitions to and from the levels in an ion, but they are usually less important than electron–ion collisions; we will discuss their contribution to the total line emission in Section 4.9.

4.2.1 Electron–ion inelastic collisions

An electron–ion inelastic collision can be described as

$$X_i^{+m} + \mathrm{e}(E_1) \rightarrow X_j^{+m} + \mathrm{e}(E_2), \tag{4.1}$$

where i and j are the initial and final levels of the ion X^{+m}, with energies E_i and E_j respectively, and E_1 and E_2 are the initial and final energies of the free incident electron. When $E_i < E_j$, the ion X^{+m} is *collisionally excited* to a level with higher energy, while the free electron loses an amount of energy

$$E_2 - E_1 = E_i - E_j. \tag{4.2}$$

The inverse process, called *collisional de-excitation*, occurs when $E_i > E_j$, so that the ion is de-excited from a level with higher energy to a level with lower energy, while the energy of the free electron has increased according to Equation (4.2). In order to excite an ion from level i to level j, the free incident electron must have an initial energy E_1 larger than or equal to $\Delta E_{ij} = E_j - E_i$.

An ion in a plasma is subject to a number of elastic and inelastic collisions. In unit time, the number of electrons with velocity v that strike the unit area is given by

$$N_{\mathrm{coll}} = N_{\mathrm{e}}(v)\, v, \tag{4.3}$$

where $N_{\mathrm{e}}(v)$ indicates the density of electrons with velocity v. In solar plasmas, the velocities of electrons have a distribution $f(v)$ such that the number of electrons with velocity in the interval $v, v + dv$ is given by

$$N_{\mathrm{e}}(v, v + dv) = N_{\mathrm{e}} f(v)\, 4\pi v^2\, dv \Longrightarrow N_{\mathrm{coll}}\,(v, v + dv) = N_{\mathrm{e}} f\,(v)\, v \times\, 4\pi v^2\, dv, \tag{4.4}$$

where N_{e} is the total electron density. The probability that the collision will be inelastic and the ion will change state from the initial level i to a final level j is given by the *cross-section*, whose value is dependent on the velocity of the incident electron as well as on the nature of the target ion and of the type of transition between levels i and j. We will indicate the cross-section for inelastic collisions between the ion and an electron with velocity v as $\sigma_{ij}(v)$. The dimensions of $\sigma_{ij}(v)$ are cm^2, and we can qualitatively interpret the cross-section as the surface area of the ion that can be struck by the incoming electron in a classical collision in order to produce the $i \rightarrow j$ transition in the target ion.

Inelastic collisions can either excite or de-excite the target ion. We first consider the case of de-excitation; we indicate with i and j the two levels of the target ion involved in the transition, such that $E_j > E_i$, and the ion changes state from the initial level j to the final level i. The total number of inelastic collisions giving rise to the transition $j \rightarrow i$ per unit time is therefore

$$N_{\mathrm{coll}}^{ji} = N_{\mathrm{e}} \int_0^{\infty} f(v)\, \sigma_{ji}(v)\, v\, 4\pi v^2\, dv \qquad \mathrm{s}^{-1}. \tag{4.5}$$

We define the *collision rate coefficient* for inelastic, de-exciting collisions as the quantity

$$C_{ji}^{\mathrm{d}} = \int_0^{\infty} f(v)\, \sigma_{ji}(v)\, v\, 4\pi v^2\, dv \qquad \mathrm{cm}^3\ \mathrm{s}^{-1}. \tag{4.6}$$

The de-excitation cross-section can also be expressed by using a non-dimensional quantity, called *collision strength* (Ω), defined by the equation

$$\sigma_{ji} = \frac{I_{\mathrm{H}}}{E} \frac{\Omega_{ji}}{\omega_j} \pi a_0^2, \tag{4.7}$$

where a_0 is the Bohr radius and I_{H} is the hydrogen ionization potential, given by Equations (3.7) and (3.12) respectively, and ω_j is the statistical weight of the initial level j. The main property of the collision strength is its symmetry over the initial and final levels, since it can be shown that

$$\Omega_{ji}(E) = \Omega_{ij}(E' = E - \Delta E_{ij}), \tag{4.8}$$

where Ω_{ij} and Ω_{ji} are the excitation and de-excitation collision strengths respectively. The energy E' must be positive in order to have non-zero Ω_{ji} values. The energy E is measured relative to the energy of the upper level j.

In solar plasmas, the distribution of velocities is normally a Maxwell–Boltzmann distribution:

$$f(v) = \left(\frac{m}{2\pi k_{\mathrm{B}} T}\right)^{\frac{3}{2}} \exp\left(-\frac{mv^2}{2k_{\mathrm{B}} T}\right), \tag{4.9}$$

where m is the electron rest mass, and k_{B} the Boltzmann constant. By using Equations (4.7) and (4.9), and putting $E = \frac{1}{2}mv^2$ as the kinetic energy of the incoming electron associated with its velocity v relative to the target ion, the rate coefficient for collisional de-excitation in Equation (4.6) can be written as

$$\begin{aligned} C_{ji}^{\mathrm{d}} &= \frac{A}{\omega_j k_{\mathrm{B}} T^{3/2}} \int_0^\infty \Omega_{ji}(E) \exp\left(-\frac{E}{k_{\mathrm{B}} T}\right) dE, \\ A &= I_{\mathrm{H}} a_0^2 \sqrt{\frac{8\pi}{m k_{\mathrm{B}}}} \simeq 8.63 \times 10^{-6}. \end{aligned} \tag{4.10}$$

The units of A are cm^3 s^{-1} K$^{1/2}$. As before, we can define the *effective collision strength*, or *Maxwellian-averaged collision strength*, i.e. averaged over a Maxwell–Boltzmann distribution, by

$$\Upsilon = \int_0^\infty \Omega_{ji}(E) \exp\left(-\frac{E}{k_{\mathrm{B}} T}\right) d\left(\frac{E}{k_{\mathrm{B}} T}\right) \tag{4.11}$$

so that

$$C_{ji}^{\mathrm{d}} = \frac{A}{\omega_j T^{1/2}} \Upsilon. \tag{4.12}$$

The coefficient C_{ij}^{e} for excitation from i to j can be determined from C_{ji}^{d} using the principle of detailed balance. Under equilibrium, excitations and de-excitations must balance each other, so that

$$N_i N_{\mathrm{e}} C_{ij}^{\mathrm{e}} = N_j N_{\mathrm{e}} C_{ji}^{\mathrm{d}}. \tag{4.13}$$

However, if we assume the plasma to be in thermodynamic equilibrium, the densities N_i and N_j of the ions in the levels i and j are related by the Boltzmann equation:

$$\frac{N_j}{N_i} = \frac{\omega_j}{\omega_i} \exp\left(-\frac{\Delta E_{ij}}{k_{\mathrm{B}} T}\right), \tag{4.14}$$

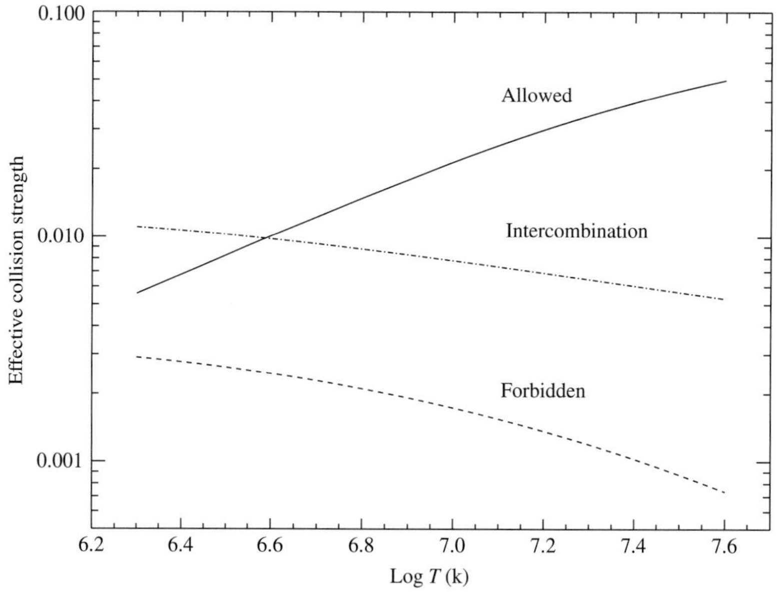

Fig. 4.1. Effective collision strengths Υ for an allowed (full line), intercombination (dash-dotted line) and forbidden (dashed line) transition in Ar XV.

where ω_j and ω_i are the statistical weights of levels j and i. Combining Equations (4.13) and (4.14), we obtain

$$
\begin{aligned}
C_{ij}^{\mathrm{e}} &= \frac{\omega_j}{\omega_i} C_{ji}^{\mathrm{d}} \exp\left(-\frac{\Delta E_{ij}}{k_{\mathrm{B}} T}\right) \\
&= \frac{A}{\omega_i T^{1/2}} \Upsilon \exp\left(-\frac{\Delta E_{ij}}{k_{\mathrm{B}} T}\right).
\end{aligned}
\tag{4.15}
$$

The effective collision strengths depend on the temperature, varying slowly with T for allowed and intercombination transitions, and decreasing at high temperatures for forbidden transitions. Examples of effective collision strengths for allowed, intercombination, and forbidden collision strengths are shown in Figure 4.1.

Even though Equation (4.15) has been derived under the assumptions of thermodynamic equilibrium, this relation is general and can be applied for non-equilibrium situations.

4.2.2 Dielectronic collisions

The levels above the ionization threshold of an ion can be excited in two different ways: electron–ion collisional excitation and dielectronic collisions. Electron collisional excitation occurs from levels below the ionization threshold in the same way as described in the previous section.

Dielectronic collisional excitation of a level j of an ion X^{+m} above the ionization threshold is an intrinsically different process. It is summarized in Figure 4.2. The process occurs when a free electron approaches an ion X^{+m+1}. If we indicate by i the state of the target ion X^{+m+1}, the excitation through dielectronic recombination can be described as

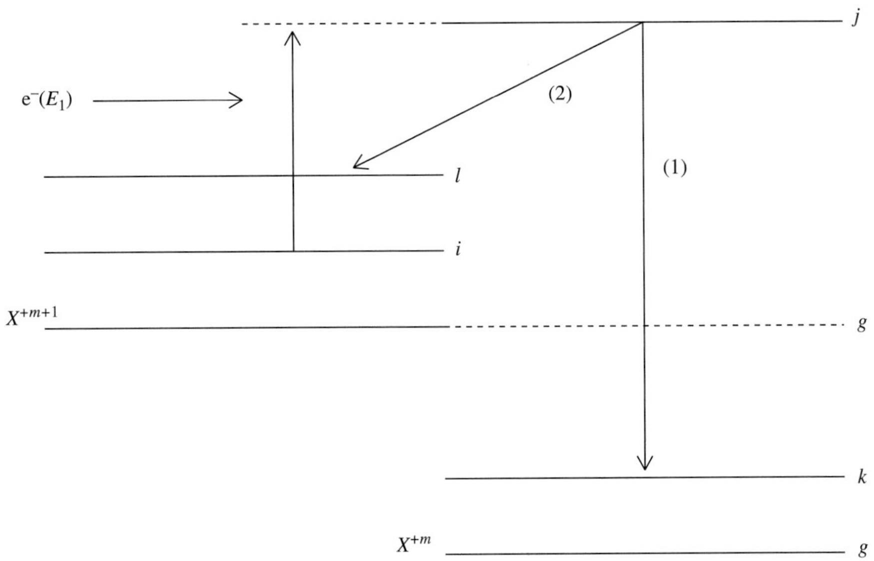

Fig. 4.2. Excitation (2) of the ion X^{+m+1} and dielectronic recombination (1) to ion X^{+m} by means of a dielectronic collision with a free electron $e^-(E_1)$. The ground level of ion X^{+m+1} corresponds to the first ionization energy of the ion X^{+m}.

$$X_i^{+m+1} + \mathrm{e}(E_1) \rightarrow X_j^{+m}, \tag{4.16}$$

where level j lies above the ionization threshold of the ion X^{+m}. The free electron is captured by the target ion, which recombines temporarily into the ion X_j^{+m} in the highly excited level j. This transition occurs only when the energy of the incoming electron relative to the ground state of the recombining ion X^{+m+1} is $E_1 = E_j - E_i$, so no radiation is emitted. Level j can be de-excited in two different ways:

(1) through a radiative transition, to another level k of the X^{+m} ion, usually below the ionization threshold: i.e. $X_j^{+m} \rightarrow X_k^{+m} + h\nu_{jk}$; or
(2) through *autoionization*, i.e. the electron is released from the recombined X^{+m}, leaving it in a new state l of the next stage of ionization: $X_k^{+m} \rightarrow X_l^{+m+1} + \mathrm{e}(E_2)$.

Process (1) is described by the usual Einstein coefficients for spontaneous radiation. Process (2) is regulated by the autoionization rate A^{a} introduced in Section 3.4.5, and involves no emission of radiation. The final state l of the ion X^{+m+1} can be different from the initial state i; the energy of the outgoing electron is changed by the quantity $E_2 - E_1 = E_i - E_l$.

The rate coefficient of dielectronic capture of a free electron is described by a collisional coefficient, C^{diel}, which can be determined from the principle of detailed balance. If we indicate by $N\left(X_j^{+m}\right)$ and $N\left(X_i^{+m+1}\right)$ the densities of ions X^{+m} and X^{+m+1} in levels j and i respectively, with level j lying above the ionization threshold of ion X^{+m}, the principle of detailed balance requires that under equilibrium

$$N\left(X_i^{+m+1}\right) N_{\mathrm{e}} C_{ij}^{\mathrm{diel}} = N\left(X_j^{+m}\right) A_{ji}^{\mathrm{a}}. \tag{4.17}$$

If we assume the plasma to be in thermodynamic equilibrium, the relative population of levels i and j follows the Saha–Boltzmann equation,

$$\frac{N\left(X_j^{+m}\right)}{N\left(X_i^{+m+1}\right)} = \frac{h^3}{2(2\pi m k_B T)^{3/2}} \frac{\omega_j}{\omega_i} \exp\left(-\frac{E_j - E_i}{k_B T}\right) N_e. \tag{4.18}$$

Combining Equations (4.17) and (4.18), the dielectronic collisional excitation rate coefficient can be expressed as

$$C_{ij}^{\text{diel}} = \frac{h^3}{2(2\pi m k_B T)^{3/2}} \frac{\omega_i}{\omega_j} A_{ji}^{a} \exp\left(-\frac{E_j - E_i}{k_B T}\right) \text{ cm}^3\,\text{s}^{-1}. \tag{4.19}$$

Although Equation (4.19) was derived under the assumption of thermodynamic equilibrium, it is also valid when this assumption does not hold.

4.2.3 *Atomic codes for electron–ion collisions*

The calculation of cross-sections, collision strengths, and collision rate coefficients for collisional excitation by a free electron requires the use of computer programs by which the energy level structure of the target ion and the ion–electron interaction can be determined. Several approximations with increasing accuracy and complexity have been developed over the years, leading to the creation of comprehensive atomic codes. Among these are the University College London Distorted Wave (DW) code (Eissner (1998)), the Queen's University Belfast R-matrix (RMATRX) code (Berrington *et al.* (1995)), the Hebrew University–Lawrence Livermore Atomic Code (HULLAC: Bar-Shalom *et al.* (1988)), and the Flexible Atomic Code (FAC: Gu (2003)). It is beyond the scope of this book to review them, but the reader is referred to the review by Bautista (2000) for further details.

Collision rates for dielectronic recombination can be calculated from the autoionization rates obtained using the atomic codes discussed in Chapter 3.

4.3 Level populations

The probability that an ion will be in an excited level depends on the interplay between the excitation and de-excitation mechanisms. In equilibrium, these two processes balance each other. For energy levels below the ionization threshold, electron impact excitation is the main contributor to level excitation, while spontaneous radiative decay and collisional de-excitation are the main de-excitation mechanisms. In the solar upper atmosphere, plasmas are usually optically thin and radiation fields are usually sufficiently weak to allow us to neglect de-excitation through induced emission.

The ions making up a plasma will in general be in different excitation states. We define the *population* of a given level i of an ion X^{+m} as the fraction of ions X^{+m} in the plasma that are in level i, i.e. $N\left(X_i^{+m}\right)/N\left(X^{+m}\right)$. The sum of the populations of all the ion's energy levels is unity.

4.3.1 *The two-level atom*

In order to determine level populations, we first consider a two-level atom with both levels below the ionization threshold: the ground level g and an excited level i. In equilibrium, the probability of finding the ion in the level i will be found by solving the equation of balance between excitation and de-excitation to and from levels i and g:

Table 4.1. *Values of A_{ig} (s^{-1}) as a function of atomic number for one allowed (Allow.), one intercombination (Intc.), and one forbidden (Forb.) transition in the carbon isoelectronic sequence. Powers of ten are indicated in parentheses.*

Transition	Type	N II	Ne V	Ca XV	Fe XXI
$2s^22p^2\,{}^3P_0 - 2s2p^3\,{}^3D_1$	Allow.	2.15(+8)	7.12(+8)	4.25(+9)	1.17(+10)
$2s^22p^2\,{}^3P_1 - 2s2p^3\,{}^5S_2$	Intc.	7.17(+1)	2.23(+3)	2.04(+6)	3.42(+7)
$2s^22p^2\,{}^3P_0 - 2s^22p^2\,{}^3P_1$	Forb.	2.08(−6)	1.67(−3)	9.56(+1)	6.54(+3)

$$N_{\rm g} N_{\rm e} C^{\rm e}_{{\rm g}i} = N_i \left(N_{\rm e} C^{\rm d}_{i{\rm g}} + A_{i{\rm g}} \right), \tag{4.20}$$

where A_{ig} is the Einstein coefficient for spontaneous emission from i to g and $N_{\rm e}$ is the electron density. We can determine the population of level i relative to g as

$$\frac{N_i}{N_{\rm g}} = \frac{N_{\rm e} C^{\rm e}_{{\rm g}i}}{N_{\rm e} C^{\rm d}_{i{\rm g}} + A_{i{\rm g}}}. \tag{4.21}$$

The relative importance of the radiative rate A_{ig} and of the collision de-excitation term $N_{\rm e} C^{\rm d}_{ig}$ determines the dependence of the relative populations of levels i and g on the temperature and density of the plasma.

The collisional de-excitation coefficients $C^{\rm d}_{ig}$ can have a broad range of values, generally between 10^{-15} and 10^{-7} cm^3 s^{-1}. The electron density in the solar atmosphere decreases with distance from the photosphere, from $\sim 10^{11}$ cm^{-3} to $\sim 10^7$ cm^{-3} (e.g. large-scale coronal loops), and is up to 10^{13} cm^{-3} at the start of some solar flares. Therefore the rate $N_{\rm e} C^{\rm d}_{ig}$ never exceeds $\sim 10^5$–10^6 s^{-1}.

The value of A_{ig} depends on the type of transitions originating from the upper level, being large for optically allowed transitions, smaller for intercombination transitions and very small for forbidden transitions. The value of A_{ig} also depends on the atomic number Z of the ion. Examples of A-values for allowed, intercombination, and forbidden transitions in the carbon isoelectronic sequence are given in Table 4.1.

Equation (4.21) admits three cases:

$$\begin{aligned} &(1) \quad A_{i{\rm g}} \ll N_{\rm e} C^{\rm d}_{{\rm g}i}, \\ &(2) \quad A_{i{\rm g}} \gg N_{\rm e} C^{\rm d}_{{\rm g}i}, \\ &(3) \quad A_{i{\rm g}} \simeq N_{\rm e} C^{\rm d}_{{\rm g}i}. \end{aligned} \tag{4.22}$$

In Case (1), the relative population $N_i/N_{\rm g}$ is given by the ratio of the two collision rate coefficients $C^{\rm d}_{ig}$ and $C^{\rm e}_{gi}$, so that, using Equation (4.15), we find that the populations of the two-level ion follow the Boltzmann equation. The populations of the two levels are comparable when $\Delta E_{gi} \simeq k_{\rm B} T$, and their relative value depends on the temperature T. The level i for which case (1) applies is called a *metastable level*. However, this case is rather uncommon in the solar atmosphere, since electron densities are usually rather small, and it occurs only among levels in the ground configuration of ions with low atomic numbers, where only forbidden transitions can occur and the A_{ig} values are small.

Case (2) is much more common, since it describes the situation of an ion in a low-density plasma whose upper level i is depopulated by allowed transitions. Equation (4.21) becomes

$$\frac{N_i}{N_g} = \frac{C^e_{gi}}{A_{ig}} N_e \ll 1. \tag{4.23}$$

Therefore level i is much less populated than the ground level, so that nearly all the ions are in the ground state. This case is called the *coronal model approximation*. In the past it was the first approximation to be adopted for the calculation of level populations. Under this approximation, the relative population of level i is directly proportional to the electron density and also temperature through the temperature dependence of C^e_{gi}.

In case (3), the relative level population N_i/N_g is given by Equation (4.21), and it depends on both the electron density and temperature; the detailed dependence must be evaluated on a case-by-case basis.

4.3.2 The multi-level atom

Equation (4.20) can be generalized to the calculation of the population of an excited level i in the more realistic case of an atom with a number of excited levels $N > 1$. The excitation and de-excitation processes are the same, but now level i can also be excited from, and de-excited to, the other excited levels in the ion. However, the populations of these additional levels must also be calculated. In this case, Equation (4.20) needs to be modified to include the additional excitation and de-excitation to and from the other excited levels, and it becomes:

$$\underbrace{\sum_{j>i} N_j N_e C^d_{ji}}_{\text{E1}} + \underbrace{\sum_{j<i} N_j N_e C^e_{ji}}_{\text{E2}} + \underbrace{\sum_{j>i} N_j A_{ji}}_{\text{E3}} =$$

$$N_i \left(\underbrace{\sum_{j<i} N_e C^d_{ij}}_{\text{D1}} + \underbrace{\sum_{j>i} N_e C^e_{ij}}_{\text{D2}} + \underbrace{\sum_{j<i} A_{ij}}_{\text{D3}} \right). \tag{4.24}$$

The left side of the equation includes all the processes that excite level i, the right side includes all the de-excitation processes from level i; their equality is required by the equilibrium assumption. The terms in Equation (4.24) have the following meanings:

(1) E1: collisional de-excitation to level i from higher energy levels j;
(2) E2: collisional excitation to level i from lower energy levels j;
(3) E3: spontaneous radiative decay to level i from higher energy levels j (*cascades*);
(4) D1: collisional de-excitation from level i to lower energy levels j;
(5) D2: collisional excitation from level i to higher energy levels j;
(6) D3: spontaneous radiative decay from level i to lower energy levels j.

Equation (4.24) must be spelled out for each of the N levels in the ion, so that we have a system of N equations, the solution of which provides the population distribution of all levels in the ion. To this system the following condition must be added:

$$\sum_i N_i = 1 \tag{4.25}$$

using the definition of level populations.

The three cases in Equation (4.22) can also be discussed for a multi-level atom. In Case (1), we can neglect the radiative decay rate for all levels, and it is possible to show that all level populations follow the Boltzmann distribution. However, such a case does not usually apply to the solar corona, since for each level, with the only exceptions of the levels of the ground configuration and very few more levels in excited configurations, there is at least one allowed radiative transition that cannot be neglected.

In Case (2), collisional de-excitation can be neglected for all levels. This means that their populations are very small, because for each level $A_{ij} \gg N_e C^d_{ij}$ so that, since C^d_{ij} is of the same order as C^e_{ji}, $A_{ij} \gg N_e C^e_{ji}$. This means that the ground level is by far the most populated level in the ion, so that collisional excitation from the ground level is much more important than collisional excitation from any other excited level; in fact the latter process can be neglected. Radiative decay to an excited level from levels with higher energies can also be neglected, because the populations of the latter are small so that their total contribution to the excitation of level i is much smaller than collisional excitation from the ground level. Therefore, each excited level becomes disconnected from all the other excited levels, except for radiative decays towards lower energy levels, while excitation to level i occurs only by excitation from the ground level. The system of Equations (4.24) then reduces to N equations similar to the case of the one-level ion:

$$N_g N_e C^e_{gi} = N_i \sum_{j<i} A_{ji}, \tag{4.26}$$

so that the relative population of level i is given by

$$\frac{N_i}{N_g} = \frac{C^e_{gi}}{\sum_{j<i} A_{ji}} N_e \ll 1, \tag{4.27}$$

which is equivalent to Equation (4.23). The level populations are very small, and are proportional to the electron density. They also depend on temperature through C^e_{gi}. This case is the coronal model approximation applied to multi-level ions; in the solar atmosphere, this approximation holds for all the ions in the H-like, Li-like, and Na-like isoelectronic sequences, except for the lightest elements. Equation (4.27) holds also if the upper level is populated via spontaneous radiative decays from upper levels following collisional excitation, but only if these higher levels are collisionally excited from the ground level only. If this condition holds, Eqaution (4.27) is valid, but the excitation rate C^e_{gi} needs to be modified including these other contributions. An example is given in Section 5.2.

Case (3) requires the full solution of the system of Equations (4.24), since now each excited level is connected to one or more other excited levels. In general, relative level populations will depend on both temperature and density, although the detailed dependence needs to be calculated on a case-by-case basis. This situation is common in the solar atmosphere, where the levels in the ground configuration and a few other levels in excited configurations may be metastable. An example of level populations is given in Figure 4.3 for the 10 lowest levels of Mg VI as a function of the electron density. At very low density ($N_e < 10^5$ cm^{-3}) the ground level is the only one significantly populated, and the ion follows the coronal model approximation. For densities in the $10^5 < N_e < 10^9$ cm^{-3} range, a few metastable levels in the ground configuration start to be significantly populated, and for higher densities their relative populations follow the Boltzmann distribution. The levels belonging to the $2s2p^4$

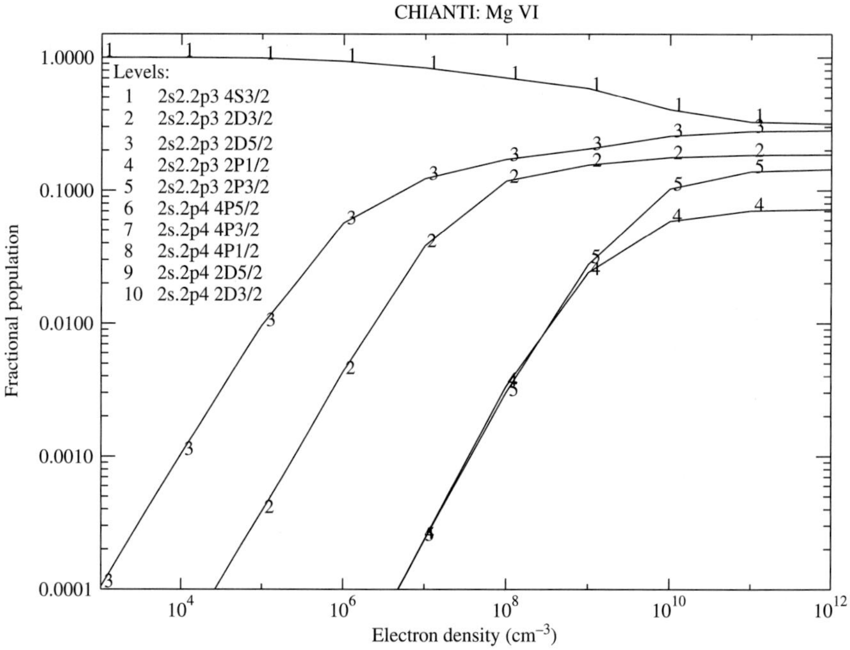

Fig. 4.3. Level populations for the 10 lowest levels of Mg VI. Populations of levels 6 to 10 are always smaller than 10^{-4}.

configuration have populations lower than 10^{-4} at all densities, since all these levels can be depopulated by allowed transitions, so they are off the scale of Figure 4.3.

The statistical equilibrium equations for levels of ion X^{+m} above the ionization threshold must also take into account autoionization and excitation through dielectronic recombination. Since the autoionization rates are large, these levels can be considered to be in the coronal model approximation. However, while direct collisional excitation usually occurs from the ground level g of the ion X^{+m}, dielectronic capture occurs from the ground level g$'$ of the ion X^{+m+1}, so that

$$N_g N_e C^e_{gi} + N_{g'} N_e C^{diel}_{g'i} = N_i \left(A^a_{tot} + \sum_{j<i} A_{ji} \right), \tag{4.28}$$

where A^a_{tot} is the total autoionization rate for level i. In order to solve Equation (4.28) it is necessary to take into account the relative populations of ions X^{+m} and X^{+m+1}, and hence include ionization and recombination (see next section). However, such an approach, although formally correct, greatly complicates the calculations. A simpler, more efficient, and yet accurate way of dealing with the populations of such levels can be developed by considering that the two collisional processes are completely independent from each other, and even involve two different target ions, so that their contributions to the total population of the level i are totally uncorrelated. We can therefore separate the population of level i in the sum of the populations given by the two processes, and calculate each of them separately:

$$N_i = N^e_i + N^{diel}_i, \tag{4.29}$$

where N_i^{e} is the portion of the population given by direct collisions and N_i^{diel} is the amount given by dielectronic recombination. These two populations can be determined by solving each of the following equations:

$$N_{\mathrm{g}} N_{\mathrm{e}} C_{\mathrm{g}i}^{\mathrm{e}} = N_i^{\mathrm{e}} \left(A_{\mathrm{tot}}^{\mathrm{a}} + \sum_{j<i} A_{ji} \right)$$

$$\Longrightarrow \frac{N_i^{\mathrm{e}}}{N_{\mathrm{g}}} = N_{\mathrm{e}} \frac{C_{\mathrm{g}i}^{\mathrm{e}}}{\left(A_{\mathrm{tot}}^{\mathrm{a}} + \sum_{j<i} A_{ji} \right)}, \tag{4.30}$$

$$N'_{\mathrm{g}} N_{\mathrm{e}} C_{\mathrm{g}'i}^{\mathrm{diel}} = N_i^{\mathrm{diel}} \left(A_{\mathrm{tot}}^{\mathrm{a}} + \sum_{j<i} A_{ji} \right)$$

$$\Longrightarrow \frac{N_i^{\mathrm{diel}}}{N'_{\mathrm{g}}} = N_{\mathrm{e}} \frac{C_{\mathrm{g}'i}^{\mathrm{diel}}}{\left(A_{\mathrm{tot}}^{\mathrm{a}} + \sum_{j<i} A_{ji} \right)}. \tag{4.31}$$

It is important to note that the population N_i^{diel} is related to the ground level of the ion X^{+m+1} rather than to X^{+m}, to which the level i belongs. Usually $N_i^{\mathrm{diel}} > N_i^{\mathrm{e}}$ so that the total population of the level i depends on the abundance of the ion X^{+m+1}: this property has very important consequences for temperature diagnostics, as we will see in Chapter 5.

When dielectronic recombination dominates, introducing Equation (4.19) into Equation (4.31) gives

$$\frac{N_i^{\mathrm{diel}}}{N'_{\mathrm{g}}} = \frac{h^3}{2(2\pi m k_{\mathrm{B}} T)^{3/2}} \frac{\omega_i}{\omega_j} \exp\left(-\frac{\Delta E_{\mathrm{g}'i}}{k_{\mathrm{B}} T} \right) \frac{A_{i\mathrm{g}'}^{\mathrm{a}}}{A_{\mathrm{tot}}^{\mathrm{a}} + \sum_{j<i} A_{ji}} N_{\mathrm{e}}. \tag{4.32}$$

The ratio $A_{i\mathrm{g}'}^{\mathrm{a}} / \left(A_{\mathrm{tot}}^{\mathrm{a}} + \sum_{j<i} A_{ji} \right)$ represents the probability that the ion X^{+m} autoionizes back into level g′ of the ion X^{+m+1}. The dielectronic-dominated level population is strongly dependent on the temperature.

4.4 Ionization and recombination

When an ion is subject to some perturbation from outside causing one of its bound electrons to leave the ion and to become free, the ion undergoes an *ionization*. When the opposite occurs and a free electron is captured by an ion into a bound state, the ion undergoes a *recombination*. The interplay between ionizations and recombinations determines the fractional abundance of all stages of ionization of each element in a plasma. Since the strength of spectral lines emitted by particular ions depends on the ion fractional abundance, the ability to predict the spectrum relies on the calculation of the relative abundances of all stages of ionization of all elements and hence on the knowledge of their ionization and recombination rates.

Ionizations and recombinations can occur via interactions of an ion with an external radiation field, or with one or more free electrons, or between two ions. The main ionization and recombination mechanisms can be summarized as follows:

(1) interaction with the ambient radiation field;
(2) collisions with free electrons involving levels *below* the ionization threshold;

(3) collisions with free electrons involving levels *above* the ionization threshold;
(4) transfer of one electron from one colliding ion to the target ion, or vice versa.

Each of these mechanisms gives rise to an ionization–recombination pair of processes, each being the inverse of the other:

(1)	Photoionization	Radiative recombination
(2)	Collisional ionization	Three-body recombination
(3)	Excitation–autoionization	Dielectronic recombination

Charge transfer between two ions is simultaneously an ionization and a recombination process. However, charge transfer only provides non-negligible contributions to the state of ionization of the plasma at low temperatures, and is expected to be negligible in the solar transition region and corona (Arnaud & Rothenflug (1985), Bryans *et al.* (2006)), so it will not be discussed here.

4.4.1 Photoionization and radiative recombination

Photoionization and radiative recombination take place as a result of the interaction between an ion and the ambient radiation field. The two processes can be summarized as

$$\begin{aligned} X_i^{+m} + h\nu &\rightarrow X_j^{+m+1} + \mathrm{e}(E) \qquad \textit{Photoionization} \\ X_j^{+m+1} + \mathrm{e}(E) &\rightarrow X_i^{+m} + h\nu \qquad \textit{Radiative recombination} \end{aligned} \tag{4.33}$$

where X_i^{+m} and X_j^{+m+1} are the ions corresponding to the mth and $(m+1)$th stages of ionization of the element X in their states i and j respectively, $h\nu$ is the energy of the radiation, and $E = \frac{1}{2}mv^2$ is the kinetic energy of the free electron with speed v. During photoionization, the absorption of a photon of energy $h\nu$ larger than the ionization threshold level i provides one bound electron with enough energy to escape from the ion with kinetic energy E, leaving behind an atomic system in the level j of the next higher stage of ionization. Recombination is the opposite process. The energies of the free electron and of the photon are fixed by the relation

$$h\nu = E_j - E_i + \frac{1}{2}mv^2, \tag{4.34}$$

where E_j and E_i are the energies of the j and i levels relative to the ground level of the ion X^{+m}. Since the velocity v of the free electron can in principle have all possible positive values, the photons involved in radiative recombination and photoionization can have all possible energies down to a minimum energy $E_0 = h\nu_0 = E_j - E_i$. Below the frequency ν_0 the radiation field is unable to photoionize the ion, so that the emission or absorption spectrum resulting from this mechanism is null at $\nu < \nu_0$. The cut-off wavelength $\lambda_0 = c/\nu_0$ is called the *recombination edge*.

Photoionization is usually described by making use of a photoionizing cross-section σ^{ph}, which can be qualitatively interpreted as the size of the target ion in a collision between an incoming photon of energy $h\nu$ and the target ion resulting in a photoionization. If we indicate by $\rho(\nu)$ the photon energy density of the radiation field, so that the radiation energy dE in the unit volume and in the frequency interval between ν and $\nu + d\nu$ is given by

$$dE = \rho(\nu)\, d\nu, \tag{4.35}$$

the number of photons in the interval $d\nu$ around a given frequency ν is given by

$$N(\nu) = \frac{\rho(\nu)}{h\nu} d\nu. \tag{4.36}$$

Thus the total number of photoionizations per unit volume and time is given by

$$\frac{dN_{ij}^{\mathrm{ph}}}{dV\,dt} = c\frac{\rho(\nu)}{h\nu} N\big(X_i^{+m}\big)\sigma_{ij}^{\mathrm{ph}}\, d\nu, \tag{4.37}$$

where c is the speed of light, $\sigma_{ij}^{\mathrm{ph}}$ is the cross-section for the photoionization from the level i of the ion X^{+m} to the level j of the ion X^{+m+1}, and $N\big(X_i^{+m}\big)$ is the number density of target ions in the initial state i. The cross-section $\sigma_{ij}^{\mathrm{ph}}$ is summed over all possible combinations of angular momenta of the total system and of the resulting free electron.

Radiative recombination is the inverse process of photoionization, and can also be described using a cross-section σ^{rr}. Indicating by N_{e} the number density of free electrons, the number of radiative recombinations per unit time and volume is given by

$$\frac{dN_{ij}^{\mathrm{rr}}}{dV\,dt} = N\Big(X_j^{+m+1}\Big)\sigma_{ij}^{\mathrm{rr}} N_{\mathrm{e}} f(v) v\, 4\pi v^2\, dv, \tag{4.38}$$

where $f(v)$ is the distribution of free electron velocities and $\sigma_{ij}^{\mathrm{rr}}$ is the cross-section for recombination from the level j of the ion X^{+m+1} to the level i of the ion X^{+m}. The radiative recombination cross-section can be expressed as a function of the photoionization cross-section. Their relation can be determined from the principle of detailed balance in the case of thermodynamic equilibrium, where

$$\frac{dN_{ij}^{\mathrm{rr}}}{dV\,dt} = \frac{dN_{ij}^{\mathrm{ph}}}{dV\,dt}. \tag{4.39}$$

Under thermodynamic equilibrium, electron velocities have a Maxwell–Boltzmann distribution given by Equation (4.9), and the radiation energy density is given by

$$\rho(\nu) = \frac{8\pi h\nu^3}{c^3}\frac{1}{\exp(h\nu/k_{\mathrm{B}}T) - 1} \tag{4.40}$$

and the rate of photoionization must be corrected by a factor $\big(1 - e^{-h\nu/k_{\mathrm{B}}T}\big)$ to take into account stimulated emission. Combining this factor with Equations (4.9), (4.39), and (4.40), we have

$$\sigma_{ij}^{\rm rr} = \frac{2mv^2}{hc^2v^2}\left(\frac{2\pi k_{\rm B}T}{m}\right)^{3/2}\frac{1-\exp\left(-h\nu/k_{\rm B}T\right)}{\exp\left(h\nu/k_{\rm B}T\right)-1}$$

$$\times\exp\left(\frac{mv^2}{2k_{\rm B}T}\right)\left[\frac{1}{N_{\rm e}}\frac{N\left(X_i^{+m}\right)}{N\left(X_j^{+m+1}\right)}\right]\sigma_{ij}^{\rm ph}$$

$$= \frac{2mv^2}{hc^2v^2}\left(\frac{2\pi k_{\rm B}T}{m}\right)^{3/2}\left[\frac{1}{N_{\rm e}}\frac{N\left(X_i^{+m}\right)}{N\left(X_j^{+m+1}\right)}\right]\exp\left(-\frac{E_j-E_i}{k_{\rm B}T}\right)\sigma_{ij}^{\rm ph}, \qquad (4.41)$$

where Equation (4.34) has been used to simplify the exponentials, and it has also been differentiated to give

$$dv = \frac{h}{mv}d\nu. \qquad (4.42)$$

The relative population between the levels of the two ions involved is given by the Saha–Boltzmann distribution in Equation (4.18). Combining Equations (4.18) and (4.41) we obtain the relation between the radiative recombination and photoionization cross-sections:

$$\sigma_{ij}^{\rm rr} = \frac{\omega_i}{\omega_j}\frac{h^2\nu^2}{m^2c^2v^2}\sigma_{ij}^{\rm ph}, \qquad (4.43)$$

where ω_i and ω_j are the statistical weights of levels i and j. This relation involves physical properties of the ion and of the incident electron only, and therefore it is valid also outside thermodynamic equilibrium. The total photoionization and recombination rates can then both be expressed as a function of $\sigma_{ij}^{\rm ph}$. Using the Maxwell–Boltzmann distribution of velocities, the total number of recombinations from X_j^{+m+1} to X_i^{+m} can be expressed as

$$\frac{dN_{ij}^{\rm rr}}{dV\,dt} = N_{\rm e}N\left(X_j^{+m+1}\right)\frac{\omega_i}{\omega_j}\int_0^\infty\sqrt{\frac{2}{\pi m}}\frac{h^2\nu^2}{c^2}\sigma_{ij}^{\rm ph}\frac{\exp(-mv^2/2k_{\rm B}T)}{(k_{\rm B}T)^{3/2}}v\,dv, \qquad (4.44)$$

so that by using Equations (4.34) and (4.42), and with the new variable $u = h\nu/k_{\rm B}T$, we obtain

$$\frac{dN_{ij}^{\rm rr}}{dV\,dt} = N_{\rm e}N\left(X_j^{+m+1}\right)\frac{\omega_i}{\omega_j}\sqrt{\frac{2}{\pi}}\frac{\exp\left(\frac{E_j-E_i}{k_{\rm B}T}\right)}{c^2}\left(\frac{k_{\rm B}T}{m}\right)^{3/2}$$

$$\int_{\frac{E_j-E_i}{k_{\rm B}T}}^\infty u^2e^{-u}\sigma_{ij}^{\rm ph}\,du. \qquad (4.45)$$

The total number of recombinations from X^{+m+1} to X^{+m} can be obtained by summing Equation (4.45) over all levels i and j. If we define the *total radiative recombination rate* $\alpha^{\rm rr}$ as

$$\frac{dN^{\rm rr}}{dV\,dt} = N_{\rm e}N\left(X^{+m+1}\right)\alpha^{\rm rr}, \qquad (4.46)$$

the summation over all levels i and j gives

$$\alpha^{\rm rr} = \frac{1}{c^2}\sqrt{\frac{2}{\pi}}\left(\frac{k_{\rm B}T}{m}\right)^{3/2}\sum_i\sum_j\frac{N\left(X_j^{+m+1}\right)}{N\left(X^{+m+1}\right)}\exp\left(\frac{E_j-E_i}{k_{\rm B}T}\right)$$

$$\times\frac{\omega_i}{\omega_j}\int_{\frac{E_j-E_i}{k_{\rm B}T}}^{\infty}u^2e^{-u}\sigma_{ij}^{\rm ph}du \qquad {\rm cm}^3\,{\rm s}^{-1}, \tag{4.47}$$

where $N(X_j^{+m+1})/N(X^{+m+1})$ is the relative population of level j.

The total photoionization rate can be obtained from Equation (4.38) using similar arguments, noting that for a given ambient radiation field $\rho(\nu)$

$$\frac{dN_{ij}^{\rm ph}}{dV\,dt} = \int_{\nu_0}^{\infty} cN\left(X_i^{+m}\right)\sigma_{ij}^{\rm ph}\frac{\rho(\nu)}{h\nu}d\nu. \tag{4.48}$$

If we define the *total photoionization rate coefficient* $\alpha^{\rm ph}$ as

$$\frac{dN^{\rm ph}}{dV\,dt} = N\left(X^{+m}\right)\alpha^{\rm ph}, \tag{4.49}$$

it takes the form

$$\alpha^{\rm ph} = c\sum_i\sum_j\frac{N\left(X_i^{+m}\right)}{N\left(X^{+m}\right)}\int_{\nu_0}^{\infty}\sigma_{ij}^{\rm ph}\frac{\rho(\nu)}{h\nu}d\nu \qquad {\rm s}^{-1}. \tag{4.50}$$

4.4.2 Collisional ionization and three-body recombination

The processes of collisional ionization and three-body recombination involve collisions between the target ion and free electrons, according to

$$X_i^{+m} + {\rm e}_1(E_1) \rightarrow X_j^{+m+1} + {\rm e}_1(E_2) + {\rm e}_2(E_3) \qquad \textit{Collisional ionization} \tag{4.51}$$

$$X_j^{+m+1} + {\rm e}_1(E_2) + {\rm e}_2(E_3) \rightarrow X_i^{+m} + {\rm e}_1(E_1) \qquad \textit{Three-body recombination} \tag{4.52}$$

where in Equation (4.51) $\rm e_1$ is the incident electron and $\rm e_2$ is the electron that escapes from the target ion following ionization. In order for these two processes to take place, the energies of the free electrons involved, and those of the atomic levels i and j (E_i and E_j respectively) must satisfy the following requirements:

$$\begin{cases} E_j - E_i = E_1 - E_2 - E_3, \\ E_1 = \frac{1}{2}mv^2 > E_j - E_i, \\ E_2 < E_1 - \left(E_j - E_i\right). \end{cases}$$

The first requirement is an expression of the conservation of energy, while the second and the third are necessary for the incident electron to ionize the target ion and the ejected electron to have non-zero kinetic energy.

The number of ionizations following inelastic collisions with free electrons can be obtained by introducing the collisional ionization cross-section $\sigma_{\rm ci}^{ij}(E_1, E_2)$. This can be qualitatively interpreted as the size of the target ion struck by a free electron with energy E_1 for an ionization to occur with the incident electron having a final kinetic energy in the range E_2 to

$E_2 + dE_2$: since an electron of energy E_3 is ejected from the target ion, the final energy of the incident electron can have all possible values between 0 and $E_1 - (E_j - E_i)$. The number of ionizations from incident electrons per unit volume and time, with initial energy E_1 and final energy in the range E_2 to $E_2 + dE_2$, is given by

$$\frac{dN_{ij}^{\mathrm{ci}}(E_1, E_2)}{dV\,dt} = N_{\mathrm{e}} N\left(X_i^{+m}\right) v\, f(v)\, \sigma_{ij}^{\mathrm{ci}}(E_1, E_2)\, 4\pi v^2\, dv\, dE_2, \tag{4.53}$$

where $E_1 = mv^2/2$. To determine the total collisional ionization rate between X_i^{+m} and X_j^{+m+1}, regardless of the initial and final energies of the incident electron, we must integrate Equation (4.53) over both E_1 and E_2. If the distribution of electron velocities $f(v)$ is Maxwell–Boltzmann, the total collisional ionization rate per unit volume and time is given by

$$\frac{dN_{ij}^{\mathrm{ci}}}{dV\,dt} = N_{\mathrm{e}} N\left(X_i^{+m}\right) \sqrt{\frac{8k_{\mathrm{B}}}{m\pi}}\, T^{1/2} \int_{I_{ij}}^{\infty} \left(\frac{E_1}{k_{\mathrm{B}}T}\right) \exp\left(-\frac{E_1}{k_{\mathrm{B}}T}\right) d\left(\frac{E_1}{k_{\mathrm{B}}T}\right) \times \int_0^{E_1 - I_{ij}} \sigma_{\mathrm{ci}}^{ij}(E_1, E_2)\, dE_2 \tag{4.54}$$

$$= N_{\mathrm{e}} N\left(X_i^{+m}\right) C_{ij}^{\mathrm{ci}}, \tag{4.55}$$

where C_{ij}^{ci} is the collisional ionization rate coefficient and $I_{ij} = E_j - E_i$. The number of collisional ionizations from ion X^{+m} to ion X^{+m+1}, regardless of the initial and final atomic levels i and j, can be obtained by summing Equation (4.55) over all levels i and j. If we define the *total collisional ionization rate coefficient* α^{ci} as

$$\frac{dN^{\mathrm{ci}}}{dV\,dt} = N_{\mathrm{e}} N\left(X^{+m}\right) \alpha^{\mathrm{ci}}, \tag{4.56}$$

then the rate coefficient α^{ci} is

$$\alpha^{\mathrm{ci}} = \sqrt{\frac{8k_{\mathrm{B}}}{m\pi}}\, T^{1/2} \sum_i \sum_j \frac{N\left(X_i^{+m}\right)}{N\left(X^{+m}\right)} \int_{I_{ij}}^{\infty} \left(\frac{E_1}{k_{\mathrm{B}}T}\right) \exp\left(-\frac{E_1}{k_{\mathrm{B}}T}\right) d\left(\frac{E_1}{k_{\mathrm{B}}T}\right) \times \int_0^{E_1 - I_{ij}} \sigma_{ij}^{\mathrm{ci}}(E_1, E_2)\, dE_2, \tag{4.57}$$

where α^{ci} is in cm^3 s^{-1}. The three-body recombination rate from X_j^{+m+1} to X_i^{+m} can be obtained through the principle of detailed balance. If we indicate with C_{ji}^{3b} the three-body recombination rate coefficient, the number of three-body recombinations per unit volume and time can be written as

$$\frac{dN_{ji}^{\mathrm{3b}}}{dV\,dt} = N_{\mathrm{e}} N\left(X_j^{+m+1}\right) C_{ji}^{\mathrm{3b}}. \tag{4.58}$$

The principle of detailed balance requires that

$$N_{\mathrm{e}} N\left(X_j^{+m+1}\right) C_{ji}^{\mathrm{3b}} = N_{\mathrm{e}} N\left(X_i^{+m}\right) C_{ij}^{\mathrm{ci}}, \tag{4.59}$$

so that

$$C_{ji}^{\mathrm{3b}} = \frac{N\left(X_i^{+m}\right)}{N\left(X_j^{+m+1}\right)} C_{ij}^{\mathrm{ci}} = \frac{h^3}{2(2\pi m k_{\mathrm{B}} T)^{3/2}} \frac{\omega_i}{\omega_j} \exp\left(-\frac{E_i - E_j}{k_{\mathrm{B}}T}\right) C_{ij}^{\mathrm{ci}} N_{\mathrm{e}}, \tag{4.60}$$

where the Saha–Boltzmann distribution (Equation (4.18)) has been used. Defining the *total three-body recombination rate coefficient* α^{3b} as

$$\frac{dN^{3b}}{dV\,dt} = N_e N\left(X^{+m+1}\right)\alpha^{3b}, \tag{4.61}$$

Equation (4.60) leads to

$$\alpha^{3b} = \frac{h^3}{2(2\pi m k_B T)^{3/2}} N_e \sum_j \sum_i \frac{\omega_i}{\omega_j} \exp\left(-\frac{E_i - E_j}{k_B T}\right) \frac{N\left(X_j^{+m+1}\right)}{N\left(X^{+m+1}\right)} C_{ij}^{ci}, \tag{4.62}$$

where α^{3b} is in cm^3 s^{-1}.

4.4.3 Excitation–autoionization and dielectronic recombination

The processes of excitation–autoionization and dielectronic recombination involve levels above the ionization threshold. These two processes can be summarized as

$$X_i^{+m+1} + e(E) \rightarrow X_j^{+m} \rightarrow X_l^{+m} + h\nu \qquad \textit{Dielectronic recombination}$$

$$X_l^{+m} + e(E_1) \rightarrow X_j^{+m} + e(E_2) \rightarrow X_i^{+m+1} + e(E_2) + e'(E)$$

$$\textit{Excitation-autoionization} \tag{4.63}$$

where E_1 and E_2 are the kinetic energies of the incident electron before and after the excitation–autoionization process, the energy of level j lies above the ionization threshold of the ion X^{+m}, and e$'$ is the electron that escapes from the ion X^{+m} with kinetic energy E.

The process of dielectronic recombination involves the dielectronic capture of an electron described in Section 4.2.2: an incoming electron is captured by the target ion X_i^{+m+1} into the level j above the ionization threshold of ion X^{+m} via a radiationless transition provided that $E = E_j - E_i$. The recombined ion is in an unstable state, and it can either autoionize, i.e. the captured electron is released from the ion X^{+m} via another radiationless transition, or it can radiatively decay into a level below ionization threshold of the ion X^{+m}. The recombination process is complete only if the radiative decay takes place.

The number of dielectronic captures per unit time and volume is given by

$$\frac{dN_{ij}^{diel}}{dV\,dt} = N_e N\left(X_i^{+m+1}\right) C_{ij}^{diel}, \tag{4.64}$$

where C_{ij}^{diel} rate coefficient is given by Equation (4.19). The probability P_{jl} that the ion X^{+m} will decay radiatively from the level j to a level l below the ionization threshold is given by

$$P_{jl} = \frac{A_{jl}^r}{\sum_l A_{jl}^r + A_j^a}, \tag{4.65}$$

where A_{jl}^r are the Einstein coefficients for spontaneous emission from j to l and their summation is carried out over all bound levels l of the ion X^{+m}, while A_j^a is the total autoionization rate from level j. The total probability of radiative stabilization of the ion X_j^{+m} is given by $P = \sum_l P_{jl}$, so that the total number of dielectronic recombinations from the ion X^{+m+1} to X^{+m} per unit time and volume will be given by

$$\frac{dN^{diel}}{dV\,dt} = N_e \sum_j \sum_i N\left(X_i^{+m+1}\right) C_{ij}^{diel} \sum_l P_{jl}, \tag{4.66}$$

so that, combining Equations (4.19) and (4.65), we have

$$\frac{dN^{\rm diel}}{dV\,dt} = N_{\rm e}\frac{h^3}{2(2\pi m k_{\rm B}T)^{3/2}}\sum_j\sum_i N\left(X_i^{+m+1}\right) \times\frac{\omega_i}{\omega_j}\exp\left(-\frac{E_j-E_i}{k_{\rm B}T}\right)A^{\rm a}_{ij}\frac{\sum_l A^{\rm r}_{jl}}{\sum_l A^{\rm r}_{jl}+A^{\rm a}_j}. \tag{4.67}$$

Defining the *total dielectronic recombination rate coefficient* $\alpha^{\rm dr}$ as

$$\frac{dN^{\rm diel}}{dV\,dt} = N_{\rm e}N\left(X^{+m+1}\right)\alpha^{\rm dr}, \tag{4.68}$$

Equation (4.67) gives

$$\alpha^{\rm dr} = \frac{h^3}{2(2\pi m k_{\rm B}T)^{3/2}}\sum_j\sum_i\frac{\omega_i}{\omega_j}\frac{N\left(X_i^{+m+1}\right)}{N\left(X^{+m+1}\right)}\exp\left(-\frac{E_j-E_i}{k_{\rm B}T}\right) A^{\rm a}_{ij}\frac{\sum_l A^{\rm r}_{jl}}{\sum_l A^{\rm r}_{jl}+A^{\rm a}_j}, \tag{4.69}$$

where $N(X_i^{+m+1})/N(X^{+m+1})$ is the relative population of level i, and $\alpha^{\rm dr}$ is in cm^3 s^{-1}.

The inverse process is given by autoionization, as described in Section 4.2.2. In order for it to take place, it is necessary that the ion X^{+m} is first excited to a level j above the ionization threshold by some other process. In the solar atmosphere, the most common process is inelastic collisions with free electrons, although in some cases photoexcitations by X-ray radiation may occur. Because collisional excitation must precede autoionization, the entire process is called excitation–autoionization.

The number of excitations per unit volume and time to level j due to inelastic collisions from a level l below ionization threshold is given by

$$\frac{dN^{\rm exc}_{lj}}{dV\,dt} = N_{\rm e}N\left(X_l^{+m}\right)C^{\rm e}_{lj}, \tag{4.70}$$

where $C^{\rm e}_{lj}$ is the collisional excitation rate coefficient defined in Equation (4.15) in the case of a Maxwell–Boltzmann velocity distribution. The total probability that the excited ion X_j^{+m} will autoionize is given by

$$P_j = \frac{A^{\rm a}_j}{\sum_l A^{\rm r}_{jl}+A^{\rm a}_j}, \tag{4.71}$$

so that the total number of ionizations per unit volume and time resulting from the excitation–autoionization process from X^{+m} to X^{+m+1} is given by

$$\frac{dN^{\rm ea}}{dV\,dt} = N_{\rm e}\sum_l\sum_j C^{\rm e}_{lj}P_jN\left(X_l^{+m}\right) = \frac{A}{T^{1/2}}N_{\rm e}\sum_l\sum_j\frac{\Upsilon_{lj}(T)}{\omega_j}\exp\left(-\frac{\Delta E_{lj}}{k_{\rm B}T}\right)\frac{A^{\rm a}_j}{\sum_l A^{\rm r}_{jl}+A^{\rm a}_j}N\left(X_l^{+m}\right), \tag{4.72}$$

where use has been made of Equation (4.15). If we define the *total excitation–autoionization rate coefficient* α^{ea} from X^{+m} to X^{+m+1} as

$$\frac{dN^{\mathrm{ea}}}{dV\,dt} = N_{\mathrm{e}} N\left(X^{+m}\right)\alpha^{\mathrm{ea}}, \tag{4.73}$$

then Equation (4.72) gives

$$\alpha^{\mathrm{ea}} = \frac{A}{T^{1/2}} \sum_{l} \sum_{j} \frac{\Upsilon_{lj}(T)}{\omega_j} e^{-\Delta E_{lj}/k_{\mathrm{B}}T} \frac{A^{\mathrm{a}}_{j}}{\sum_l A^{\mathrm{r}}_{jl} + A^{\mathrm{a}}_{j}} \frac{N\left(X^{+m}_{l}\right)}{N\left(X^{+m}\right)}, \tag{4.74}$$

where $N(X^{+m}_{l})/N(X^{+m})$ is the relative population of level l and α^{ea} is in cm^3 s^{-1}.

4.5 Ion populations and ionization equilibrium

The abundance of each ionization state can be calculated by solving the system of equations that regulate the interplay between ionizations and recombinations for each ionization state of a given element. In the solar atmosphere, ionizations and recombinations usually occur only between adjacent stages of ionization, and the processes of charge transfer, photoionization, and three-body recombination can be neglected, mostly because of the low density and of the lack of radiation fields sufficiently strong to compete with collisional processes in ionizing ions. Charge transfer may, however, be important at temperatures smaller than 25 000 K for a few ions (Arnaud & Rothenflug (1985)).

The time evolution of the populations of each of the N stages of ionization of the element X can be described by the set of N equations

$$\begin{aligned}\frac{dN_i}{dt} &= N_{\mathrm{e}}\left[N_{i-1}\left(\alpha^{\mathrm{ci}} + \alpha^{\mathrm{ea}}\right) + N_{i+1}\left(\alpha^{\mathrm{rr}} + \alpha^{\mathrm{dr}}\right)\right] \\ &\quad - N_{\mathrm{e}} N_i\left(\alpha^{\mathrm{rr}} + \alpha^{\mathrm{dr}} + \alpha^{\mathrm{ci}} + \alpha^{\mathrm{ea}}\right),\end{aligned} \tag{4.75}$$

with the additional condition that

$$\sum_{i=1}^{N} N_i = N(X), \tag{4.76}$$

where $N(X)$ is the total number density of element X.

Solving the system of Equations (4.75) can be quite involved, and the solution depends on the boundary and initial conditions of the plasma being studied; such a solution also requires the availability of a large number of ionization and recombination rates. Moreover, this system of equations also depends on the populations of each level of each ionization stage, because the ionization and recombination rates depend on the relative level populations. This means that the system of equations for ion populations and for level populations are interconnected. However, in most conditions the time scales for ionization and recombination are much larger than those for level excitation and de-excitation, so that the two processes can be treated separately and the level populations can be calculated independently using the system of Equations (4.24).

When the plasma is in equilibrium, the time derivatives can be set to zero. In this case the system of Equations (4.75) is reduced to

$$N_{i-1}\left(\alpha^{\rm ci}+\alpha^{\rm ea}\right)+N_{i+1}\left(\alpha^{\rm rr}+\alpha^{\rm dr}\right)=N_i\left(\alpha^{\rm rr}+\alpha^{\rm dr}+\alpha^{\rm ci}+\alpha^{\rm ea}\right). \qquad (4.77)$$

It is important to note that in the case of equilibrium the electron density cancels out from both sides of Equation (4.77), so that unless the ionization and recombination rate coefficients depend on density, the ion populations resulting from Equation (4.77) are only a function of temperature.

Calculating the ionization and recombination rate coefficients can be quite difficult, given the large number of atomic levels that might be involved in each process. However, several approximate, simplified formulas have been developed. They can be parameterized by analytical fits that can be used to solve the system of Equations (4.75). As an example, Shull and Steenberg (1982) parameterized the rate coefficients for the four relevant processes in the solar atmosphere as follows:

$$\alpha^{\rm ci} = A_{\rm ci}T^{1/2}\left(1+a_{\rm ci}\frac{T}{T_{\rm ci}}\right)^{-1}\exp\left(-\frac{T_{\rm ci}}{T}\right), \qquad (4.78)$$

$$\alpha^{\rm ea} = A_{\rm ea}T^{-1/2}\exp\left(-\frac{T_{\rm ea}}{T}\right), \qquad (4.79)$$

$$\alpha^{\rm rr} = A_{\rm rr}\left(\frac{T}{10^4}\right)^{-\chi_{\rm rr}}, \qquad (4.80)$$

$$\alpha^{\rm dr} = A_{\rm dr}T^{-3/2}\exp\left(-\frac{T_{\rm dr1}}{T}\right)\left(1+B_{\rm dr}\exp\left(-\frac{T_{\rm dr2}}{T}\right)\right), \qquad (4.81)$$

where the parameters $A_{\rm ci}$, $a_{\rm ci}$, $T_{\rm ci}$, $A_{\rm ea}$, $T_{\rm ea}$, $A_{\rm rr}$, $\chi_{\rm rr}$, $A_{\rm dr}$, $T_{\rm dr1}$, $B_{\rm dr}$, and $T_{\rm dr2}$ are given by Shull and Steenberg (1982) for each ion.

Analytical fits to the ionization and recombination rate coefficients have also been used by a number of authors to calculate ion populations, the most recent being Arnaud and Rothenflug (1985) for the most abundant elements in the Universe; from Mazzotta *et al.* (1998) and Bryans *et al.* (2006) for all elements with $Z \leq 28$; and from Arnaud and Raymond (1992) for Fe ions. The differences between these calculations lie in the increasingly accurate approximations and analytical fits used to calculate ionization and recombination rates, relying both on laboratory measurements and on increasingly sophisticated calculations of such rates. All these calculations assume that ionization and recombination rates are not density sensitive. This assumption holds if the electron density is sufficiently small so that the effects of ionizations and recombinations to and from excited levels can be neglected: this is called the *zero-density* approximation. However, when the density is sufficiently high, excited levels need to be taken into account, since level populations are density dependent so that the rates and the resulting ion abundances will also be density dependent. Also, electron density can influence the dielectronic recombination rate and hence ion abundances. Some of these cases have been discussed by Jordan (1969), Summers (1972), and Summers (1974). An example of the ion populations as a function of temperature is given in Figure 1.12, where the ion populations for Fe from Mazzotta *et al.* (1998) are shown.

In the solar corona it is usually assumed that the plasma is in ionization equilibrium. However, this assumption may not hold during dynamic phenomena such as explosive events and flares. Also, in the solar wind far from the disk the electron density is so low that ionization and recombination are very slow processes, so that the outflowing wind plasma maintains the same ion abundance as in the locations where it originated. In this case, ion abundances are decoupled from the local temperatures, and the ionization states are said to be *frozen in*.

4.6 Line radiation

The solar atmosphere is in most cases optically thin, so that all the photons emitted along the line of sight eventually reach the observer. Under these conditions, the number of photons in a spectral line will be given simply by the sum of all contributions of the plasma along the line of sight.

The power emitted by a volume element dV by the transition $j \rightarrow i$ in the ion X^{+m} and observed at a distance d by a detector with collecting area S is given by

$$dE = \frac{S}{4\pi d^2} N\left(X_j^{+m}\right) A_{ji} h\nu_{ij}\, dV \quad \text{erg s}^{-1}, \tag{4.82}$$

where ν_{ij} is the frequency, A_{ji} the Einstein coefficient for spontaneous emission, and $N(X_j^{+m})$ is the density of emitters. Since the plasma is optically thin, the total observed power is given by the sum of the power emitted by each volume element dV along the line of sight, according to

$$E = \frac{S}{4\pi d^2} \int_V N\left(X_j^{+m}\right) A_{ji} h\nu_{ij}\, dV. \tag{4.83}$$

We define *flux* of the line as the power passing through the unit area at distance d; from Equation (4.83), the flux is given by

$$F = \frac{E}{S} = \frac{1}{4\pi d^2} \int_V N\left(X_j^{+m}\right) A_{ji} h\nu_{ij}\, dV \quad \text{erg cm}^{-2}\ \text{s}^{-1}. \tag{4.84}$$

The quantity F is commonly referred to as flux in astrophysics and solar physics, and we use this term here. However, as indicated in Chapter 2 (Section 2.5.1), the term irradiance should strictly be used.

The density of emitters can be expressed as a function of the electron density using the relation

$$N\left(X_j^{+m}\right) = \frac{N\left(X_j^{+m}\right)}{N\left(X^{+m}\right)} \frac{N\left(X^{+m}\right)}{N(X)} \frac{N(X)}{N(H)} \frac{N(H)}{N_e} N_e, \tag{4.85}$$

where

(1) $N(X_j^{+m})/N(X^{+m})$ is the relative level population (Section 4.3 describes how this can be evaluated);
(2) $N(X^{+m})/N(X)$ is the relative ion population (see Section 4.5);
(3) $N(X)/N(H)$ is the abundance of the element X relative to hydrogen;
(4) $N(H)/N_e$ is the abundance of hydrogen relative to the free electron density (this is $\sim$0.83 for regions of the solar atmosphere with temperature higher than 10^5 K).

Using Equation (4.85), we can define the *contribution function* $G(T, N_e)$ as

$$G(T, N_e) = \frac{N\left(X_j^{+m}\right)}{N\left(X^{+m}\right)} \frac{N\left(X^{+m}\right)}{N(X)} \frac{N(X)}{N(H)} \frac{N(H)}{N_e} \frac{A_{ji}}{N_e} h\nu_{ij} \quad \text{erg cm}^3\ \text{s}^{-1}, \tag{4.86}$$

so that the flux observed at distance d can be written as

$$F = \frac{1}{4\pi d^2} \int_V G(T, N_e) N_e^2 \, dV . \tag{4.87}$$

The contribution function includes all the atomic physics involved in the process of line formation and, once all the relevant rates are available, it can be calculated theoretically as a function of electron density and temperature and of the element abundance by solving the systems of Equations (4.24) and (4.75) or (4.77) to determine level and ion populations, and by adopting a value for the element abundances.

Examples of the contribution functions for Fe XXI lines in the EUV range are given in Figure 4.4 as a function of electron temperature (left) and density (right): they are fairly sharply peaked with temperature owing to their dependence on the ion abundance, and are density sensitive in the $10^{10} - 10^{13}$ cm^{-3} density range. The peak temperature of the contribution function as well as the range of density sensitivity are typical of each ion; different lines within the same ion may have different dependence on the electron density: this property is of great importance for density diagnostics, as described in Chapter 5.

The quantity $N_e^2 \, dV = d(\mathrm{EM})$ is called the *emission measure* (EM) of the volume element dV, and has units cm^{-3}. It is proportional to the number of free electrons ($N_e \, dV$) and to the electron density in dV, and so depends on the physical conditions of the emitting plasma. The total emission measure EM of the plasma is given by

$$\mathrm{EM} = \int_V N_e^2 \, dV . \tag{4.88}$$

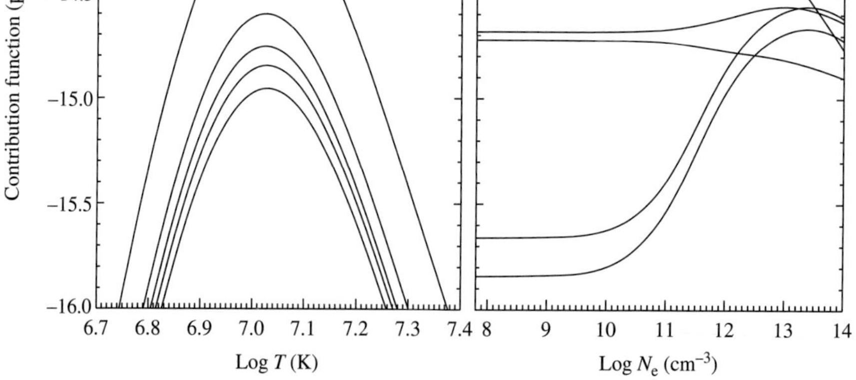

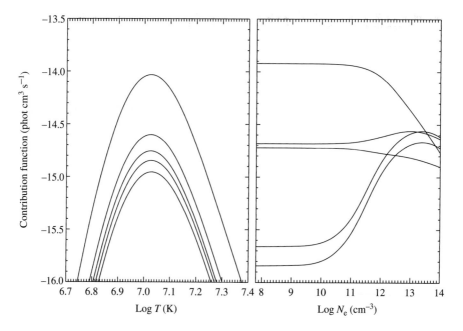

Fig. 4.4. Contribution function for Fe XXI lines in the 90–130 Å range as a function of electron temperature at $N_e = 10^9$ cm^{-3} (*left*) and as a function of electron density, at $T = 1.6 \times 10^7$ K (*right*).

Equation (4.87) can be cast as an integral over density and temperature, provided a relationship exists between the volume on one hand and temperature and density on the other. Following Brown *et al.* (1991), the volume along the line of sight can be divided into several volume elements dV such that in each of them the temperature and density of the plasma lie in the $T + dT$ and $N_e + dN_e$ range. We indicate with S_n and S_T the surfaces where the electron density and the temperature respectively are constant: usually in a plasma such surfaces will not be coincident but will intersect along a line $L_{N_e,T}$. The element volume dV_j can then be expressed as

$$dV_j = dL^j_{N_e,T} \frac{dN_e \, dT}{|\nabla N_e| \;\; |\nabla T| \sin\theta^j_{N_e,T}}, \tag{4.89}$$

where $\theta_{N_e,T}$ is the local angle between the vectors ∇N_e and ∇T representing the gradients of N_e and T perpendicular to the surfaces S_n and S_T. In a plasma, there will be many disjoint volume elements dV_j where the electron density and temperature lie in the $T+dT$ and N_e+dN_e ranges; using Equation (4.89) we can define the emission measure differential in N_e and T as

$$\psi(T, N_e) = \sum_j \int_{L^j_{N_e,T}} \frac{N_e^2 \, dL^j_{N_e,T}}{|\nabla N_e| \;\; |\nabla T| \sin\theta^j_{N_e,T}}, \tag{4.90}$$

so that Equation (4.87) can be rewritten as

$$F = \frac{1}{4\pi d^2} \int_T \int_{N_e} G(T, N_e)\psi(T, N_e) \, dT \, dN_e. \tag{4.91}$$

The quantity $\psi(T, N_e)$ can be considered a measure of the distribution of the plasma along the line of sight as a function of electron temperature and density. However, the dependence of $\psi(T, N_e)$ on N_e is practically impossible to determine from observations since the sensitivity of the contribution function $G(T, N_e)$ to N_e is small so that the inversion of Equation (4.91) provides very uncertain results along the density direction (see Section 5.6). For this reason, in most studies a more limited definition of $\psi(T, N_e)$ has been used, in which only the distribution of plasma as a function of temperature is considered, and the dependence on N_e ignored. The emitting volume dV where the electron temperature is confined in the $T, T + dT$ range is expressed as

$$dV = \frac{dT}{|\nabla T|}, \tag{4.92}$$

where, unlike in Equation (4.89), $|\nabla T| = dT/dV$ indicates differentiation over volume, so that the emission measure differential in temperature $\varphi(T)$ (also called *differential emission measure*, or *DEM*) is defined as

$$\varphi(T) = \frac{N_e^2}{|\nabla T|} \quad \text{cm}^{-3}\ \text{K}^{-1}. \tag{4.93}$$

Equation (4.87) can therefore be written as

$$F = \frac{1}{4\pi d^2} \int_T G(T, N_e)\varphi(T) \, dT. \tag{4.94}$$

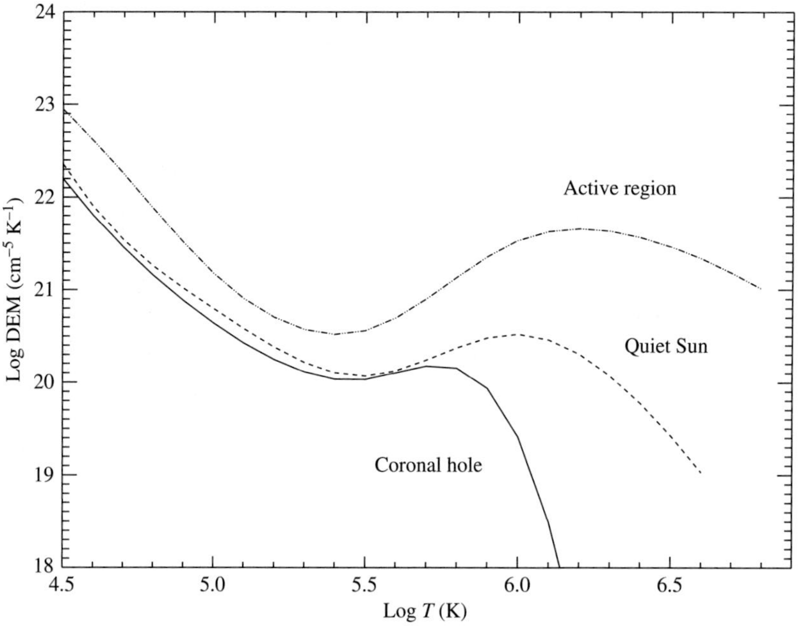

Fig. 4.5. Typical DEM curves for coronal holes, quiet Sun and active regions.

Although $\varphi(T)$ loses any information along the density direction, it can be more easily measured from observations than the more generalized function $\psi(T, N_e)$ by inverting Equation (4.94). Although such inversion still poses several problems, many techniques have been developed to carry it out, as described in Section 5.6.

Typical DEM curves for coronal hole, quiet Sun, and active regions are shown in Figure 4.5: all curves have similar shapes at log $T \leq 5.6$, but differ sharply at higher temperatures, where the amount of material with log $T \geq 5.8$ greatly increases in the quiet Sun and even more in active regions.

When the plasma is isothermal with temperature T_c, the DEM can be described using a δ function:

$$\varphi(T) = \text{EM}\delta(T - T_c). \tag{4.95}$$

4.7 Continuum radiation

Continuum radiation in the UV, EUV, and X-ray ranges is generated by the interaction between free electrons and ions. If we consider the interaction between one incoming electron and one ion X^{+m} in the ground state, the final state of the electron can be of two kinds:

(1) the electron is still free: $e(E_1) + X^{+m} \Longrightarrow e(E_2) + X^{+m} + h\nu$;
(2) the electron is captured by the ion, which recombines: $e(E_1) + X^{+m} \Longrightarrow X_i^{+m-1} + h\nu$.

In these expressions, E_1 and E_2 are the initial and final kinetic energies of the free electron, and i is the final level of the recombined ion X^{+m-1} where the incident electron is bound. In both cases radiation is emitted: in the first case it is called *free–free radiation* (or *bremsstrahlung*), in the second case it is called *free–bound radiation*. The frequency of the emitted radiation depends on the initial kinetic energy of the electron and on its final state:

$$h\nu = E_1 - E_2 \qquad \text{free–free}, \tag{4.96}$$

$$h\nu = E_1 - E_i \qquad \text{free–bound}, \tag{4.97}$$

where E_i is the energy of the atomic state in which the electron is bound when recombination occurs. Note that in Equation (4.97) the energy has been taken relative to the energy of the ground level of the ion X^{+m}, so that E_i is negative. Since the incident free electrons can have all possible energies, the total radiation originating from a large number of free–free and free–bound processes will be emitted at all frequencies, resulting in a continuum spectrum. Note, however, that in the case of free–bound continuum the radiation can be emitted only down to a limit frequency ν_c such that $h\nu_c = E_I - E_i$, where E_I is the ionization threshold of the ion X^{+m}. The case $\nu = \nu_0$ corresponds to the limit case where the velocity of the incident electron is zero. This generates an edge in the free–bound spectrum. Each level i in the recombined ion X_i^{+m-1} generates its own edge. The shape of the free–bound spectrum of a plasma with many different elements and ions shows therefore many jumps, as shown in Figure 4.6, where the edges due to recombination into the ground level of many ions are apparent. Figure 4.6 also shows the free–free spectrum of a plasma at three different temperatures (0.5, 1.0 and 10 MK).

In this section we give expressions for the free–free and free–bound continuum emission in their classical approximation, together with quantum mechanical correction factors known as *Gaunt factors*.

A third process – two-photon emission – also generates continuum radiation below 2000 Å, but in the solar atmosphere the two-photon continuum is usually much weaker than free–free and free–bound emission, so that it will not be considered further in this book.

4.7.1 Free–free radiation

In the classical picture, the electron–ion interaction generating the free–free emission consists of a free electron approaching an ion with charge Ze and accelerated by the Coulomb force. We will consider only the case where the incident electron velocity is non-relativistic. Since the electron is a charged particle, the acceleration results in the emission of radiation, according to the formula

$$\frac{dR}{dt} = \frac{2e^2}{3c^3}\,|a|^2, \tag{4.98}$$

where R is the emitted energy and a is the electron acceleration. In order to determine the acceleration, we first assume that the energy lost by the electron through radiation is negligible relative to its kinetic energy, so that the electron energy can be assumed to be constant. Under this assumption, the electron's orbit around the ion is hyperbolic, with the ion in one of its foci (see Figure 4.7). If we describe the electron motion using the spherical coordinates centred on the ion, with the angle θ taken from the major axis of the hyperbolic orbit, and indicate with θ_0 the angle between the direction of the initial velocity v of the electron with the major axis of the hyperbola, we have

$$a = -\frac{Ze^2}{mr^2}, \tag{4.99}$$

$$\frac{1}{r} = \frac{1}{p\tan\theta_0}\left(1 - \frac{\cos\theta}{\cos\theta_0}\right), \tag{4.100}$$

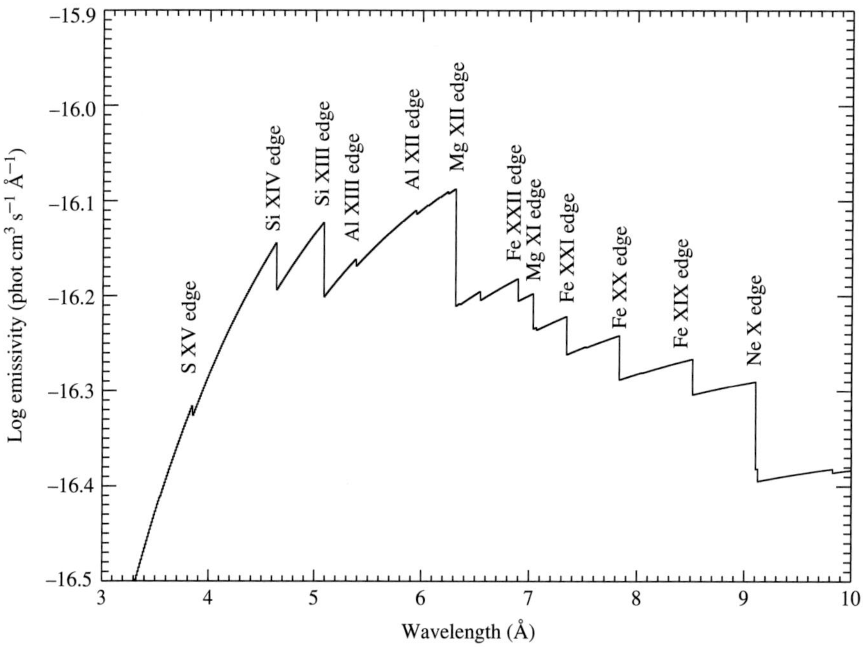

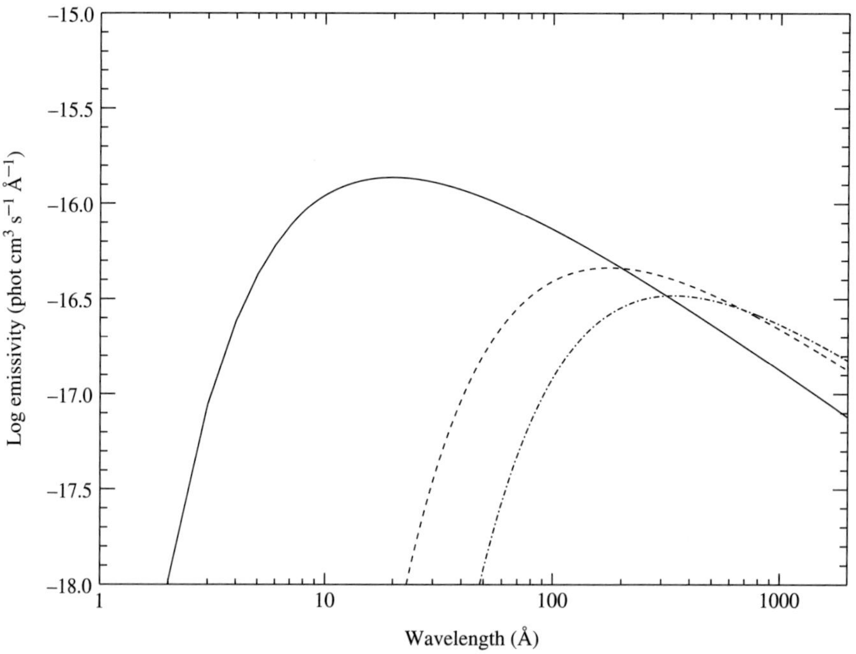

Fig. 4.6. (*Top*) Free–bound flux as a function of wavelength for a solar coronal plasma with temperature of 10^7 K, and unit emission measure and distance, in photon cm^3 s^{-1} Å^{-1}. (*Bottom*) Free–free flux as a function of wavelength for three different temperatures: 5×10^5 K (dash-dotted line), 10^6 K (dashed line) and 10^7 K (full line).

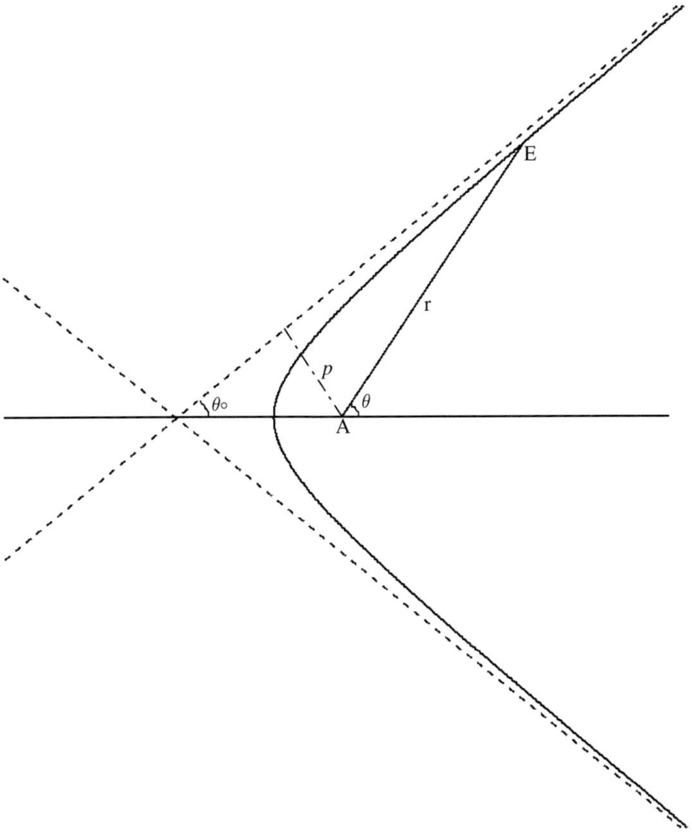

Fig. 4.7. Geometry of the ion–electron interaction. The electron is located at the point E at a distance r from the ion at A, and follows the hyperbolic trajectory given by the full line curve.

where r is the electron–ion distance, p is the distance between the ion and the initial direction of motion of the incoming electron. The angle θ_0 measures the amount of deviation of the electron trajectory from the initial one, and it is related to the initial velocity v and the distance p by the relation (Rutherford, 1911):

$$\tan\theta_0 = \frac{mpv^2}{Ze^2}, \tag{4.101}$$

where m is the electron mass. We will assume that $\tan\theta_0 \ll 1$, corresponding to the case in which the electron's final trajectory is very close to the initial one, but the direction of motion is opposite. Such an assumption simplifies the problem of calculating the spectral distribution of the emitted radiation, and allows us to obtain relations that can be used as a starting point for including quantum mechanical corrections. Under this assumption, $\sin\theta_0 \ll 1$ and $\cos\theta_0 \simeq 1$, so that the emitted energy can be calculated as

$$\frac{1}{r} = \frac{1 - \cos\theta}{p \tan\theta_0}, \tag{4.102}$$

$$R = \int \frac{2Z^2 e^6}{3m^2 c^3} \frac{dt}{r^4}. \tag{4.103}$$

The integral involving R can be solved by using the fact that the angular momentum of the electron relative to the ion is constant,

$$r^2 \frac{d\theta}{dt} = pv, \tag{4.104}$$

so that the total radiated energy can be calculated as

$$R = \frac{8Z^4 e^{10}}{3m^4 c^3 p^5 v^5} \int_0^\infty \frac{dz}{\left(1 + z^2\right)^3} = \frac{2\pi Z^4 e^{10}}{m^4 c^3 p^5 v^5}, \tag{4.105}$$

where the integral variable z is defined by the relation $\cos\theta = -\left(1 - z^2\right)/\left(1 + z^2\right)$. However, we are also interested in the spectrum of the emitted radiation. Following Kramers (1923), the amount of energy emitted in the interval $\nu \to \nu + d\nu$ is given by

$$dR(p, \nu) = 4\pi^2 \frac{Z^2 e^6}{c^3 m^2 p^2 v^2} P(\gamma)\, d\nu \qquad \text{with} \qquad \gamma = 2\pi\nu \frac{m^2 p^3 v^3}{Z^2 e^4}, \tag{4.106}$$

where the function $P(\gamma)$ is shown in Figure 4.8 and it represents the relative amount of radiation emitted in the $\gamma \longrightarrow \gamma + d\gamma$ interval. The emitted energy $dR(p, \nu)$ depends on the impact parameter p. In order to determine the total free–free emission of a plasma, we first need to calculate the total radiation emitted in the $\nu \to \nu + d\nu$ interval by N electrons approaching the target ion with all possible values of the impact parameter p. With N_e the density of electrons, the number N of electrons approaching the ion in the time interval dt with impact parameter in the $p \longrightarrow p + dp$ range is given by $N = N_\mathrm{e} v\, 2\pi p\, dp\, dt$. The total emitted radiation per unit time and frequency is then given by

$$dR(\nu) = N_\mathrm{e}\, dt\, d\nu \int_0^\infty 4\pi^2 \frac{Z^2 e^6}{c^3 m^2 p^2 v} P\left(2\pi\nu \frac{m^2 p^3 v^3}{Z^2 e^4}\right) 2\pi p\, dp. \tag{4.107}$$

By considering that

$$\frac{dp}{p} = \frac{1}{3} \frac{d\gamma}{\gamma} \qquad \text{and} \qquad \int_0^\infty \frac{P(\gamma)}{\gamma}\, d\gamma = \frac{4}{\pi\sqrt{3}}, \tag{4.108}$$

the total energy emitted per unit time and frequency is given by

$$\frac{dR(\nu)}{dt\, d\nu} = \frac{32}{3\sqrt{3}} \pi^2 \frac{Z^2 e^6}{m^2 v c^3} N_\mathrm{e}. \tag{4.109}$$

If we consider a plasma having free electrons all with the same non-relativistic velocity v, and ions with the same charge Ze, with densities given by N_e and N_i respectively, the emitted free–free radiation energy per unit time t and volume V is

$$\frac{dR(\nu)}{dt\, d\nu\, dV} = \frac{32\pi^2}{3\sqrt{3}} \frac{Z^2 e^6}{m^2 v c^3} N_e N_i. \tag{4.110}$$

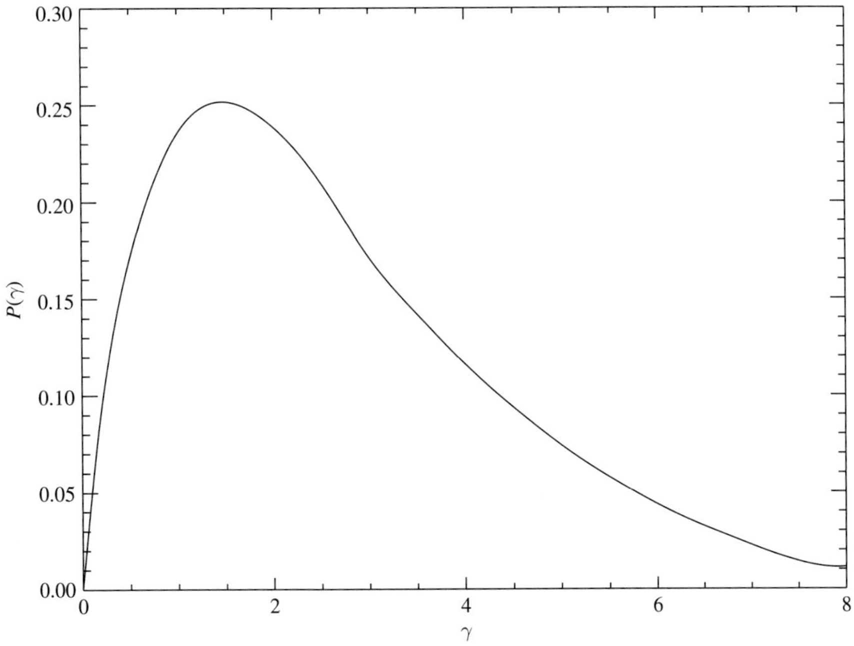

Fig. 4.8. Function $P(\gamma)$ in Equation (4.106). From Kramers (1923).

Equation (4.110) is generally corrected by the relaxation of the assumptions used in Kramers' (1923) treatment with a Gaunt factor g_{ff}:

$$\frac{dR(\nu)}{dt\,d\nu\,dV} = \frac{32\pi^2}{3\sqrt{3}}\frac{Z^2e^6}{m^2vc^3}N_{\mathrm{e}}N_{\mathrm{i}}g_{\mathrm{ff}}. \tag{4.111}$$

The Gaunt factor g_{ff} can be calculated with quantum mechanical models and depends on both the initial and final energies of the incident electron or, alternatively, on the initial energy and on the frequency of the emitted radiation. Since Kramers' (1923) paper, a number of authors have dealt with the Gaunt factor g_{ff} using a variety of different methods and approximations, and considering different régimes of the frequency ν and of the electron velocity v. Such results are reviewed by Brussaard and van de Hulst (1962), where the non-relativistic case is considered and where also references to the relativistic case can be found. Usually the Gaunt factor has values between 0.1 and 10, and is most often close to unity.

To determine the energy emitted from a plasma characterized by a distribution of the incident electron velocities $f(v)$, it is necessary to integrate over all velocities the total emitted energy in Equation (4.111) weighted by the electron velocity distribution $f(v)$. In the solar atmosphere, the electron velocity distribution is usually Maxwell–Boltzmann and is given by Equation (4.9), so that

$$\frac{dR(\nu)}{dt\,d\nu\,dV} = \frac{32}{3}\pi\sqrt{\frac{2}{3}\pi}\frac{Z^2e^6}{m^{1/2}k_{\mathrm{B}}^{3/2}c^3}T^{-3/2}N_{\mathrm{e}}N_{\mathrm{i}}\int_{\sqrt{\frac{2h\nu}{m}}}^{\infty} v \times \exp\left(-\frac{mv^2}{2k_{\mathrm{B}}T}\right)g_{\mathrm{ff}}\,dv, \tag{4.112}$$

where the lowest integration limit is given by the requirement that the incident electron energy $mv^2/2$ be larger than the emitted radiation energy $h\nu$. By changing the integration variable into

$$u = \frac{mv^2}{2k_{\mathrm{B}}T} - \frac{h\nu}{k_{\mathrm{B}}T} \qquad \Longrightarrow \qquad v\,dv = \frac{k_{\mathrm{B}}T}{m}\,du, \tag{4.113}$$

we obtain

$$\frac{dR(\nu)}{dt\,d\nu\,dV} = \frac{32}{3}\pi\sqrt{\frac{2}{3}\pi}\frac{Z^2e^6}{m^{3/2}k_{\mathrm{B}}^{1/2}c^3}\frac{1}{T^{1/2}}N_{\mathrm{e}}N_{\mathrm{i}}\exp\left(-\frac{h\nu}{k_{\mathrm{B}}T}\right)\langle g_{\mathrm{ff}}\rangle \tag{4.114}$$

where

$$\langle g_{\mathrm{ff}}\rangle = \int_0^{\infty} g_{\mathrm{ff}}e^{-u}\,du \tag{4.115}$$

is the *velocity-averaged free–free Gaunt factor*. The total frequency-averaged energy emitted by the free–free radiation can be calculated with

$$\begin{aligned}\frac{dR}{dt\,dV} &= \frac{32}{3}\pi\sqrt{\frac{2}{3}\pi}\frac{Z^2e^6}{m^{3/2}k_{\mathrm{B}}^{1/2}c^3}T^{-1/2}N_{\mathrm{e}}N_{\mathrm{i}}\int_0^{\infty} < g_{\mathrm{ff}} > \exp\left(-\frac{h\nu}{k_{\mathrm{B}}T}\right)d\nu\\ &= \frac{32}{3}\pi\sqrt{\frac{2\pi k_{\mathrm{B}}}{3m}}\frac{Z^2e^6}{mhc^3}N_{\mathrm{e}}N_{\mathrm{i}}T^{1/2}\overline{\langle g_{\mathrm{ff}}\rangle},\end{aligned} \tag{4.116}$$

where $\overline{\langle g_{\mathrm{ff}}\rangle}$ is the *frequency-averaged free–free Gaunt factor* given by

$$\overline{\langle g_{\mathrm{ff}}\rangle} = \int_0^{\infty}\langle g_{\mathrm{ff}}\rangle e^{-x}\,dx, \qquad x = \frac{h\nu}{k_{\mathrm{B}}T}. \tag{4.117}$$

The total emission from a plasma composed of many elements and ions can be obtained by summing Equation (4.116) over the contributions of each ion i in the plasma:

$$\frac{dR}{dt\,dV} = \frac{32}{3}\pi\sqrt{\frac{2\pi k_{\mathrm{B}}}{3m}}\frac{e^6}{mhc^3}N_{\mathrm{e}}T^{1/2}\sum_i Z_i^2N_{\mathrm{i}}\overline{\langle g_{\mathrm{ff}}\rangle}, \tag{4.118}$$

where Z_i is the atomic number of the ion i.

4.7.2 Free–bound radiation

In this section we will describe a simplified model for free–bound emission. The difference between free–free and free–bound emission consists of the incident electron being captured by the target ion in the latter process. The loss of energy of the electron is compensated for by the emission of radiation whose energy is equal to the energy lost by the electron.

Free–bound Gaunt factor

While a classical treatment of the free–free emission process, although approximate, is still possible, the recombination process is a problem that must be treated quantum mechanically since it involves the structure of the target ion in a much more intimate way than the free–free process. However, a full quantum mechanical description of the free–bound emission is beyond the scope of this book, so we will limit ourselves to outlining a semi-classical model for free–bound radiation, making use of a few concepts of atomic physics outlined in Chapter 3, which allows us to derive the correct equations for the radiated free–bound emission.

The main difference between a free–free and a free–bound transition is that in the former the outgoing electron could in principle have all possible kinetic energies in the interval between 0 and the initial value $mv^2/2$. Therefore, the amount of energy lost by the electron through radiation has a continuous spectrum of values. On the other hand, the jump in energy experienced by an electron when it recombines into an atomic level n is fixed once the initial velocity v is set, and it is given by

$$\Delta E = \frac{1}{2}mv^2 - E_n, \tag{4.119}$$

where E_n is the energy of the final atomic level that the electron is bound to relative to the first ionization threshold of the ion. The emitted radiation from the recombination of one electron is therefore monochromatic.

To quantify the amount of energy lost by an electron through recombination using a semi-classical model, we will consider the processes of free–free and free–bound emission as two possible outcomes of the same collisional interaction between the incoming electron and the target ion. We will indicate the cross-section of such an interaction by σ. We define the cross-section for the free–free scattering between free electrons and one ion from Equation (4.109) as

$$\frac{dR(\nu)}{dt} = \frac{32}{3\sqrt{3}}\pi^2 \frac{Z^2 e^6}{m^2 v c^3} N_{\mathrm{e}} d\nu = h\nu\, \sigma_{\mathrm{ff}}(\nu)\, N_{\mathrm{e}}\, v, \tag{4.120}$$

so that $\sigma_{\mathrm{ff}}(\nu)\, d\nu$ represents the probability of the electron being de-excited to a free state with energy smaller than the initial one by a quantity $h\nu$ through the emission of a photon of frequency ν. We will assume that the cross-section of the recombination process is the same as in Equation (4.120) but two important differences must be taken into account. First, the recombined electron must obey the Pauli exclusion principle, so it cannot recombine into a state already occupied by a bound electron: if the final energy E_n corresponds to a degenerate level, the cross-section must then be proportional to the fraction of states in the degenerate level unoccupied by other bound electrons. Second, since the energy lost by the recombining electron is fixed, the $d\nu$ must be proportional to the number of energy levels of the target ion with energy close to E_n.

Following Kramers (1923), we assume that the target ion is hydrogenic, so that Equation (4.119) takes the form

$$E_n = -2\pi^2 \frac{Z^2 e^4 m}{h^2 n^2}, \tag{4.121}$$

$$h\nu = \frac{1}{2}mv^2 + 2\pi^2 \frac{Z^2 e^4 m}{h^2 n^2}. \tag{4.122}$$

The hydrogenic energy level is degenerate, and the number of levels having the same energy E_n is $2n^2$ (see Chapter 3). If we indicate with n_{free} the number of levels unoccupied by other bound electrons, the free–bound cross-section will be proportional to $n_{\text{free}}/2n^2$. The number of levels with energy included within an interval $h\,d\nu$ from E_n can be determined in the case where n is large. In fact, for large values of n, the energies of the hydrogenic levels become very close and form a pseudo-continuum, so that it is possible to treat the principal quantum number n as a continuous variable and differentiate Equation (4.122); with the initial electron energy fixed, we have

$$d\nu = -4\pi^2 \frac{Z^2 e^4 m}{h^3 n^3} dn. \tag{4.123}$$

Such an approximation is inadequate for the lowest energy levels, whose energies are well separated. We will discuss this case later. Using Equations (4.120) and (4.123), and setting $dn = 1$, the free–bound cross-section can then by written as

$$\begin{aligned}\sigma_{\text{fb}}(v, \nu, n) &= \sigma_{\text{ff}}(\nu)\, \frac{n_{\text{free}}}{2n^2} \\ &= \frac{128}{3\sqrt{3}}\pi^4 \frac{Z^4 e^{10}}{m v^2 c^3 h^4 \nu}\frac{1}{n^3}\frac{n_{\text{free}}}{2n^2}.\end{aligned} \tag{4.124}$$

This cross-section can be interpreted as the area of the target ion that can be struck by the incoming electron in order to have a recombination into the level with energy E_n. The energy emitted per unit time and volume by a number N_{rec} of recombinations of many electrons with the same initial velocity v interacting with many target ions with density N_{i} in the volume element dV is given by

$$N_{\text{rec}} = N_{\text{e}} v \sigma_{\text{fb}}(v, \nu) dt\; N_{\text{i}}\, dV, \tag{4.125}$$

$$dR = N_{\text{rec}} h\nu = \frac{128}{3\sqrt{3}}\pi^4 \frac{Z^4 e^{10}}{m v c^3 h^3}\frac{1}{n^3}\frac{n_{\text{free}}}{2n^2} N_{\text{e}} N_{\text{i}}\, dV dt. \tag{4.126}$$

The above model is based on semi-classical arguments and is therefore approximate. As with free–free radiation, full quantum mechanical calculations have been carried out by a number of authors, and their results have been expressed in terms of a correction factor, called the *free–bound Gaunt factor* g_{fb}, which multiplies the right-hand side of Equation (4.126). This factor is a function of the initial electron energy and it has been calculated for all levels in the hydrogenic case. Results have shown that g_{fb} is close to unity in all cases. This indicates that the approximation made in Equation (4.123), although very gross for the hydrogenic levels with small n, still provides reasonably accurate results at any n. Values of g_{fb} have been calculated for example by Karzas and Latter (1961).

The total energy emitted per unit volume and time by a plasma characterized by a Maxwell–Boltzmann distribution of velocities is given by

$$\frac{dR}{dV dt} = \frac{128}{3\sqrt{3}}\pi^4 \frac{Z^4 e^{10}}{m v c^3 h^3}\frac{1}{n^3}\frac{n_{\text{free}}}{2n^2} N_{\text{e}} N_{\text{i}} g_{\text{fb}} \left(\frac{m}{2\pi k_{\text{B}} T}\right)^{3/2} \exp\left(-\frac{m v^2}{2 k_{\text{B}} T}\right) 4\pi v^2\, dv. \tag{4.127}$$

By using Equations (4.121) and (4.122), and noting that $v\,dv = h\,d\nu/m$, the total emitted energy per unit time, volume, and frequency becomes

$$\frac{dR}{dV\,dt\,d\nu} = \frac{128}{3}\pi^3\sqrt{\frac{2\pi}{3m}}\frac{Z^4 e^{10}}{c^3 h^2}\frac{1}{(k_B T)^{3/2}}\frac{1}{n^3}\frac{n_{\rm free}}{2n^2}\exp\left(-\frac{h\nu}{k_B T}\right)\exp\left(-\frac{E_n}{k_B T}\right)g_{\rm fb}N_{\rm e}N_{\rm i}. \tag{4.128}$$

It is important to note that, although the recombination of a single free electron with a given initial velocity provides radiation at a monochromatic frequency, the presence in the plasma of free electrons with all possible velocities makes the free–bound emission from a plasma a continuum radiation.

By summing over all the ions in the plasma and over all the levels in each ion, the total emitted energy per unit time, volume, and frequency is

$$\begin{aligned}\frac{dR}{dV\,dt\,d\nu} &= \frac{128}{3}\pi^3\sqrt{\frac{2\pi}{3m}}\frac{e^{10}}{c^3 h^2}\frac{1}{(k_B T)^{3/2}}\exp\left(-\frac{h\nu}{k_B T}\right)N_{\rm e}\\ &\times\sum_{i,n}\frac{1}{n^3}\frac{n_{\rm free}}{2n^2}\exp\left(-\frac{E_n}{k_B T}\right)Z_i^4 N_{\rm i}g_{\rm fb}.\end{aligned} \tag{4.129}$$

Free–bound continuum and photoionization cross-section

Another approach for calculating free–bound radiation uses the relation between the photoionization and radiative recombination cross-sections (Equation (4.43)). Since free–bound radiation is generated during a recombination process as described in Section 4.4.1, the energy emitted by recombinations of free electrons with velocity v into level i of the recombined ion X^{+m} per unit time and volume is given by

$$\frac{dE}{dV\,dt} = h\nu\frac{dN_{ij}^{\rm rr}}{dV\,dt} = h\nu N_{\rm e}N\left(X_j^{+m+1}\right)\sigma_{ij}^{\rm rr}f(v)v\,4\pi v^2\,dv, \tag{4.130}$$

where ν is the radiation frequency, $dN_{ij}^{\rm rr}/dV\,dt$ is the number of recombination per unit time and volume, $f(v)$ is the distribution of velocities of the incident electrons, $\sigma_{ij}^{\rm rr}$ is the cross-section for recombination $X_j^{+m+1} + e(v) \to X_i^{+m} + h\nu$, and $N\left(X_j^{+m+1}\right)$ is the density of target ions in level j.

In the solar atmosphere $f(v)$ is the Maxwell–Boltzmann distribution given by Equation (4.9), so that combining Equation (4.130) with Equations (4.34) and (4.43) the energy emitted per unit time, volume, and frequency is given by

$$\frac{dE}{dV\,dt\,d\nu} = N\left(X_j^{+m}\right)N_{\rm e}\sqrt{\frac{2m}{\pi k_B}\frac{h^4}{k_B m^2 c^2}}\frac{\omega_i}{\omega_j}\frac{\nu^3}{T^{3/2}}\exp\left(-\frac{h\nu - E_{ij}}{k_B T}\right)\sigma_{ij}^{\rm ph}, \tag{4.131}$$

where $E_{ij} = E_j - E_i$ is the difference between the energies of the levels j and i, whose statistical weights are g_j and g_i respectively, and $\sigma_{ij}^{\rm ph}$ is the photoionization cross-section for the transition $X_i^{+m} + h\nu \to X_j^{+m+1} + e(v)$.

The total energy emitted by a plasma per unit time, volume, and frequency can be obtained by summing over all levels i and j, as well as over all ions of all elements in the plasma:

$$\frac{dE}{dV\,dt\,d\nu} = \sqrt{\frac{2m}{\pi k_B}}\frac{h^4}{k_B m^2 c^2}\frac{\nu^3}{T^{3/2}}\exp\left(-\frac{h\nu}{k_B T}\right)N_e N(H)$$
$$\times \sum_{i,j}\sum_{X,m}\frac{N\left(X_j^{+m+1}\right)}{N\left(X^{+m+1}\right)}\frac{N\left(X^{+m+1}\right)}{N(X)}\frac{N(X)}{N(H)}\frac{\omega_i}{\omega_j}\exp\left(\frac{E_{ij}}{k_B T}\right)\sigma_{ij}^{\text{ph}}, \tag{4.132}$$

so that the total free–bound emission depends on element abundances $N(X)/N(H)$ and level populations of the recombining ion.

Usually, the recombining ion is assumed to be in the coronal model approximation, so that its populations are negligible for all levels except the ground level, for which $N\left(X_j^{+m+1}\right)/N\left(X^{+m+1}\right) = 1$.

Equation (4.132) has the advantage of dispensing with the use of the approximations in the semi-classical treatment outlined above and of the Gaunt factors, and relies on the photoionization cross-sections, which are calculated by atomic codes such as those listed in Section 3.7.

4.8 Spectral line profiles

In Chapter 3 we saw how the radiative emission from a transition between two atomic levels of an ion takes place at a fixed frequency $\nu_{ij} = \left(E_j - E_i\right)/h$, where E_j and E_i are the energies of the upper and lower levels involved in the transition, giving rise to a *monochromatic* spectral line with flux F_0. However, when observed by an instrument, the flux of each line is always distributed over a limited wavelength range, so that the line is not monochromatic:

$$F(\nu) = F_0\varphi(\nu) \qquad \text{where} \qquad \int_0^\infty \varphi(\nu)\,d\nu = 1. \tag{4.133}$$

The function $\varphi(\nu)$ that describes the distribution of the flux is called the *line profile*. Usually this distribution is peaked around the *central frequency* ν_{ij}. As we will see in Chapter 7, spectrometers introduce an instrumental spectral line profile caused by the finite resolution of their optical components. When this instrumental profile is deconvolved from the measured line profile, the spectral line is still not monochromatic. Other broadening mechanisms intrinsic to the emitting source give rise to the spectral line profile.

Lines are broadened by a variety of processes, the most common ones being:

(1) *natural* line profile, due to the intrinsic energy width of atomic energy levels;
(2) *Doppler* line profile, due to motions of the emitting ion along the line of sight;
(3) *pressure* line profile, due to the interaction between the emitting ion and the other particles present in the plasma along the line of sight.

Also, opacity can significantly affect the shape of a line profile, since absorption will be more efficient (and hence remove more photons) at frequencies where the line profile is larger, and smaller at frequencies where the line profile is smaller.

Doppler-broadened profiles are caused by a variety of motions: random thermal motions, regulated by the plasma distribution of velocities, rotation, turbulence, waves and bulk motions of portions of the plasma along the line of sight, and many other types of motions.

Pressure-broadened profiles can be due to many different processes, namely: linear and quadratic Stark effect, due to the electric fields of the nearby particles; resonance broadening, due to the interaction with an identical particle that can induce energy exchange processes; and van der Waals forces.

All these processes provide spectral profiles that depend on the electron density, temperature, or both, and on the particle motions, and can therefore be used as diagnostic tools for the measurement of these quantities in the emitting plasmas.

In the solar atmosphere, line widths are dominated by Doppler broadening, and only in cases of very strong lines can the effects of other types of broadening be observed far from the central wavelengths. Pressure broadening is usually negligible because of the low density of the plasma, while opacity broadening is usually unimportant since the solar atmosphere plasma is optically thin in the transition region and corona. Natural broadening is also negligible in the solar atmosphere.

4.8.1 Thermal Doppler broadening

When an emitting ion is moving with velocity v along the line of sight (with $v \ll c$, c = speed of light), its emission will be shifted by the *Doppler effect*: if ν_0 (λ_0) is the frequency (wavelength) of the radiation at rest, the radiation will be observed at a frequency (wavelength):

$$\nu = \nu_0\left(1 + \frac{v}{c}\right) \quad \text{or} \quad \lambda = \lambda_0\left(1 - \frac{v}{c}\right), \tag{4.134}$$

so that if $v > 0$ (when the ion is moving towards the observer) the frequency will increase and the wavelength will decrease, and vice versa if the ion is moving away from the observer ($v < 0$). The profile of a naturally broadened spectral line emitted by an ion is given by

$$\varphi(\nu) = \frac{1}{\pi} \frac{\frac{\Gamma}{4\pi}}{(\nu - \nu_0)^2 + \left(\frac{\Gamma}{4\pi}\right)^2}, \tag{4.135}$$

where $\nu_0 = \left(E_j - E_i\right)/h$ and $\Gamma = \sum_{k<j} A_{jk} + \sum_{k<i} A_{ik}$; this profile is called a *Lorentzian profile*. When Doppler motions are present, the ion emits at a shifted frequency $\nu_0' = \nu_0 + \nu_0 v/c$, so that its profile is

$$\varphi(\nu) = \frac{1}{\pi} \frac{\frac{\Gamma}{4\pi}}{\left(\nu - \nu_0 - \nu_0\frac{v}{c}\right)^2 + \left(\frac{\Gamma}{4\pi}\right)^2}. \tag{4.136}$$

In most plasmas in the solar atmosphere, the ions move according to the Maxwell–Boltzmann distribution of velocities, so that the cumulative Doppler effect due to thermal motions along the line of sight of all the ions in the plasma is given by

$$\varphi(\nu) = \frac{1}{\pi}\int_{-\infty}^{+\infty} \frac{\frac{\Gamma}{4\pi}}{\left(\nu - \nu_0 - \nu_0\frac{v}{c}\right)^2 + \left(\frac{\Gamma}{4\pi}\right)^2} f_l(v)dv$$
$$= \frac{1}{\sqrt{\pi}\Delta\nu} H(a, b),$$
$$H(a, b) = \frac{a}{\pi}\int_{-\infty}^{+\infty} \frac{e^{-x^2}}{(b-x)^2 + a^2} dx, \qquad (4.137)$$

where $a = \Gamma/(4\pi\Delta\nu)$, $b = (\nu - \nu_0)/\Delta\nu$, $\Delta\nu = (\nu_0/c)\sqrt{2k_B T_i/M}$, $x = \sqrt{M/2kT}$, and $f_l(v)$ is the Maxwell–Boltzmann distribution of velocities along the line of sight, given by

$$f_l(v) = \sqrt{\frac{M}{2\pi k_B T_i}} \exp\left(-\frac{Mv^2}{2k_B T_i}\right) \qquad (4.138)$$

where M and T_i are the ions' mass and temperature. The function $H(a, b)$ is called the *Voigt profile*. When the Doppler broadening dominates over the natural broadening, as is always the case in the solar atmosphere, the Voigt profile can be approximated as

$$H(a, b) = \begin{cases} e^{-b^2} & b < 1, \\ a/\sqrt{\pi} b^2 & b \gg 1, \end{cases} \qquad (4.139)$$

where $b < 1$ corresponds to the frequencies close to ν_0 (line *core*), and $b \gg 1$ to its *wings*. The exponential term dominates at the line core, while the second term, due to the Lorentzian profile, dominates in the line wings. However, except for very strong lines, the spectral line profiles can be approximated with Gaussians e^{-b^2} since the wings are very faint and are usually not detected. The line profile can then be written

$$\varphi(\nu) = \frac{1}{\sqrt{\pi}\Delta\nu} \exp\left[-\left(\frac{\nu - \nu_0}{\Delta\nu}\right)^2\right] \qquad (4.140)$$

and it is usually characterized by the *full width at half maximum* (FWHM), defined as the interval between the two frequencies where the line profile is half its peak value:

$$\text{FWHM} = \frac{\nu_0}{c}\sqrt{4\ln 2\,\frac{2k_B T_i}{M}}. \qquad (4.141)$$

When the line profile is expressed as a function of wavelength, using similar arguments we have

$$\varphi(\lambda) = \frac{1}{\sqrt{\pi}\Delta\lambda} e^{-\left(\frac{\lambda-\lambda_0}{\Delta\lambda}\right)^2}, \qquad \text{FWHM} = \frac{\lambda_0}{c}\sqrt{4\ln 2\,\frac{2k_B T_i}{M}}, \qquad (4.142)$$

with $\Delta\lambda = (\lambda_0/c)(2k_B T_i/M)^{1/2}$.

Spectral line widths have been measured in the spectrum of the solar atmosphere for many years, and in most cases they are found to be Doppler broadened. However, measured line widths are always larger than expected from the thermal broadening alone, indicating the presence of other motions that preserved the Gaussian shape of the line profiles but increased the width. These motions might be caused by many possible mechanisms, although no

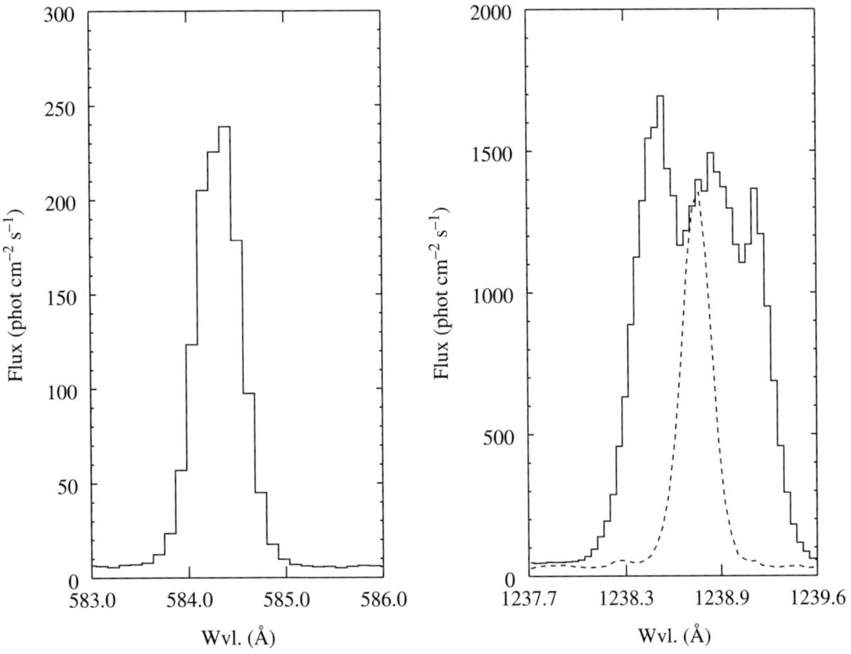

Fig. 4.9. Typical spectral line profiles in the solar atmosphere. (*Left*) Gaussian profile of He I 584.34 Å, from the quiet Sun. (*Right*) Doppler motion distortion of the profile of N V 1238.82 Å from an explosive event; the unshifted, quiet Sun profile of this line is also shown (dashed line). Courtesy *SOHO* SUMER team.

model has yet been able to predict their value. These additional motions can be described by a *non-thermal velocity* $v_{\rm nth}$, so that the FWHM of the resulting line profile can be expressed as

$$\mathrm{FWHM} = \frac{\lambda_0}{c}\sqrt{4\ln 2\left(\frac{2k_{\rm B}T_{\rm i}}{M} + v_{\rm nth}^2\right)}, \tag{4.143}$$

where $v_{\rm nth}$ is used as a parameter to describe the non-thermal dynamics in theoretical models.

When bulk motions along the line of sight are present, the line profile can be either further broadened, or distorted, or shifted, or a combination of these effects. The most common case is the distortion of the line profile at rest, sometimes accompanied by a shift of the central wavelength of the thermal profile if the bulk motions involve most of the emitting plasma.

Figure 4.9 shows examples of the Gaussian and distorted line profiles from observations carried out on the Sun.

4.9 Other excitation mechanisms of spectral lines

4.9.1 Photoexcitation from photospheric black-body radiation

An ion can be excited from level i to level j via the absorption of a photon whose energy equals the difference in the energy of the two levels: $h\nu_{ij} = |E_j - E_i|$. In most cases photoexcitation can be neglected as it is less efficient than excitation via inelastic collisions, but there are a few cases when this process must be taken into account.

If an ion is immersed in a radiation field with average intensity $J_{\nu_{ij}}$, the total number of photoexcitations will be given by

$$N_{\mathrm{abs}} = N(X_i^{+m}) B_{ij} J_{\nu_{ij}}, \tag{4.144}$$

where B_{ij} is the Einstein coefficient for absorption, given by Equation (3.112) or (3.116). A similar expression can be used for induced emission:

$$N_{\mathrm{em}} = N(X_j^{+m}) B_{ji} J_{\nu_{ij}}. \tag{4.145}$$

These processes can be accounted for in the equation for statistical equilibrium (4.24) by adopting a generalized form for the Einstein coefficients. In fact, we can define a new coefficient A'_{ij} as

$$A'_{ij} = \begin{cases} A_{ij} W(R) \frac{g_j}{g_i} \frac{c^2}{2h\nu_{ij}^3} J_{\nu_{ij}} & i > j \\ A'_{ij} = A_{ji}(1 + \frac{c^2}{2h\nu_{ij}^3} W(R) J_{\nu_{ij}}) & i < j \end{cases} \tag{4.146}$$

where A_{ji} and A_{ij} are the Einstein coefficients for spontaneous radiation and $W(R)$ is the radiation dilution factor which accounts for the weakening of the radiation field at distances R from the source centre. For a uniform spherical source with radius R_{s}, the dilution factor is given by

$$W(R) = \frac{1}{2} \left(1 - \sqrt{1 - \frac{R_{\mathrm{s}}^2}{R^2}} \right). \tag{4.147}$$

In the solar atmosphere, stimulated emission is negligible while photoexcitation from photospheric black-body radiation can be important as an excitation mechanism for the states within the ground configuration of many ions. This mechanism has important consequences for level populations and plasma diagnostics at low densities, when collisional excitation is less effective, since it provides an additional excitation mechanism that can alter the overall population of the ion.

An example of the effects of photoexcitation by photospheric black-body radiation on the populations of the levels of the ground term of Si IX is shown in the left panel of Figure 4.10, where populations of these levels were calculated including (dashed line) and neglecting (full line) this process; black-body radiation has been calculated assuming $T = 5770$ K and a distance of $1.05 R_{\odot}$. Photoexcitation provides significant contributions at densities in the $10^7 - 10^8$ cm^{-3} range, typical of the off-limb solar corona, and is the dominant excitation process at densities below 10^7 cm^{-3}.

In the right panel of Figure 4.10 the flux ratio of two Si IX lines calculated as a function of the electron density neglecting (full line) and including (dashed line) photoexcitation is shown, to demonstrate the importance of this process for plasma diagnostics.

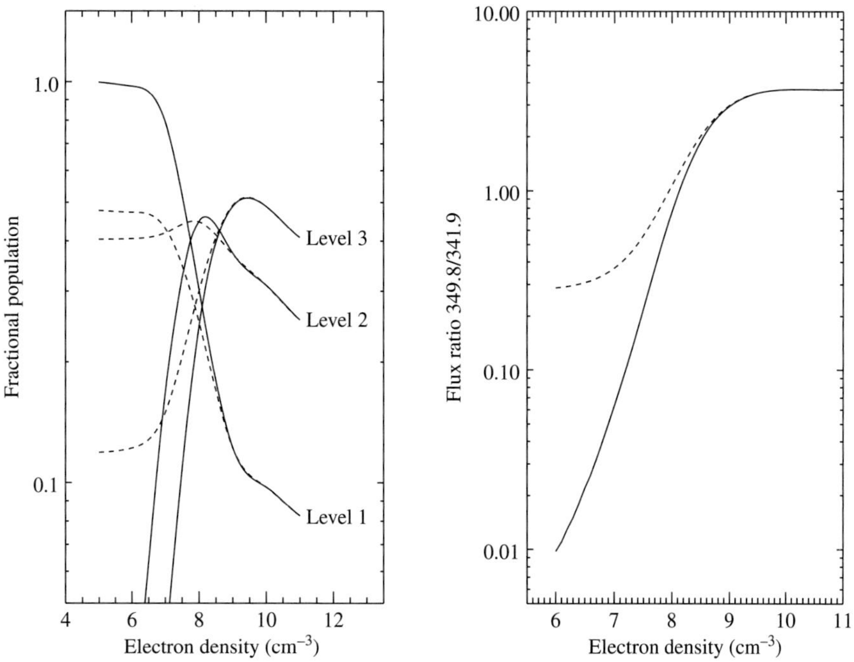

Fig. 4.10. (*Left*) Populations of the levels of the ground term of Si IX. (*Right*) Flux ratio of the two Si IX lines 349.8/341.9. The curves are calculated including (dashed line) and neglecting (full line) photoexcitation from solar photospheric radiation. Black-body radiation has been calculated assuming $T = 5770$ K and a distance of 1.05 $R_\odot$.

4.9.2 *X-ray fluorescence lines in flares*

Spectra in the vicinity of the Fe XXV resonance line (transition $1s^2\,{}^1S_0 - 1s2p\,{}^1P_1$) at 1.850 Å show the presence of inner-shell Fe lines near 1.94 Å during X-ray flares. They are due to the removal of one of the $1s$ electrons in neutral Fe atoms in the photosphere, with the vacancy in the $1s$ shell then being filled by an electron in the $2p$ shell – one may speak of a 'hole' transition $[1s] - [2p]$ in Fe II. The excess energy created by this transition may appear as an Auger electron after the rearrangement of the remaining electrons, leaving the Fe atom with two vacancies, or as an emitted photon. As with Ly-α transitions in hydrogenic ions, a doublet of lines results, depending on the spin of the $2p$ electron. The wavelengths of the two lines are 1.936 Å (Kα_1, due to $[1s_{1/2}] - [2p_{3/2}]$) and 1.940 Å (Kα_2, due to $[1s_{1/2}] - [2p_{1/2}]$), and like the H-like ion Ly-α lines have a flux ratio Kα_1:Kα_2 of 2:1.

To remove a $1s$ electron from the electron array of neutral Fe requires an energy of 7.11 keV, known as the K ionization energy for Fe. Having been ionized, the probability of photon emission, as opposed to an Auger electron ejection, is called the fluorescence yield (written ω_K). Its value strongly increases with Z; for Fe it is $\omega_K = 0.33$ (Fink *et al.* (1966)), i.e. one-third of K-shell-ionized Fe atoms relax by the $2p$ electron filling the $1s$ hole with emission of Kα line radiation and two-thirds undergo rearrangement of their electrons and ejection of Auger electrons. Either an electron or photon with energy > 7.11 keV can remove the $1s$ electron. Relatively few electrons exist at this energy in Maxwell–Boltzmann distributions at typical flare temperatures, but there are many in non-thermal electron distributions that are present during the impulsive hard X-ray bursts at the onset of an X-ray flare. Electrons with power-law energy distributions are deduced from the fact that the hard X-ray emission during such bursts has a power-law photon spectrum, $(h\nu)^{-n}$. This was the basis for calculations by Phillips and Neupert (1973) of the amount of Kα emission excited by electron K-shell ionization during energetic flares.

Fluorescence or photon excitation of photospheric Fe is also possible, and was suggested by Doschek *et al.* (1971) and others. Calculations by Bai (1979) showed that there would as a result be a large centre-to-limb variation in the Fe Kα line intensity – flares near Sun centre were predicted to show the most intense Fe Kα emission, those near the limb hardly any. This appears to be the case observationally (Chapter 10, Section 10.9), clearly indicating that the dominant excitation mechanism for the Fe Kα lines is fluorescence, i.e. K-shell ionization by photons with energies $h\nu > 7.11$ keV, mostly from thermal distributions. Photons from an X-ray-emitting flare plasma a few thousand km above the photosphere are incident on the photosphere, ionizing the neutral Fe atoms there to produce the observed Kα emission. The flux of Fe Kα photons is (Bai (1979))

$$F(\mathrm{K}\alpha) = \frac{\Gamma f(\theta)}{4\pi(1+\alpha)(\mathrm{AU})^2} \int_{7.11}^{\infty} I_{\mathrm{obs}}(h\nu)\, d(h\nu) \tag{4.148}$$

where Γ is the ratio of Fe Kα photons emitted by the photosphere to photons with $h\nu >$ 7.11 keV emitted by the X-ray flare plasma incident on the photosphere, $f(\theta)$ expresses the angular dependence of the emission (θ is the flare's angular distance from the Sun's centre), α is an albedo correction for Compton-scattered photons, and I_{obs} is the observed X-ray emission from the flare. The angular term $f(\theta)$ is not far from a cosine dependence when the X-ray flare height above the photosphere h is zero, i.e. $f(\theta) = 0$ for $\theta = 90°$, but for $h > 0$, the value of $f(90°)$ becomes non-zero. Bai's calculations of the various factors (Γ, $f(\theta)$, α) entering into the expression for the fluorescence Fe Kα emission were done by Monte Carlo procedures. They are illustrated by the curves in Figure 10.11.

The vacancy in the K shell following ionization of Fe atoms may be filled by a $3p$ rather than a $2p$ electron, i.e. a $[1s] - [3p]$ hole transition, so forming the Kβ line feature at 1.762 Å. The branching ratio of $K\alpha/K\beta$ emission is 0.88/0.11 (Bambynek *et al.* (1972)), so the Fe Kβ line feature is much weaker than Kα. As the excitation mechanism is the same as for the Kα lines, a centre-to-limb variation of the Kβ line feature's flux is predicted. Also, fluorescence may create an L-shell vacancy in Fe atoms rather than one in the K shell, with the vacancy filled by an electron in the M shell.

Corresponding fluorescence lines occur in elements other than Fe, but the steep dependence of the fluorescence yield ω on Z means that their fluxes are insignificant for lighter elements such as S or Ca.

4.9.3 Resonant scattering

In plasma above the photosphere where the density is sufficiently low, a spectral line can be emitted by an ion that absorbs a photon of the same line coming from the direction of the Sun, and then re-emits it in another direction. This process is called *resonant scattering*. Like the absorption of photospheric black-body radiation, this process enhances the flux of lines that otherwise would be too faint to be observed. Also, resonant scattering alters both the flux and the profile of a spectral line, and can be used to measure the temperature of the absorbing ions or their velocity distribution (see Chapter 5).

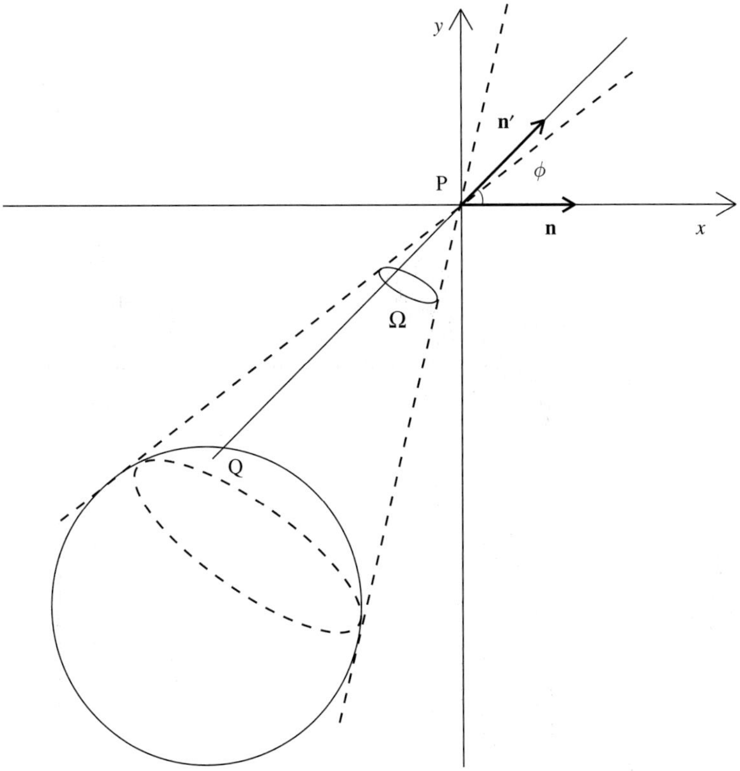

Fig. 4.11. Resonant scattering of disk radiation by a coronal ion at P, towards the observer along the direction **n**.

To derive the flux of a spectral line excited by resonant scattering we will assume that the frequency of the absorbed and emitted radiation is the same, and in addition that the ion can absorb radiation only at a single frequency, i.e. its absorption profile is a Dirac δ function. Figure 4.11 describes the resonant scattering of a line emitted by the solar disk, seen by an ion at P under a solid angle Ω: $\mathbf{n}'$ is the direction of the radiation emitted by a point Q on the solar disk incident on the ion at P, and $\mathbf{n}$ indicates the direction from P to the observer. The flux $F(\Omega)$ per unit solid angle emitted by the point Q on the disk, with profile $\varphi(\nu_1)$, is seen by an ion at P moving with velocity $\mathbf{v}$ at a frequency ν' shifted from the original frequency ν_1 because of the Doppler effect:

$$\nu' = \nu_1\left(1 - \frac{\mathbf{v}\cdot\mathbf{n}'}{c}\right). \tag{4.149}$$

This radiation will be absorbed by the ion at P only if $\nu' = \nu_0$, where ν_0 is the rest frequency associated to the atomic transition. If we indicate by $N\left(X^{+m}\right) f(\mathbf{v})\, d\mathbf{v}$ the number of absorbers with velocity in the $\mathbf{v}$, $\mathbf{v} + d\mathbf{v}$ range, where $f(\mathbf{v})$ is the ion velocity distribution function, the total number of absorptions of radiation coming within a solid angle $d\Omega$ per unit volume and time is

$$N_{\rm abs} = N\left(X^{+m}\right)\alpha_{\rm res} \tag{4.150}$$

with

$$\alpha_{\rm res} = B\ F(\Omega)\ d\Omega\ f(\mathbf{v})\ d\mathbf{v} \int_0^\infty d\nu_1\ \varphi(\nu_1)\ \delta\left[\nu_1\left(1 - \frac{\mathbf{v}\cdot\mathbf{n}'}{c}\right) - \nu_0\right], \tag{4.151}$$

where B is the Einstein absorption coefficient. In the solar corona, resonant scattering is an important excitation mechanism when the electron density is low, and the populations of the excited levels relative to the ground level are very low. In this case level populations can be calculated using the coronal model approximation, so that absorptions occur only when the ion is in the ground level. If we neglect for simplicity collisional excitation, and use resonant scattering (Equation (4.151)) in place of electron excitation in Equation (4.27), the total population $N_i\left(X^{+m}\right)$ of the excited level i is

$$N_i\left(X^{+m}\right) = N\left(X^{+m}\right)\frac{\alpha_{\rm res}}{A_{\rm tot}}, \tag{4.152}$$

where $A_{\rm tot} = \sum_{j<i} A_{ij}$ is the total decay rate by spontaneous emission. If A_i is the Einstein coefficient for spontaneous radiation of the absorbed line and $p(\phi)$ the coefficient taking into account the anisotropy of resonant scattering emission (where ϕ is the angle between $\mathbf{n}'$ and $\mathbf{n}$), the number of photons emitted along the direction $\mathbf{n}$ towards the observer is given by

$$N(\nu_0) = N\left(X_i^{+m}\right)p(\phi)A_i = N\left(X^{+m}\right)\alpha_{\rm res}\,p(\phi)\frac{A_i}{A_{\rm tot}}. \tag{4.153}$$

The angular coefficient $p(\phi)$ depends on the transition being considered. For transitions of the type ${}^2S - {}^2P$ we have (Noci *et al.* (1987)):

$$p(\phi) = \begin{cases} 1/4\pi & {}^2S_{1/2} - {}^2P_{1/2}, \\ (7 + 3\cos^2\phi)/32\pi & {}^2S_{1/2} - {}^2P_{3/2}, \\ (11 + 3\cos^2\phi)/48\pi & {}^2S - {}^2P. \end{cases} \tag{4.154}$$

In the frame of the observer, the radiation emitted at frequency ν_0 is shifted because of the motion of the ion along the direction $\mathbf{n}$, and it is observed at frequency

$$\nu = \nu_0\left(1 + \frac{\mathbf{v}\cdot\mathbf{n}}{c}\right). \tag{4.155}$$

The total flux of the line seen by the observer at frequency ν is given by integrating Equation (4.153) over the solid angle Ω, the velocity $\mathbf{v}$, and along the direction x of the observer.

By using Equation (4.151) we obtain

$$\begin{aligned} F(\nu) = h\nu\ B\frac{A_i}{A_{\rm tot}} \int_{-\infty}^{+\infty} dx \frac{N\left(X^{+m}\right)}{N(X)}\frac{N(X)}{N(H)}\frac{N(H)}{N_e}N_e \\ \times \int_\Omega d\Omega\ F(\Omega)\ p(\phi) \int_0^\infty d\nu_1\varphi(\nu_1)\int_\infty d\mathbf{v}\,f(\mathbf{v}) \\ \times\,\delta\left[\nu_1\left(1 - \frac{\mathbf{v}\cdot\mathbf{n}'}{c}\right) - \nu_0\right]\delta\left[\nu - \nu_0\left(1 + \frac{\mathbf{v}\cdot\mathbf{n}}{c}\right)\right], \end{aligned} \tag{4.156}$$

where we also use the relation

$$N\left(X^{+m}\right) = \frac{N\left(X^{+m}\right)}{N(X)} \frac{N(X)}{N(H)} \frac{N(H)}{N_e} N_e. \tag{4.157}$$

Equation (4.156) shows that the scattered light depends on the ion velocity distribution $f(\mathbf{v})$, on the ion and element abundances and on the electron density along the line of sight, and on the profile $\varphi(\nu_1)$ of the incident radiation.

The integral over the velocity distribution can be interpreted as the effective absorption profile of the absorbing ion for the transition with frequency ν_0, given by Doppler shifts due to their thermal or bulk motions. This absorption profile can be defined as

$$\psi(\nu_1, \nu) = \int_\infty d\mathbf{v}\, f(\mathbf{v})\, \delta\left[\nu_1\left(1 - \frac{\mathbf{v}\cdot\mathbf{n}'}{c}\right) - \nu_0\right] \delta\left[\nu - \nu_0\left(1 + \frac{\mathbf{v}\cdot\mathbf{n}}{c}\right)\right] \tag{4.158}$$

so that the quantity

$$D(\nu) = \int_0^\infty d\nu_1\, \varphi(\nu_1)\, \psi(\nu_1, \nu), \tag{4.159}$$

called the *Doppler dimming term*, can be considered to be proportional to the total probability that the disk flux is absorbed by the ion. This quantity depends on the velocity distribution of the ion, but in case of outflow it also depends on the outflow velocity v_r: this property is of great importance for the measurement of the solar wind speed (see Chapter 5).

When the electron collisional excitation is non-negligible, resonant scattering only provides a portion of the line flux, and before using this line for plasma diagnostics it is necessary to separate the two components of the line flux given by collisional and resonant scattering excitation.

As an example, Equation (4.156) can be used to calculate the flux ratio of the two very strong lines $1s\,^2S_{1/2} - 2p\,^2P_{1/2,3/2}$ emitted by Li-like ions. Such lines, commonly observed in the solar corona, can be used as tracers of the importance of resonant scattering in the corona, of the electron density of the plasma, and of the velocity of the solar wind (see Chapter 5). In collisionally dominated plasmas, the flux ratio between these two lines is

$$\frac{F_{^2S_{1/2}-^2P_{3/2}}}{F_{^2S_{1/2}-^2P_{1/2}}} = 2, \tag{4.160}$$

while in plasmas dominated by resonant scattering, Equation (4.156) gives

$$\frac{F_{^2S_{1/2}-^2P_{3/2}}}{F_{^2S_{1/2}-^2P_{1/2}}} = 4, \tag{4.161}$$

the integrals in Equation (4.156) being identical since both lines are emitted by the same ion, the values of A_i and A_{tot} are approximately the same, $B(^2P_{3/2}) = 2B(^2P_{1/2})$ and $F(\Omega)(^2P_{3/2}) = 2F(\Omega)(^2P_{1/2})$ since the incident radiation from the solar disk is emitted by collisionally dominated plasmas. In plasmas where both collisions and scattering are important, the $F(^2S_{1/2}) - (^2P_{3/2})/F(^2S_{1/2}) - (^2P_{1/2})$ ratio lies between 2 and 4, and its measured value can be used to determine the fractions of the measured flux of each line due to collisional and to resonant scattering.

Resonant scattering also alters the profile of the line emitted along the line of sight from that of the incident spectral line. This is due to the fact that the distribution of velocities of the *excited* ions is different from that of the non-excited ions. In fact, motions along the direction $\mathbf{n}'$ of the incident radiation shift the frequency of the latter in the rest frame of the ion, outside the peak of the profile $\varphi(\nu_1)$, so that the excitation rate $\alpha_{\rm res}$ will be smaller than in the case where the ion moves perpendicular to $\mathbf{n}'$. The velocity distribution of the excited ions is therefore different from that of the non-excited ions, and will cause the profile of the line emitted along the direction of the line of sight (x axis in Figure 4.11) to be distorted from the incident one. The distortion is larger the smaller the angle ϕ between $\mathbf{n}$ and $\mathbf{n}'$ is: the effect will be largest when $\phi = 0$, and null when $\phi = \pi/2$. However, the integration along the line of sight of the resonantly scattered lines encompasses all values of ϕ between 0 and π, so that departures from the incident line profiles (a Gaussian) are usually moderate.

4.9.4 *Proton–ion collisional excitation*

Inelastic collisions between protons and ions can be important contributors to the populations of the levels of the ground multiplet of many ions when the plasma temperature increases. Such collisions can be described in the same way as electron–ion collisions are described at the beginning of this chapter. If $\sigma_{ij}^{\rm p}$ is the cross-section for excitation from level i to level j of the ion X^{+m}, the number of exciting collisions $dN_{ij}^{\rm p}$ over a target ion per unit time will be given by

$$\frac{dN_{ij}^{\rm p}}{dt} = N_{\rm p} f(v)\sigma_{ij}^{\rm p} v 4\pi v^2\, dv, \tag{4.162}$$

where $N_{\rm p}$ is the density of protons, $f(v)$ the proton velocity distribution, and v the incident proton velocity. In order for an excitation to take place, the energy of the incident proton must be larger than the energy difference $\Delta E = E_j - E_i$ between the two atomic levels, so that the proton velocity must be larger than a minimum value $v_{\rm min} = \sqrt{2\Delta E/m}$.

The total number of excitations per unit time and volume can be obtained by integrating over the proton velocities. If $N(X_i^{+m})$ is the density of target ions in the level i, we have

$$\begin{aligned} \frac{dN_{ij}^{\rm exc}}{dV\,dt} &= N\big(X_i^{+m}\big) N_{\rm p} \int_{v_{\rm min}}^{\infty} f(v)\sigma_{ij}^{\rm p} v 4\pi v^2\, dv \\ &= N\big(X_i^{+m}\big) N_{\rm p} C_{ij}^{\rm p}, \end{aligned} \tag{4.163}$$

where $C_{ij}^{\rm p}$ is the *proton collisional excitation rate coefficient*. Usually the distribution of proton velocities is assumed to be Maxwell–Boltzmann, characterized by the proton temperature $T_{\rm p}$. In the case of thermal plasmas, $T_{\rm p}$ is equal to the electron temperature.

Seaton (1964) showed that collision excitation rates for proton–ion collisions were comparable to or greater than those for electron–ion collisions when $\Delta E \ll k_{\rm B}T$, where ΔE is the energy difference of the two levels in the transition. For this reason, only transitions between fine-structure levels, whose ΔE is small, can be significantly affected by proton collisional excitation. For most atomic systems, the only transitions affected by proton excitation that are important for the level population balance are those within levels of the ground multiplet, since this is the only multiplet with significant population, while proton excitation can be neglected for all other transitions.

Proton excitation rates have an increased importance in very hot plasmas such as in flares, while for cooler plasmas, such as in coronal holes and quiet Sun, their contribution can be neglected even for the levels in the ground term.

4.9.5 Ionization and recombination

Ionization and recombination processes can be of great importance for the calculation of level populations and line fluxes, and their effect is due both to direct recombination or ionization into an excited level, and to cascades from higher-energy levels where a recombination or an ionization has occurred. Including these two processes in the level population calculation can be extremely complex since they require the coupling of the system of equations that governs ionization and recombination with the system of equations that governs level populations.

However, if the total population of the excited levels of an ion can be described by the coronal model approximation, the effects of ionization and recombination can be accounted for in a relatively straightforward way, since they can be treated as a correction to the case where populations are calculated neglecting them.

Without ionization and recombination, the population of the excited level i of the ion X^{+m} under the coronal model approximation is given by (Equation (4.27)):

$$N_g N_e C^e_{gi} = N_i \sum_{j<i} A_{ij} \qquad \Longrightarrow \qquad \frac{N_i}{N_g} = \frac{N_e C^e_{gi}}{\sum_{j<i} A_{ij}}, \tag{4.164}$$

where C^e_{gi} is the electron collisional rate coefficient from the ground level. When ionization and recombination make significant contributions, Equation (4.164) must be modified to include the rate coefficients for ionization (α^i_{ion}) and recombination (α^i_{rec}) to level i:

$$N_g N_e \left(n_m C^e_{gi} + n_{m-1}\alpha^i_{ion} + n_{m+1}\alpha^i_{rec} \right) = N_i \sum_{j<i} A_{ij} n_m, \tag{4.165}$$

where n_{m-1}, n_m, n_{m+1} are the ion fractions for the ions X^{m-1}, X^m, and X^{m+1} respectively, while α^i_{ion} and α^i_{rec} are the effective ionization and recombination rates, which take into account both direct ionization and recombination to level i, and radiative cascade contributions from higher levels. Ionization and recombinations *from* level i are neglected. The population of the excited level can then be expressed as

$$\frac{N_i}{N_g} = \frac{N_e C^e_{gi}}{\sum_{j<i} A_{ij}} \times \mathcal{K} \tag{4.166}$$

where the correction $\mathcal{K}$ is given by

$$\mathcal{K} = 1 + \frac{n_{m-1}\alpha^i_{ion} + n_{m+1}\alpha^i_{rec}}{n_m C^e_{gi}}. \tag{4.167}$$

The correction $\mathcal{K}$ is temperature sensitive and can be large when the collisional excitation rate is small or when the abundance of the ion m is much smaller than the abundance of one of the adjacent ions. Correction factors can be as high as 100, but the largest values are typical of levels with low collisional excitation rates, whose populations are small and therefore do not give rise to observed lines. Corrections to populations of levels emitting observed lines are normally within a factor 2, and are directly propagated into line fluxes.

This method of including the effects of ionization and recombination cannot be applied to ions where collisional excitations occur not only from the ground level but also from other excited levels. If this occurs, Equation (4.165) does not hold, and level populations must be calculated in a more complete way. An example is given by the He-like ions, whose populations require a more complete treatment of the effects of recombination, as shown by Gabriel and Jordan (1973).

5

Plasma diagnostic techniques

5.1 Introduction

In this chapter we will review the main diagnostic techniques that allow the measurement of the physical properties of optically thin plasmas. These techniques can be used to measure the electron and ion temperatures, electron density, the thermal structure of the plasma, its chemical composition and ionization state, its dynamics, and the velocity distribution of its electrons. Most, but not all, of the techniques we describe assume that the emitting plasma is (i) thermal; (ii) homogeneous and isothermal; and (iii) in collisional ionization equilibrium. The effects of the relaxation of one or more of these assumptions will be discussed when possible, although the effects of departures from ionization equilibrium are difficult to study because of the complexity of evaluating ion abundances in the dynamic environment of out-of-equilibrium plasmas.

Diagnostic techniques require atomic databases and spectral codes to be used. These codes are necessary to provide theoretical estimates of line and continuum fluxes as a function of the physical parameters of the plasma, whose comparison with observations provides the measurement of such parameters. In this chapter, we also briefly review the main spectral codes available in the literature. The diagnostic techniques we discuss can be used with either fluxes (radiances) or intensities (irradiances), but for clarity we will mostly use fluxes.

5.2 Electron density diagnostics

5.2.1 Line flux ratios

Measuring electron densities using the flux ratio of two spectral lines is the most popular method. For a non-isothermal plasma, the flux ratio of two lines emitted by the same ion is given by Equation (4.94):

$$\frac{F_{ij}}{F_{kl}} = \frac{\int_T G_{ij}(T, N_e)\, \varphi(T)\, dT}{\int_T G_{kl}(T, N_e)\, \varphi(T)\, dT}, \tag{5.1}$$

where $\varphi(T)$ is the differential emission measure (DEM) and the $G(T, N_e)$ functions are the contribution functions of the two lines having transitions ij and kl. Since the ion and the element abundances are the same for the two lines, the integrals only depend on the populations of the upper levels i and k of the lines, and on the Einstein coefficients for spontaneous radiation A_{ij} and A_{kl}.

The density sensitivity of the population ratio can be due to the competing importance of collisional and radiative de-excitation from upper levels, or the density sensitivity of

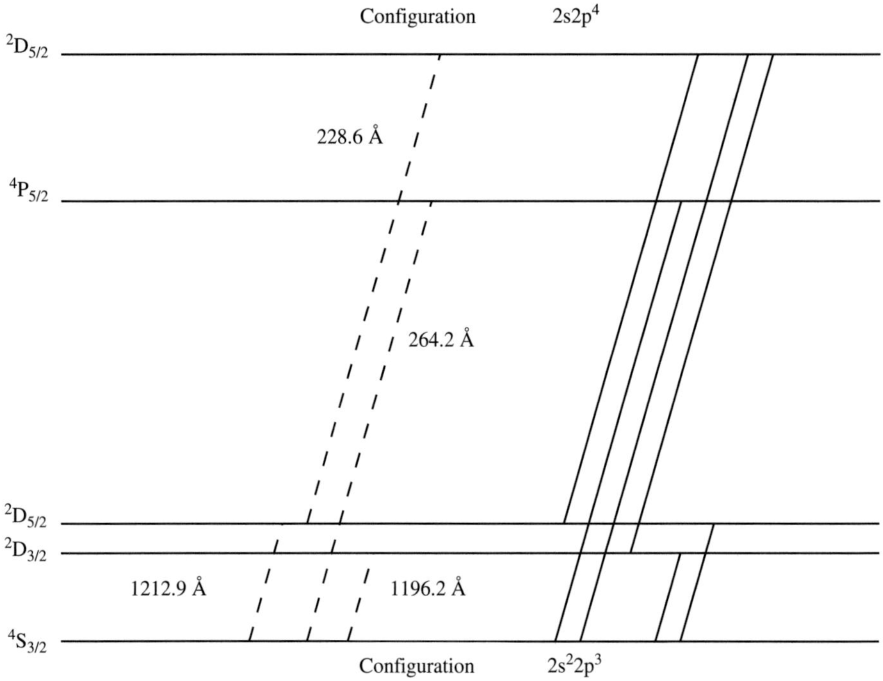

Fig. 5.1. S X levels and transitions for four spectral lines. Full lines: transitions for collisional excitation. Dashed lines: radiative transitions that generate the lines.

the population of the lower level (or levels) from which the upper level is populated. An example is shown in Figure 5.1, where the atomic levels and transitions of two line pairs of S X (λ1196/λ1212 and λ228/λ264) are shown (here λ1196 indicates the line with wavelength at 1196 Å).

Case 1: Ratios involving forbidden lines

The density sensitivity of two forbidden lines arises from the relative importance of collisional and radiative de-excitation. The upper levels of the transitions for the two lines are metastable: they are populated by inelastic collisions from the ground level and from radiative cascades from higher energy levels, and can decay radiatively only via forbidden transitions with small A-values to a lower level belonging to the same configuration. In the cases where radiative cascades from higher levels can be neglected, the population of these metastable levels is given by (see Section 4.3)

$$\frac{N_i}{N_{\mathrm{g}}} = \frac{N_{\mathrm{e}}\, C^{\mathrm{e}}_{gi}}{\sum_{j<i} A_{ij} + N_{\mathrm{e}} \sum_{j<i} C^{\mathrm{d}}_{ij}}, \tag{5.2}$$

where g is the ground level. If the electron density is sufficiently small, collisional de-excitation is negligible:

$$N_{\mathrm{e}} \sum_{j<i} C^{\mathrm{d}}_{ij} \ll \sum_{j<i} A_{ij} \quad \longrightarrow \quad \frac{N_i}{N_{\mathrm{g}}} = \frac{C^{\mathrm{e}}_{gi}}{\sum_{j<i} A_{ij}} N_{\mathrm{e}}. \tag{5.3}$$

If collisional de-excitation dominates over spontaneous emission, then

$$N_e \sum_{j<i} C^d_{ij} \gg \sum_{j<i} A_{ij} \quad \longrightarrow \quad \frac{N_i}{N_g} = \frac{C^e_{gi}}{\sum_{j<i} C_{ij}}. \tag{5.4}$$

Level populations depend linearly on density when $N_e \sum_{j<i} C_{ij} \ll \sum_{j<i} A_{ij}$, and are constant otherwise. The transition between these two régimes occurs at a critical density $N^{cr} = \sum_{j<i} A_{ij} / \sum_{j<i} C_{ij}$. Different atomic levels within the same ion have different critical densities. The flux ratio of two lines emitted by levels with different critical densities N^{cr}_1 and N^{cr}_2 are proportional to density in the $N^{cr}_1 < N_e < N^{cr}_2$ range. Electron density from an observed ratio can be estimated from comparing the ratio with the theoretical ratio as a function of N_e. An example of ratios of this kind for N-like sequence ions is shown in Figure 5.2.

The same argument can be repeated when the metastable levels are also populated via radiative decays from higher levels. Most contributions from higher energy levels come from allowed transitions, so that the populations of the levels from which these decays originate can be described using the coronal model approximation. If collisional excitation of these higher levels is from the ground level only, then Equations (5.2) to (5.4) can still be used, provided that the collisional excitation rate coefficient C^e_{gi} is replaced by a corrected rate coefficient that includes cascades following collisional excitation that, to a first approximation, can be written as:

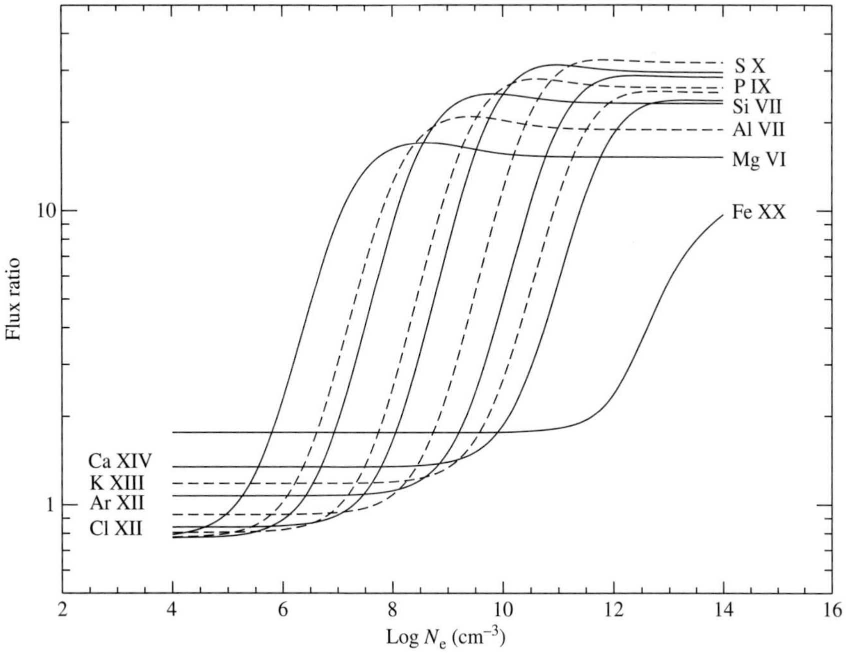

Fig. 5.2. Flux ratio $(2s^2\,2p^3\,{}^4S_{3/2} - 2s^2\,2p^3\,{}^2D_{3/2})/(2s^2\,2p^3\,{}^4S_{3/2} - 2s^2\,2p^3\,{}^2D_{5/2})$ for two forbidden transitions as a function of N_e along the N-like isoelectronic sequence.

$$(C^{\rm e}_{gi})' = C^{\rm e}_{gi} + \sum_{j>i} C_{gj} \frac{A_{ji}}{\sum_{k<j} A_{jk}}. \tag{5.5}$$

More complicated situations where collisional excitation is from other excited levels need to be evaluated on a case-by-case basis by solving the full system of equations for level populations.

Density-sensitive ratios involving forbidden lines can be found mostly in the 500–2000 Å region.

Case 2: Ratios involving allowed lines

This case applies to flux ratios of two allowed lines, such as the S X λ228/λ264 ratio in Figure 5.1. Since these lines are allowed, their A-values are larger than the collisional de-excitation rates at all densities typical of the solar atmosphere, so that the populations of their upper levels are proportional to the electron density and to the population of the lower level (or levels) from which the collisional excitation proceeds. For the S X λ228/λ264 ratio, the population $N(^4P_{5/2})$ of the upper level $^4P_{5/2}$ is mainly excited from the ground level and therefore is proportional to the population of $N_{\rm g}$ and to $N_{\rm e}$. At low densities, the ground level population is unity, so that $N(^4P_{5/2}) \propto N_e$, but as other excited levels begin to be significantly populated, $N_{\rm g}$ decreases, so that $N(^4P_{5/2}) \propto N_e^{\beta}$ where $0 < \beta < 1$.

The population $N(^2D_{5/2})$ at low densities is given mainly by excitation from the ground level, but as the density increases the populations of the ^2D doublet in the ground configuration become larger, and since the collisional excitation rate from these levels to the $^2D_{5/2}$ level is larger than from the ground level, collisional excitation from the latter loses importance relative to excitation from the ^2D doublet. Since the populations of the ^2D levels increase with density, $N(^2D_{5/2}) \propto N_e^{\alpha}$ with $\alpha > 1$. Therefore, the λ228/λ264 flux ratio is proportional to $N_{\rm e}^{\alpha-\beta}$. The value of $\alpha - \beta$ can be high so that some flux ratios vary by more than an order of magnitude in limited density ranges.

Ratios between allowed lines can be found mostly at wavelengths shorter than 400 Å.

Case 3: Ratios involving allowed and forbidden lines

Flux ratios of allowed and forbidden lines can also be used to measure the electron density of a plasma, sometimes to a high degree of accuracy. Well-known examples are found in the X-ray region, where density-dependent line ratios occur in He-like ion spectra, with allowed and forbidden lines formed by the de-excitation of the $1s2p$ and $1s2s$ configurations respectively.

The atomic structure and the main processes that populate and depopulate these levels are summarized in Figure 5.3. Their relevance for plasma diagnostics was given for the first time by Gabriel and Jordan (1969). In the solar atmosphere, photoexcitation from the $1s2s\,^3S_1$ level to the $1s2p\,^3P_{0,1,2}$ levels by UV radiation is negligible for elements heavier than C (though this might not be true in stellar atmospheres), so inelastic collisions from the ground level are the dominant excitation process for all levels except $1s2s\,^3S_1$, which is mainly populated by radiative cascades from the $1s2p\,^3P$ levels. Additional contributions are made by direct recombination into each level, and from cascades from $n > 2$ levels following collisional excitation and recombination into these higher levels. As shown in Figure 5.3, four prominent lines are emitted in He-like ion spectra, with labels according to the notation of Gabriel (1972) (see also Table 10.1): an allowed line with transition $1s^2\,^1S_0 - 1s2p\,^1P_1$ (w);

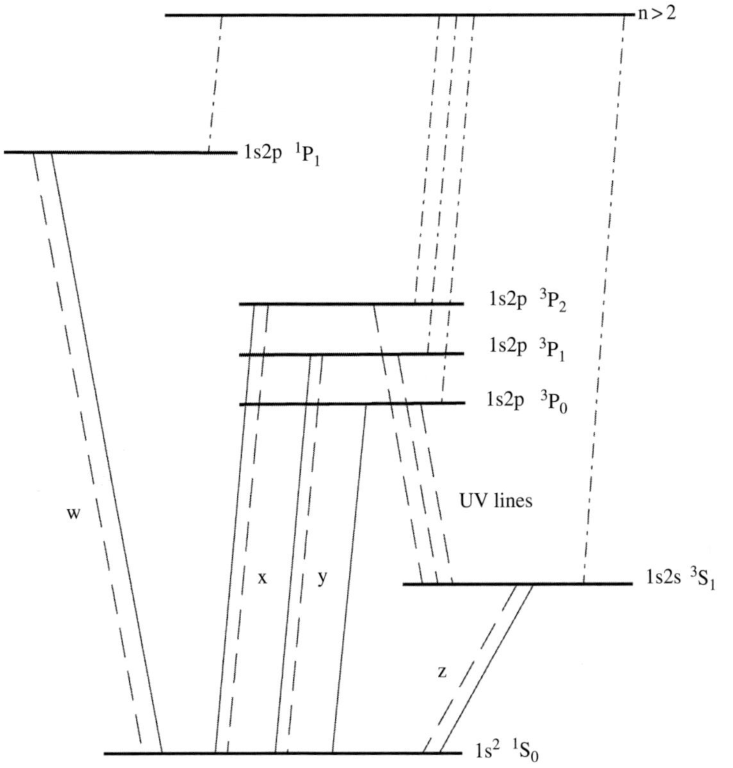

Fig. 5.3. Excitation of lines from the $n = 2$ configurations in He-like ion spectra. Full lines indicate collisional excitations, dashed lines indicate spontaneous radiative decays. Dot-dashed lines indicate direct recombination into each level and radiative cascades from $n > 2$ levels. Level $1s2s\,^1S_0$ has been omitted as it is not involved in the transitions of interest.

two intercombination lines with transitions $1s^2\,^1S_0 - 1s2p\,^3P_{1,2}$ (x and y); and a forbidden line with transition $1s^2\,^1S_0 - 1s2s\,^3S_1$ (z). Gabriel and Jordan (1969) demonstrated that the ratio $R = (x + y)/z$ depends strongly on N_e for densities above a minimum value N_{min}, and slightly on electron temperature, while the ratio $G = (x + y + z)/w$ is temperature sensitive but does not depend on density. The value of N_{min} depends on the ion and is discussed in Chapter 6.

Ratios involving lines from high energy levels

A third group of lines whose flux ratios are in principle density sensitive includes allowed lines emitted by levels in high-energy configurations. These flux ratios behave in the same way as those described in the previous section except that their excitation energies ΔE are large. The excitation rate coefficient (see Equation (4.15)) is proportional to $\exp[-(\Delta E/k_B T)]/T^{1/2}$ so that when $\Delta E \sim k_B T$ even small changes in the plasma temperature can cause large variations in collisional excitation rates. This strong temperature dependence can be a problem if lines emitted by energy levels with different excitation energy are used, since their flux ratios are dependent both on electron density and temperature.

Dielectronic satellite line spectra

The population of the level j of the ion X^{+m} above ionization threshold excited by a dielectronic collision is given by Equation (4.31), derived assuming that the target ion of the dielectronic collision is in the ground state. This is so for the dielectronic satellites of Li-like ions, excited from the ground state of He-like ions. However, for dielectronic satellites of ions such as those in the C-like, N-like, or O-like ion sequences, there are several levels within the ground configuration of the B-like, C-like, and N-like ions from which these satellites were excited; thus the C-like ion has a ground configuration $1s^2 2s^2 2p^2$, and the five possible levels are (in order of increasing energy) $^3P_0$, $^3P_1$, $^3P_2$, $^1D_2$, and $^1S_0$. At low densities, only the $^3P_0$ level is populated, but at increasing densities other levels have significant populations. The case of Fe dielectronic satellites is of some interest, since they are relatively strong compared with lighter elements and appear in the decay stages of solar flares. Thus, Fe XX satellites are formed by the dielectronic excitation of C-like Fe, which at higher densities may be any of the five levels of the ground configuration. The Fe XX dielectronic satellite spectrum, which covers a very small spectral region (1.90–1.91 Å), should therefore vary with N_e. The Fe XX spectrum changes substantially from densities expected in solar flares ($N_e \sim 10^{11}$ cm^{-3}) to those in tokamak devices used for fusion research ($N_e \sim 10^{13}$ cm^{-3}). Other families of Fe satellites, particularly Fe XIX and Fe XXI, are also density dependent, covering the spectral range 1.89–1.92 Å. Figure 5.4 shows theoretical spectra for these ions calculated by Phillips *et al.* (1983). Note that at higher temperatures, the Fe XXI satellites may be confused by the presence of an Fe XXIV dielectronic satellite (transition $1s^2\,2p\,^2P_{1/2} - 1s\,2s^2\,^2S_{1/2}$, labelled p by Gabriel (1972)).

The observation of the density-dependent Fe satellites in solar flare spectra requires high-resolution, high-sensitivity spectrometers, though the fact that the satellites occur over very small wavelength ranges makes them ideal for following density changes during the decay stages of flares since possibly uncertain instrumental sensitivities are unlikely to vary over such ranges. In Section 10.5, details are given of the available flare observations.

Ratios involving lines from different ions

Flux ratios of lines from different ions can also be used to measure N_e, sometimes providing useful diagnostic pairs. The main advantage in using lines from different ions is that line pairs with high density dependence can be found in strong lines very close in wavelength, making them easy to observe at high time cadence with spectrometers while minimizing instrument sensitivity uncertainties. However, these ratios present several sources of additional uncertainty, such as temperature sensitivity, due to ion abundances, as well as uncertainties in the adopted ion fractions and element abundances (if the lines are emitted by different elements).

Ratios of the collisional and resonantly scattered components of spectral lines

In the outer corona, where the electron density is low and resonant scattering is important, the electron density can be measured using the relative strength of the collisional and resonantly scattered components of the same spectral line, if plasma bulk velocities are negligible. From Equation (4.87), the flux of a collisionally excited spectral line is proportional to $\int N_e^2\,dx$ (where x is the line-of-sight path length), while Equation (4.156) shows that the flux of a resonantly scattered line is proportional to $\int N_e\,dx$. The flux ratio of these two components can therefore provide an estimate of N_e along the line of sight.

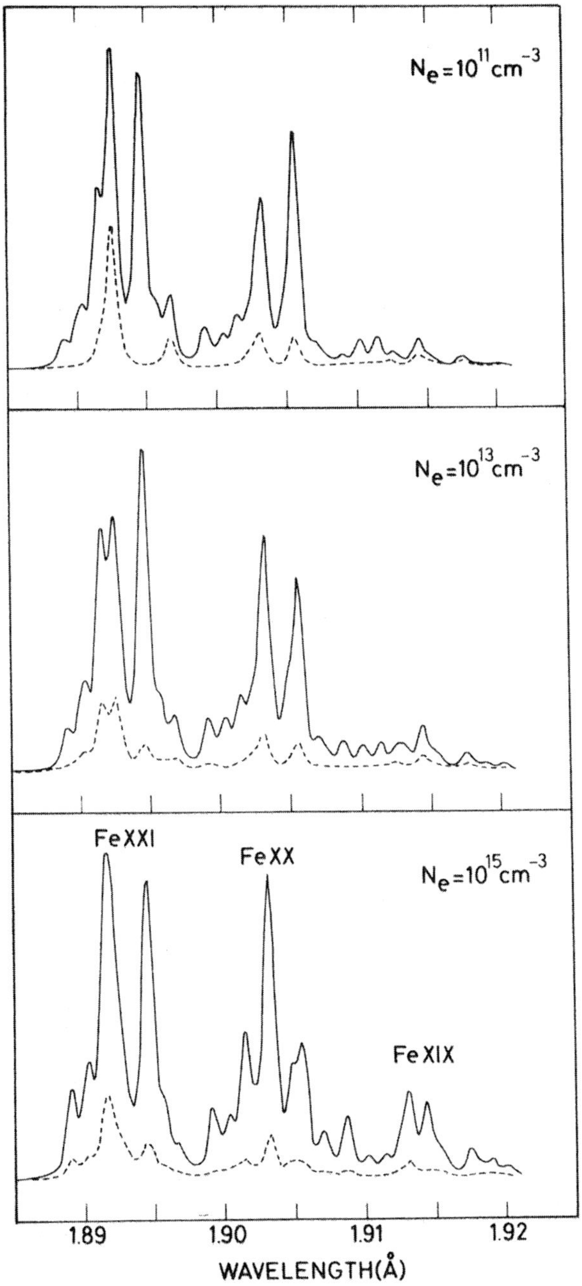

Fig. 5.4. Calculated spectra of Fe XIX–XXI lines between 1.885 and 1.921 Å for three electron densities, plotted on a linear intensity scale. The solid lines include both dielectronic and inner-shell excitation of the satellites, dashed lines those from collisional excitation alone. The assumed electron temperature is 10^7 K. From Phillips *et al.* (1983).

However, both the collisional and the scattered components must be formed in the same plasma, and bulk motions must either be negligible or known accurately, so that the Doppler dimming effect (see Section 5.9.2) can be taken into account when calculating the predicted flux ratio as a function of density. Also, the two components are blended together so that some technique must be used to separate them.

5.2.2 *L-function method*

The L-function method as a density diagnostic was introduced by Landi and Landini (1997) and uses a set of two or more lines from the same ion. From its definition in Equation (4.86), the contribution function $G_i(\log T, N_e)$ of a line i can be written as

$$G_i(\log T, N_e) = \frac{K}{N_e} h_i(\log T, N_e)\, g(\log T), \tag{5.6}$$

where the constant K includes the Einstein coefficient, the element abundance, and the $N(\mathrm{H})/N_e$ ratio, $h_i(\log T, N_e)$ is the level population, and $g(\log T)$ is the fractional ion abundance, i.e. $N(X^{+m})/N(X)$ for the ion X^{+m}. The function $h_i(\log T, N_e)$ is linear with $\log T$ to a good approximation. The level population can therefore be expressed using the first two terms of a Taylor series around a temperature $\log T_0$ as

$$h_i(\log T, N_e) = h_i\,(\log T_0, N_e) + \left(\frac{\partial h_i}{\partial \log T}\right)_{\log T_0} (\log T - \log T_0). \tag{5.7}$$

From Equation (4.94), the flux F_i of an observed spectral line emitted by a plasma with density $N_e^\star$ is then given by

$$\begin{aligned} F_i =& \frac{1}{4\pi d^2} \frac{K}{N_e^\star} h_i(\log T_0, N_e^\star) \int \varphi(\log T)\, g(\log T)\, dT \\ &+ \frac{1}{4\pi d^2} \frac{K}{N_e^\star} \left(\frac{\partial h_i}{\partial \log T}\right)_{\log T_0} \int \varphi(\log T)\, g(\log T)\, (\log T - \log T_0)\, dT, \end{aligned} \tag{5.8}$$

where d is the distance between the observer and the source. If we choose the temperature T_0 such that

$$\log T_0 = \frac{\int \varphi(\log T)\, g(\log T)\, \log T\, dT}{\int \varphi(\log T)\, g(\log T) dT}, \tag{5.9}$$

the observed flux simplifies to

$$F_i = \frac{1}{4\pi d^2} \frac{K}{N_e^\star} h_i(T_0, N_e^\star) \int \varphi(\log T)\, g(\log T)\, dT, \tag{5.10}$$

where the dependence of the flux on the temperature and the electron density appears as two independent terms. Equation (5.10) allows us to define, for each line i, a new function (the L-function) as

$$\begin{aligned} L_i(N_e) &= \frac{F_i}{G_i(\log T_0, N_e)} \\ &= \frac{1}{4\pi d^2} \frac{N_e}{N_e^\star} \frac{h_i(\log T_0, N_e^\star)}{h_i(\log T_0, N_e)} \int \frac{g(\log T)}{g(\log T_0)} \varphi(\log T)\, dT, \end{aligned} \tag{5.11}$$

so that

$$L_i(N_e^\star) = \frac{1}{4\pi d^2} \int \frac{g(\log T)}{g(\log T_0)} \varphi(\log T)\, dT. \tag{5.12}$$

The quantity $L_i(N_e^\star)$ is the same for all lines of the same ion since it depends only on the ion fraction and the plasma DEM. However, since the dependence of $L_i(N_e)$ on the electron density is different for each spectral line, Equation (5.12) can be used to measure the electron density. If we plot all the L-functions of a set of observed lines of the same ion as a function of density together, all the curves will meet in the same point $[N_e^\star, L_i(N_e^\star)]$, giving the density of the emitting source. An example is given in Figure 5.5, where the L-functions for several Fe XIII lines observed by the *SOHO* CDS instrument in an active region indicate a density of $\log N_e = 9.2 \pm 0.2$ (N_e in cm^{-3}).

When the observed spectral feature includes many individual lines of the same ion, the L-function can be defined as

$$\begin{aligned} L(N_e^\star) &= \frac{F}{\sum_j G_j(\log T_0, N_e)} \\ &= \frac{1}{4\pi d^2} \frac{N_e \sum_j h_j(\log T_0, N_e^\star)}{N_e^\star \sum_j h_j(\log T_0, N_e)} \int \frac{g(\log T)}{g(\log T_0)} \varphi(\log T)\, dT \end{aligned} \tag{5.13}$$

and will cross the same point as all the other lines.

This diagnostic method has the advantage over the line flux ratio method in that it does not assume an isothermal plasma, eliminating a source of uncertainty from the measured

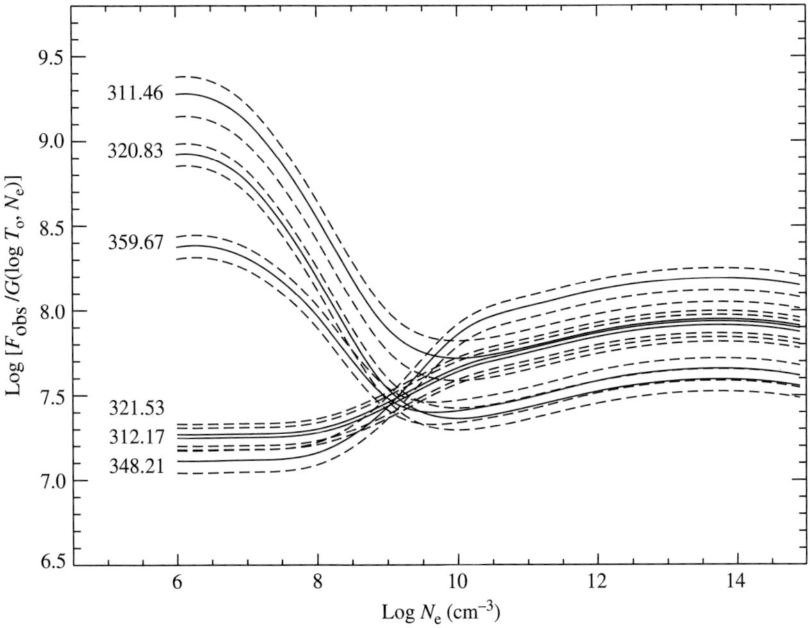

Fig. 5.5. Application of the L-function method to *SOHO* CDS observations of Fe XIII lines from an active region. The line at 311.46 Å is blended with a Cr XII line. Dashed lines indicate the uncertainties in the L-functions.

electron density, as well as making simultaneous use of all the observed lines of the same ion. The latter is important when one or more lines in the observed spectrum of the ion are either blended with unidentified lines from other ions, or when inaccurate atomic data are used. The line flux ratio technique is unable to distinguish which of the two lines has problems related to atomic physics, and from many flux ratios it is not easy to identify which line is at fault. The L-function method allows us immediately to understand where the problems are: in Figure 5.5 the L-function of the 311.46 Å line is higher than the crossing point of all other curves, due to a blending Cr XII line. However, a disadvantage is that the DEM function $\varphi(\log T)$ of the plasma must be known before the L-function method can be applied.

5.2.3 Densities from emission measure values

From isothermal plasmas

The flux of a line emitted by an isothermal plasma homogeneous in density is proportional to the emission measure of the plasma,

$$F = \frac{1}{4\pi d^2} G(T, N_e)\,\mathrm{EM}, \qquad \mathrm{EM} = \int_V N_e^2\, dV, \tag{5.14}$$

where $G(T, N_e)$ is the contribution function (Equation (4.86)) and V is the volume of the emitting plasma. The volume emission measure EM can be used to infer the electron density if some assumptions about the volume and geometry of the emitting region are made. If the emitting plasma is homogeneous and fills all the volume, the electron density is

$$N_e = \sqrt{\frac{\mathrm{EM}}{V}} = \sqrt{\frac{4\pi d^2}{V} \frac{F}{G(T, N_e)}}. \tag{5.15}$$

In general, the emitting plasma is likely to be far from homogeneous. Only an estimate of the average density $\overline{N_e}$ can be obtained,

$$F = \frac{1}{4\pi d^2} G(T, \overline{N_e}) \int_V N_e^2\, f\, dV \quad \approx \quad \frac{G(T, \overline{N_e})}{4\pi d^2}\, \overline{N_e^2}\, f\, V, \tag{5.16}$$

where f is the fraction of the volume occupied by the emitting plasma, called the filling factor, and $G(T, \overline{N_e})$ the contribution function at the average density $\overline{N_e}$. When the electron density is known from line flux ratios, Equation (5.16) can be used to derive the plasma filling factor f, giving an indication of the extent to which the plasma is structured.

From multithermal plasmas

If the plasma is not isothermal, some additional approximations must be introduced, since the line flux in Equation (5.16) must be modified to allow for the fact that T and V are related through the differential emission measure (see Equations (4.90) and (4.93)). Usually the contribution function is assumed to be equal to some fraction $b < 1$ of its peak value, so that

$$F = \frac{b}{4\pi d^2} G(T_M, N_e)\, f\, \mathrm{EM} \quad \approx \quad \frac{b}{4\pi d^2} G(T_M)\, \overline{N_e^2}\, f V, \tag{5.17}$$

where T_M is the temperature at which the contribution function reaches its maximum value. Also, in this case, it is better to use density-insensitive lines. Equation (5.17) can be used to measure the plasma filling factor if N_e is known from some other means.

From the continuum

Both the free–free and free–bound continuum can be used to derive N_e since they can be used to measure the emission measure of the plasma. From Equations (4.118) (free–free continuum) and (4.129) (free–bound continuum), the continuum emission can be expressed as

$$\frac{dR_{ff}}{dt} = F_{ff}(T, a, b)\,\mathrm{EM}, \tag{5.18}$$

$$\frac{dR_{fb}}{d\nu dt} = F_{fb}(T, a, b)\,\mathrm{EM}, \tag{5.19}$$

where the adopted set of ion fractional abundances are a and the element abundances b. Equations (5.18) and (5.19) allow us to determine the emission measure EM from measured continuum emission, to be used for density estimates as described above.

Underlying assumptions

Using the EM method above requires knowledge of the geometrical volume V of the emitting plasma, its filling factor f, and temperature T (if isothermal). Thus the temperature must be either known or assumed: a common assumption is $T = T_M$, where T_M is the temperature at which the contribution function maximizes. The geometrical volume of the emitting plasma can be inferred from images made by instruments with narrow-band filters sensitive to temperature similar to T_M, or using raster scans of lines formed at temperatures similar to T_M. If no image of the emitting plasma is available, additional assumptions on the volume V must be made.

In general, measurements of electron density from the emission measure are much less precise than those obtained from line ratios, but may be useful if no density-sensitive line pairs are available.

5.2.4 Lyman series in H_I

The profiles of neutral hydrogen (H I) lines are determined by not only thermal motions (Section 4.8) but also the interaction with the electric fields of ions and free electrons in the plasma, or the Stark effect, and by collisions between neutral H atoms and free electrons. This broadening depends on N_e, and is responsible for the blending of the lines of the H I series with principal quantum number n larger than a limiting value n_S. The value of n_S for the last line in a given series (Lyman, Balmer, Paschen) that can be distinguished as a separate spectral feature depends on the electron density; lines with $n > n_S$ will be unresolved, forming a pseudo-continuum (see Figure 3.1). The measurement of the electron density from the observed n_S value was first developed for the Balmer series of lines in the near UV, but was extended to the Lyman and Paschen series by Kurochka (1970) and Kurochka and Maslennikova (1970). Here we are concerned with the Lyman series, lying between 911 Å and 1215 Å.

Kurochka (1970) and Kurochka and Maslennikova (1970) assumed two subsequent lines of the series to be unresolved when their half-widths $\Delta\lambda_{1/2}$ were as large as half the distance between the two lines. Using this criterion, and considering line broadening due to the linear Stark effect from ion electric fields and impacts with free electrons as two independent processes, Kurochka (1970) derived the broadening of an H I line as a function of N_e as

$$\Delta\lambda_{1/2} = k_{n,m} F_0 (4 + 0.8\gamma), \tag{5.20}$$

where

$$k_{n,m} = 5.5 \times 10^{-5} \frac{n^4 m^4}{n^2 - m^2},$$
$$F_0 = 1.25 \times 10^{-9} N_e^{2/3},$$
$$\gamma = 8.4 \times 10^6 \frac{N_e^{1/3}}{T_e^{1/2}} \frac{I_{n,m}}{n^2 - m^2} \log\left(\frac{4 \times 10^{13}\, T_e^2}{N_e I_{n,m}}\right),$$

and n and m are the principal quantum numbers of the upper and lower levels involved in the transition respectively. The function $I_{n,m}$ was calculated by Minaeva (1968) and Minaeva and Sobel'man (1968) for the Lyman ($m = 1$), Balmer ($m = 2$), and Paschen ($m = 3$) series; in the Lyman series,

$$I_{n,1} = \frac{9}{4}(n^4 - 3n^2). \tag{5.21}$$

From Equation (5.20), the largest value n_S for which a spectrometer can resolve two adjacent lines in the Lyman series can be obtained. The value of n_S depends mostly on N_e and on the spectrometer's resolving power, so N_e can be measured. Kurochka and Maslennikova (1970) provided an approximated formula to calculate the dependence of n_S on the electron density as

$$\log N_e = 22 - 7.0 \times \log(n_S + \delta), \tag{5.22}$$

where δ is a correction depending on the quality of the observed spectrum. They extended Equation (5.20) to the case where Doppler broadening also provides a significant contribution to the total H I line widths, obtaining

$$\Delta\lambda_{1/2} = \begin{cases} \sqrt{[k_{n,m} F_0 (4 + 0.8\gamma)]^2 + \Delta\lambda_D^2} & \beta_D \le 15, \quad (5.23) \\ 0.8 k_{n,m} F_0 \gamma + 0.833 \Delta\lambda_D & \beta_D \ge 15, \quad (5.24) \end{cases}$$

where

$$\beta_D = \frac{\Delta\lambda_D}{k_{n,m} F_0}, \qquad \Delta\lambda_D = \frac{\lambda_0}{c}\sqrt{\frac{2k_B T}{M} + v_{nth}^2}, \tag{5.25}$$

and T and M are the hydrogen atom temperature and mass, λ_0 the rest wavelength of the line, and v_{nth} is the non-thermal velocity. Diagrams of n_S versus log N_e for Doppler-broadened lines are shown in Figure 5.6.

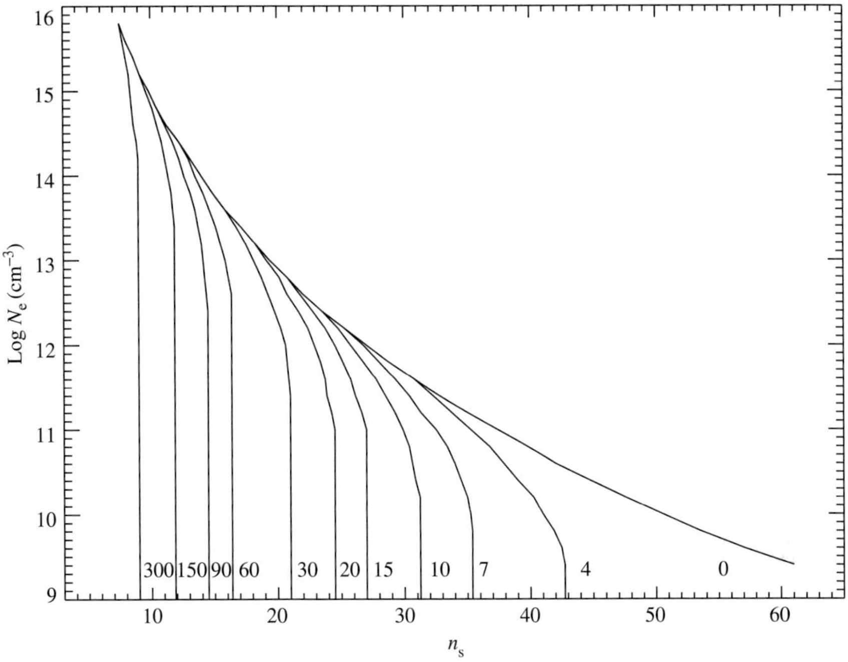

Fig. 5.6. Dependence of n_s on the electron density at different non-thermal velocities (in km s^{-1}, indicated at the side of each curve) for the H I Lyman series.

5.3 Electron temperature diagnostics

5.3.1 *Line flux ratios*

The most widely used tools for temperature diagnostics are flux ratios of two lines or one line and the continuum. These can be grouped into four classes, according to the lines that are used: (i) ratios of lines of the same ion; (ii) ratios of lines of different ions; (iii) ratios involving dielectronic satellite lines; and (iv) ratios involving one line and free–free continuum.

Ratios of lines from the same ion

The temperature sensitivity of flux ratios of two lines of the same ion is mostly due to the exponential factor in the collisional excitation rate coefficient. According to Equation (4.15), this rate coefficient is given by the product of the effective collision strength Υ and the factor $\exp(-\Delta E/k_B T)/T^{1/2}$, where ΔE is the line excitation energy, i.e. the energy difference between the lower and the upper levels. If we assume for the sake of simplicity that the upper levels of the two lines are excited from only one lower level (as for example in the coronal model approximation), the ratio of the two line fluxes is

$$\frac{F_1}{F_2} \propto \frac{\Upsilon_1(T)}{\Upsilon_2(T)} \exp[-(\Delta E_1 - \Delta E_2)/k_B T]. \tag{5.26}$$

The ratio of the two effective collision strengths is not strongly temperature dependent unless the two lines are emitted by transitions of two different types (e.g. a forbidden and an allowed line), while the exponential factor is very strongly temperature dependent when $k_B T \ll (\Delta E_1 - \Delta E_2)$, and decreases its temperature sensitivity when the temperature approaches a limiting value $T_m = (\Delta E_1 - \Delta E_2)/k_B$. For this reason, ratios of this kind should involve lines with very different excitation energies. This condition is usually met by lines with the upper levels of their transitions belonging to different configurations. The limitations of these ratios lie in the fact that, unless the difference in excitation energy is large relative to the plasma temperature, their temperature dependence is not strong and so there may be large uncertainties in the measured temperature. Also, both lines must be density insensitive or at least depend on the electron density in the same way, so that the ratio itself is independent of density. However, under equilibrium conditions these ratios are independent of ion and element abundances.

Ratios of lines from different ions

The temperature sensitivity of flux ratios of lines from different ions is mostly due to the different temperature dependence of the abundances of the ions emitting the line pair; this sensitivity is usually very strong, independent of the excitation energy of the two lines. Usually such ratios are of lines emitted by ions of the same element to avoid uncertainties in element abundances. Such ratios have stronger temperature sensitivity than ratios of lines of the same ion, but they are subject to the uncertainties in the ion fraction calculations. Also, any departures from equilibrium will lead to misleading results. The line ratios should not have any dependence on N_e.

Ratios involving dielectronic satellite lines

The ratio of a dielectronic satellite line emitted by an ion X^{+m} and the resonance line emitted by the parent ion X^{+m+1}, generally in the X-ray region, is a commonly used way of measuring temperatures of active regions or flares. The details of the method were first given by Gabriel (1972). The most well-known ratios involve Li-like ion satellites generally having transitions of the type $1s^2\,2p - 1s\,2p\,np$ to He-like ions for which the resonance line transition is $1s^2\,{}^1S_0 - 1s2p\,{}^1P_1$ (line w). Satellites with $n = 2$ are the most intense, and occur over a small wavelength region to the long-wavelength side of line w. He-like satellites with transitions like $1s\,2l - nl\,2p$ ($l = s,\ p$) also occur on the long-wavelength side of the Ly-α line of H-like ions; again the $n = 2$ satellites are the most intense. Such ratios are useful not only because of their strong temperature sensitivity but also because dielectronic satellite lines are close in wavelength to the resonance lines of the parent ion. This makes them very suitable for finding the temperature of rapidly changing flares, since they can be observed with rapidly scanning flat crystal spectrometers or with bent crystal spectrometers having limited wavelength ranges (Section 7.5). Another advantage of these ratios is that they are independent of the ion fractions since the population of the upper level of a satellite line is determined by the dielectronic capture of a free electron by the parent ion X^{+m+1}, so that the flux of the dielectronic satellite line is proportional to the ion abundance of the parent ion.

The flux ratio of the dielectronic satellite to the resonance line is given by combining Equations (4.15), (4.19), (4.23), and (4.32):

$$\frac{F_{\rm diel}}{F_{\rm res}} = K\frac{\exp[(\Delta E_{\rm res} - \Delta E_{\rm diel})/k_B T]}{T}, \tag{5.27}$$

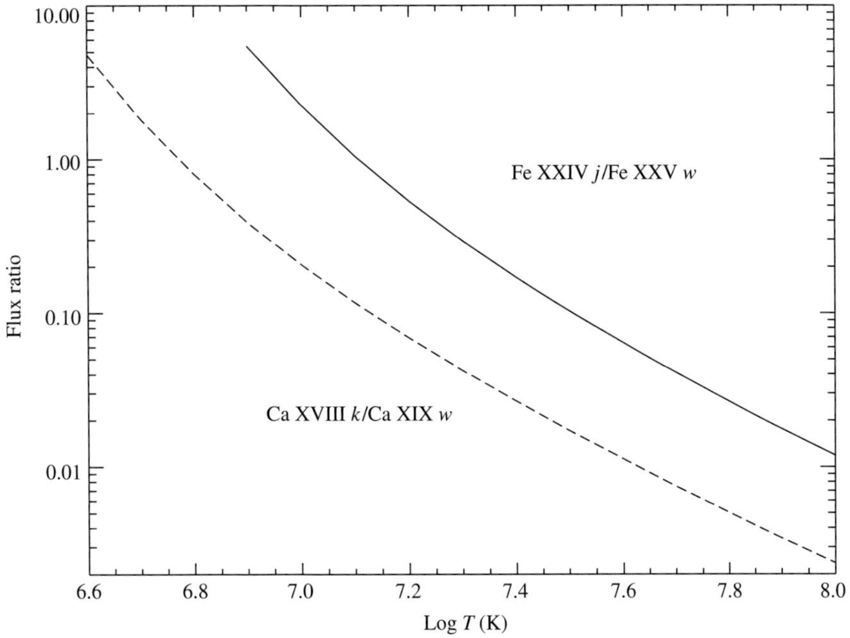

Fig. 5.7. Dielectronic satellite-to-resonance line flux ratios: (*full line*) Fe XXIV satellite j to Fe XXV line w; (*dashed line*) Ca XVIII satellite k to Ca XIX line w.

where the constant K includes physical constants and atomic factors, $\Delta E_{\rm diel}$ is the energy of the transition involved in the emission of the dielectronic line, and $\Delta E_{\rm res}$ is the excitation energy of the allowed line. Since the excitation energies $\Delta E_{\rm res}$ and $\Delta E_{\rm diel}$ are similar, $F_{\rm diel}/F_{\rm res}$ varies approximately as T^{-1}. The ratio is independent of density. Examples of two satellite-to-resonance line flux ratios are shown in Figure 5.7. For Fe spectra, used to measure the higher-temperature parts of X-ray flares, the flux ratio of the most intense Fe XXIV satellite line j (transition $1s^2\,2p\,^2P_{3/2} - 1s2p^2\,^2D_{5/2}$) at 1.8655 Å to the Fe XXV resonance line w (transition $1s^2\,^1S_0 - 1s2p\,^1P_1$) is a suitable and frequently used means of determining $T_{\rm e}$. For Ca spectra, the Ca XVIII satellite line j is blended with the Ca XIX z line ($1s^2\,^1S_0 - 1s2s\,^3S_1$) but satellite line k ($1s^2\,2p\,^2P_{1/2} - 1s2p^2\,^2D_{3/2}$) is available for measuring temperature through the k/z ratio.

A disadvantage of using dielectronic satellites near He-like ion resonance lines w having transitions $1s^2\,2p - 1s\,2p^2$ is that many lines are crowded in a small wavelength range, and high-resolution spectrometers are needed to resolve the important satellites unambiguously. Satellites near He-like ion lines with transitions $1s^2\,^1S_0 - 1s\,np\,^1P_1$ ($n \geq 3$) have a distinct advantage in that the wavelength region to the long-wavelength side of the He-like ion lines is much less crowded – there is, for example, no equivalent of the $1s^2\,^1S_0 - 1s2s\,^3S_1$ (z) line, and the intercombination line with transition $1s^2\,^1S_0 - 1snp\,^3P_1$ ($n \geq 3$) is very weak compared with the $1s^2\,^1S_0 - 1s\,np\,^1P_1$ line. The satellites form a single spectral feature that is very prominent at relatively small temperatures. They were observed in laboratory plasmas some years ago (e.g. Feldman *et al.* (1974)), but it was not until their observation in RESIK solar flare spectra (Sylwester *et al.* (2005)) that their use as flare temperature indicators was realized. In RESIK spectra, Si and S satellites were observed in numerous flares, and with atomic data from the UCL codes their temperature sensitivity was calculated and compared with RESIK data (Phillips *et al.* (2006b)). Of particular interest is the case of the satellite feature at 1.593 Å (7.78 keV) near the Fe XXV $1s^2\,{}^1S_0 - 1s\,3p\,{}^1P_1$ line at 1.573 Å (7.88 keV), to date never observed with high-resolution spectrometers, which would make an excellent temperature diagnostic for the hottest part of solar flares (or indeed of other high-temperature, non-solar sources). These lines make up the bulk of the Fe/Ni 8.0 keV line feature seen by the *RHESSI* spacecraft (Section 10.2) for flare temperatures up to 30 MK. Future crystal spectrometers observing this spectral region would be likely to make a significant contribution to our knowledge of hot flare kernels occurring at the earliest stages of flare development.

5.3.2 Narrow-band filter ratios

Ratios of fluxes in two neighbouring spectral regions observed with narrow-band filters, e.g. in imaging telescopes, can be used as approximate determinations of temperature. Often filter passbands are chosen so that strong, isolated lines are included. This applies to the imaging EUV instruments EIT on *SOHO* and *TRACE* and the X-ray telescopes such as the SXT on *Yohkoh* and XRT on *Hinode*. Figure 5.8 (left) shows the temperature response of the 171 Å and 195 Å filters of the *TRACE* satellite, as calculated by Phillips *et al.* (2005b). The ratios of the emission in the 171 Å filter to that in the 195 Å filter can be used to infer temperature under the assumption of isothermal plasma.

The main advantage of narrow-band filter ratios is that they can provide temperature measurements for large areas of the solar atmosphere observed simultaneously. Scanning

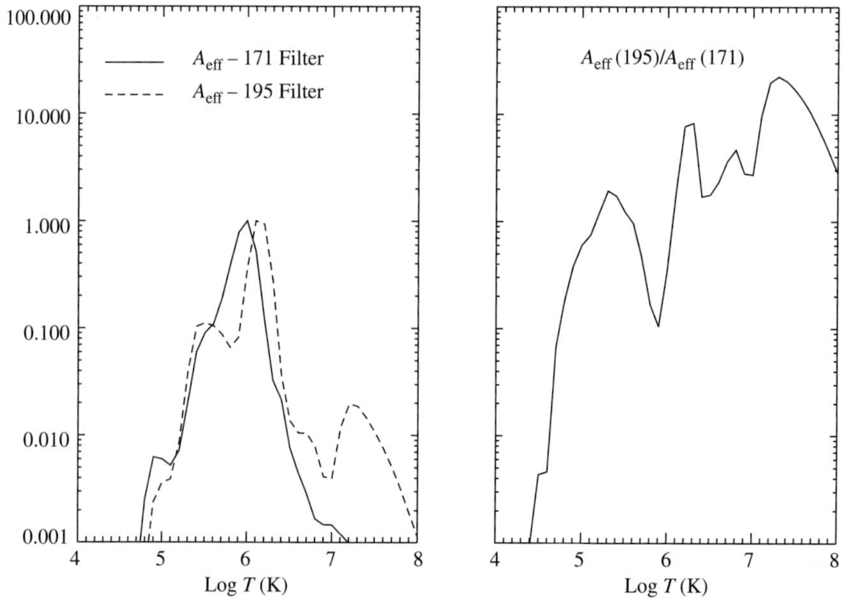

Fig. 5.8. (*Left*) Normalized effective areas A_{eff} for the *TRACE* 171 Å and 195 Å channels as a function of temperature. (*Right*) Effective area ratio $A_{eff}(195)/A_{eff}(171)$ as a function of temperature. Effective areas are taken from Phillips *et al.* (2005b).

spectrometers like the CDS on *SOHO*, by contrast, need to scan the same areas by repeatedly stepping the entrance slit, thus requiring long observation times over which the emitting region may have significantly changed. The main disadvantage of narrow-band filter ratios is that they are often non-monotonic, so that a given ratio results in more than one temperature. Figure 5.8 (right) shows how three temperatures correspond to a single filter ratio of between 0.1 and 1.0, though not all these temperatures will be realistic depending on the features being observed. For multithermal plasmas, the method may be even less reliable, as Weber *et al.* (2005) showed. For an emitting plasma that is multithermal over a broad enough temperature range, the ratio of filter fluxes tends to the ratio of the integral of their response functions. In the calculated flux ratios, assumed element abundances influence the ratio between two different filters.

5.3.3 *Flux fall-off with height in the solar atmosphere*

When the observed plasma is in hydrostatic equilibrium, an estimate of the electron temperature can be obtained from the flux decrease as a function of height above the solar limb. In hydrostatic equilibrium, the electron pressure p decreases exponentially according to

$$p(h) = p_0 \exp\left(-\int_0^h \frac{dz}{H(z)}\right), \qquad H(h) = \frac{k_B\, T(h)}{\mu g}, \tag{5.28}$$

where μ is the mean particle mass, h is the height above the photosphere, g is the acceleration due to gravity at the solar surface, and p_0 the pressure corresponding to $h = 0$. The quantity $H(z)$ is the pressure scale height. If the plasma is isothermal, as for example in the off-disk quiet solar corona, the perfect gas law can be used to determine the density decrease as a function of height as

$$N_e = N_{e0}\, e^{-h/H} \quad \longrightarrow \quad \mathrm{EM} = N_{e0}^2\, e^{-2h/H} dV \quad \longrightarrow \quad F = F_0\, e^{-2h/H}, \tag{5.29}$$

where N_{e0} is the density at height $h = 0$ and H is calculated using the coronal temperature, and F, F_0 are the line fluxes at heights h and 0. If the line flux is measured as a function of distance from the solar limb, by fitting its dependence on the distance it is possible to measure the scale height H and to derive from it the temperature T. Such a technique has been applied by several authors (for example, Landi *et al.* (2006b)).

5.3.4 *Broad-band resolution X-ray spectra*

Spectra in X-rays ($\lambda < 20$ Å, or $E > 0.6$ keV) with broad-band resolution can also be used for the measurement of the electron temperature of solar plasmas, even though individual spectral lines cannot be resolved.

Temperature from broad-band spectra

In some solar flare X-ray spectra with broad-band resolution, groups of lines from different ions may be observed as a single, unresolved feature superimposed on a thermal continuum. The total flux of this feature can be used to estimate the electron temperature if expressed as a fraction of the continuum flux. We may define the *equivalent width* of the line feature as the width of a portion of the continuum at the feature's wavelength whose flux is equal to the flux of the line feature. Knowing the details of the individual lines making up the line feature, the variation of the equivalent width with temperature can be determined assuming the abundance of relevant elements. Comparison of measured and predicted equivalent

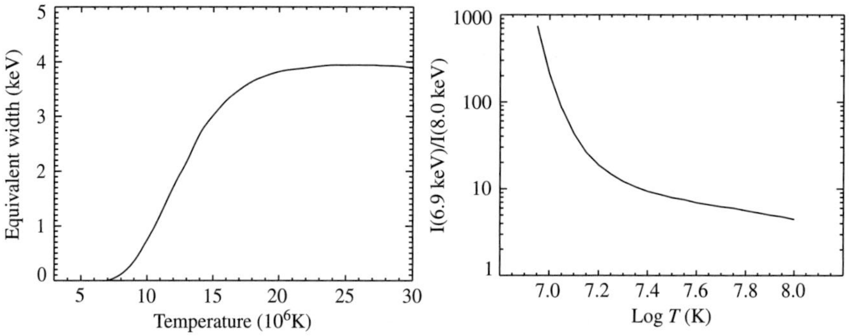

Fig. 5.9. (*Left*) Equivalent width of the unresolved group of lines observed by *RHESSI* at 6.7 keV ($\simeq$ 1.85 Å), as a function of electron temperature. (*Right*) Ratio (photon units) of the Fe line feature observed by *RHESSI* at 6.7 keV ($\simeq$ 1.85 Å), mainly formed by Fe XXV and its satellites, and the Fe/Ni line feature at 8.0 keV ($\simeq$ 1.55 Å), formed by Fe XXV, Fe XXVI and Ni ion lines. A coronal abundance of Fe (Feldman & Laming (2000)) is assumed. From Phillips (2004).

widths should then give temperature. This has been done for the case of the well-known Fe line feature (Phillips (2004)) at ~1.9 Å (6.7 keV), which is prominent in solar flare spectra from a number of instruments including the detectors on the *RHESSI* spacecraft. The line feature is made up of Fe XXV and Fe XXVI resonance lines and numerous dielectronic satellites. Atomic data for many of these lines were used to find the equivalent width of the 6.7 keV Fe line feature shown in Figure 5.9 (left). For coronal abundances of Fe (Feldman & Laming (2000)), the maximum equivalent width of the Fe line feature is about 4 keV in energy units.

At higher energies (~8 keV, wavelengths of ~1.55 Å), a weaker line feature occurs in *RHESSI* flare spectra, made up mostly of Fe XXV $1s^2 - 1snp$ ($n > 2$) lines and (at higher temperatures) Fe XXVI $1s - np$ ($n > 2$) lines. A small proportion of this feature is due to lines of He-like and H-like Ni ions, and for this reason this line feature has become known as the Fe/Ni feature. The flux ratio of the Fe line and Fe/Ni line features should also give temperature, as a check on the Fe line equivalent width value. Figure 5.9 (right) gives this ratio as a function of T. The value of the Fe abundance does not affect the ratio.

Continuum

The energy radiated per unit time, volume and frequency in the free–free and free–bound continuum is given by Equations (4.114) and (4.129), and can be summarized as

$$\frac{d R_{\rm ff}(\nu)}{dt\, dV\, d\nu} = C\, T_{\rm e}^{-1/2} \exp(-h\nu/k_{\rm B}T) \sum_i Z_i^2 \langle g_{\rm ff}\rangle N_{\rm e}\, N_{\rm i}, \tag{5.30}$$

$$\frac{d R_{\rm fb}(\nu)}{dt\, dV\, d\nu} = C'\, T_{\rm e}^{-3/2} \sum_i K_i' \exp[-(h\nu + E_i)/k_{\rm B}T]\, g_{\rm fb}\, N_{\rm e}\, N_{\rm i}, \tag{5.31}$$

where $N_{\rm e}$ and $N_{\rm i}$ are the electron and ion densities, $Z_{\rm i}$ is the ion atomic number, $\langle g_{\rm ff}\rangle$ and $g_{\rm fb}$ (both dependent on $T_{\rm e}$) are, respectively the velocity-averaged free–free Gaunt factor and the free–bound Gaunt factor, E_i is the ionization energy for the ion i, and the values of the constants C, C', and K' can be calculated from Equations (4.114) and (4.129). Both continua provide significant contributions at flare temperatures ($T \leq 3 \times 10^7$ K). Using Equations (5.30) and (5.31), the total continuum emission can be calculated as a function of wavelength (or energy) and of temperature: comparison with the observed continuum spectrum will indicate which value of T_e provides best agreement.

This method has the disadvantage that the predicted total emission depends on the absolute abundances of the elements (O to Ni) emitting the free–bound continuum: incorrect assumed abundances lead to errors in the total continuum and hence the estimated temperature. Observing the continuum in a wide wavelength (or energy) range minimizes uncertainties in the measured temperature.

5.3.5 Lyman-α electron scattering

The electron temperature of coronal plasmas can also be measured by using the Thomson scattering of solar disk radiation by free electrons in the solar corona. This method is most useful when no suitable line pairs are available for the application of the flux ratio technique, and when the plasma density is low enough for the coupling between electrons and ions to be weakened, leading to departures of the ion and electron temperatures (ion temperatures can be estimated from line widths). Following Withbroe *et al.* (1982), if $F(\lambda_s, \omega, \mathbf{n_s})$ is the photon flux in a spectral line per unit wavelength in the direction $\mathbf{n_s}$, the number of photons scattered by coronal electrons per unit time and volume in the solid angle $d\omega$ is

$$\frac{dN}{dV\,dt} = N_e f(\mathbf{v})\,\sigma\, F(\lambda_s, \omega, \mathbf{n_s})\, d\lambda_s\, d\omega\, 4\pi\, \mathbf{v}^2\, d\mathbf{v}, \tag{5.32}$$

where the velocity of the scattering electrons is between v and $v + dv$, $f(v)$ is the electron velocity distribution, $\mathbf{n_s}$ is the unit vector in the direction of the incident photon, and σ is the Thomson scattering differential cross-section of the electron, given by

$$\sigma = \frac{e^4}{m^2 c^4} = 7.94 \times 10^{-26}\ \mathrm{cm}^2\ \mathrm{sr}^{-1}. \tag{5.33}$$

(The total Thomson cross-section is $(8\pi/3)\sigma = 6.65 \times 10^{-25}$ cm^2.) The fraction of incident photons scattered along the direction $\mathbf{n}$ of the observer at a given wavelength λ is given by (Withbroe *et al.* (1982))

$$\begin{aligned}\frac{dN(\lambda)}{dV\,dt} = \frac{1}{4\pi}\frac{dN}{dV\,dt}\left[\frac{3}{4}\left(1 + (\mathbf{n}\cdot\mathbf{n_s})^2\right)\right]\delta\left[\left(\lambda_s + \frac{\lambda_s}{c}\mathbf{v}\cdot\mathbf{n_s}\right)\right.\\ \left. - \left(\lambda - \frac{\lambda}{c}\mathbf{v}\cdot\mathbf{n}\right)\right],\end{aligned} \tag{5.34}$$

where $\mathbf{n_s}$ and $\mathbf{n}$ are, respectively, the unit vectors of the directions of the incident radiation and the line of sight, and $\mathbf{v}$ is the electron velocity vector. The Dirac δ functions specify which wavelengths λ_s of the incident radiation are scattered by an electron to provide radiation at wavelength λ towards the observer. The term $\frac{3}{4}\left(1 + (\mathbf{n}\cdot\mathbf{n_s})^2\right)$ is the intensity modulation factor taking into account the anisotropy of Thomson scattering.

The total Thomson-scattered flux emitted towards the observer can be obtained by combining Equations (5.32) and (5.34) and integrating over the electron velocity, the wavelength of the incident radiation, the solid angle, and the line of sight:

$$F(\lambda) = \frac{3\sigma}{16\pi}\int_0^{\infty} dx\, N_e \int_{\omega} d\omega\left(1 + (\mathbf{n}\cdot\mathbf{n_s})^2\right)\int_0^{+\infty} d\lambda_s\, F(\lambda_s, \omega, \mathbf{n_s})$$
$$\times\int_{-\infty}^{+\infty} f(v)\,\delta\left[\left(\lambda_s + \frac{\lambda_s}{c}\mathbf{v}\cdot\mathbf{n_s}\right) - \left(\lambda - \frac{\lambda}{c}\mathbf{v}\cdot\mathbf{n}\right)\right]4\pi v^2\, d\mathbf{v}. \quad (5.35)$$

The spectral profile of the emerging radiation will thus be determined by the electron velocity distribution and its parameters and by the scattering geometry. If the electron velocities have a Maxwell–Boltzmann distribution, T_e can be determined from the line width. Also, the dependence of the line profile on $f(v)$ provides a tool for determining departures of the electron velocity distribution from the assumed one (normally a Maxwell–Boltzmann distribution).

An example of the profile of a Thomson-scattered line is given in Figure 5.10 (left) obtained from the calculations given by Withbroe *et al.* (1982): Figure 5.10 shows that the line width of the Thomson-scattered line can be as large as 100 Å or more, and is strongly dependent on T_e. The effects of plasma flows and turbulence on the line profile in Figure 5.10 were neglected, since the thermal speed (~7000 km s^{-1}) of the electrons is so large. This emission blends

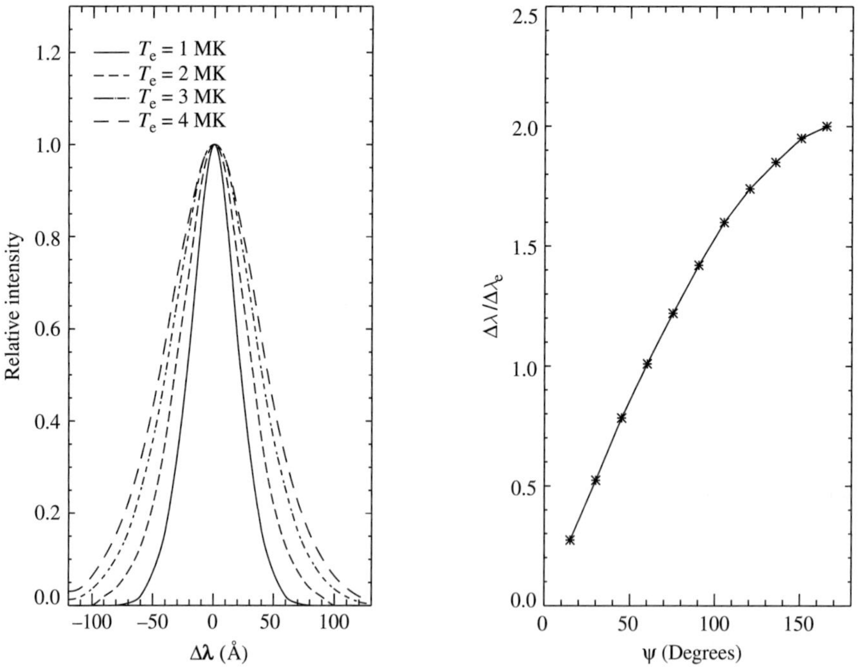

Fig. 5.10. (*Left*) Normalized profiles of the electron-scattered component of H I Ly-α, calculated at different values of the electron temperature. (*Right*) Width of the electron-scattered profile of H I Ly-α as a function of the angle between the line of sight and the direction between Sun centre and the emitting plasma element. $\Delta\lambda_e$ is the electron thermal Doppler width (see text), $\Psi = 90$ degrees is the plane of the sky. From Withbroe *et al.* (1982).

with the local line emission and the resonant-scattered emission of the line, but the width of Thomson-scattered line emission is so large that it can be easily separated from the latter two components whose widths are a few Å at most. The dependence of the Thomson-scattered radiation on the angle Ψ between the line of sight and the direction between Sun centre and the emitting plasma element is shown in Figure 5.10, as the ratio $\Delta\lambda/\Delta\lambda_e$ where $\Delta\lambda$ is the line width and $\Delta\lambda_e = \lambda/c\sqrt{2k_B T_e/m}$ is the electron thermal Doppler width. Figure 5.10 shows that geometry must be taken into account when interpreting Thomson-scattered emission: Withbroe *et al.* (1982) estimated that uncertainties in T_e resulting from the geometry can be as high as 25%.

The flux of the Thomson-scattered radiation depends on the incident flux of a spectral line and on the local electron density. However, only the strongest lines can provide enough Thomson-scattered intensity to be observed, so that this diagnostic technique has been applied mostly to the H I Ly-α line.

5.3.6 Problems in electron temperature determination

All the diagnostic techniques outlined in this section share the same assumption of the plasma being isothermal. If the plasma is multithermal, all techniques only provide an approximate estimate of the temperature of the plasma that emits the spectral features considered, and a differential emission measure analysis (Section 5.6) must be used to determine the thermal structure of the plasma. For spectral lines formed in plasmas with similar temperatures, the temperature measurement obtained from their flux ratios approximates the actual temperature. This is usually true for ratios of lines emitted by the same ion and, to a lesser extent, of lines emitted by adjacent stages of ionization of the same element.

5.4 Ion temperature diagnostics

5.4.1 UV spectra

As shown in Section 4.8, the FWHM of an optically thin spectral line broadened by thermal motions and by additional Gaussian distributed non-thermal motions is given by

$$\Delta\lambda = \frac{\lambda_0}{c}\sqrt{4\ln 2\left(\frac{2k_B T_i}{M} + v_{nth}^2\right)}, \tag{5.36}$$

where $\Delta\lambda$ is the FWHM, v_{nth} is the non-thermal velocity, T_i is the ion temperature, M is the ion mass, and λ_0 is the line centroid rest wavelength. Equation (5.36) sets upper limits on both v_{nth} and T_i:

$$v_{nth} \le \frac{c}{\lambda_0}\sqrt{\frac{1}{4\ln 2}}\,\Delta\lambda, \tag{5.37}$$

$$T_i \le \frac{M}{8k_B\ln 2}\left(\frac{c}{\lambda_0}\Delta\lambda\right)^2, \tag{5.38}$$

where the equalities are for $T_i = 0$ (Equation (5.37)) and $v_{nth} = 0$ (Equation (5.38)). To set a lower limit on the ion temperature it is necessary to make some assumptions about non-thermal velocities.

Measurement of H and He ion temperatures

H and He lines are emitted by coronal plasmas when the emission measure is large because the fractional abundance of neutral H, and neutral or once-ionized He in ~1 MK plasmas, though small, is not negligible. The profiles of the coronal H and He lines are dominated by thermal Doppler broadening rather than non-thermal broadening since the H and He masses are very small. For these lines, Equation (5.36) can be simplified by neglecting $v_{\rm nth}$ so that $T_{\rm i}$ can be directly measured from $\Delta\lambda$. In dense active regions the ion temperatures are equal to the electron temperatures, and can exceed 2×10^6 K. In such cases, the velocities $v_{\rm th}$ caused by random thermal motions are much larger than typical values of $v_{\rm nth}$: for example, $v_{\rm th} = 190$ km s^{-1} and 90 km s^{-1} at $T_{\rm e} = 2 \times 10^6$ K for H and He respectively. The dependence of the H and He line widths on the ion temperature is shown in Figure 5.11.

The measurement of $T_{\rm H}$ and $T_{\rm He}$ in active region and flare off-limb plasmas has the advantage that the line width is easy to measure because it is very large (~1 Å and 0.3 Å for H and He respectively), thus minimizing instrumental broadening effects. However, when observed outside the solar disk, these lines may be too weak to be detected, while in disk spectra their profile is not Gaussian.

Measurement of other ion temperatures

Tu *et al.* (1998) developed a technique that assumes only that the non-thermal velocity be the same for all ions, allowing the measurement of $T_{\rm i}$ when many lines of several

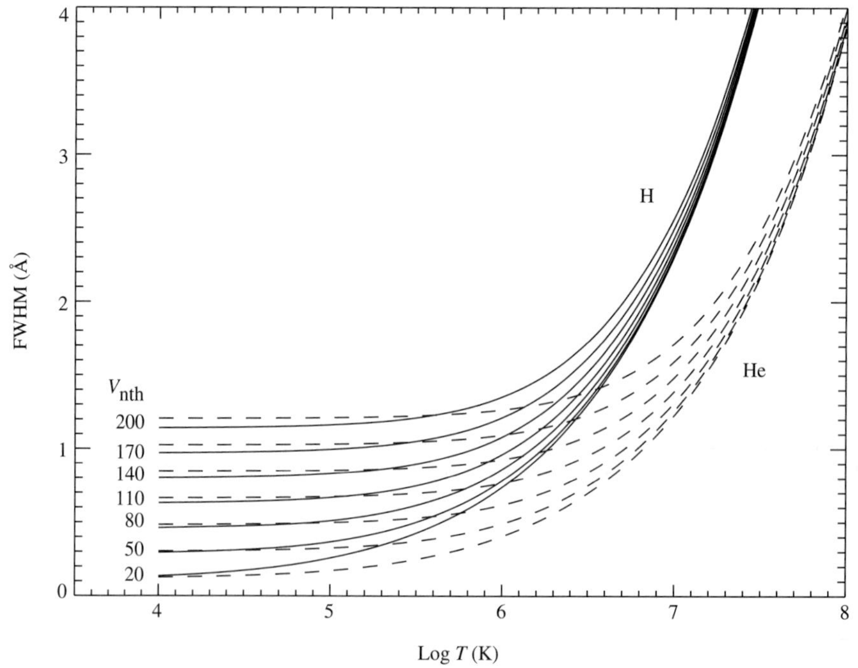

Fig. 5.11. Dependence of the FWHM of the H I lines at 1025.7 Å and of the He II lines at 1084.9 Å on the electron temperature. The curves have been calculated adopting several values of the non-thermal velocity $v_{\rm nth}$ (in km s^{-1}).

elements are observed. According to this technique, we first calculate the maximum values v_{max} of v_{nth} for each line of each ion using Equation (5.37). Since this technique assumes that v_{nth} is the same for all ions, the maximum value for non-thermal velocities compatible with all lines, which we indicate by v_M, corresponds to the minimum value of the set of v_{max}. The velocity v_M can be used for all ions to determine a minimum value for T_i:

$$T_i \geq \frac{M}{2k_B}\left(\frac{c^2}{4\ln 2}\left(\frac{d\lambda}{\lambda_0}\right)^2 - v_M^2\right). \tag{5.39}$$

This value, together with Equation (5.38), provides the allowed range of T_i for each ion compatible with the set of observed line widths and the uniform v_{nth} assumption. This method can also be used to test the commonly used assumption that $T_i = T_e$: for example, Tu *et al.* (1998) applied it to coronal hole plasmas observed close to the limb and found that T_i was significantly higher than T_e.

5.4.2 Hydrogen atom and proton temperatures in the solar wind

In regions where the electron density is low, the H I Ly-α line at 1215.67 Å provides one of the best tools for measuring the proton and hydrogen atom temperature. When observed outside the limb, the H I Ly-α line is made up of two components given by scattering of chromospheric radiation: (i) resonant scattering by local hydrogen atoms; and (ii) Thomson scattering by local electrons (Section 5.3.5). The FWHM of the resonant-scattered radiation is ~1 Å, while the FWHM of the Thomson-scattered radiation is ~50 Å, so these two components can be easily separated and used to measure the hydrogen atom temperature and the local electron temperature respectively.

The profile of a resonant-scattered line can be computed numerically from the considerations in Section 4.9.3. It is very sensitive to the hydrogen atom temperature T_H. An example is shown in Figure 5.12 where the calculation of the line profiles has been taken from Withbroe *et al.* (1982). Figure 5.12 shows that an increase of the hydrogen atom temperature from 1×10^6 K to 2×10^6 K causes the line width to increase by 0.34 Å, easily measurable with a high-resolution spectrometer. Also, the line profile is close to Gaussian.

In solar wind regions where the coronal expansion time is larger than the lifetime of the hydrogen atom (usually at low altitudes, i.e. $R < 2.5 R_\odot$), the proton temperature T_p is similar to the hydrogen atom temperature, so that the width of the H I Ly-α line can also be used to measure T_p.

The above assumes that the non-thermal velocity v_{nth} is zero. If this is not true, the Ly-α profiles in Figure 5.12 only provide the kinetic temperature $T_k = T_H + m_H v_{nth}^2/2k_B$. However, the non-thermal velocities in the low corona have been measured to be ~30 km s^{-1}, much smaller than the hydrogen atom thermal speeds (~160 km s^{-1} at $T_H = 1.5 \times 10^6$ K), so their effect on the measured T_H in the low corona is not large.

This diagnostic method is subject to two additional sources of uncertainty. Radial flows of hydrogen atoms in both directions along the line of sight may provide additional line broadening and thus lead to an overestimation of T_H and T_p. Also, according to Section 4.9.3, the profile of the resonant-scattered Ly-α line depends on the angle between the line of sight and the vector from the emitting plasma element and Sun centre. The cumulative uncertainties given by geometrical effects were estimated by Withbroe *et al.* (1982) to be around 10%,

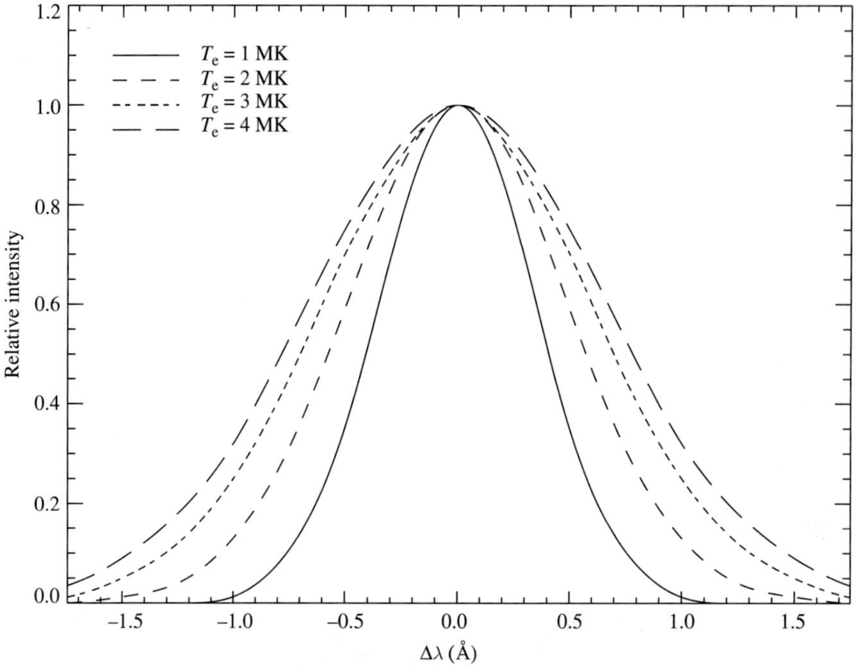

Fig. 5.12. Calculated normalized profiles of the resonantly scattered component of H I Ly-α. From Withbroe *et al.* (1982).

while the increase in the FWHM due to radial motions was estimated to be 13% when $v = 400$ km s^{-1}; less when outflow velocities are smaller.

5.5 Emission measure diagnostics

The emission measure (EM) of a plasma can be measured both in isothermal and multithermal plasmas. In the latter case, the total EM of a plasma is obtained by integrating the differential emission measure (DEM) function $\varphi(T)$ or the density distribution in space (see Section 4.6):

$$\text{EM} = \int_T \varphi(T)\, dT = \int_{T, N_e} \psi(T, N_e)\, dT\, dN_e \quad \text{or} \quad \text{EM} = \int_V N_e^2\, dV. \tag{5.40}$$

The total EM of a multithermal plasma can be determined only after the DEM or spatial density distribution of the plasma has been estimated by some independent means. If the plasma is isothermal, the EM can be directly measured from line or continuum fluxes.

5.5.1 The EM loci method

The measured flux of a spectral line emitted by an isothermal plasma can be derived from Equations (4.94) and (4.95) as

$$F = \frac{1}{4\pi d^2} G(T_c)\, \text{EM}, \tag{5.41}$$

where d is the distance between the observer and the source, and $G(T_c)$ is the contribution function of the line, calculated at the plasma temperature T_c. Equation (5.41) allows us to define the function

$$\mathrm{EM}(T) = 4\pi d^2 \frac{F}{G(T)}. \tag{5.42}$$

The function $\mathrm{EM}(T)$, also called *EM loci*, strongly depends on the electron temperature because of the contribution function (e.g. Figure 4.4), but when $T = T_c$, $\mathrm{EM}(T) = \mathrm{EM}$. The emission measure diagnostic technique consists of displaying all the $\mathrm{EM}(T)$ curves calculated from the fluxes of a set of observed lines from many different ions in the same plot as a function of temperature: all the curves will meet in the same point (T_c, EM), unless their contribution functions have atomic physics inaccuracies, or the lines are blended with lines from other ions. The single crossing point determines simultaneously the plasma temperature and emission measure, and its presence implies an isothermal plasma. If the plasma is multithermal, no common crossing point can be identified. An example of this diagnostic technique is given in Figure 5.13 (left), where the application to the spectra of a quiet Sun region is shown. Other examples and further details can be found in Jordan *et al.* (1987), Landi *et al.* (2002), and references therein.

This method can only be applied when the fluxes of lines emitted by two or more ions are available, since the $\mathrm{EM}(T)$ curves of lines of the same ion are coincident within experimental uncertainties. The $\mathrm{EM}(T)$ curves are generally calculated assuming ionization equilibrium, so that results obtained using this method are affected by departures from equilibrium, as well as by uncertainties in the ion fractions themselves. In order to minimize the effects of uncertainties in the element abundances, it is best to use a set of lines from different ions of the same element: in this case, the measured EM will be insensitive to inaccuracies in the relative element abundances except for a constant shift. Otherwise, if lines from different elements are used, inaccurate adopted abundances may lead to wrong values of the measured T and EM, or even prevent the definition of the common crossing point of the $\mathrm{EM}(T)$ curves, leading to a false conclusion that the plasma is multithermal. When using a large number of

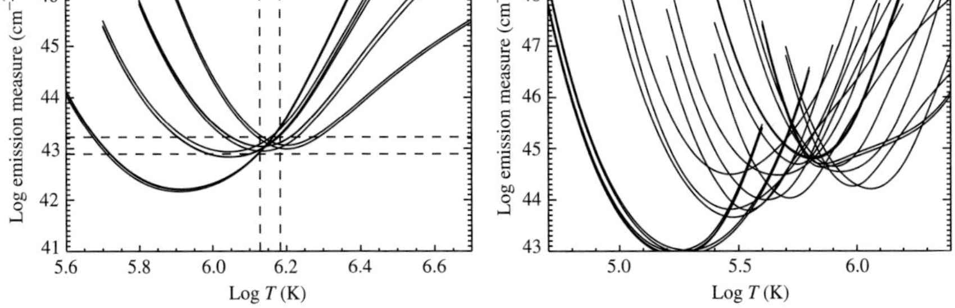

Fig. 5.13. Emission measure analysis using fluxes of Si VIII to Si XII lines. (*Left*) Isothermal plasma from the quiet solar corona: the common crossing point is indicated by the dashed lines at $\log T = 6.16 \pm 0.03$ and $\log \mathrm{EM} = 43.05 \pm 0.15$. (*Right*) multithermal plasma: no single crossing point can be defined.

lines, with several lines emitted by each element, vertical shifts of the crossing points given by lines of the same element can be used to determine abundances (Section 5.7).

5.5.2 *Line flux analysis*

If the flux of only one line is available, the EM can be determined directly from Equation (5.42), provided a measurement of the plasma temperature is available from some other method; otherwise only a lower limit to the EM can be determined. The use of a single line for EM measurements provides results affected by the same uncertainties as the EM analysis method discussed in the previous section, and also by the uncertainties in the measured T value. The latter source of error is particularly important as the contribution function has such a strong temperature dependence: even small errors in T can provide large variations in $G(T)$, unless the plasma temperature is close to the peak temperature of the $G(T)$ function.

5.5.3 *Continuum flux analysis*

Equation (5.42) can also be applied to the continuum flux to measure the plasma emission measure in the same way and with the same limitations as described in the previous section. However, at short wavelengths (<50 Å) the continuum emission strongly depends on temperature, so that T_e must be measured independently before using the continuum to determine the emission measure. By contrast, at longer wavelengths the continuum emission only slightly depends on temperature, so that inaccuracies in the assumed value of T_e have smaller effects on the measured EM than if individual line fluxes are used.

5.6 Differential emission measure diagnostics

5.6.1 *Fundamental limitations of DEM analysis*

The measurement of the emission measure differential in temperature and density $\psi(T, N_e)$ consists of inverting Equation (4.91),

$$F_i = \frac{1}{4\pi d^2} \int_T \int_{N_e} G(T, N_e)\, \psi(T, N_e)\, dT\, dN_e, \tag{5.43}$$

for a set of lines whose flux F_i has been observed, and whose contribution functions $G(T, N_e)$ are known. Usually, there are two sources of uncertainties involved in the inversion: those in the measured flux, and those in the atomic physics underlying the calculation of the $G(T, N_e)$ function. Moreover, the nature of the inversion process itself poses some additional problems. For a review of the difficulties associated with the inversion of Equation (5.43), see Jefferies *et al.* (1972), Craig & Brown (1976), Brown *et al.* (1991), Judge *et al.* (1997), and references therein.

If uncertainties in the contribution function and observed fluxes are negligible, the nature of the $G(T, N_e)$ function already poses a fundamental limitation to the determination of $\psi(T, N_e)$: from Figure 4.4 it is easy to see that the contribution functions $G(T, N_e)$ have a considerable width ΔT_G in temperature, and are only slightly dependent on the electron density. The temperature width implies that in the inversion of Equation (5.43) it is impossible to determine variations in temperature of $\psi(T, N_e)$ on scales smaller than ΔT_G. The slight dependence of $G(T, N_e)$ on density causes the derived $\psi(T, N_e)$ to have very large uncertainties in density. This is because changes of $\psi(T, N_e)$ in the temperature direction faster than ΔT_G, or changes in the density direction only cause small changes in the observed flux obtained from Equation (5.43). This means that (i) the $\psi(T, N_e)$ function resulting from the inversion of Equation (5.43) has large uncertainties, especially in density; and (ii) many different $\psi(T, N_e)$ functions can equally reproduce the observed fluxes F_i, so that the solution is non-unique. The presence of uncertainties in the contribution functions or in the observed fluxes F_i only magnifies these problems.

Also, the number of lines available to invert Equation (5.43) is an important factor. Because of its limited temperature width, the contribution function can only sample the $\psi(T, N_e)$ function along a temperature range $\Delta T \approx \Delta T_G$, so that many lines are necessary to cover the full span of temperature where $\psi(T, N_e)$ is defined. A small set of lines that do not cover the full temperature domain of the $\psi(T, N_e)$ function undersamples $\psi(T, N_e)$ in temperature, thus leading to a more imprecise determination of the solution, since no information is provided at certain temperature ranges. On the other hand, the availability of too many lines causes oversampling of the temperature domain of $\psi(T, N_e)$, and may lead to conflicting results from lines whose contribution functions overlap in temperature and which are affected by atomic physics or blending problems.

Because of the large uncertainties in the determination of $\psi(T, N_e)$ in the density direction, most diagnostic techniques have been developed to determine the DEM $\varphi(T)$, defined by Equation (4.93) as

$$\varphi(T) = \frac{N_e^2}{|\nabla T|} \quad \Longrightarrow \quad F = \frac{1}{4\pi d^2} \int_V G(T, N_e)\, \varphi(T)\, dT. \tag{5.44}$$

We now briefly review some of the methods for determining $\varphi(T)$. In all cases, the plasma electron density N_e is assumed to be known, or the $G(T, N_e)$ curves are assumed to be independent of N_e.

5.6.2 Iterative techniques

A crude approximation to $\varphi(T)$ can be obtained by setting the contribution function to a constant fraction of its peak value at $T = T_M$ inside a certain temperature interval ΔT around T_M, and zero outside:

$$G(T, N_e) = K'\, G(T_M, N_e) \longrightarrow F = \frac{1}{4\pi d^2} K'\, G(T_M, N_e) \int_{\Delta T} \varphi(T)\, dT, \tag{5.45}$$

so that the integral of $\varphi(T)$ over ΔT can be readily measured from the observed F value. Repeating this measurement for many lines with different T_M allows us to sample and measure the EM as a function of T. However, this technique, first introduced by Pottasch (1964), only provides rough estimates of the integral of $\varphi(T)$ around T_M and not $\varphi(T)$ itself.

A better method of determining $\varphi(T)$ without formally inverting Equation (5.44) consists of an iterative technique that starts from a known, arbitrary initial $\varphi_0(T)$. This method was introduced by Withbroe (1975), by choosing an arbitrary function $\varphi_0(T)$ to calculate the predicted flux $F_{0,l}$ of a line l as

$$F_{0,l} = \frac{1}{4\pi d^2} \int_T G_l(T, N_e) \varphi_0(T)\, dT. \tag{5.46}$$

The ratio of $F_{0,l}$ and the measured flux F_l gives an estimate of the correction to $\varphi_0(T)$, and can be used to improve the latter by defining a new function $\varphi_1(T)$ as

$$\varphi_1(T) = \varphi_0(T)\frac{\sum_l (F_l/F_{0,l})W_l(T)}{\sum_l W_l(T)}, \tag{5.47}$$

where

$$W_l(T) = G_l(T, N_e)\varphi_0(T)\frac{\int_T G_l(T, N_e)\varphi_0(T)\, dT}{\int_T [G_l(T, N_e)\varphi_0(T)]^2\, dT} \tag{5.48}$$

is a weighting function that ensures that the correction $F_l/F_{0,l}$ of each line to $\varphi_0(T)$ is applied most strongly to the temperatures where the line is formed. Once $\varphi_1(T)$ is calculated, it can be used as an initial DEM function in Equation (5.46) to calculate $F_{1,l}$ and $F_l/F_{1,l}$ and repeat the procedure, until at iteration i all the ratios $F_l/F_{i,l}$ are unity to within uncertainties.

This technique requires that the temperature range over which the DEM is defined be sampled by many lines. However, lines with overlapping $G_l(T, N_e)$ may provide conflicting corrections at similar temperatures, so that sharp artificial variations to the next trial function are introduced: in this case, such corrections are smoothed over T. An advantage of this procedure is that Equations (5.47) and (5.48) ensure that the correction provided by each line is not associated with any arbitrary temperature but contributes to all temperatures where the line is formed.

A slightly different version of this iterative technique was developed by Landi and Landini (1997), who provided a self-consistent definition of the temperature at which the corrections to an initial arbitrary DEM are applied. Landi and Landini (1997) defined the true DEM of the plasma $\varphi(\log T)$ as

$$\varphi(\log T) = \omega(\log T)\, \varphi_0(\log T), \tag{5.49}$$

where $\varphi_0(\log T)$ is the initial approximated DEM, and $\omega(\log T)$ is a smoothly varying correction function. Landi and Landini (1997) expanded the function $\omega(\log T)$ in a Taylor series around a temperature T_l as

$$\omega(\log T) = \omega(\log T_l) + \left(\frac{\partial \omega}{\partial \log T}\right)_{T_l} (\log T - \log T_l), \tag{5.50}$$

so that the flux of line l can be written as the sum of two contributions:

$$\begin{aligned} F_l =& \frac{1}{4\pi d^2}\, \omega(\log T_l) \int G_l(\log T, N_e)\, \varphi_0(\log T)\, dT \\ &+ \frac{1}{4\pi d^2}\left(\frac{\partial \omega}{\partial \log T}\right)_{T_l} \int (\log T - \log T_l)\, G_l(\log T, N_e)\, \varphi_0(\log T)\, dT. \end{aligned} \tag{5.51}$$

Landi and Landini (1997) defined $\log T_l$ as

$$\log T_l = \frac{\int G_l(\log T, N_e)\, \varphi_0(\log T)\, \log T\, dT}{\int G_l(\log T, N_e)\, \varphi_0(\log T)\, dT}, \tag{5.52}$$

so that Equation (5.51) simplifies to

$$F_l = \frac{1}{4\pi d^2}\, \omega(\log T_l) \int G_l(\log T, N_e)\, \varphi_0(\log T)\, dT. \tag{5.53}$$

Thus, each line defines its own value of $\log T_l$ and allows the measurement of the correction $\omega(\log T_l)$ to the initial DEM, $\varphi_0(\log T)$, at $T = T_l$ from Equation (5.53) and its measured flux F_l. The function $\omega(\log T)$ is then interpolated from the set of measured $\omega(\log T_l)$ and used in Equation (5.49) to repeat the procedure using the new $\varphi_1(\log T) = \varphi_0(\log T)\ \omega(\log T)$ as initial trial function in Equations (5.50) to (5.53). This procedure is repeated until the correction function $\omega(\log T)$ is unity within uncertainties.

This method has the advantage of providing a self-consistent definition of the temperature associated to each line's correction, which takes into account both the contribution function of the line and the shape of the DEM. However, it requires an interpolation among the $\omega(\log T_l)$ values, and when several lines provide $\omega(\log T_l)$ at similar values of $\log T_l$, a smoothing of the corrections from these lines is required to avoid instabilities and sharp oscillations.

5.6.3 *Maximum entropy techniques*

Equation (5.44) can be discretized in temperature T so that it can be analyzed numerically, by dividing the temperature range in a series of n intervals $T_1, T_2, \ldots, T_n$ sufficiently small so that the DEM can be considered constant in each of them. Equation (5.44) can then be written as

$$F_s^{\text{th}} = \sum_{i=1}^{N} K_{s,i}\varphi_i, \tag{5.54}$$

where F_s^{th} is the flux of line s predicted by the discretized DEM, φ_i is the constant value of the DEM in the temperature interval i, and

$$K_{s,i} = \frac{1}{4\pi d^2}\int_{T_i}^{T_{i+1}} G_s(T)\, dT \tag{5.55}$$

is the kernel. Equation (5.54) can be solved, and the φ_i values determined, by an iterative technique that minimizes the quantity

$$X^2 = \sum_{i=1}^{N} \frac{(F_s^{\text{th}} - F_s)^2}{\sigma_s^2}, \tag{5.56}$$

where σ_s are the uncertainties of the observed flux values F_s. Equation (5.56) can be put in the form of a linear system of equations (one equation for each observed line)

$$\frac{d(X^2)}{d\varphi_j} = 2\sum_{s=1}^{N} \frac{K_{s,j}(\sum_i K_{s,i}\varphi_i - F_s)}{\sigma_s^2}. \tag{5.57}$$

However, this linear system of equations provides non-unique and oscillatory solutions, which may also be negative and therefore unphysical. In order to avoid these effects, a further condition is imposed, that the entropy of the distributions φ be a maximum, so that the new quantity to be optimized in place of Equation (5.56) is

$$H = S + \alpha X^2 \qquad \text{and} \qquad S = \sum_i \varphi_i \ln \varphi_i, \tag{5.58}$$

where α is a Lagrangian multiplier. The quantity S is called the *entropy* of the function φ_i. The initial values of φ_i can be chosen by the user. The advantage of this method is to provide fast convergence of $\varphi(T)$, and the use of the maximum entropy constraint ensures positivity of the resulting DEM and allows the distribution to be found that best satisfies the observed data. However, this method has the disadvantage of being sensitive to errors in atomic physics quantities affecting the kernels K. McIntosh (2000) devised a modified maximum entropy method where line flux ratios rather than individual absolute fluxes were considered:

$$R_{ij} = \frac{F_i}{F_j} = \frac{\int_T G_i(T)\,\varphi(T)\,dT}{\int_T G_j(T)\,\varphi(T)\,dT} \tag{5.59}$$

and a revised X^2 was introduced involving both the observed and theoretical uncertainties $\sigma_{\rm obs}$ and $\sigma_{\rm th}$ of each ratio:

$$X^2 = \sum_i \frac{(R^i_{\rm obs} - R^i_{\rm th})^2}{\sigma^2_{\rm obs} + \sigma^2_{\rm th}}, \tag{5.60}$$

where $R^i_{\rm obs}$ and $R^i_{\rm th}$ are the observed and estimated values of the ith ratio. The solution $\varphi(T)$ is found using a maximum entropy method:

$$H = \lambda X^2 + \phi(\varphi) \qquad \text{with} \qquad \phi(\varphi) = \int_T \frac{\varphi}{W} \log \frac{\varphi}{W}\,dT \tag{5.61}$$

with W a user-defined function. The resulting DEM $\varphi(T)$ is scaled to absolute values by calculating the absolute fluxes F_s. McIntosh (2000) showed that this approach is much more robust against uncertainties in the atomic physics underlying the calculation of the contribution functions, so effectively removing one of the sources of uncertainties in the DEM determination problem.

Thompson (1990) also showed that regularization techniques such as the maximum entropy method may cause the solution to be oversmoothed where the DEM has large values, and undersmoothed where it has low values. This problem is more likely to happen when the DEM curve values span many decades, as DEM curves measured in the solar atmosphere usually do. Thompson (1990) suggested an *adaptive smoothing* method to avoid this problem, consisting of rescaling the DEM and the kernels so that the dynamic range covered by the solution of the inversion technique is within one order of magnitude. In this way, the maximum entropy techniques described above can avoid over- and undersmoothing. Following Thompson (1990), the DEM is written as $\varphi(T) = \varphi_0(T)\,\varphi'(T)$, with the arbitrary DEM $\varphi_0(T)$ satisfying the equation

$$F^0_s = \int_T G(T)\,\varphi_0(T)\,dT \tag{5.62}$$

for all lines, where F^0_s is a crude estimate of the emitted flux of line s, whose difference from the observed value is smaller than one order of magnitude. Equation (5.44) can then be written as

$$F'_s = \int_T \left(\frac{G_s(T)\,\varphi_0(T)}{F^0_s} \right) \varphi'(T)\,dT = \int_T G'(T)_s\,\varphi'(T)\,dT \tag{5.63}$$

where $F'_s = F_s / F^0_s$ and the range of values of $\varphi'(T)$ is much reduced. The maximum entropy or another inversion method can be applied to Equation (5.63) to obtain a solution $\varphi'(T)$ (and hence the final $\varphi(T)$) more accurately than in the case where the DEM was determined directly.

A proper choice of $\varphi_0(T)$ is important for the success of this technique. Since it is not necessary for $\varphi_0(T)$ to be accurately determined, it may be selected by approximating $G(T)$ with an average $\overline{G}$ over the temperature range where the value of the latter is greater than half of the peak, and calculating

$$\varphi_0^s(T) = 4\pi d^2 \frac{F_s}{\overline{G}} \tag{5.64}$$

from all the lines in the dataset, and then interpolating the $\varphi_0^s(T)$ in the temperature range where the DEM is sought.

5.6.4 Monte Carlo techniques

Kashyap and Drake (2000) developed a technique that utilized a statistical approach based on a Bayesian statistical formalism, which allows the determination of the most probable DEM curve that reproduces the observed line fluxes. This technique can be applied to problems with a large set of unknown parameters, even in the case where abundances are unknown or the $\psi(T, N_e)$ is sought. Also, this method relaxes the smoothness assumption commonly adopted by inversion or iterative techniques, by confining smoothing only to local scales whose lengths are tied to the widths of the contribution functions. The heart of this technique relies on the application of the Bayes theorem to line fluxes, stating that the probability $P(X, F)$ of obtaining a set of observed line fluxes $F = (F_1, F_2, ..., F_n)$ from a DEM characterized by a set of parameters $X = (X_1, X_2, ..., X_m)$ is given by

$$P(X, F) = P(X)\left[\Pi_{i=1,n} P(X, F_i)/P(F)\right], \tag{5.65}$$

where $P(F)$ is a normalization factor, $P(X)$ is an a-priori probability of the set of parameters X, and $P(X, F_i)$ is the probability of obtaining the observed flux F_i with the set of parameters X and has been defined by Kashyap and Drake (2000) as

$$P(X, F_i) = \exp\left[-((F_i - F_i^{\text{th}})/\sqrt{2}\sigma_i)^2\right], \tag{5.66}$$

where F_i^{th} is the flux of line i predicted from Equation (5.43) using the DEM described by the set of parameters X, and σ_i is the uncertainty of the observed flux F_i.

To determine the best set of parameters X that provide the maximum probability $P(X, F)$, Kashyap and Drake (2000) adopted a Markov-chain Monte Carlo approach. In this approach, the set of parameters X that describes an initial trial DEM is varied step by step; in each step only one of the parameters in the set X is varied and the others are left unchanged, and the change introduced to the varied parameter only depends on the set of parameters X of the previous step. The new set of parameters X' has a different probability $P(X', F)$ from the previous one, and is accepted or rejected according to the Metropolis algorithm (Metropolis *et al.* (1953)), based on the change of probability P. This algorithm consists of generating a random number u, such that $0 \le u < 1$, and a function $A(X, X')$ defined as

$$A(X, X') = \min\left[1, \frac{P(X', F)}{P(X, F)}\right]. \tag{5.67}$$

The new set of parameters is accepted if $u < A(X, X')$, and rejected otherwise. In this way, not only is a new set of parameters X' with greater probability than the previous one always accepted (since $A(X, X') = 1$), but also a new set with a smaller probability has a finite chance of being selected. This latter property helps finding the best distribution of parameters X by moving the solution out of local maxima.

The system is implemented by first choosing the ranges of temperature over which the DEM can be evaluated, as those where $G(T)$ is non-zero. Then an initial set of parameters X that describes the DEM is varied and the variations are selected or rejected according to the Metropolis algorithm. At the end of the run, the distribution of the values of the DEM obtained in all steps is used to determine the confidence interval at each temperature, thus providing an estimate of the uncertainty of the DEM at all temperatures.

The DEM is smoothed over intervals L calculated from the log T width of each contribution function by correlating the latter with a so-called Mexican hat function with parameter δ:

$$M(\log T, \delta) = \frac{1}{\delta}\left[1 - \left(\frac{\log T}{\delta}\right)^2\right] \exp\left[-\left(\log T/\sqrt{2}\delta\right)^2\right]. \tag{5.68}$$

The value of δ for which the response of the contribution function to $M(\log T, \delta)$ is maximum determines the local length $L = 2\delta$ over which the DEM is smoothed.

The Markov-chain Monte Carlo method offers several advantages. First, it is possible to show (Neal (1993)) that as the number of steps increases such a method converges to a fixed final set of parameters X_f regardless of the initial set X_i chosen. Also, the final set of parameters X_f does not depend on the initial one. Another advantage is that this method retains as much temperature structure of the DEM function as possible, by smoothing the DEM over the smallest possible interval. Also, this method provides confidence limits on the most probable DEM, even if a rigorous treatment of uncertainties is not carried out. This method is well suited to cases where the number of parameters is large, although in some cases convergence is slow.

5.7 Element abundance diagnostics

The abundance of an element can only be measured relative to the abundance of another element, because both line and continuum radiation are directly proportional to the abundance of the element or elements that are emitting them. Traditionally, the *absolute abundance* of an element is defined relative to the abundance of hydrogen, since the latter is by far the most abundant element in the solar atmosphere. Abundances are often expressed as logarithmic values on a scale in which $H = 12.0$.

5.7.1 *Relative abundances*

Line flux ratios

The most direct method of measuring the relative abundance of two elements X and Y is to compare the observed ratio of the fluxes F_X and F_Y of lines emitted by each of the two elements with a theoretical ratio as a function of the relative abundance. Following Equations (4.86) and (4.87), the theoretical ratio can be written as

$$\frac{F_X}{F_Y} = \frac{A(X)}{A(Y)} b(T, N_\mathrm{e}), \qquad b(T, N_\mathrm{e}) = \frac{\int_V G'_X(T)\, N_\mathrm{e}^2 dV}{\int_V G'_Y(T)\, N_\mathrm{e}^2\, dV}, \tag{5.69}$$

where $A(X) = N(X)/N(\mathrm{H})$ and $A(Y) = N(Y)/N(\mathrm{H})$ are the absolute abundances of the elements X and Y, i.e. relative to hydrogen, and $G'_X(T)$ and $G'_Y(T)$ are the contribution functions of the lines calculated assuming unit abundance.

There are several disadvantages in using this technique. First, the two ions must be formed at very similar temperatures so that it is still reasonable to assume that both lines are emitted by the same plasma. Second, the plasma electron temperature and density must be determined with some precision before measuring the relative abundance; temperature is especially important since ion fractions are very sensitive to T_e. Third, the theoretical $b(T, N_e)$ function is subject to the uncertainties in the ion fractions, and in the assumption that the plasma is in ionization equilibrium.

EM and DEM analysis

When many spectral lines are available, it is possible to apply the diagnostic techniques described in Sections 5.5 and 5.6 to determine the EM or the DEM of the emitting plasma using only the lines emitted by ions of the same element. The resulting EM or DEM is directly proportional to the absolute abundance of the element used. The ratio of the EM or DEM curves from two different elements provides a direct measurement of their relative abundance. An example is given in Figure 5.14. This method has the advantage of being less dependent on the ion fractions for each element, since lines from many ions are used at the same time, and can be applied to isothermal and multithermal plasmas alike by determining the EM or DEM respectively. However, when only a few lines of two elements can be observed, this technique leads to very large uncertainties.

5.7.2 *Absolute abundances*

The absolute abundance of an element X can be measured by comparing the flux of one or more lines of this element with those of an H I line or with free–free continuum

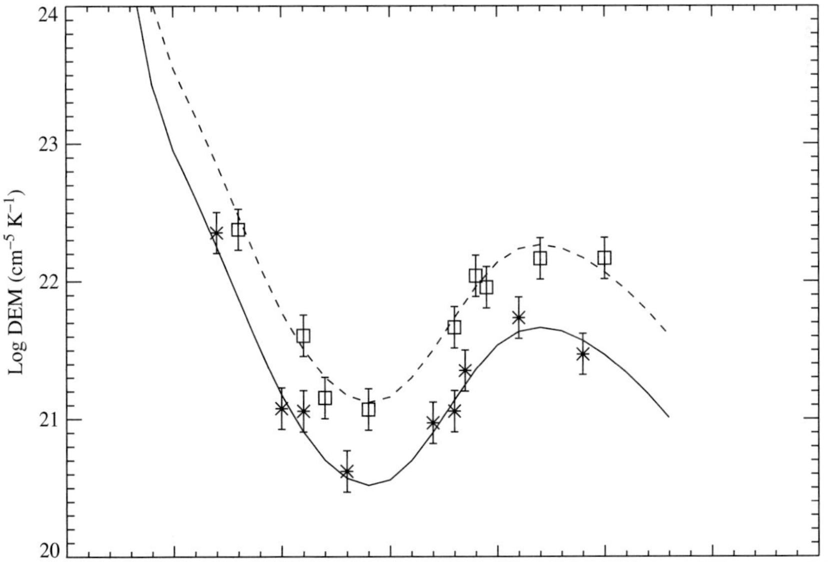

Fig. 5.14. Example of relative abundance diagnostics using the DEM analysis: stars and squares correspond to lines due to two different elements.

radiation; the latter can be used because hydrogen is by far the most abundant element in the solar atmosphere so that free–free radiation is mostly due to the interaction of free electrons with protons.

Line flux ratios

Equation (5.69) can in principle be applied to a flux ratio involving a H I line to measure absolute abundances, but H I lines in the solar atmosphere are emitted mostly by chromospheric plasmas and the most intense members of the series are optically thick. Also, since chromospheric plasmas are not isothermal, only lines formed at the same temperature as H I can be used to measure absolute element abundances, but these lines too are optically thick. These difficulties can be overcome in off-disk observations of intense active regions or flares, when the temperature is a few million degrees, and the emission measure is significantly larger than in the quiet Sun. In these cases, the fraction of neutral H is very small but non-zero, and the EM is still high enough to provide observable line fluxes for the brightest lines in the H I Lyman series. These lines are emitted locally by the same plasma that emits lines from highly ionized species, whose flux ratios with H I lines can be used to determine the absolute abundances of these elements. For example, Feldman *et al.* (2005a) used such a technique for the measurement of the absolute abundance of He in post-flare plasmas, using He II/H I flux ratios of lines observed at around 1000 Å. Since the H I lines are often fairly weak, care must be taken to remove the contributions of other nearby lines and also, if present, instrument-scattered light.

Line flux ratios of resonantly scattered lines in the solar wind

The flux ratio of lines formed by resonant scattering is also proportional to the element abundances of the emitting elements, as shown by Equation (4.156). Since the H I Ly-α line at 1215.6 Å is one of the most intense lines emitted by the solar wind plasma, it can be used to measure the absolute abundance of other elements present in solar wind spectra. If the element and ion abundances are constant along the line of sight, the ratio between the observed flux F_X of a line of element X to the Ly-α flux is given by

$$\frac{F_X}{F(\text{Ly-}\alpha)} = \frac{N(X)}{N(\text{H})} R, \qquad (5.70)$$

where R can be calculated from Equation (4.156) assuming unit abundances. In order to use this technique, the wind velocity v_r, the ion abundance, and the velocity distribution f of both H and the element X must be known in the radial direction to calculate R.

Free–free continuum

Ratios of the flux of a spectral line and that of the nearby free–free continuum are widely used for measuring absolute element abundances. This technique has been applied extensively to X-ray spectra of solar flares, where the free–free emission is very apparent. However, the X-ray free–free emission is contaminated by the free–bound continuum mostly from He-like and H-like ions of O to Ni, so that it is necessary to subtract this emission from the total continuum. This can be done if the temperature of the emitting plasma and the absolute abundances of the elements emitting most of the free–bound emission are known. Failure to account for the free–bound emission can lead to large underestimations of the absolute abundances of the element emitting the line being used. The relative importance

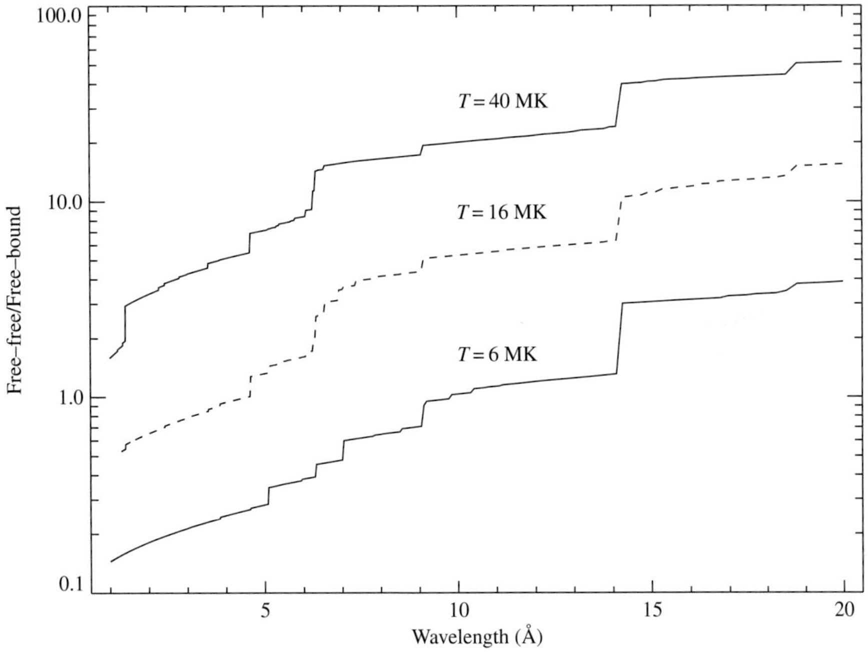

Fig. 5.15. Ratio of free–free and free–bound radiation at 6 MK, 16 MK and 40 MK, corresponding to hot active regions, average-size flare and a bright stellar flare. Free–bound emission has been calculated using the coronal abundances from Feldman (1992) and the CHIANTI database.

of free–free and free–bound radiation depends both on the electron temperature and the wavelength: see Figure 5.15.

Free–free emission can also be observed in the off-disk solar corona during the impulsive and decay phases of flares, and even in non-flaring, hot active regions, although in the latter it is much less intense. If no chromospheric plasma is present along the line of sight, this emission, when observed in the UV, can be used to measure absolute element abundances from flux ratios with lines emitted by highly ionized ions so long as the temperature is known. The advantage of the method is that in the UV range the free–free emission depends on $T_e^{1/2}$, since the exponential term in Equation (4.114) is nearly unity for wavelengths $\lambda > 720$ Å when $T_e = 2 \times 10^6$ K (and for even shorter wavelengths when the temperature is higher). Thus, even if T_e is not well known, the free–free emission can still be predicted with accuracy. The temperature of the emitting plasma must still be known accurately to predict the emissivity of the line being used.

Broad-band resolution spectra

Absolute abundances can be measured using the continuum when the instrumental spectral resolution is only moderate. The line-to-continuum ratio technique can be applied to fluxes from two different portions of the spectrum, one dominated by the free–free continuum, and the other by lines of the same element. In Section 5.3.4, it was seen that *RHESSI* broad-band resolution spectra in the 1.5–2 Å (6–8 keV) range give much information about the

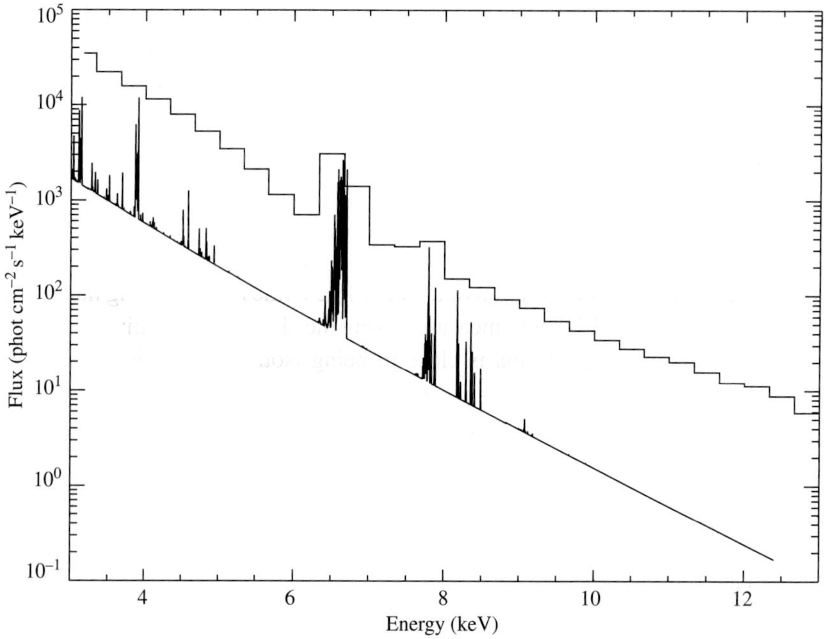

Fig. 5.16. *RHESSI* spectrum of the decay phase of a solar flare observed on 2006 April 26 (histogram) compared with a synthetic spectrum from an isothermal plasma with $T = 14$ MK calculated using the CHIANTI code (Dere *et al.* (1997)). The CHIANTI spectrum is displaced down by an order of magnitude for clarity. At this temperature, the Fe line feature at 6.7 keV is made up mostly of Fe XXV lines and Fe XXIV satellites, and the weaker Fe/Ni line feature at 8.0 keV is mostly made up of Fe XXV $1s^2 - 1snp$ $(n > 2)$ lines and nearby Fe XXIV satellites. Courtesy the *RHESSI* team.

temperature and emission measure of flares. Knowing the temperature from the slope of the *RHESSI* continuum, we can find the abundance of iron from the flux of the 1.85 Å (6.7 keV) line feature that is nearly always present in *RHESSI* flare spectra. This is of some importance for examining the FIP effect and its dependence on flare type (see Chapter 11).

The Fe 6.7 keV line feature is made up of resonance lines of He-like and H-like Fe (Fe XXV and Fe XXVI) and dielectronic satellite lines nearby. This is illustrated by Figure 5.16 which shows an observed flare *(RHESSI)* spectrum compared with a synthesized spectrum from the CHIANTI code. Adding together the contribution functions of all the lines gives the contribution function of the whole 6.7 keV feature, as is illustrated in Figure 5.9, based on atomic calculations on individual lines by Phillips (2004). All the lines are due to Fe, so the ratio of the 6.7 keV line feature to the continuum leads to the abundance of Fe if the flare temperature is known independently. This may be obtained most accurately from the slope of the continuum, observed by *RHESSI*, or less certainly from the ratio of the Fe line feature and the Fe/Ni line feature seen at 8.0 keV. A convenient means of doing this for a number of flares is to use the Fe line feature's equivalent width (Section 5.3.4) which reaches a maximum of about 4 keV for coronal abundances of Fe. Comparison of the observed and calculated Fe line equivalent widths gives the iron/hydrogen abundance ratio for flares.

5.8 Ion abundances

5.8.1 Flux ratios

Ion abundances can sometimes be estimated from solar spectra to give a useful check on the accuracy of ionization equilibrium calculations. This has been done using solar flare X-ray spectra by measuring the ratios of collisionally excited dielectronic satellite lines to resonance lines in He-like Ca and Fe ion spectra, with the electron temperature independently and simultaneously measured from the ratios of dielectronically formed dielectronic satellites to resonance lines (Antonucci *et al.* (1984)). Alternatively, broad-band X-ray flare spectra may be used to measure the variation of the Fe 6.7 keV line feature during flares with electron temperature, again independently measured using the slope of the continuum. Either method is accurate if the emitting plasma is close to being isothermal but is of limited use if the emitting plasma is multithermal.

5.9 Plasma dynamics diagnostics

The effects of plasma motions on spectral lines can be detected from the Doppler effect by shifts of the line's centroid wavelength, or by distortions or broadening of the line profile. Direct observation of motions of structures in the solar atmosphere can also be detected from spatial shifts of well-resolved structures in a sequence of images recorded consecutively.

5.9.1 Doppler shifts

The measurement of the rest wavelength of the centroid of a line profile is the most direct way of measuring the velocity $v_{\rm los}$ along the line of sight of the emitting plasma. According to the Doppler formula for non-relativistic velocities,

$$\frac{\Delta\lambda}{\lambda_0} = -\frac{v_{\rm los}}{c}, \tag{5.71}$$

where λ_0 is the rest wavelength of the line and $\Delta\lambda$ is the displacement of the line centroid: positive values of $\Delta\lambda$ imply that the source is receding from the observer (*red shifts*), negative values that the source is approaching the observer (*blue shifts*). Equation (5.71) provides the average velocity along the direction of the observer (line-of-sight velocity) but does not provide information on motions perpendicular to it (plane-of-sky velocity), so that the total velocity or its direction cannot be measured with Equation (5.71) alone. Sometimes, in solar spectra, many different velocity components may be present along the same line of sight, giving rise to distorted profiles of the spectral lines (Figure 5.17).

The velocity vector can only be determined from both the measured Doppler line-of-sight velocities and plane-of-sky velocities from high-cadence imaging. Imaging with a cadence of a few seconds can be achieved more easily with instruments having narrow-band filters, but the presence of more than one line emitted by ions formed at different temperatures within the filter's bandpass can make the detection of real motions in the plane of the sky ambiguous. An imaging spectrometer can remove this ambiguity since it is able to measure individual line fluxes which allow true motions in the plane of the sky to be distinguished from other effects. Although current imaging spectrometers have a slower imaging cadence than imagers with narrow-band filters, their ability to monitor large areas is sufficiently fast if a wide slit and a short exposure time are used.

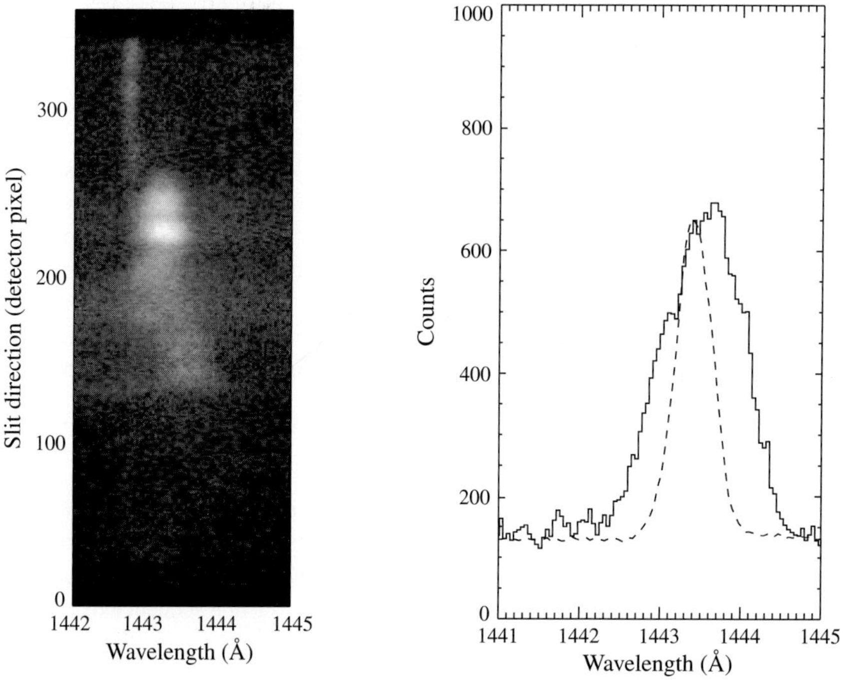

Fig. 5.17. (*Left*) SUMER slit image in the 1442–1445 Å wavelength range during the onset of a moderate flare: the bright Fe XX 721.5 Å line in second order at 1443 Å is broadened and distorted. (*Right*) Spectrum of the Fe XX 721.5 Å summed over pixels 120–200 along the slit direction (full line); the dashed line is a Gaussian spectrum at the rest wavelength normalized to the peak of the Fe XX line. Courtesy *SOHO* SUMER team.

5.9.2 *Solar wind velocities*

Doppler dimming

The flux of a resonantly scattered line is proportional to the flux of the incident radiation at the frequency of the line ν_0, according to Equation (4.156). However, radial motions of the scattering plasmas, such as those due to an outflowing solar wind, can alter substantially the amount of radiation being scattered: the incident radiation is red-shifted in the ion's reference frame so that only the wings of its profile can be absorbed by the ion, leading to a decrease of the line flux emitted by the ion. This effect can be described by defining a Doppler dimming term, by analogy with Equation (4.159):

$$D(v_r) = \int_0^\infty \varphi(\nu_1 - \delta\nu)\, \psi(\nu_1, \nu_0)\, d\nu_1, \tag{5.72}$$

where $\delta\nu = \nu_1 v_r/c$, v_r is the outflow velocity, $\varphi(\nu_1)$ is the line profile, and $\psi(\nu_1, \nu_0)$ the ion absorption profile, defined by Equation (4.158).

The quantity $D(v_r)$ depends on the outflow velocity v_r. Figure 5.18 illustrates the value of $D(v_r)$ for the Ly-α transitions in H I and He II, and the O VI line at 1031.9 Å: $D(v_r)$ was calculated by Withbroe *et al.* (1982) assuming an isothermal corona with $T_e = 1.5 \times 10^6$ K, a Maxwell–Boltzmann distribution of velocities, and a Gaussian $\varphi(\nu_1)$ profile. Because of

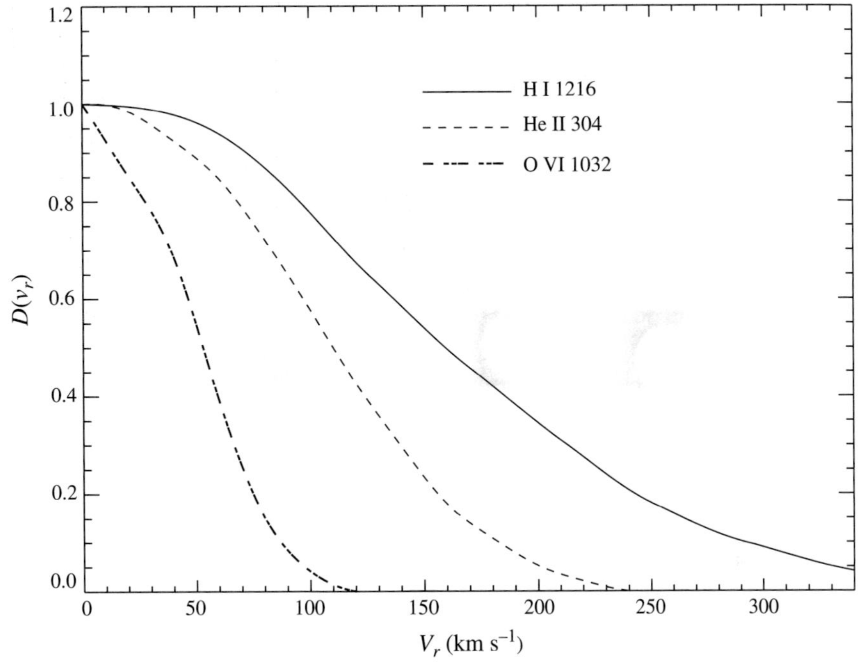

Fig. 5.18. Doppler dimming term for H I 1215.6 Å (H I Ly-α), He II 303.7 Å (He II Ly-α), and O VI 1031.9 Å calculated as a function of the outflow velocity v_r. From Withbroe *et al.* (1982).

their broader profiles, lighter ions have a slower decrease of $D(v_r)$ so that they can sample the red-shifted line profile at higher velocities, thus extending the diagnostic capabilities of $D(v_r)$ at higher radial speeds. For example, H I can be used from 60 to 300 km s^{-1}, while O VI lines cease to be formed through resonant scattering at velocities larger than 100 km s^{-1}.

There are several methods of measuring the Doppler dimming term $D(v_r)$ that make use of radiation below 2000 Å. One method consists of measuring the ratio of the resonantly scattered ($F_{\rm res}$) and the electron scattered ($F_{\rm Thom}$) flux components of a line. From Equations (4.156) and (5.35), this ratio is proportional to

$$\frac{F_{\rm res}}{F_{\rm Thom}} \propto \frac{N(X^{+m})}{N(X)} \frac{N(X)}{N({\rm H})} D(v_r), \tag{5.73}$$

where the relative abundance $N(X)/N({\rm H})$ is unity if an H I line is used, and $N(X^{+m})/N(X)$ is the fractional ion abundance. From the comparison of measured and predicted values $D(v_r)$ can be measured and an estimate of v_r obtained. The main disadvantage of this technique is that the flux of spectral lines formed by electron scattering is small; in fact, it has been observed for H I Ly-α and He II Ly-α (Breeveld *et al.* (1988)). Also, Equation (5.73) is oversimplified and, in reality, assumptions on the ion velocity distribution and abundance along the line of sight must be made, and Equations (4.156) and (5.35) must be fully considered.

Another method of determining $D(v_r)$ consists of measuring the flux ratio of the two lines in the $2s\,^2S - 2p\,^2P$ doublet in Li-like ions, and between those in the $3s\,^2S - 3p\,^2P$ doublet in Na-like ions. Both lines are excited by collisions and resonant scattering. The ratio of these two components is

$$\frac{F_{\mathrm{res}}}{F_{\mathrm{coll}}} \propto D(v_r)\frac{\int_{-\infty}^{+\infty} N_{\mathrm{e}}\, dx}{\int_{-\infty}^{+\infty} N_{\mathrm{e}}^2\, dx}, \tag{5.74}$$

where F_{coll} is the flux produced by electron–ion collisional excitation and F_{res} is the flux produced by resonant scattering. The fractions of each component can be measured from the observed ratio of the two lines in the doublet and the fact that the ratio of the collisional components of the doublet is 2 and the ratio between the resonantly excited components is 4, as described in Section 4.9.3.

Equation (5.74) is also oversimplified: to obtain the full expression, it is necessary to use Equations (4.86), (4.87), and (4.156), the measured flux and profile of the incident radiation, and make assumptions about the electron density, temperature, velocity distribution and abundance of the emitting ion along the line of sight.

Doppler-enhanced scattering

For each ion, the Doppler dimming term becomes zero when the outflow velocity is sufficiently large to shift the incident line profile far from the rest frequency. If a second, strong line is present at shorter wavelengths close to the line considered, the reverse of Doppler dimming – Doppler-enhanced scattering – can occur if the outflow velocities are large enough to shift the profile of this neighbouring line to the rest frequency of the first line. As the velocity increases, the Doppler dimming term starts to increase with velocity, reaches a new maximum and then decreases again. This property can be used in connection with the techniques described in the previous section to extend their range of application.

The calculation of the Doppler dimming term $D(v_r)$, taking into account the photoexcitation of nearby lines, depends on many factors such as the ion temperature and velocity distribution, as well as the measured profile and flux of the additional line. An example of this is given in Figure 5.19. More details on the Doppler-enhanced scattering can be found in Noci *et al.* (1987), who first developed this technique.

This technique has been applied to the O VI doublet ratio $\lambda 1031.9/\lambda 1037.6$, since the 1037.6 Å line is very close to two strong C II chromospheric lines at 1036.3 Å and 1037.0 Å, which can increase resonant scattering at velocities in the 150–400 km s^{-1} range.

5.9.3 Electron velocity distribution

5.9.3.1 *Electron velocity distribution in the solar wind*

The electron velocity distribution in the solar wind can be studied using the profile of the electron scattered radiation of a spectral line. According to Equation (5.35), the scattered line profile strongly depends on the electron velocity distribution, offering a possible means of measuring it. However, the electron-scattered light is distributed over tens of Å, so that it is difficult to distinguish it from the background, and so any measurement has large uncertainties. Moreover, electron scattered radiation depends on the geometry of the scattering process (see Figure 5.10), so that a model of the geometrical distribution of the scattering electrons, of the source of the incident radiation, and of the position of the observer are needed for measuring the electron velocity distribution using this method.

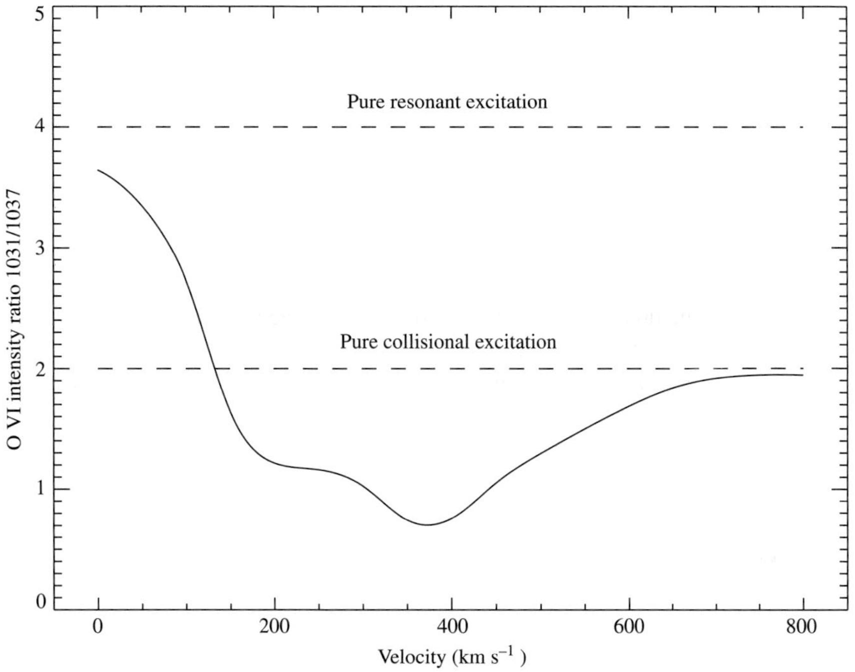

Fig. 5.19. Example of the flux ratio of the two O VI lines at 1031.9 Å and 1037.6 Å as a function of outflow velocity. From Cranmer *et al.* (1999). The contributions of the two C II lines at 1036.3 Å and 1037.0 Å to resonant scattering cause the O VI ratio to have values lower than 1.

Non-thermal tails of high-energy electrons

Normally for solar atmospheric plasmas, the assumption of Maxwell–Boltzmann distributions of particle speeds $f(v)$ (see Equation (4.9)) is fully justified since for particle densities $\gtrsim 10^8$ cm^{-3} departures from such distributions are rapidly smoothed out. Any departures from a Maxwell–Boltzmann distribution affect ionization, recombination, and excitation rate coefficients, which in turn affect ionization fractions and level populations and therefore line fluxes. In practice only large departures will produce measurable changes in line fluxes.

Lines emitted by levels with excitation energy $E_{\rm exc}$ much larger than the average thermal energy $E_{\rm th}$ can be used to measure deviations of $f(v)$ from an assumed distribution. In fact, only a small portion of thermal electrons can excite these levels, so the presence of additional non-thermal high-energy electrons with energy larger than $E_{\rm exc}$ provides substantial contributions to the populations of such levels. For levels of He-like ions, the additional non-thermal electrons have little effect on ion fractions, since these are near unity for wide temperature ranges. Therefore, the populations of these levels and the emission of the lines they generate can easily be calculated as a function of energy E and number N of the additional non-thermal electrons, and comparison with observations can yield the values of these parameters. The electron temperature and emission measure (or differential emission measure) of the emitting plasma must be known independently.

Dielectronic satellites with different excitation energies can enable departures from Maxwell–Boltzmann distributions to be examined (Gabriel & Phillips (1979)). The reason is that dielectronic satellites are excited by electrons with a single energy, whereas lines that are collisionally excited arise from the excitation by electrons in the Maxwell–Boltzmann distribution above a threshold energy. This diagnostic method has been investigated for flare plasmas particularly, for which electron temperatures can rise to ~ 20 MK or more. At such temperatures, the Fe XXV resonance line w at 1.850 Å and associated Fe XXIV dielectronically formed satellites are intense. Line w is formed by collisional excitation of Fe^{+24} ions with electrons having energies above the threshold energy 6.70 keV. The prominent dielectronically formed satellites j and k (at 1.8655 Å and 1.8631 Å) are due to transitions $1s^2\,2p\,^2P - 1s\,2p^2\,^2D$, i.e. with one of the $2p$ electrons being the non-participating or spectator electron. They are excited by electrons with energy 4.69 keV. Families of dielectronic satellites having a $3p$ spectator electron exist and are spectroscopically distinguishable on the long-wavelength side of line w. They include the blended dielectronic satellite feature $d13$ and $d15$, due to transitions $1s^2\,3p\,^2P - 1s\,2p\,3p\,^2D$ (Bely-Dubau *et al.* (1979)), which are excited by electrons with energy 5.82 keV.

Figure 5.20 illustrates the level diagram for $Fe^{+24} + e^-$ and the recombined ion Fe^{+23}. In this figure, the free electron exciting the Fe XXIV satellites j and $d13 + d15$ is assumed to have a Maxwell–Boltzmann distribution with $T_e = 20$ MK. Any non-thermal distribution

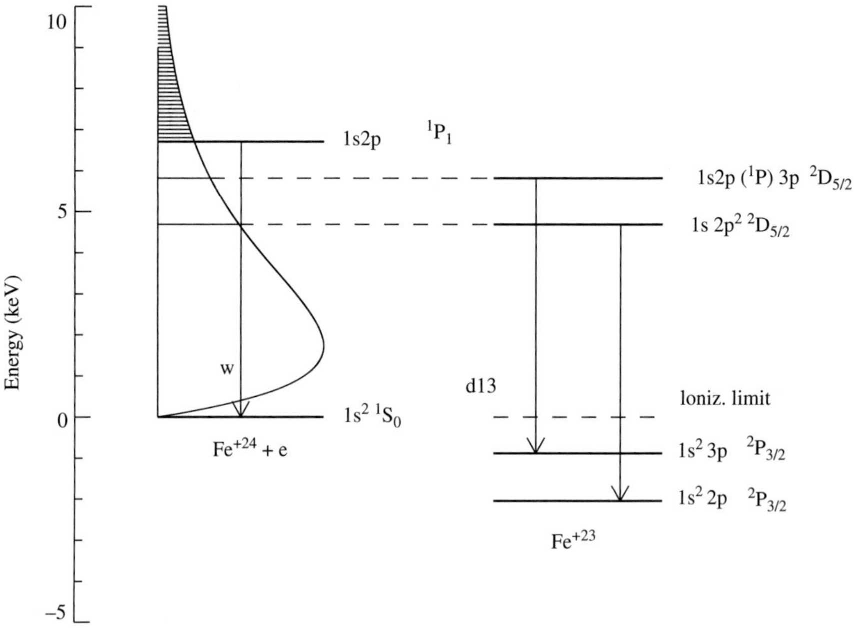

Fig. 5.20. Energy level diagram for Fe^{+24} (*left*) and $Fe^{+23} + e^-$ (*right*). Fe XXV line w is excited by electrons with energies above 6.70 keV, the Fe XXIV satellite j is excited by mono-energetic electrons with $E = 4.69$ keV, and the Fe XXIV satellite $d13$ is excited by mono-energetic electrons with $E = 5.82$ keV. The free electron is illustrated in the left part of this figure having a Maxwell–Boltzmann distribution with $T_e = 20$ MK.

added to this Maxwell–Boltzmann distribution will add to the excitation of line w but if (as is commonly assumed to be the case) there is a cut-off energy, e.g. of $\sim$ 10 keV, in the non-thermal electron distribution, there will be negligible additional excitation to the two satellite lines. There should therefore be a difference in estimated temperatures derived from the j/w ratio or the $(d13 + d15)/w$ line ratio.

5.10 Magnetic field diagnostics

The measurement of the strength and direction of the magnetic field B in the corona is of major importance for modelling virtually any phenomenon in the solar upper atmosphere, and yet no technique has been developed involving radiation in the UV and X-ray wavelength ranges. Here we list some of the techniques used or under development that may lead to possible advances in the measurement of coronal magnetic fields.

5.10.1 *Zeeman effect*

The main method for measuring the magnetic field strength B makes use of the Zeeman effect, consisting of the removal of the degeneracy of atomic energy levels with respect to M, the quantum number of the operator $\mathbf{J_z}$ projection of the operator $\mathbf{J}$ of the total angular momentum of the ion (Section 3.4.4), when an ion is in a magnetic field B. The splitting of these otherwise degenerate levels breaks down the transition that connects them into a multiplet of separate lines whose wavelengths differ from the original by amounts $\Delta\lambda$ (in Å) approximately given by

$$\Delta\lambda = 4.67 \times 10^{-13}\, B\, \lambda_0^2, \tag{5.75}$$

where B is the magnetic field strength (gauss) and λ_0 is the wavelength (in Å) of the transition when $B = 0$. For lines below 2000 Å and a magnetic field of 100 G, $\Delta\lambda = 0.19$ mÅ, far less than the thermal Doppler-broadened width of a spectral line. In order to be measurable at 2000 Å, $\Delta\lambda$ must be at least 0.1 Å, so that B should be at least 54 kG, much larger than what is thought to be the coronal field strength. Only at infrared wavelengths does the Zeeman effect provide reliable measurement of coronal magnetic fields of the order of 1000 G.

5.10.2 *Radio and EUV combined observations*

Solar EUV radiation has been used in conjunction with radio observations at 4.87 GHz and 8.45 GHz by Brosius *et al.* (2002) to measure the vector magnetic field over a $2' \times 2'$ area above an active region, finding field strengths of 1000–1500 G at different altitudes. The EUV line fluxes were used to determine the DEM of the emitting plasma for each pixel of the field of view, to be used mainly in the calculation of the free–free optical thickness in the radio frequency domain. Since this diagnostic technique uses radio emission, we will not describe it here in detail as it is outside the scope of this book. Details are given by Brosius *et al.* (2002) and references therein.

5.10.3 *Line flux ratios*

The presence of an external magnetic field around an ion has the effect of resolving the degeneracy of the magnetic sublevels of a particular level and of mixing them with sublevels of a neighbouring level provided they have the same magnetic quantum number and parity. This means that the wavefunction of one level can include, in the presence of an external magnetic field, components from the wavefunctions of neighbouring levels. In general, the magnitude of these components is very small, so that they will have negligible consequences for all levels. However, when one atomic level has either zero radiative decay rate or a very small one, magnetic field mixing can provide this level with a finite decay rate and therefore allow the emission of a forbidden line that otherwise could not be emitted.

This effect has been studied by Beiersdorfer *et al.* (2003), who showed that the level $2s^2\,2p^5\,3s\,^3P_0$ in Ne-like ions interacted, in the presence of an external magnetic field, with the level $2s^2\,2p^5\,3s\,^1P_1$. Decay from the $^3P_0$ level to the ground level $2p^6\,^1S_0$ is strictly forbidden, but the small contributions of the $^1P_1$ wavefunction to the $^3P_0$ wavefunction result in the $^3P_0 - {}^1S_0$ transition having a small, non-negligible, Einstein coefficient for spontaneous radiation, giving rise to a line that has been observed in laboratory plasmas in the isoelectronic sequence S VII, Ar IX, and Fe XVII. The flux of the $^1S_0 - {}^3P_0$ line depends on the magnetic field, while the nearby lines emitted by the other levels from the $2s^2\,2p^5\,3s$ configuration are insensitive to it. Therefore, the flux ratio of the $^1S_0 - {}^3P_0$ line and one of the nearby lines can be used to measure the magnetic field.

The minimum strength of the magnetic field that can be measured with these flux ratios depends on the atomic weight of the ion, being smaller for lighter ions, and on the electron densities of the emitting plasmas. At densities typical of the solar atmosphere, the range of magnetic field sensitivity is from a few thousand to a few tens of thousand gauss. These fields are much larger than coronal field strengths so it is unlikely that Ne-like line ratios can be used to measure solar magnetic fields; however, the technique can be applied to other types of ions so that future studies may lead to this method being applicable to the solar atmosphere.

5.10.4 *Hanle effect*

The application of the Hanle effect to measure the strength and direction of the coronal magnetic field has been proposed by Bommier and Sahal-Brechot (1982) and Fineschi *et al.* (1999). This method exploits the change in linear polarization of spectral line components formed by resonant scattering, induced by the presence of magnetic fields. It has never been tested observationally, but has been proposed for building advanced UV coronagraphs. Its sensitivity is best for magnetic fields with strengths in the 1–60 G range, when applied to H I Ly-α, Ly-β, Ly-γ, and O VI 1031.9 Å lines.

5.11 Spectral codes for solar corona research

The application of the diagnostic techniques described in this chapter to high- and low-resolution spectra, as well as the use of the measured physical properties to calculate synthetic spectra, need a vast amount of atomic data, which are necessary to calculate level populations, ion abundances, and line and continuum contribution functions whose comparison with observations provide the diagnostics measurements we need. Chapter 4 describes the physical quantities needed for the calculation of the line and continuum contribution functions:

energy levels, transition wavelengths, quantum numbers;
ionization potentials;
Einstein coefficients;
electron–ion and proton–ion collision excitation rate coefficients;
ionization and recombination rate coefficients;
free–free and free–bound Gaunt factors.

These quantities have been published in the literature for a large number of ions and for the continuum by many authors, and they have been collected in a few spectral codes to make them easily accessible to users.

The main spectral codes freely available for research in the solar upper atmosphere below 2000 Å are:

CHIANTI (Dere *et al.* (1997), Landi *et al.* (2006a))
APEC/APED (Smith *et al.* (2001))
MEKAL (Mewe *et al.* (1995))

Each of these codes can also be applied to astrophysical objects other than the Sun. Other spectral codes are available for more general astrophysical research. Details of each spectral code can be found in the papers quoted above.

6

Ultraviolet and X-ray emission lines

6.1 Introduction

In this chapter we describe the solar spectrum between 1 Å and 2000 Å, and briefly introduce some applications of the diagnostic techniques described in Chapter 5 to line and continuum emission emitted by the solar atmosphere.

The 1–2000 Å range is here divided into six intervals, roughly corresponding to the spectral ranges covered by different types of spectrometers (see Chapter 7); this division also takes into account the different nature of spectral lines and atomic systems belonging to each of the intervals.

The six spectral ranges we will discuss are:

(1) 1–7 Å: this range is observed by crystal spectrometers, and includes lines emitted by H-like and He-like ions of Si to Ni and their dielectronic satellites during flares or from bright active regions.
(2) 7–23 Å: this range is observed by crystal and grazing-incidence spectrometers, and includes lines emitted primarily by H-like and He-like ions of O to Al and lines from configurations with principal quantum number $n \geq 3$ in Fe XVII to Fe XXIV and Ni XIX to Ni XXVI spectra emitted by active regions and flares.
(3) 23–170 Å: this range is observed by grazing-incidence spectrometers and mostly includes lines from $n \geq 3$ configurations in Li-like to Cl-like ions emitted at $T \approx 1$–5 MK, and allowed lines from $n = 2$ configurations in Fe XVIII to Fe XXIII and Ni XV to Ni XXVI emitted by active regions and flares.
(4) 170–630 Å: this range is observed by grazing- and normal-incidence spectrometers and includes allowed lines from $n = 2$ configurations of Li-like to Mg-like ions emitted over a wide temperature range, and allowed lines from $n = 3$ configurations in Fe IX to Fe XVII spectra.
(5) 630–2000 Å (off-disk spectrum): this range is observed by normal-incidence spectrometers. When observed outside the solar disk at heights larger than $1.05 R_{\odot}$, it is dominated by allowed, forbidden and intercombination lines from $n = 2$ configurations and allowed lines between $n = 3$ configurations in coronal ions formed at $T \geq 0.7$ MK.
(6) 630–2000 Å (disk spectrum): when observed on the solar disk, this range includes mostly chromospheric and transition region lines from all the solar abundant elements, the Lyman continuum, and a small number of bright coronal lines.

A complete list of all lines observed in the solar spectrum in the 1–2000 Å range is given in Appendix 2.

6.2 The 1–7 Å range

The 1–7 Å spectral range is dominated by emission from bright active regions and flares. The spectrum observed in this range is composed of free–free and free–bound continua, lines from H-like and He-like ions from Si to Ni, and satellite lines formed by dielectronic recombination and inner-shell collisional excitation. An annotated theoretical solar spectrum in this wavelength range for a temperature of 20 MK is shown in Figure 6.1.

The relative importance of free–free and free–bound continuum emission in this range depends on wavelength and temperature, as seen in Figure 5.15. The observed continuum can be used to derive temperature and emission measure (EM), and with line fluxes for measuring absolute element abundances, though the relative contribution of free–bound and free–free continua must first be known. For many instruments the continuum emission is difficult to measure because of instrumental background.

Lines from H-like ions of Si to Ni are emitted by allowed $1s\ ^2S - np\ ^2P\ (n \geq 2)$ transitions and their contribution functions are density insensitive. The H-like line fluxes can be used for emission measure, differential emission measure (DEM), and abundance measurements. Flux ratios of lines with transitions having different upper-level principal quantum number n in the same ion are temperature sensitive up to a certain maximum temperature that depends on Δn and on the atomic number Z.

Helium-like ions from Si to Ni emit allowed, intercombination, and forbidden lines that are important for measuring temperature and density in solar flare plasmas (Section 5.2.1). Since the work of Gabriel and Jordan (1969), the R and G ratios have been extensively used to measure temperature and density, and have been applied to flare and active region

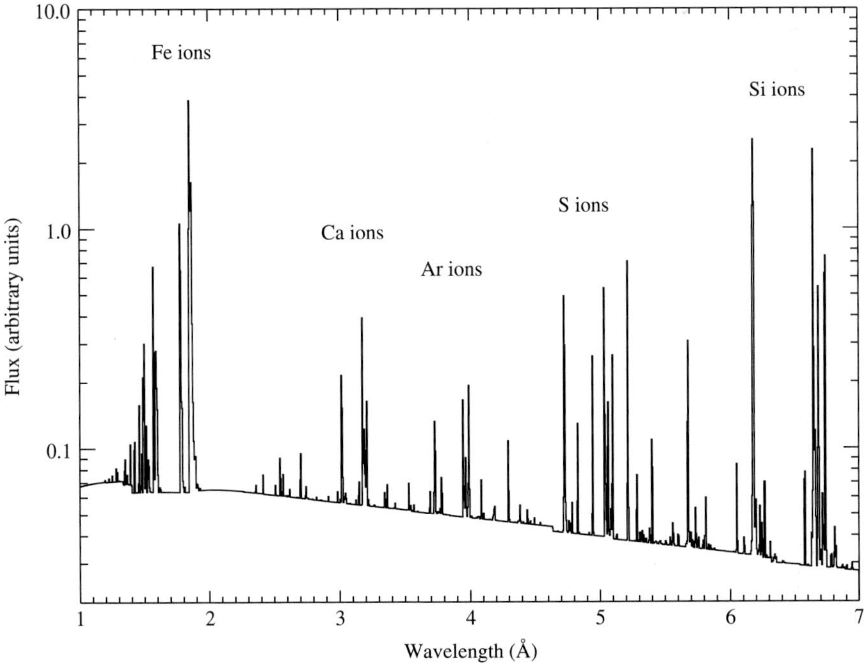

Fig. 6.1. Spectrum in the 1–7 Å spectral range, calculated with the CHIANTI database, with a temperature of 20 MK and with coronal abundances.

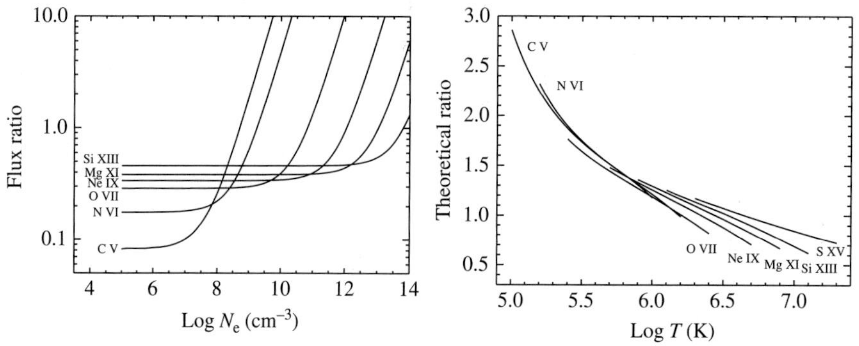

Fig. 6.2. Plasma diagnostics using the He-like ions. (*Left*) Density from the $R = (x + y)/z$ flux ratio. These curves do not include the effect of dielectronic satellites. (*Right*) Temperature from the $G = (x + y + z)/w$ flux ratio. The ratios are calculated with the CHIANTI database.

X-ray spectra. These ratios are shown in Figure 6.2. The lower limit on density ($N_{\rm min}$) is $\sim 10^{12}$ cm^{-3} for Mg and increases with the atomic charge Z; this value may be greater than flare or active region densities so that the use of the ratio R for density diagnostics in this spectral range is limited. Satellite lines from configurations above the ionization threshold with $n \geq 3$ in Li-like ions provide significant blending to the He-like ion lines at relatively low temperatures when their fluxes are larger. Such blends must be considered when calculating the R and G ratios as a function of temperature or density.

The numerous dielectronic satellite lines occurring in this spectral region are useful for measuring the temperature of active regions or flares (Section 5.3.1). Many satellites with transitions of the form $1s^2\,nl - 1s\,2p\,nl$ occur on the long-wavelength side of resonance lines of He-like ions (transitions $1s^2\,{}^1S_0 - 1s2p^1\,P_1$). Equivalent satellites (transitions $1s\,nl - 2p\,nl$) occur on the long-wavelength side of the Ly-α lines of H-like ions. In the case of Fe the satellites form a single prominent feature (called J, at 1.792 Å), the flux of which relative to that of the Fe XXVI Ly-α lines (at (1.778, 1.784 Å) is temperature dependent, a fact that has been used to obtain the temperature of the flare superhot component (Section 10.4). Other related satellite lines occur as single line features on the long-wavelength side of He-like ion lines with transitions $1s^2\,{}^1S_0 - 1s\,np^1\,P_1$ ($n \geq 3$), and again the flux ratios are temperature dependent. This applies to Si and S lines in this wavelength range, observed with e.g. the RESIK spectrometer on *CORONAS-F*.

6.3 The 7–23 Å range

The 7–23 Å range includes H-like and He-like lines from O to Al, as well as their satellites, and a large number of lines emitted by configurations with principal quantum number $n \geq 3$ in Fe XVII to Fe XXIV and Ni XIX to Ni XXVI, with Fe lines predominating because of the higher abundance of Fe. H-like and He-like lines like those in the 1–7 Å range also occur, but with smaller-Z elements, so that the value of the minimum electron density, $N_{\rm min}$, above which the R ratio is sensitive to N_e is much smaller than for Si or heavier elements. Thus R is a suitable means of measuring flare or active region densities in O VII, Mg XI, and Al XII spectra – see Figure 6.2. The Ne IX lines at 13.45–13.70 Å are

Table 6.1. *Density-sensitive Fe lines in the 7–23 Å range.*

Ion	Line ratio[a]	log T_{max}^{b}	log N_e (max)[c]
Fe XVIII	11.456/11.420	6.78	11
Fe XVIII	11.456/11.525	6.78	11
Fe XVIII	14.125/14.206	6.78	11
Fe XVIII	14.340/14.206	6.78	11
Fe XVIII	14.562/14.584	6.78	11
Fe XX	12.982/12.812	6.92	12
Fe XX	12.982/12.827	6.92	12
Fe XX	12.982/12.845	6.92	12
Fe XXI	9.482/9.587	7.00	11
Fe XXI	12.327/12.395	7.00	11
Fe XXI	12.422/12.395	7.00	11
Fe XXI	12.691/12.822	7.00	11
Fe XXII	8.169/8.090	7.05	12
Fe XXII	9.075/8.977	7.05	12
Fe XXII	11.921/11.936	7.05	12
Fe XXII	12.053/12.201	7.05	12

[a] Line wavelengths in Å.
[b] T_{max} in K.
[c] N_e in cm^{-3}.

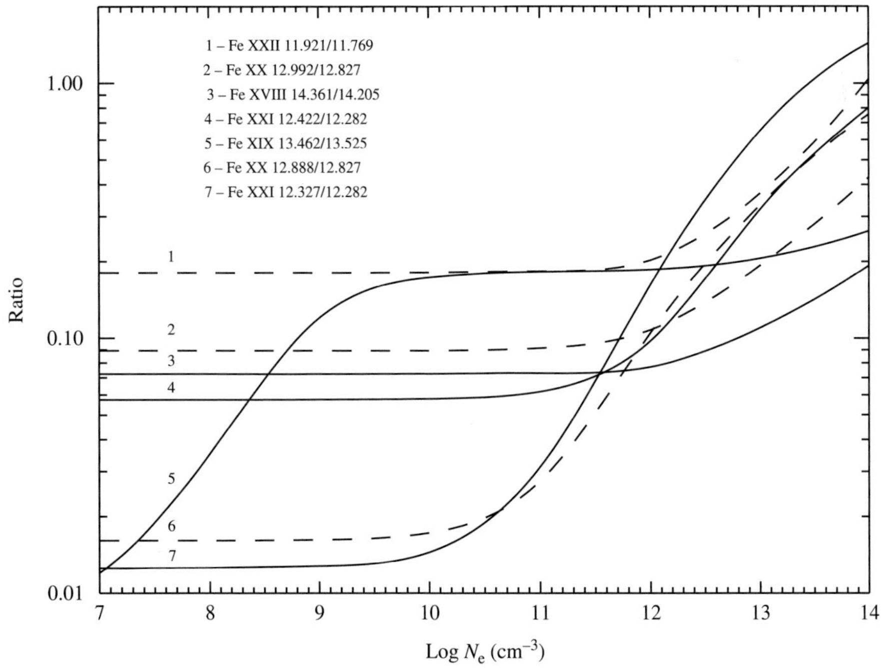

Fig. 6.3. Density-sensitive ratios from Fe lines in the 7–23 Å range. Ratios are calculated with the CHIANTI database.

also density sensitive but the presence of Fe XIX lines makes interpretation of spectra difficult unless techniques to unscramble the lines are used (Bhatia *et al.* (1989)).

The Fe XVII to Fe XXIV lines occurring in this range are useful for diagnosing flare or active regions. The fluxes of the stronger lines can be used to determine the DEM of the emitting plasma or the temperature if an isothermal approximation is valid. Also, several groups of lines are density sensitive, as is indicated in Figure 6.3 and Table 6.1. Such lines cover temperature ranges appropriate for hot active regions and flares.

Figure 6.4 shows the 10–18 Å spectrum observed during a flare decay (temperature ~6 MK) and the 10–18 Å spectrum calculated for a temperature appropriate near the maximum of a moderate flare (16 MK). The observed flare decay spectrum is dominated by Fe XVII lines at 15–17 Å, and O VII, O VIII, Ne IX, Ne X, Mg XI, Mg XII lines. At higher temperatures, the spectrum is dominated by Fe XXI to Fe XXIII lines between 10 and 13 Å.

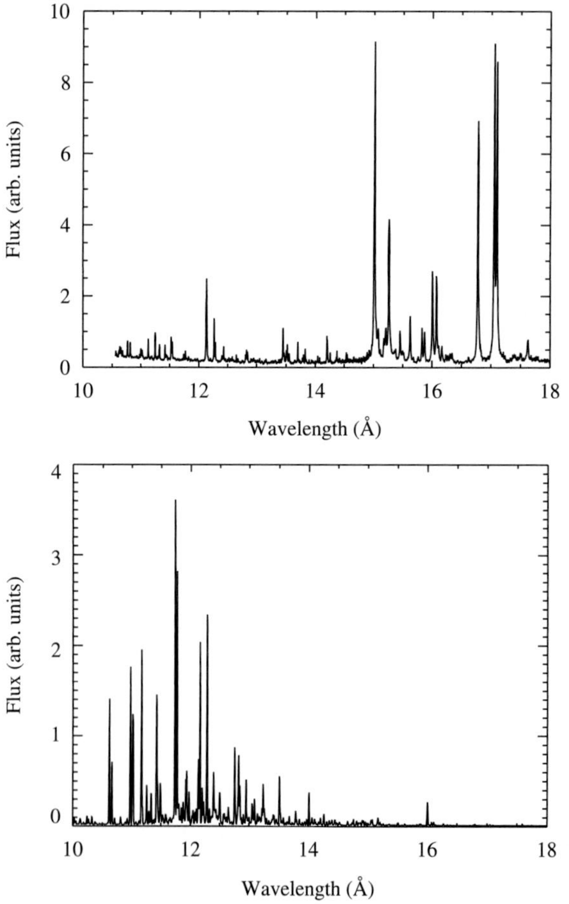

Fig. 6.4. Solar spectrum in the 10–18 Å range for $T_e \approx 6$ MK (*top*, observed with *SMM*/FCS) and a spectrum calculated with the CHIANTI database with a temperature of 16 MK (*bottom*). The spectrum at $T_e \sim 6$ MK is dominated by Fe XVII lines at 15 Å and 17 Å, while the spectrum at $T_e = 16$ MK is dominated by lines lines from Fe XXI to Fe XXIV between 10 and 13 Å.

6.4 The 23–170 Å range

The 23–170 Å spectral range has not been studied in much detail in part because other spectral ranges that include stronger lines with similar or better diagnostic capabilities have been preferred. In quiescent and active plasmas, this wavelength range is dominated by allowed lines having transitions from configurations with principal quantum number $n \geq 3$ in Li-like to Mg-like ions, and by lines with transitions among $n = 3$ configurations in Ni X to Ni XVI. These lines allow non-flaring plasmas to be studied, but are weaker than those in the 170–630 Å wavelength range. In flares, strong Fe XVIII–Fe XXIII lines with allowed transitions between levels of the $n = 2$ configurations dominate the spectrum. Unfortunately, thus far only one instrument with moderate spectral resolution has obtained flare spectra in the 66–171 Å range (Kastner *et al.* (1974)). The *Extreme Ultraviolet Explorer* (Bowyer & Malina (1991)) mission included a slitless spectrometer whose short wavelength channel included lines down to 70 Å, and its observations provided high-quality spectra allowing cool star coronae to be studied.

The Ni X–Ni XVI lines in this range include several density-sensitive lines, and also allow the DEM to be determined over the temperature range 0.7–4 MK, appropriate to coronal holes, quiet Sun and active regions. These ions have been neglected in the past because much stronger lines emitted by equivalent Fe ions (Fe VIII to Fe XIV) in the 170–230 Å wavelength range give the same information.

The 23–170 Å range includes a large number of lines of Li-like to Mg-like ions of the most abundant elements that are emitted by transitions involving configurations with $n \geq 3$ over a temperature range from the transition region to flares. Some lines are temperature sensitive or density sensitive. Figure 6.5 shows examples for Mg IX and Si XI.

Lines in the Fe XVIII to Fe XXIII spectra are emitted by allowed transitions within the $n = 2$ configurations and dominate 80–150 Å flare decay spectra, with temperatures $\gtrsim$ 6 MK. A spectrum appropriate for flares between 90 Å and 140 Å as calculated from the CHIANTI database is shown in Figure 6.6. Some lines of Fe XIX to Fe XXII are very sensitive to density and so are suitable for the decay stages of flares. They have also been used to study the atmospheres of cool stars using *Extreme Ultraviolet Explorer* observations. Examples of such ratios are given in Figure 6.7. Density-insensitive lines of Fe XVIII to Fe XXIII in this range can be used as input for obtaining the DEM of flares. Equivalent lines of Ni XX to Ni XXV occur in the 80–120 Å range, but are weaker than the Fe lines.

Lines of H-like and He-like C and N are also found in this wavelength range, below 45 Å. They are sensitive to densities down to 10^9–10^{10} cm^{-3}, much lower than equivalent lines of O and heavier elements. Calculation of the R and G ratios must include photoexcitation due to the chromospheric radiation field as well as collisional excitation.

6.5 The 170–630 Å range

The 170–630 Å wavelength range is rich in strong spectral lines suitable for plasma diagnostics and has been used to study the solar atmosphere with many rocket and satellite missions. This wavelength range is also easily accessible to grazing-incidence and normal-incidence spectrometers, and includes lines from the solar chromosphere, transition region, and corona under all conditions (from coronal hole to flares), and also continuum emission.

The continuum is mostly free–bound emission, formed by the recombination of once- and twice-ionized helium. The He I and He II edges are at 504.3 Å and 227.8 Å respectively. These continua can only be observed for sources on the solar disk but not for coronal sources off the limb where the temperature is too high for the formation of significant amounts of

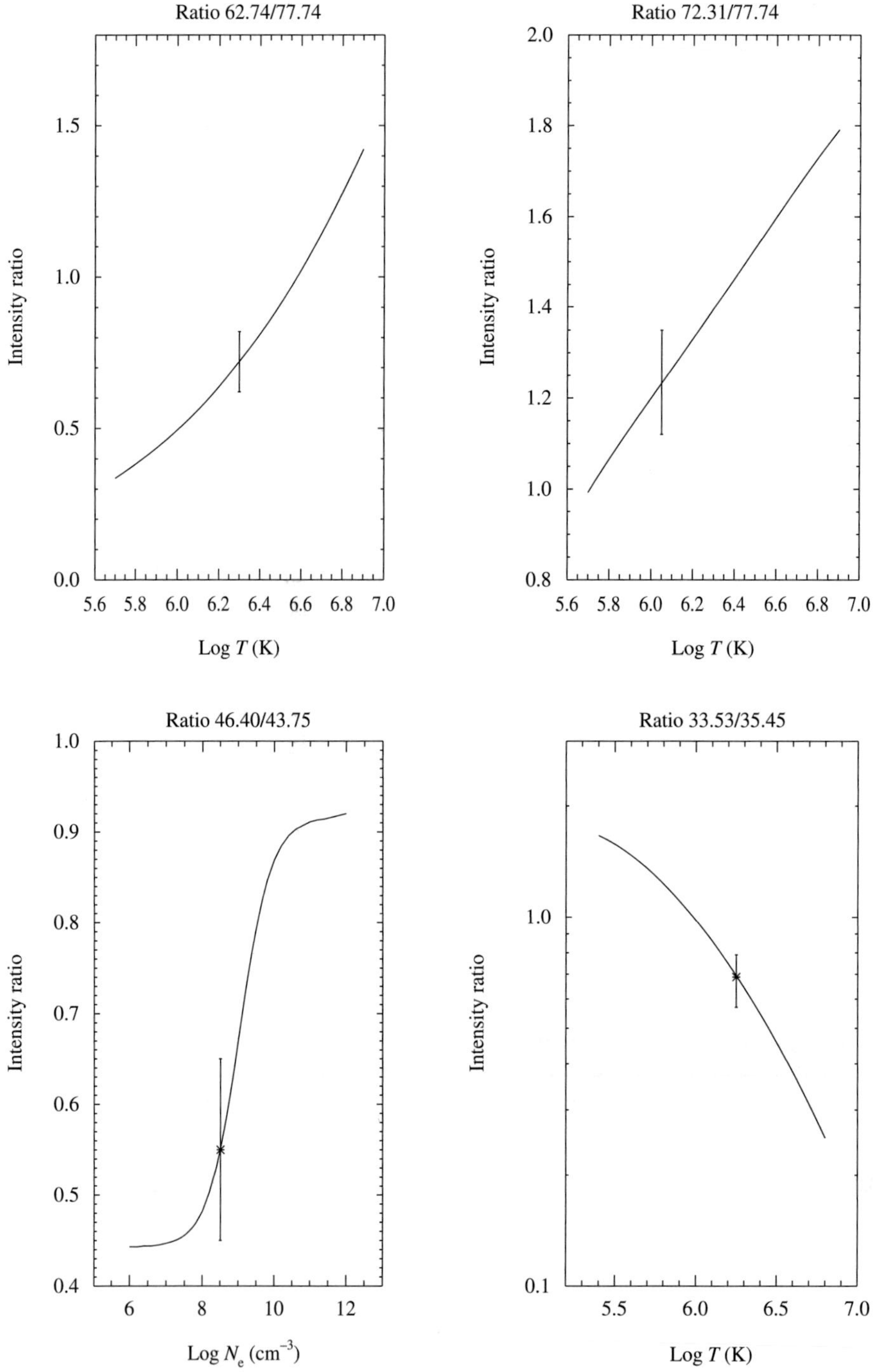

Fig. 6.5. Temperature diagnostics from Mg IX line ratios (*top*), and density and temperature diagnostics from Si XI line ratios (*bottom*), using lines from high-energy configurations in the 23–170 Å range. Observed ratios are from Acton *et al.* (1985). Predicted ratios are calculated with the CHIANTI database.

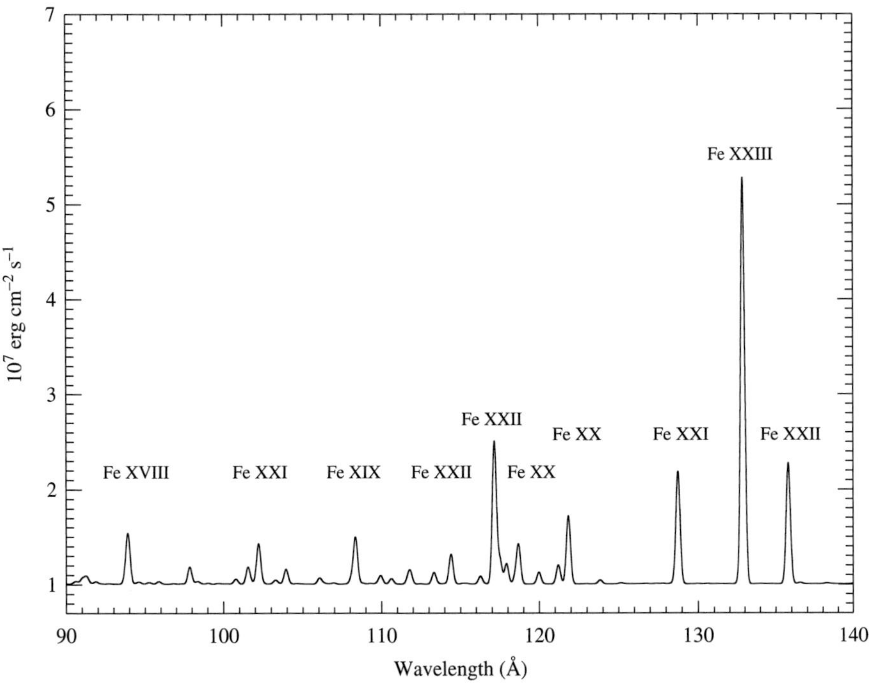

Fig. 6.6. Spectrum between 80 Å and 140 Å, calculated for solar flare temperatures using the CHIANTI database.

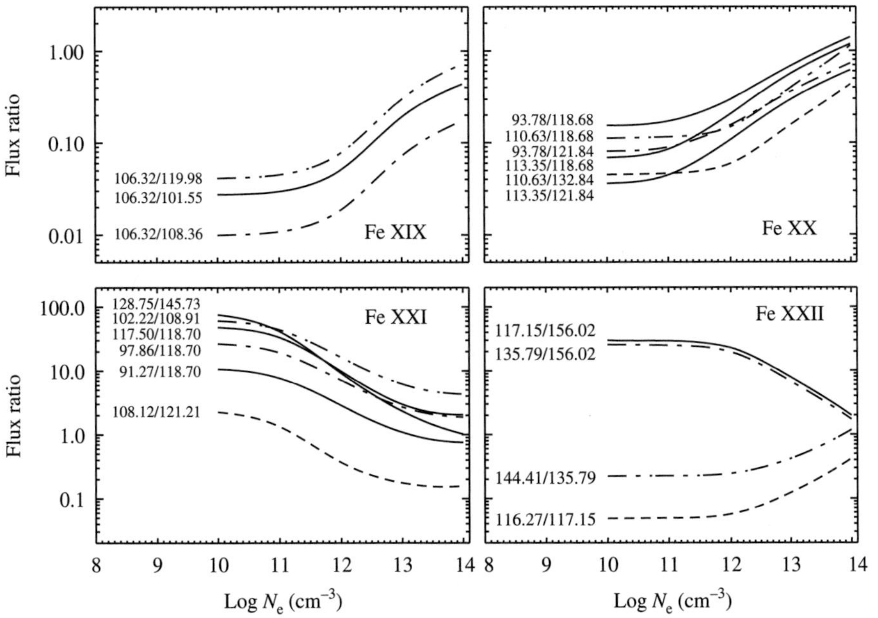

Fig. 6.7. Density-sensitive line ratios for Fe XIX to Fe XXII lines in the 23–170 Å range. The ratios were calculated using the CHIANTI database.

neutral or ionized He. Free–free emission can however be observed off-limb from solar flares and hot, dense active regions. Little attention has been paid to this emission, though it is useful for providing line-to-continuum ratios that lead to absolute abundances.

Neutral and ionized helium produce two of the strongest lines in the solar spectrum in this range. The He II Ly-α line is at 303.85 Å and is emitted mostly by the chromosphere, with a weak coronal component due to resonant scattering. The strength and characteristic temperature of this line have made it the line of choice for EUV channels of narrow-band imagers aimed at observing the lowest layers of the solar atmosphere, e.g. the EIT instrument on *SOHO*, the EUV imager (EUVI) on the two *STEREO* spacecraft, and the Atmospheric Imaging Assembly (AIA) on the future *Solar Dynamics Observatory* (*SDO*). The He I resonance line at 584.34 Å (transition $1s^2\,{}^1S_0 - 1s2p^1P_1$) is also very strong in disk spectra, as are higher members of this series. All these lines represent the coolest features in the 170–630 Å range and are valuable tools for studying the connection between chromospheric and transition region structures. However, the formation of these lines involves radiative transfer processes (not considered in Chapter 4) and so interpretation of their fluxes is not straightforward (see Andretta & Jones (1997) and references therein).

The lines occurring in the 171–630 Å range can be divided in four broad groups: (i) lines due to allowed transitions within $n = 2$ configurations in Be-like to O-like ions of elements with $Z < 20$; (ii) lines from $ns^2S - np^2P$ doublets in Li-like and Na-like ions; (iii) lines from allowed transitions within $n = 3$ configurations in Fe VIII to Fe XVII and isoelectronic ions from Ca and Ni; (iv) forbidden lines emitted by Be-like to O-like ions of Fe (Fe XIX to Fe XXIII). Allowed lines emitted by Be-like to O-like Fe and other ions constitute the bulk of the emission in the 170–630 Å wavelength range, and have been used for many diagnostic studies. Strong lines from coronal ions mostly occur below 500 Å, and the fluxes of many of them are density dependent. The most useful are lines of B-like, C-like, and N-like ions of Si and S. Important density-sensitive line ratios in this range are listed in Table 6.2. Other examples are shown in Figures 6.8, 6.9, and 6.10. Several O III lines (see Figure 6.11) and Be-like ion lines (e.g. Landi *et al.* (2001), see also Figure 6.12) are temperature sensitive. Lines in this range have also been extensively used for DEM diagnostics since they span the $4.8 \le \log T \le 6.5$ temperature range in quiet and active conditions. Lines formed at transition region temperatures include many emitted by O, Ne, Mg, and Ca ions, and their fluxes provide a probe of element abundances relevant to the FIP effect (Chapter 11). This group also includes some of the strongest lines in quiet Sun and active region spectra, particularly those with transitions $2s^2\,{}^1S_0 - 2s2p^1P_1$ in Be-like O to Ca. Many of these lines are shown in Figures 6.13 and 6.14, where the 165–310 Å range from the CDS instrument on *SOHO*, 380–450 Å from the 1989 flight of the SERTS instrument (Thomas & Neupert (1994)), and the 515–630 Å range from CDS are shown.

Li-like and Na-like ion lines with transitions ${}^2S - {}^2P$ are emitted by the corona and also include lines observed in flares. These lines are density insensitive, and so may be used in DEM and elemental abundance determinations.

Fe VIII to Fe XIV lines due to allowed transitions in $n = 3$ configurations dominate the solar spectrum in the 171–220 Å range, and provide some of the best density diagnostic information at coronal temperatures in the 171–630 Å range (Table 6.2). Also, the strength and the proximity in wavelength of many Fe ion lines formed at similar temperatures make them suitable for narrow-band filter imagers such as the EIT instrument on *SOHO*, *TRACE*, the EUVI instrument on the *STEREO* spacecraft, and the future AIA instrument on *SDO*. The

Table 6.2. *Density-dependent line pairs in the 170–630 Å spectral range.*

Ion	Ratio[a]	log T_{max} [b]	log N_{max} [c]
O IV	625.85/554.51	5.24	—
Mg V	353.09/276.58	5.44	9.0
Ne V	416.21/481.37	5.47	8.5
Mg VI	349.12/403.31	5.63	10.0
Mg VII	625.85/554.51	5.80	10.5
Fe IX	241.74/244.91	5.99	—
Si IX	258.08/292.81	6.05	11.0
Fe X	175.26/174.53	6.08	10.0
Si X	258.37/261.04	6.13	10.0
Si X	356.03/347.41	6.13	10.0
Fe XI	184.80/188.23	6.14	11.0
Fe XII	186.85/195.12	6.22	11.0
Fe XII	338.26/364.47	6.22	11.0
S XII	299.54/288.42	6.31	10.0
Fe XIII	359.64/348.18	6.31	10.0
Fe XIII	202.04/203.83	6.31	10.0
Fe XIV	219.13/211.32	6.43	11.0
Fe XIV	353.84/334.18	6.43	10.0

[a] The O IV and Fe IX flux ratios are density-dependent for a wide range of coronal densities. Other ratios are density sensitive for $N_e < N_{max}$. Line wavelengths in Å.
[b] T_{max} in K.
[c] N_e in cm^{-3}.

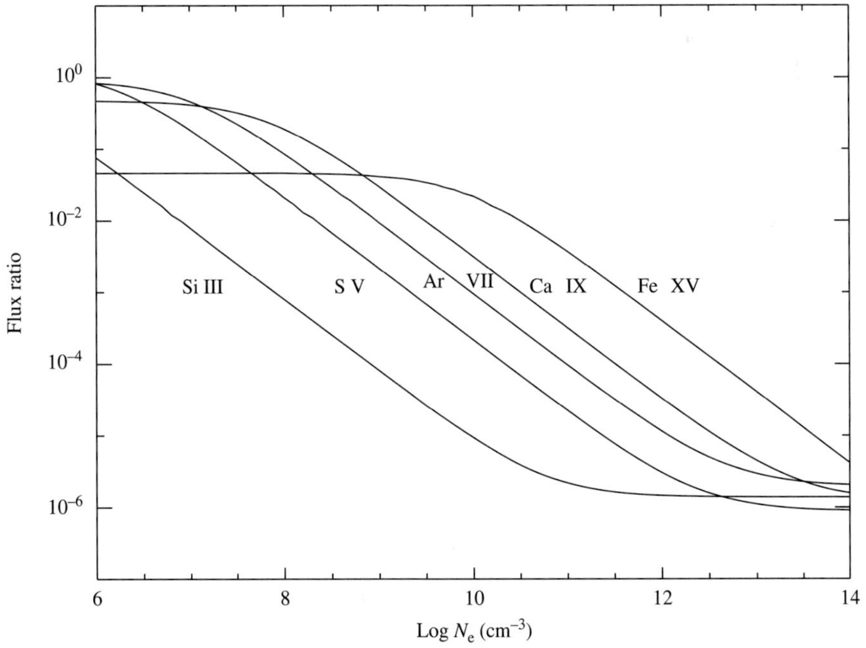

Fig. 6.8. Density-dependent Mg-like ion line pairs: Si III (1882.71/1892.03), S V (1188.28/1199.14), Ar VII (872.56/885.55), Ca IX (676.08/691.21), and Fe XV (393.98/417.26). Ratios are calculated with the CHIANTI database. Transitions are $(3s^2\,{}^1S_0 - 3s3p\,{}^3P_2)/(3s^2\,{}^1S_0 - 3s3p\,{}^3P_1)$.

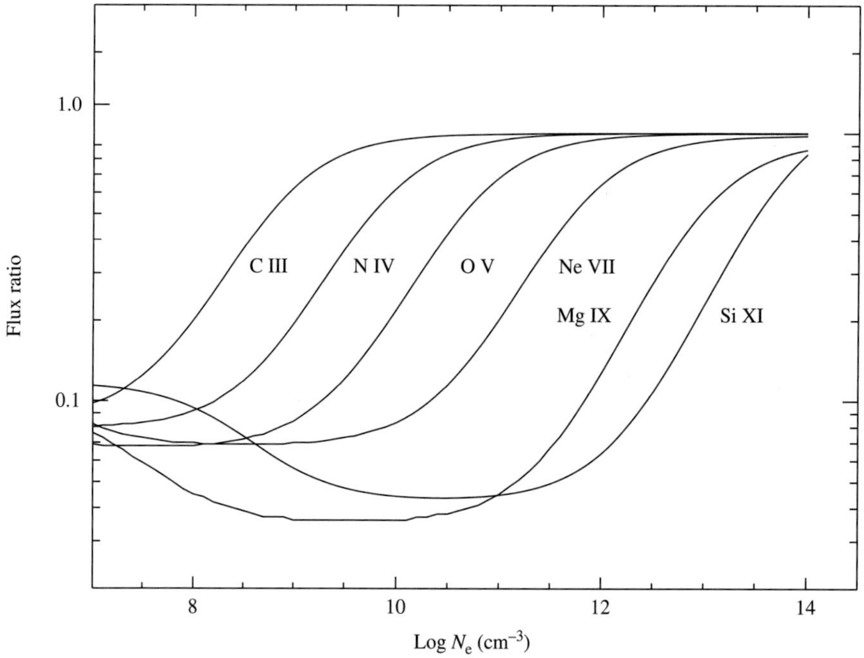

Fig. 6.9. Density-dependent Be-like ion line pairs: C III (1175.99/1175.26), N IV (923.68/922.52), O V (761.13/759.44), Ne VII (562.99/559.95), and Mg IX (445.98/441.20). Ratios are calculated with the CHIANTI database. Transitions are $(2s2p\ ^3P_1 - 2p^2\ ^3P_0)/(2s2p\ ^3P_0 - 2p^2\ ^3P_1)$.

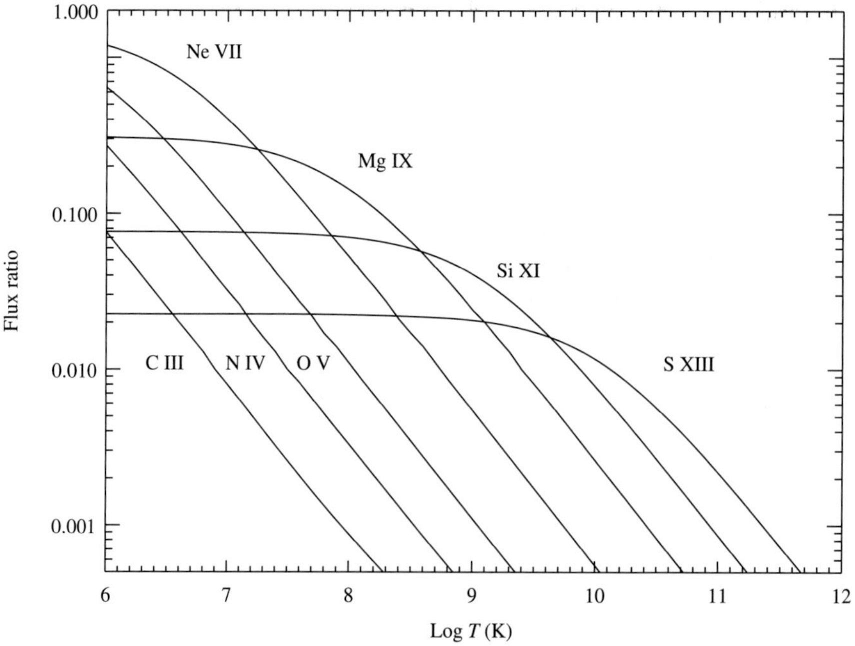

Fig. 6.10. Density-dependent Be-like ion line pairs: C III (1906.68/1908.73), N IV (1483.32/1486.50), O V (1213.81/1218.34), Ne VII (887.28/895.17), Mg IX (694.01/706.06), Si XI (563.96/580.91), and S XIII (491.46/500.34). Ratios are calculated with the CHIANTI database. Transitions are $(2s^2\ ^1S_0 - 2s2p\ ^3P_2)/(2s^2\ ^1S_0 - 2s2p\ ^3P_1)$.

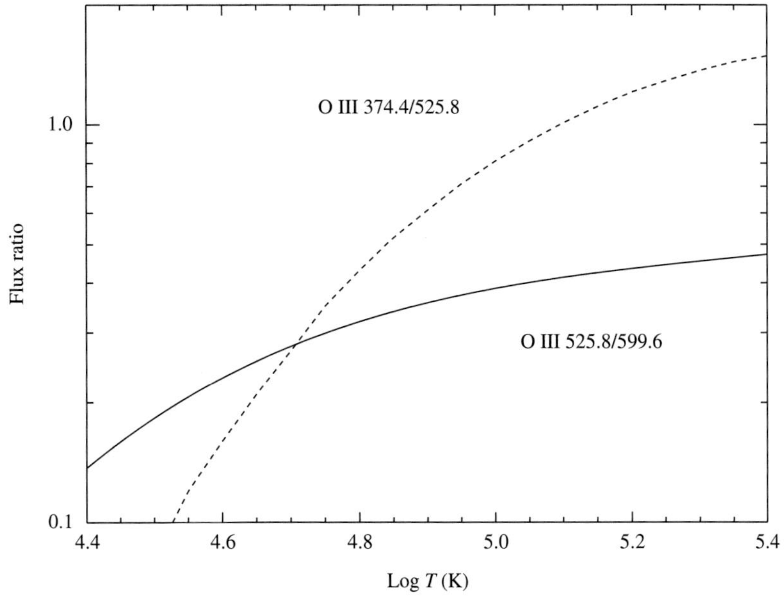

Fig. 6.11. Temperature-dependent O III line pairs (calculated in photon units). Dashed line: the theoretical ratio is increased by a factor 10. Ratios are calculated with the CHIANTI database.

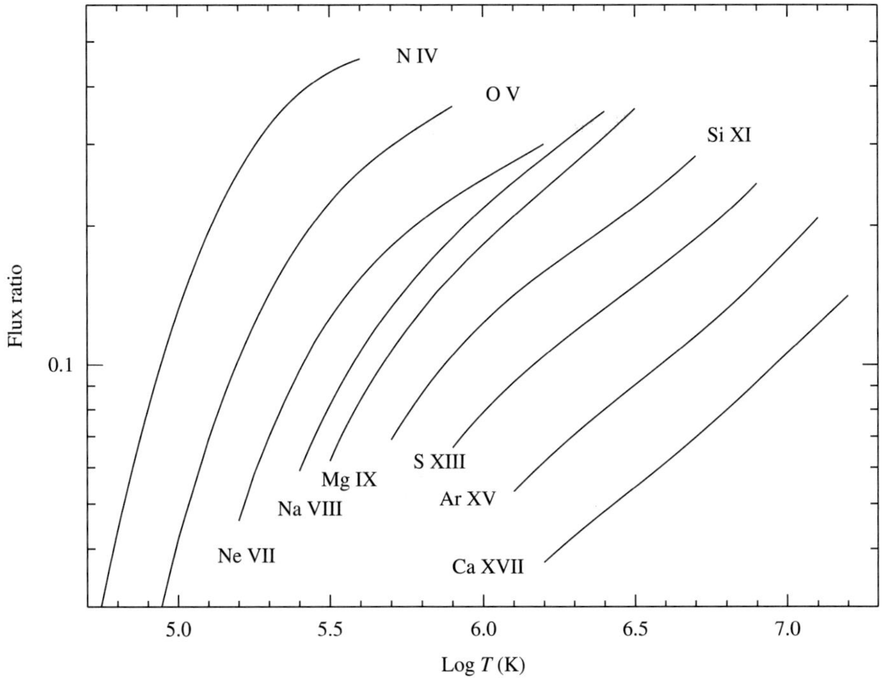

Fig. 6.12. Temperature-sensitive Be-like ion line pairs: N IV (1718.55/1486.50), O V (1371.30/1218.34), Ne VII (973.33/895.17), Na VIII (847.96/789.79), and Mg IX (749.55/706.06). Ratios are calculated with the CHIANTI database. Transitions are $(2s2_p\ ^1P_1 - 2p^2\ ^1D_2)/(2s^2\ ^1S_0 - 2s2p\ ^3P_1)$.

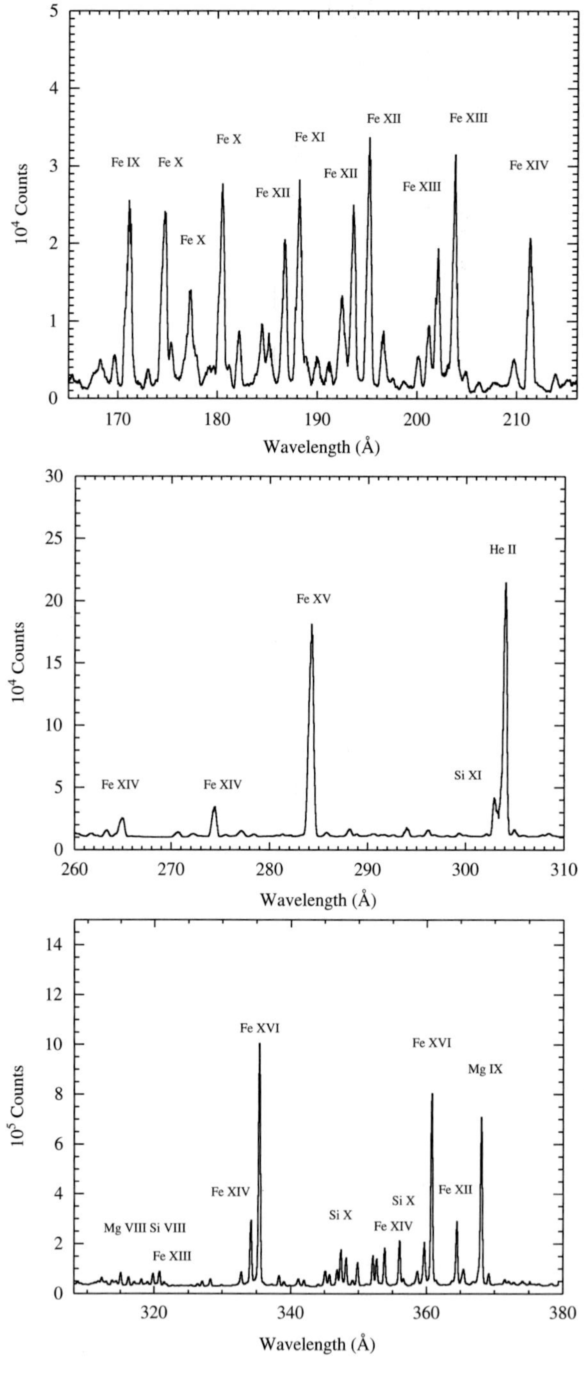

Fig. 6.13. Solar active region spectra from the CDS instrument on *SOHO*. (*Top*) 165–216 Å range, from the GIS instrument. (*Middle*) 260–310 Å range from the GIS instrument. (*Bottom*) 310–380 Å range from the NIS instrument. Courtesy *SOHO* CDS team.

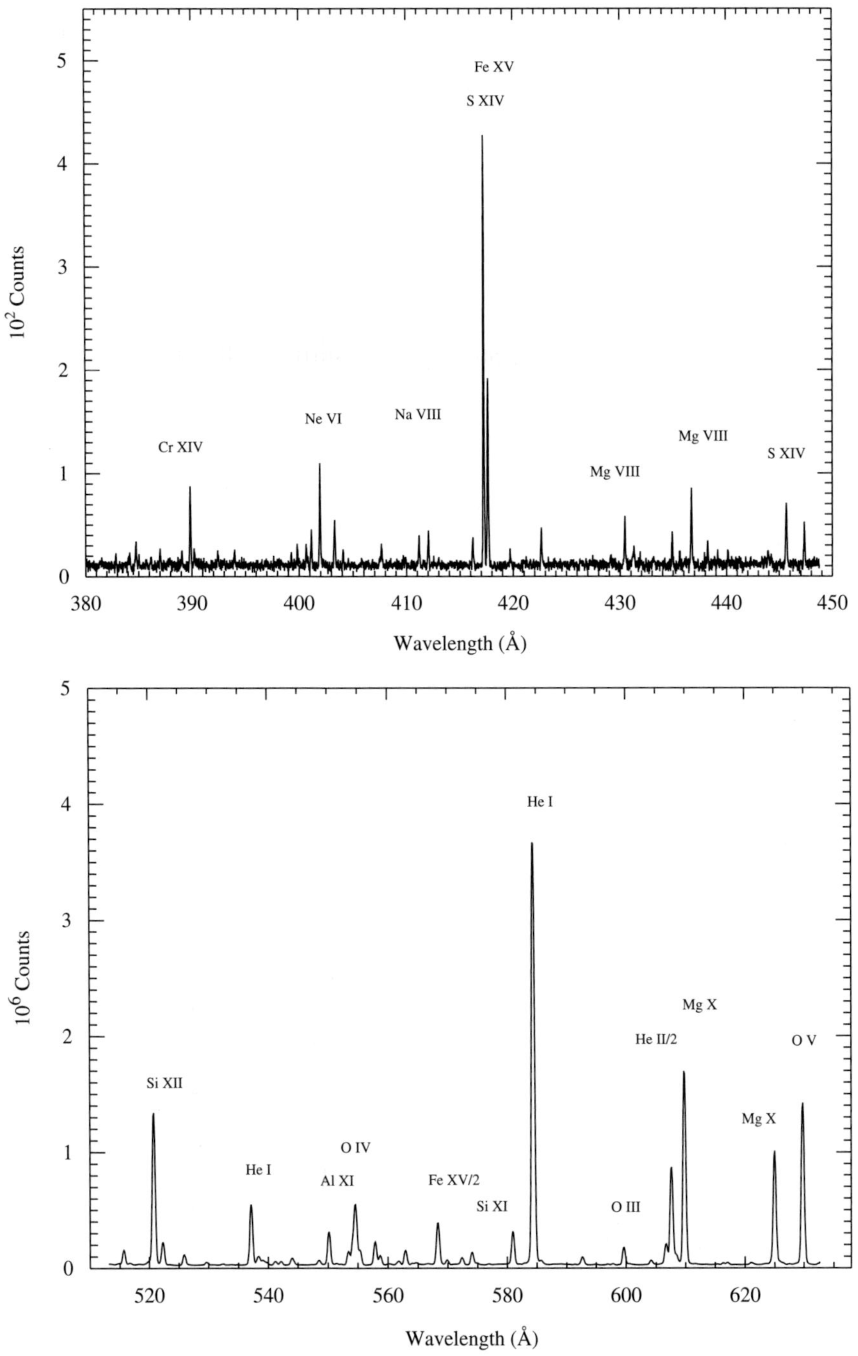

Fig. 6.14. (*Top*) Solar active region spectrum in the 380–450 Å range from the 1989 flight of SERTS. Courtesy SERTS team. (*Bottom*) Solar active region spectrum in the 515–630 Å range from the CDS/NIS instrument on *SOHO*. Courtesy *SOHO* CDS team.

wavelength region around the Fe XV $3s^2\,{}^1S_0 - 3s3p^1\,P_1$ line at 284.15 Å, which is strong in active region spectra and fairly isolated, has also been used in imagers. Figure 6.13 shows some of the Fe lines in the 165–216 Å range.

A few lines from highly ionized Fe ions excited in flares are found in the 170–630 Å range. The most important are Fe XXIV 192.03 Å and 255.11 Å, Fe XXIII 263.77 Å, the Fe XXI lines at 335.69 Å and 585.77 Å, the Fe XX lines due to transitions within the ground configuration, and the Fe XIX line at 592.24 Å. These lines have been observed in flares since the 1970s.

6.6 The 630–2000 Å range (off-limb spectrum)

The 630–2000 Å spectrum emitted by regions of the solar atmosphere above $1.05R_\odot$ includes lines emitted at temperatures $\geq 8 \times 10^5$ K as well as lines with much smaller formation temperatures, excited at least in part by resonant scattering (Section 4.9.3). The most important of the latter are the H I Ly-α and Ly-β (the most intense lines in the disk spectrum below 2000 Å), and the N V, O VI, and Ne VIII doublet $2s\,{}^2S$–$2p\,{}^2P$ lines in the Li-like sequence. Resonant-scattered lines are important for measuring the plasma dynamic state above the limb, and have provided much information on the acceleration of the solar wind. H I and He II Ly-α and other He II lines formed by collisional excitation and recombination have also been observed in the corona at low altitudes, and provide information on abundance and temperature. The Doppler thermal broadening of these lines also provides ion temperatures (Feldman *et al.* (2005a)).

Off-limb continuum emission at wavelengths $\lambda > 912$ Å in active regions and flares is mostly free–free radiation emitted by hot plasmas with large emission measure. This emission provides the means of measuring absolute coronal abundances (Feldman *et al.* (2003a)). Its weak dependence on temperature minimizes the effects of uncertainties in temperature, and its slow decrease with wavelength enables checks to be made of instrumental flux calibration (Feldman *et al.* (2005b)).

Lines from Li-like N to Na ions, and Na-like S and Ar ions can be found in this band; they can be used to find EM, relative abundances, and temperatures. Ne VIII lines are formed at $\sim 6 \times 10^5$ K, appropriate to the transition region or lower corona. Their strengths in coronal holes make them useful as tracers of the solar wind, so that the fast wind can be linked to structures in the inner solar corona. O VI lines have also been used to study the velocity of the solar wind through Doppler dimming, and together with H I lines they have provided detailed measurements of the acceleration of the fast and slow solar wind below $4R_\odot$.

In addition to the Li-like and Na-like lines, the off-disk spectrum in the 630–2000 Å range includes lines with two kinds of transitions, emitted by Be-like to F-like ions: (i) forbidden transitions within the ground configuration or between the first excited and the ground configurations; and (ii) allowed transitions between levels from highly excited $n = 3$ configurations. Forbidden transitions are limited in number, but they are generally strong, and because of the small decay rates are density sensitive in several cases. Examples are shown in Figures 6.8 (Mg-like ions), 6.10 (Be-like ions), 6.15 (O-like ions), and 6.16 (N-like ions). Also temperature can be measured from these lines (e.g. Figure 6.12 for Be-like ions). These lines can also be used for determining EM, DEM, and relative abundances (through line ratios) and absolute abundances (from line-to-continuum ratios). These lines are observed in all conditions from quiet Sun to flares.

Transitions between levels in high-energy configurations result in a large number of lines whose fluxes may be large. Flux ratios involving these lines are density sensitive in a few

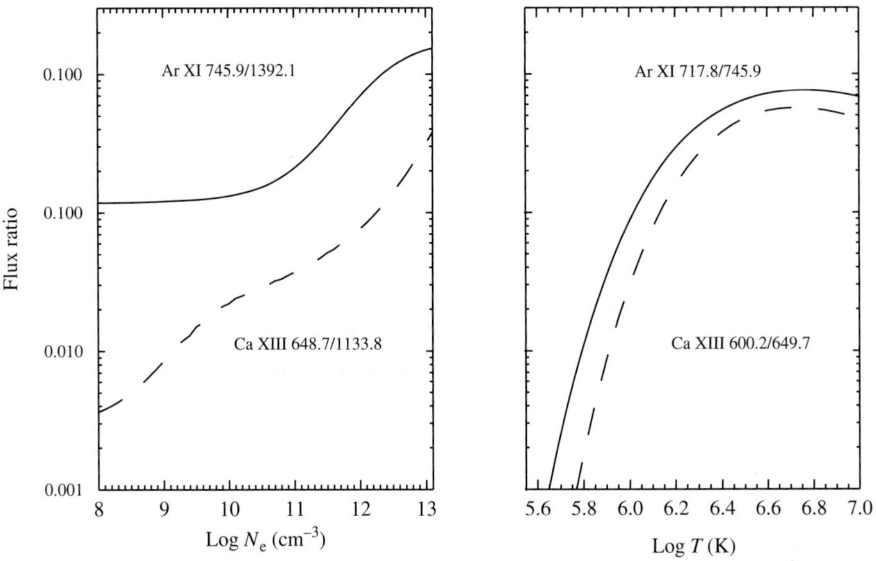

Fig. 6.15. Line flux ratios involving lines from Ar XI (full line) and Ca XIII (dashed line). (*Left*) Ratios of forbidden lines within the ground configuration. (*Right*) The 600.2 Å and 717.8 Å lines are emitted by levels from $n = 3$ configurations, while the other lines come from levels in the ground configuration. Ratios are calculated with the CHIANTI database.

cases, and are temperature sensitive for lines with different $n = 3$ configurations. What makes these $n = 3$ lines important is their proximity in wavelength to lines with forbidden transitions emitted by the same ion: the large difference in excitation energy between the $n = 3$ and the forbidden lines makes them suitable for measuring temperature. Examples are shown in Figure 6.15 for lines due to O-like ions.

A small number of forbidden lines of Fe X, Fe XI, and Fe XII also occur in this range. They can be used to determine abundance and DEM. Also, the fall-off with height above the limb of the fluxes of these lines compared with those of lines due to lighter elements like Mg, Si, and Ne has been used to detect the presence of gravitational settling. During flares, forbidden lines of Fe XVII to Fe XXIII have been observed in this range. These lines have been used to determine temperature and DEM for flares, and to measure oscillations and flows in flaring loops. Of particular importance are the two Fe XIX lines at 1118.07 Å and 1328.79 Å, since the latter line is strongly density sensitive so that the ratio of these two lines is density sensitive for $N_e > 10^8$ cm^{-3}. The Fe XVII and Fe XVIII forbidden lines at 1153.17 Å and 974.86 Å respectively are also observed in non-flaring active regions, where they provide valuable tools for studying the hottest components of active regions.

Lines from He-like ions of Ne, Na, Mg, and Si have been observed in this wavelength range, including those due to $1s2s\,^3S - 1s2p\,^3P$ transitions for Ni to Si. These lines are important because high-energy non-thermal electrons can be detected with them owing to the high energy required to excite them. Also, the fractional abundance of He-like ions is near unity over large temperature ranges, so that flux ratios of lines from He-like ions to the free–free continuum in active regions can be used to measure absolute element abundances with minimal uncertainties in ion abundances (Feldman *et al.* (2007)). This is extremely important

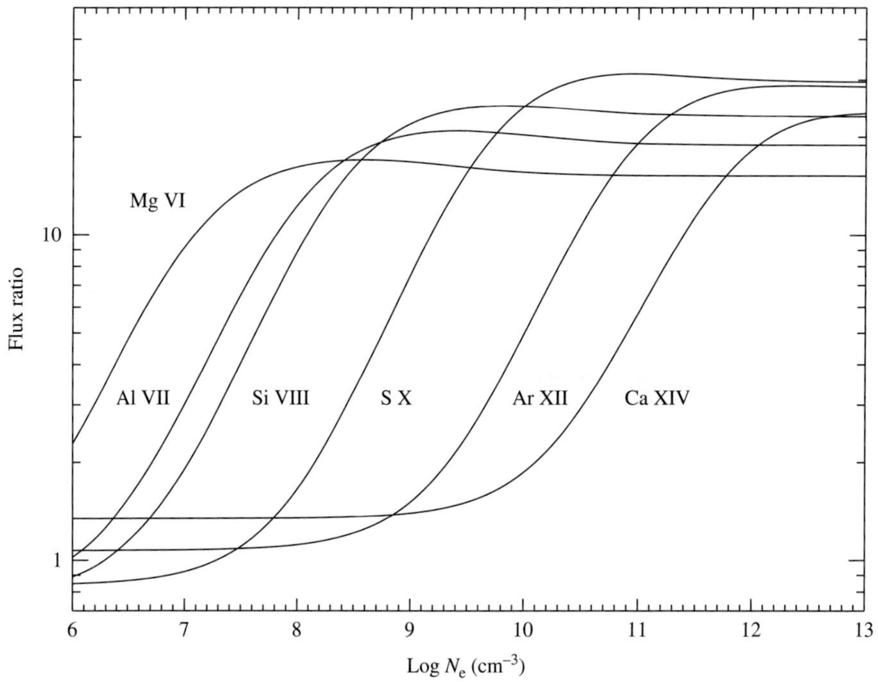

Fig. 6.16. Density-sensitive line pairs in N-like ions: Mg VI (1806.43/1806.00), Al VII (1602.93/1604.78), Si VIII (1440.51/1445.74), S X (1196.22/1212.93), Ar XII (1018.73/1054.69), and Ca XIV (880.40/943.59). Ratios are calculated with the CHIANTI database. Transitions are $(2s^2 2p^3\ ^4S_{3/2} - 2s^2 2p^3\ ^2D_{5/2}) / (2s^2 2p^3\ ^4S_{3/2} - 2s^2 2p^3\ ^2D_{3/2})$.

in the case of N, O, and Ne, because of the difficulty of measuring the absolute abundances of volatile elements in the solar atmosphere and of their importance in the calculation of the opacity of the solar interior. A list of relevant density and temperature diagnostic line ratios from lines in the 630–2000 Å range in off-disk plasmas is given in Table 6.3.

6.7 The 630–2000 Å range (disk spectrum)

The 630–2000 Å range is dominated by the H I Ly-α at 1215.67 Å emitted at chromospheric temperatures. Its very large flux has enabled instruments to image the chromosphere in fine detail. Other members of the H I Lyman series are also very strong and occur from 1025.72 Å to the long-wavelength side (920 Å) of the series limit at 911.75 Å where there is a pseudo-continuum of unresolved lines. Below the edge, the solar disk spectrum is characterized by the Lyman continuum. At wavelengths longer than 912 Å other free–bound continua from neutral elements occur, the most important of which being those of C I and Si I. Longward of these continuum edges is a forest of lines from high-*n* configurations of C I and Si I. At wavelengths longer than 1500 Å, the photospheric continuum also begins to be evident. An example of the 630–1600 Å spectrum from the SUMER instrument is shown in Figure 1.2 (upper panel).

Below 912 Å, lines from ions emitted at coronal temperatures occur. The same coronal lines observed off-disk (Section 6.6) are observed in the disk spectrum as well as those emitted by the transition region. These latter are emitted mainly by N, O, and S ions, and consist

Table 6.3. *Examples of density and temperature diagnostic line pairs in the 630–2000 Å spectral range, for off-disk plasmas.*

Ion	Ratio[a]	$\log T_{\rm max}^{b}$	$\log N_{\rm e}^{c}$ (range)
Density-sensitive line pairs			
Mg VIII	782.37/772.29	5.91	< 8
Si VIII	1440.51/1445.76	5.92	6–9
Si IX	950.16/694.69	6.05	> 11
Si IX	1822.66/1984.88	6.05	9–11
S X	1196.21/1212.97	6.14	7–10
Fe XII	1349.37/1242.00	6.22	8–12
Ar XI	745.80/1392.10	6.22	10–13
Ar XII	1018.75/1054.59	6.36	8–11
Ca XIII	648.68/1133.76	6.42	> 8
Ar XIII	656.69/1330.53	6.42	10–13
Ca XIV	880.40/943.63	6.49	10–12
Fe XIX	1328.79/1118.07	6.86	> 8
Fe XX	721.55/821.71	6.92	> 10
Fe XXI	786.03/1354.06	7.00	> 11
Temperature-sensitive line pairs			
Ne VII	973.33/895.16	5.71	
Si VII	909.43/1049.20	5.79	
Si VII	1135.46/1049.20	5.79	
Mg VIII	679.79/772.29	5.91	
Si VIII	988.21/944.47	5.92	
Si VIII	1189.51/944.47	5.92	
Mg IX	749.56/706.04	5.99	
S IX	714.80/871.73	6.02	
S IX	882.67/871.73	6.02	
S X	804.12/787.56	6.14	
S X	946.30/776.25	6.14	
Ar XI	717.86/745.80	6.22	

[a] Line wavelengths in Å.
[b] $T_{\rm max}$ in K.
[c] $N_{\rm e}$ in cm^{-3}.

of allowed lines with transitions between the $n = 2$ configurations in N and O, and $n = 3$ configurations in S. Many of these lines are sensitive to temperature and density. Examples of particular importance are the density-sensitive O V lines at around 760 Å and the temperature-sensitive O IV and O V lines at around 790 Å (Figure 6.17). Other examples of density-sensitive lines are shown in Figures 6.8 (Mg-like ions), 6.9 (Be-like ions), 6.16 (N-like ions), and 6.18 (B-like ions). The presence of many lines from consecutive stages of ionization allows the DEM to be studied from chromospheric to transition region temperatures.

Above 912 Å, with few exceptions the solar disk spectrum is crowded with lines formed at low temperatures. The coronal lines become steadily weaker and only a few of them can be distinguished. The most important among these are doublet lines from

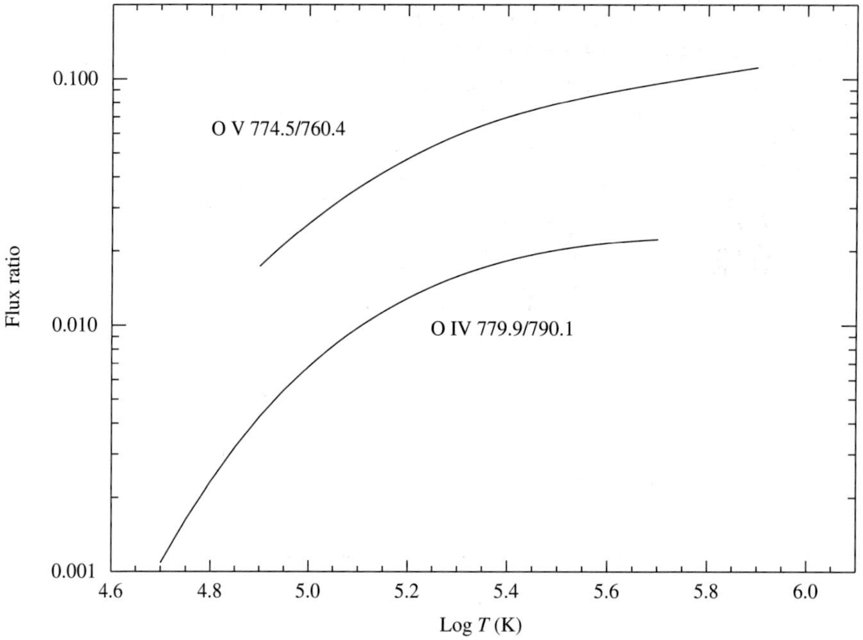

Fig. 6.17. Temperature-sensitive line pairs near 790 Å in O IV and O V. Ratios are calculated with the CHIANTI database.

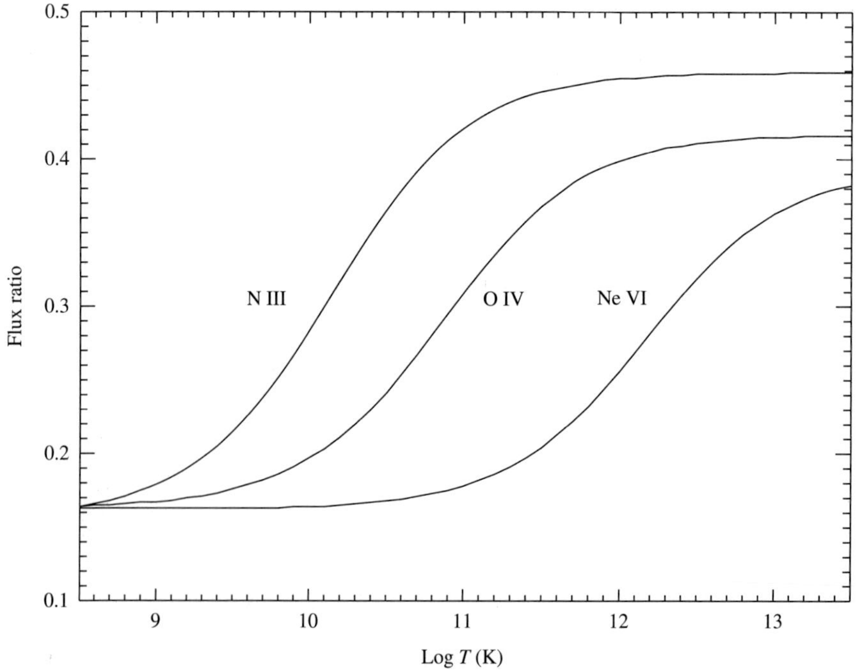

Fig. 6.18. Density-sensitive line pairs in B-like ions: N III (1754.00/1749.67), O IV (1407.38/1401.16), and Ne VI (1010.21/999.18). Ratios are calculated with the CHIANTI database. Transitions are $(2s^22p\ ^2P_{3/2} - 2s2p^2\ ^4P_{1/2})/(2s^22p\ ^2P_{3/2} - 2s2p^2\ ^4P_{5/2})$.

Li-like and Na-like ions: in addition to N v and O vi already described in Section 6.6, important lines from these sequences observed in the solar disk spectrum are the C iv doublet at around 1550 Å, extensively used by the *Skylab*, HRTS and *TRACE* instruments to observe the upper chromosphere, and the lines of Na-like Si iv and S vi lines at around 1390–1405 Å and 930–940 Å respectively. The ratios of the doublet components have been used to investigate opacity. Optical thickness is a potential problem for many density-sensitive lines longward of 912 Å. Many ratios between forbidden and intercombination lines in Be-like, B-like, Mg-like, and Al-like ions can be affected by opacity and therefore yield incorrect values of the density, if opacity is not taken into account.

The long rest wavelengths of lines of cool ions above 912 Å make them very useful for studying plasma dynamics, using both Doppler shifts and line profiles. Even small velocities of a few km s^{-1} cause shifts in the line centroid and distortions in the line profile to be readily measurable by high-resolution spectrometers, so allowing detailed studies of chromospheric and transition region dynamics including explosive events.

7

Spectrometers and imagers for observing the solar ultraviolet and X-ray spectrum

7.1 Introduction

The vast number of observations from space of solar upper atmospheric phenomena made thus far, their precision and sophistication, are due to capabilities as well as shortcomings of imaging instruments and spectrographs or spectrometers that have been used. It is the intention of this chapter to provide a general description of the instrumentation that has been commonly used in observations of the solar atmosphere in the ultraviolet and X-ray parts of the spectrum.

Temperatures of solar upper atmosphere plasmas vary by three orders of magnitude (3×10^4 K to 30 MK) and the emitted spectra span the vacuum ultraviolet (VUV, 1000–2000 Å), extreme ultraviolet (EUV, 100–1000Å) and the soft X-ray (1–100 Å) ranges. An ideal diagnostic instrument for exploring such plasmas would need to have high spectral, spatial, and temporal resolution in a wide field of view ($\sim\pm30$ arcmin) in all spectral ranges. To achieve such a goal, an instrument would need to record simultaneously data in four dimensions – two spatial, one spectral, and one temporal. Technically, the development of an instrument possessing all those properties is not feasible. This is not only because of technical and data gathering complexities, but also because the transmissive, reflective, and dispersive properties of materials needed for such an instrument are strongly wavelength dependent. No material can provide satisfactory performances at all wavelengths. To alleviate those difficulties, some instruments are constructed primarily as spectrometers and others as imaging instruments ('imagers'), each operating over limited wavelength ranges. For a review of VUV and EUV solar imagers and spectrometers, see Wilhelm *et al.* (2004).

Imagers employed in solar physics research are mostly telescopes using normal-incidence or grazing-incidence mirrors capable of imaging the Sun over narrow wavelength bands. As such, most imagers are constructed as telescopes that incorporate in their design some type of filter. Some designs incorporate filters that transmit a narrow wavelength band while rejecting the rest; other designs have telescope optics that are made to reflect fairly efficiently some wavelengths while rejecting the rest.

Spectrometers in the VUV and EUV usually consist of a telescope that images a portion of the solar upper atmosphere on the entrance aperture of the spectrometer, a dispersive element (usually an optical grating), a focusing element, and a detector. Often, the dispersive and focusing functions are incorporated in the grating design. In X-ray spectrometers a Bragg-diffracting crystal is often used as the dispersive element. Owing to the unique properties of X-ray crystals, no focusing optics are required. Spectrometers capable of operating in a raster mode, i.e. a mode in which the solar image is stepped across the aperture, can act as imagers.

Before describing several instruments as representatives of the basic classes successfully employed in solar studies, a brief review is presented on transmittance, reflectance, and dispersive properties of materials in the relevant wavelength ranges.

7.2 Transmittance properties of materials

7.2.1 *VUV transmittance of materials*

The materials lithium fluoride (LiF), magnesium fluoride (MgF_2), calcium fluoride (CaF_2), barium fluoride (BaF_2), lanthanum fluoride (LaF_3), silicon oxide (SiO_2), other types of fused silica, and sapphire (Al_2O_3) are generally chosen for windows or lenses in the VUV. The behaviour of the transmittance of these materials as a function of wavelength is similar. At wavelengths of 2000 Å and longer, the transmittance windows made from such materials may approach 90%. However, at shorter wavelengths the transmittance decreases, and below a cut-off wavelength becomes zero.

Lithium fluoride: High-quality LiF samples are transparent to radiation down to about 1050 Å. Unfortunately, LiF reacts with water vapour in the atmosphere causing it to suffer transmission loss. Since the reaction with water results in composition changes, simple cleaning will not restore the transmission of the window to its original conditions. Also, when bombarded with 1 MeV electrons, LiF becomes fairly opaque to VUV radiation if the dosage reaches values as large as 1×10^5 electrons cm^{-2}.

Magnesium fluoride, calcium fluoride, and barium fluoride: High-quality samples of MgF_2, CaF_2, and BaF_2 are transparent to radiation down to 1130 Å, 1220 Å, and 1340 Å respectively. In contrast to LiF, the solubility of these materials in water is low and as a result they suffer very little or no degradation from humidity. Although electron bombardment alters the VUV transmission of MgF_2, its effect is significantly lower than it is for LiF. Among the fluoride crystals used as VUV windows, BaF_2 has the highest resistance to degradation by electron bombardment.

Silicon oxide and sapphire: Sapphire has a cut-off wavelength of about 1450 Å, while the wavelength cut-off of various grades of fused silica can vary up to 2000 Å. The VUV transmittance of fused silica is slightly affected by electron bombardment, but sapphire is affected significantly less.

7.2.2 *EUV transmittance of thin films*

A variety of thin metal films (thicknesses of several hundred to several thousand Å) are transparent to radiation in parts of the EUV and opaque in other parts of the spectrum. The thin films most commonly used as windows or filters in the EUV are beryllium, silicon, aluminium, magnesium, germanium, tin, and indium (Hunter (1988)), as well as various plastics, e.g. kapton, vinyls, polypropylene, polystyrene, and saran. Figure 7.1 shows the transmittance as a function of wavelength of 1 μm thick beryllium foil and 800 Å thick films from each of several heavier elements. Filter structures from organic materials or from

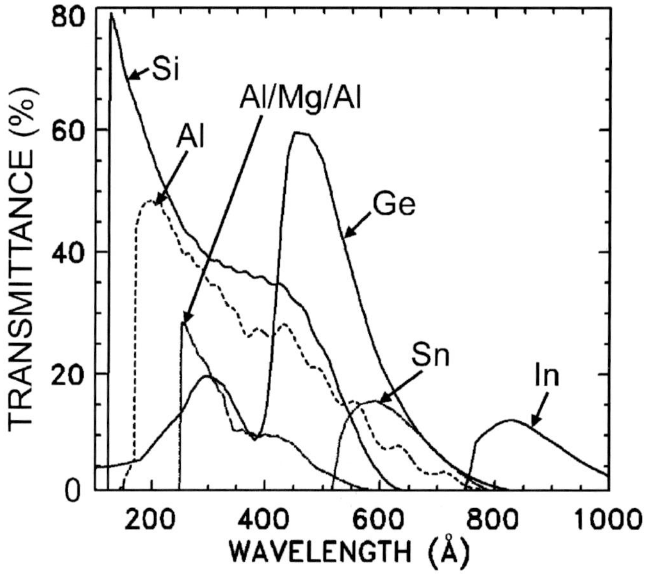

Fig. 7.1. Transmittance as a function of wavelength of common thin films used in space research. Courtesy J. Seely (NRL), data from Hunter (1988).

combinations of organic materials and metals are also used as narrow-band transmittance windows for the X-ray region (Tsuneta *et al.* (1991)).

7.3 Reflectance of VUV, EUV, and X-ray radiation from materials

7.3.1 Normal-incidence reflectance

In general, the highest spatial resolution imagers and spectrometers are designed such that incident and reflected rays are nearly normal to the optical elements they encounter. Since such instruments employ several optical elements, their total reflectance R_{tot} is the product of the reflectance R_i of all elements along the optical path, i.e.

$$R_{tot} = \prod R_i, \tag{7.1}$$

where $R_i < 1$.

Aluminium, boron carbide, and silicon carbide: Aluminium (Al) over-coated with a thin layer of magnesium fluoride is the surface of choice in the VUV. The normal-incidence reflectance of such surfaces approaches 90% at wavelengths longer than 1150 Å. At wavelengths shorter than 1150 Å, SiC and B_4C appear to be the best known reflective materials. While silicon carbide (SiC) has a higher reflectance at wavelengths longer than 750 Å, the reflectance of boron carbide (B_4C) at shorter wavelengths is much higher. For details see Figure 7.2.

High-Z elements: High-Z elements such as gold, tungsten, platinum, iridium, and several others appear to be the best normal-incidence reflectors in the 170–500 Å range. In part of this range, their reflectance reaches values as high as 25% (Figure 7.2).

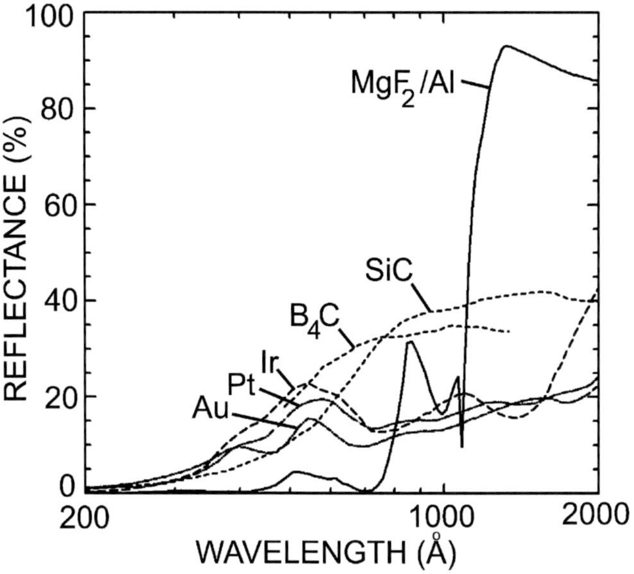

Fig. 7.2. Reflectance of coatings as a function of wavelength of MgF_2, SiC, and high-Z elements. Courtesy J. Seely (NRL).

Reflectance of multilayer surfaces: In recent years the technology of precise deposition of very thin (tens of angstroms) films has been developed. By depositing ten or more pairs of layers on very smooth surfaces, where the material of one of the pair has a very different transmittance and reflectance from the other, the reflective properties of these surfaces can be modified. There are several families of such multilayer coatings, each consisting of a few tens of pairs of materials, which can be designed to have significant reflectance (~60%) in a narrow wavelength range ($d\lambda/\lambda$ ~0.01–0.1). Outside the designed range, the reflectance falls off to very small values. Figure 7.3 shows the reflectance of some multilayer coatings. In essence, multilayer coatings operate as narrow-band filters. The multilayers of choice for the 170–350 Å range are usually made from the following pairs of materials: Si–Mo, Si–Mo_2C, or Si–B_4C. At shorter or longer wavelengths other pairs are selected. In the laboratory, multilayer surfaces have been successfully deposited to reflect radiation as short as 30 Å.

Based on the above, it is not surprising that normal-incidence spectrometers having high spectral and spatial resolution over a wide spectral range are available only for wavelengths of 500 Å and longer. Examples are the *Skylab* S082B and High Resolution Telescope Spectrometer (HRTS) instruments, both employing aluminium-coated optics, and the SUMER instrument on *SOHO*, designed with silicon carbide reflective optics. At shorter wavelengths, high spectral and spatial resolution spectrometers are designed with multilayer optics, and so operate over narrow wavelength bands. A case in point is the EIS instrument on *Hinode* operating in the 170–211 Å and 246–291 Å bands (Culhane *et al.* (2007)). The *SOHO* EIT and *TRACE* instruments are two examples of normal-incidence EUV imagers employing multilayer-coated optics.

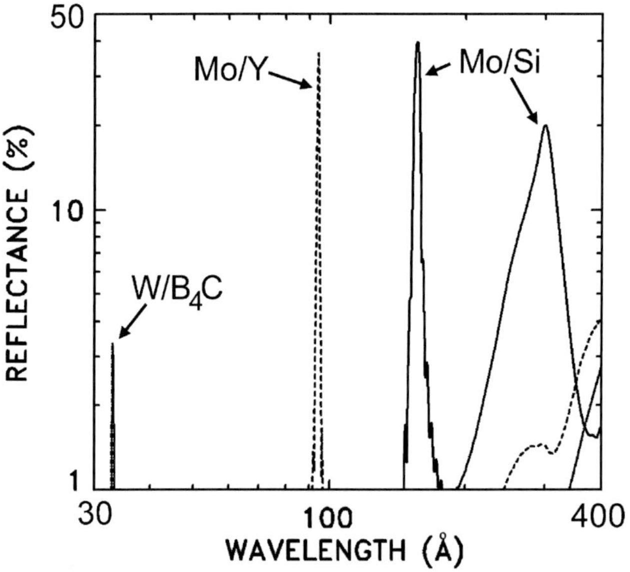

Fig. 7.3. Reflectance of some multilayer coatings. Courtesy J. Seely (NRL).

7.3.2 *Grazing-incidence reflectance*

When light is incident on an optical surface at a small (grazing-incidence) angle, the reflectance is high. For each wavelength, there is a critical grazing angle above which the reflectance diminishes. The value of this angle, θ_c, is approximately given by

$$\sin\theta_c = \lambda_{min}\left(\frac{N e^2}{mc^2\pi}\right)^{1/2}, \tag{7.2}$$

where e and m are the electron charge and mass, N the number of electrons per unit volume, and λ_{min} is the wavelength corresponding to θ_c (Samson (1967)). By inserting the values of the constants in the equations we obtain

$$\lambda_{min} = 3.33\times 10^{14}\, N^{-1/2}\, \sin\,\theta_c, \tag{7.3}$$

where λ_{min} is in Å. For aluminium, the value of $N^{1/2}$ is 8.8×10^{11} (N in cm^{-3}); for gold, platinum, and iridium, which have the largest number of electrons per unit volume, the values are a factor 2.5–2.6 larger. Thus, for aluminium, $\lambda_{min} = 6.6\times\theta^\circ$, while for platinum the value is $\lambda_{min} = 2.6\times\theta^\circ$ (θ° is θ in degrees).

Note that in order to reflect radiation at wavelengths as short as 1 Å, the angles of incidence for platinum must be less than 0.5°.

7.4 Ultraviolet spectrometers

7.4.1 *Concave spherical gratings*

Spectrometers consist of a narrow entrance slit or aperture, a dispersive element, a focusing element, and a detector on to which the dispersed radiation is focused. Owing to the low reflectance of material at short wavelengths, the dispersing and focusing elements are usually combined. The most common spectrometer design used at short wavelengths

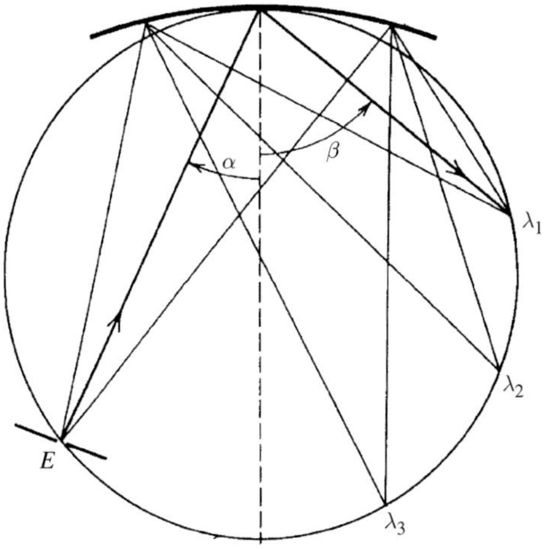

Fig. 7.4. The Rowland circle geometry where the grating radius of curvature R is twice the radius of the Rowland circle. Radiation from point E is dispersed and focused by the grating at λ_i ($i = 1, 2, 3$) along the circumference of the Rowland circle. α and β are the angles of incidence and diffraction respectively. The dashed line is the grating normal.

incorporates a concave grating of radius R mounted such that its vertex is tangential to a circle of radius $R/2$, known as the *Rowland circle.* A beam of light passing through a narrow slit on the Rowland circle and striking the grating will be dispersed and focused along the circle. Figure 7.4 illustrates the geometry.

The relation between the slit location, grating, and the image-focusing location is given by

$$m\lambda = d\,(\sin\alpha + \sin\beta), \tag{7.4}$$

where m is the spectral order, d the grating line spacing, and α and β are the angles of incidence and reflection respectively.

The angular dispersion of a Rowland circle spectrometer is

$$\frac{d\beta}{d\lambda} = \frac{m}{\cos\beta}, \tag{7.5}$$

and its resolving power RP is

$$\mathrm{RP} = \frac{\lambda}{d\lambda} = mN, \tag{7.6}$$

where N is the number of ruled grating lines exposed to the incoming radiation.

Concave gratings can be used not only to diffract but also to image radiation. In the case of a spherical grating of radius r, its focal length along the surface normal is $r/2$. Figure 7.5 illustrates this. At any other direction, the focal lengths in the tangential (f_t) and in the sagittal (f_s) directions are given by

$$f_t = \frac{r\cos^2\beta}{\cos\alpha + \cos\beta} \tag{7.7}$$

and

$$f_s = \frac{r}{\cos\alpha + \cos\beta}. \tag{7.8}$$

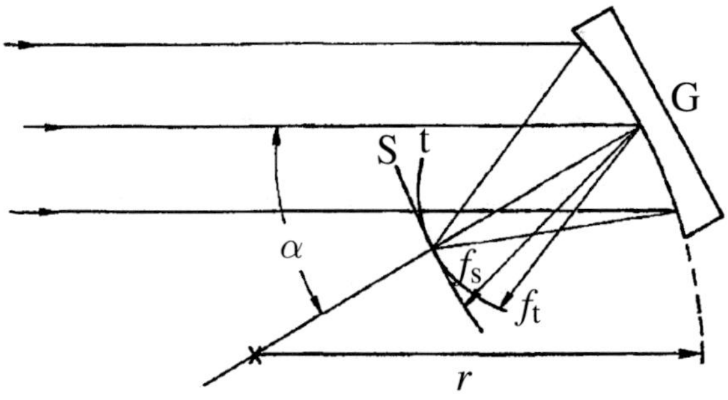

Fig. 7.5. Wadsworth mounting of concave grating. Parallel light is incident on the grating G, which has a radius of curvature r at an angle α with respect to the grating normal. f_s and f_t are the sagittal and tangential focal distances from the grating.

As seen from these equations, stigmatic imaging ($f_t = f_s$) occurs under the condition of $\beta = 0$, commonly known as the Wadsworth mount (Wadsworth (1896); Samson (1967)). Many spectrometers including the *Skylab* S082A spectroheliograph and SUMER on *SOHO* to be discussed later utilize the Wadsworth mount geometry.

7.4.2 Concave toroidal gratings

EUV stigmatic spectrographs for use in analyzing spectra that appear to emerge from a nearby location are difficult to design because of the competing requirements of a nearly grazing incidence angle to achieve a reasonable reflectivity and near-normal-incidence angles to minimize aberrations. The most easily achieved surface which permits partial correction of the aberration is a toroid. We therefore consider the image properties of a toroidal grating having radii of curvature R_s in the direction parallel to the rulings and R_t perpendicular to that direction. A point source of light located a distance r from the centre of the grating and at an angle α with respect to the normal at the grating centre will be imaged into tangential and sagittal foci for a wavelength corresponding to the diffraction angle β at distance r'_t and r'_s given by the following equations (Samson (1967)):

$$r'_t = \frac{R_t \cos^2\beta}{\cos\alpha + \cos\beta - R_t \cos^2\alpha/r} \tag{7.9}$$

and

$$r'_s = \frac{R_s}{\cos\alpha + \cos\beta - R_s/r}. \tag{7.10}$$

As a measure of the stigmatic blur, one can consider a function Δ defined to be the magnitude of the differences between r'_t and r'_s divided by the average image distance. Thus,

$$\Delta = \frac{2|r'_s - r'_t|}{r'_s + r'_t}. \tag{7.11}$$

By substituting Equations (7.9) and (7.10) into Equation (7.11), and differentiating with respect to β, it can be seen that Δ has a minimum at $\beta = 0$ regardless of the values of r. Further, if we set $r'_t = r'_s$ at $\beta = 0$, the ratio R_s/R_t ($\equiv f$) must satisfy the equations

$$r = \frac{R_s(1 - \cos\alpha)}{1 - f} \tag{7.12}$$

and

$$r' = \frac{R_s(1 - \cos\alpha)}{f - \cos^2\alpha}. \tag{7.13}$$

For real object and image distances, r and r' are greater than zero, and f must satisfy the relation

$$\cos^2\alpha < f < 1. \tag{7.14}$$

For additional details of the toroidal grating focusing properties (finite congregates), see Feldman *et al.* (1976c). The Solar Extreme-ultraviolet Research Telescope and Spectrograph (SERTS) is an example of an instrument designed according to the toroidal grating geometry concept for infinite congregates described above (Neupert *et al.* (1992)).

Spectral line profiles are influenced by plasma temperature and by non-thermal mass motions. Non-thermal mass motions (most probable velocities, ξ) of solar plasmas having temperatures of 3×10^4 K to 3×10^7 K are seldom smaller than 10 km s^{-1}. Constrained by the plasma properties, solar instruments with high spectral resolution are designed to achieve resolving powers that are equivalent to non-thermal mass motion broadenings of $\xi \sim$10 km s^{-1}. Instruments having spectral resolution $\lambda/\Delta\lambda = c/v \sim$30 000 are adequate for solar atmosphere studies.

7.4.3 Spectrometer detectors

The common high spatial resolution recording medium used in the past for the VUV and EUV ranges was special photographic plates (of any desired length) having very fine (10 µm) grain sizes. As a result, attempts were usually made to design spectrometers capable of simultaneously recording spectra over many hundreds of angstroms. One major deficiency of VUV and EUV photographic materials was their fairly limited dynamic range. To overcome this, multiple spectra with increasing exposure times were often taken at each location. Clearly, photographic plates are very useful in cases in which they could be retrieved, as was the case with the *Skylab* manned space mission, and processed on the ground. Since in typical space missions such an option is not available, an electronic recording device must be used. While photographic materials are practically unlimited in their storage capacity, present-day electronic recording devices are limited to several thousand resolution elements in both dimensions. As a result, instruments equipped with electronic devices designed to achieve a resolution of 30 000 can expose only several tens of angstroms at any one time. Examples of such instruments are SUMER on *SOHO* and EIS on *Hinode*.

7.4.4 Grating spectrometers

The *Skylab* S082B spectrometer (Figure 7.6) was a high-resolution normal-incidence spectrometer with order-sorting gratings. Being equipped with long strips of sensitive photographic film, it could record the 970–3940 Å solar spectrum in two spectral orders (970–1970 Å and 1940–3940 Å). The S082B instrument consisted of a front mirror imaging the Sun on the spectrometer entrance plane that had a $2'' \times 60''$ slit. Light entering the slit was pre-dispersed along the slit direction and imaged on an aperture by one of two gratings. One grating pre-dispersed the light so that the 970–1970 Å spectral range could pass through the aperture, while the other was aligned allowing the 1940–3940 Å range to pass. The main

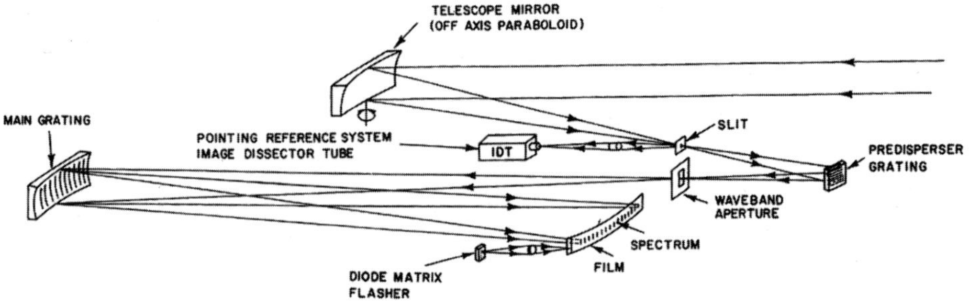

Fig. 7.6. The optical layout of the *Skylab* S082B instrument.

grating dispersed the incoming radiation in a direction perpendicular to the slit. The spectrometer resolving power in the 970–1970 Å range was $\lambda/\Delta\lambda$ ~30 000. With an accurate pointing system using the white-light limb as reference and a rigid mounting on a rotating base, the *Skylab* astronauts were able to point the spectrometer slit accurately along any solar direction (Bartoe *et al.* (1977)).

The *Skylab* S082B instrument recorded more than 8000 spectra from a variety of solar features on the disk and close to the limb. Observations close to the limb were mostly done automatically. A typical limb observation consisted of spectra at 10 predetermined locations ($-12''$, $-4''$, $-2''$, 0, $2''$, $4''$, $6''$, $8''$, $12''$, and $20''$: negative values indicate angular distances inside the limb, positive outside the limb, with the slit tangential to the limb) with accuracies of better than $\delta\theta = 0.5''$. To overcome the limited dynamic range of the photographic film, sets of three or four spectra varying in exposure time by factors of 4 were often taken at each pointing. In doing so, it was possible to derive line fluxes with ~15 % accuracy over dynamic ranges of several hundred. Copies of the entire collection of S082B spectra are deposited with the National Space Science Data Center (NSSDC).

The *SOHO* SUMER instrument (Figure 7.7) consists of a telescope mirror and a stigmatic normal-incidence spectrometer. The spectrometer was designed to image the incoming radiation on the detector with a spatial resolution of $1'' \times 1''$. It consists of an entrance slit, an off-axis paraboloid mirror collimating the light passing through the slit, a flat mirror that deflects the light onto the grating, a concave grating in a Wadsworth configuration, and two imaging detectors. The two detectors were designed to record the spectra alternately. Detector A covers the 780–1610 Å range in first order, detector B the 660–1500 Å range in first order. Second-order lines are superimposed on the first-order spectrum. Each detector has an array of 1024 (spectral) × 360 (spatial) pixels, each 26.5 μm^2 in size, and covers a spectral range of 43 Å in first order. The angular scale of a pixel at 800 Å is 1.03 arcsec, while at 1600 Å it is 0.95 arcsec. In the first-order spectrum, a pixel corresponds to 45.0 mÅ at 800 Å and 41.8 mÅ at 1600 Å. During observations, one of four slit sizes could be put in operation: 4×300 arcsec^2, 1×300 arcsec^2, 1×120 arcsec^2, or 0.3×120 arcsec^2.

The optical design of SUMER includes three normal-incidence reflections (telescope mirror, collimating mirror, and grating) and a single grazing-incidence mirror, all coated with silicon carbide, enabling it to operate with moderate efficiency in the 500–1600 Å range. The central part of the detector photo-cathode surface is coated with potassium bromide and the remainder uncoated. Since the efficiency of the detector section coated with potassium bromide is higher than the efficiency of the uncoated parts over most of the wavelength range,

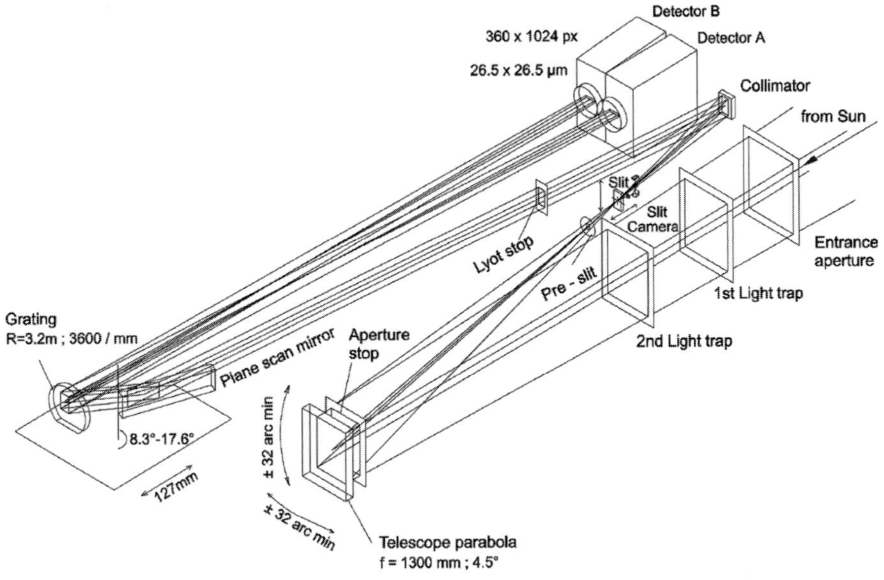

Fig. 7.7. Optical layout of the *SOHO* SUMER instrument (Wilhelm *et al.* (1995)). Courtesy the SUMER team and the Max Planck Institute for Aeronomy and Extraterrestrial Physics.

an intensity comparison between a line recorded on the potassium bromide with its intensity from the uncoated section unambiguously reveals its order. The details of the SUMER instrument and its modes of operation are described by Wilhelm *et al.* (1997).

The SUMER instrument has recorded a very large number of spectra from structures on the solar disk and as far as $1.5R_{\odot}$ above the limb. The spectra are recorded in a variety of modes, from a single frame covering a range of wavelengths ($\Delta\lambda \leq 43$ Å), to multi-frame sequences spanning the entire operational range. SUMER spectra are absolutely calibrated to better than 20%.

The Extreme-ultraviolet Imaging Spectrometer (EIS) on *Hinode* (Culhane *et al.* (2007)) is a normal-incidence instrument consisting of a telescope and a spectrometer designed to record stigmatic spectra in the 170–211 Å and 246–291 Å regions. In order to achieve a reasonably high normal-incidence reflectivity, Mo–Si multilayer coatings were applied to the two optical elements, namely, an off-axis paraboloid telescope mirror and a toroidal grating. The telescope mirror (focal length 1934 mm) is designed to create a solar image on the slit plane. Light passing the slit is dispersed and focused on the detectors by the grating ($R_s = 1183$ mm; $R_t = 1178$ mm). To attain high reflectivity in two different wavelength bands, one half of the telescope mirror and one half of the grating were coated to accommodate the shorter band, the other telescope mirror half and corresponding grating half were coated to accommodate the longer band. The efficiencies of the grating and of the coatings were measured with synchrotron radiation (Seely *et al.* (2004)).

To protect the delicate optical coatings from the intense long-wavelength solar radiation, two aluminium filters are included in the optical path, one in front of the telescope mirror and the second in front of the grating. The two CCD detectors consist of 1024 rows and 2048 columns of 13.5 µm pixels. One CCD serves the long-wavelength band and the other the short-wavelength band.

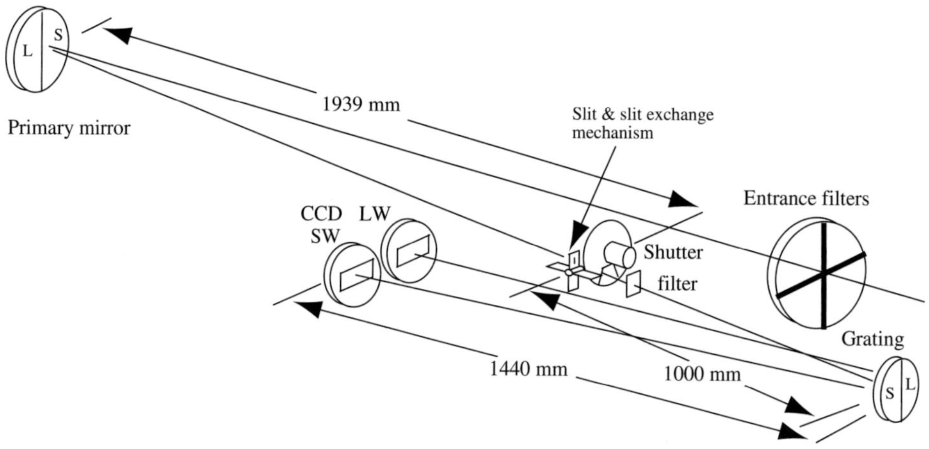

Fig. 7.8. Optical layout of the EIS instrument on *Hinode*. Courtesy the EIS team.

Coarse and fine motions of the telescope mirror cause the solar image to move on the slit plane in the solar east–west direction. The range of the fine motion is limited to 600 arcsec, and is designed to produce high spectral and spatial resolution images by rastering the mirror. The EIS instrument is equipped with four slits: two narrow slits of 1 arcsec and 2 arcsec width for forming spectral lines, and two wide slits of 40 arcsec and 266 arcsec for forming monochromatic images. Figure 7.8 is a schematic of the EIS optical layout.

Before the development of multilayer coating technology, grazing-incidence spectrometers were in essence the only practical devices to record spectra in the EUV range. Although grazing-incidence spectrometers have a high reflectance over the entire EUV range, they focus well only in the dispersion direction. Early in the space era, rocket-borne grazing-incidence spectrometers were extensively used to survey the solar spectrum. The most recent solar physics mission incorporating a grazing-incidence spectrometer is *SOHO*.

The CDS instrument on *SOHO* (Figure 7.9) is designed as a high-spatial-resolution double spectrometer device in which one half operates at normal-incidence angles and the other at grazing angles (Harrison *et al.* (1995)). To achieve this goal, a Wolter–Schwarzschild type 2 telescope feeds simultaneously a normal-incidence spectrometer (NIS) and a grazing-incidence spectrometer (GIS) which share a common entrance slit. Two portions of the incident beam are selected, one for the NIS and the other for the GIS. The NIS beam is split again into two where the parts are intercepted by normal-incidence toroidal gratings, reflected, and focused on to intensified CCD detectors. One of the CCD detectors records the 308–381 Å range and the second the 513–633 Å range. To build up large images, a plane mirror in front of the slit is scanned through a small angle in 2 arcsec steps. The NIS slit dimensions are 2×240, 4×240, and 90×240 arcsec^2. The GIS beam is intercepted by a grazing-incidence spherical grating and reflected on to four microchannel plate detectors located along the Rowland circle. The resulting GIS spectra are astigmatic. The wavelength bands they record are 151–221 Å, 256–338 Å, 393–493 Å, and 656–785 Å. By moving pinhole slits in one direction and the scan mirror in 2 arcsec increments in the plane of dispersion, monochromatic images can be formed. The dimensions of the GIS slits are 2×2, 4×4, and 8×50 arcsec^2.

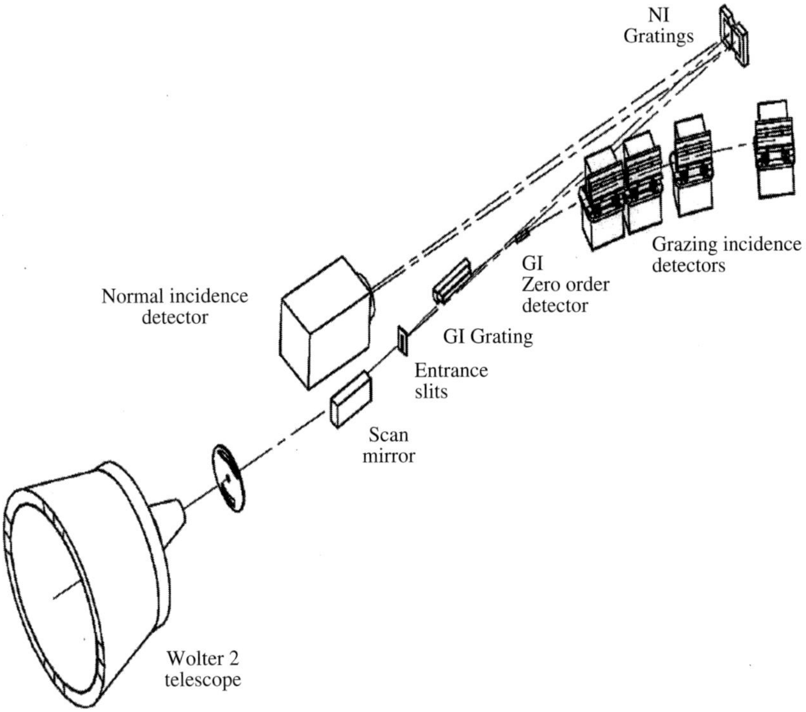

Fig. 7.9. The optical elements of the *SOHO* CDS instrument (Harrison *et al.* (1995)). Courtesy the CDS team.

7.5 X-ray spectrometers

7.5.1 *Broad-band spectrometers*

For many years, the X-ray spectra of flares have been monitored with ion chambers on some of the *Orbiting Solar Observatories* and the *Geostationary Operational Environmental Satellites* (*GOES*). The ion chambers on *GOES* have two channels covering wavelength bands nominally 0.5–3 Å and 1–8 Å. With this broad-band resolution, the temperature of the flare can be roughly estimated from the ratio of the output of the two channels. Figure 7.10 gives the wavelength dependence of the efficiencies. Each ion chamber is a sealed detector containing a xenon–argon gas mixture. X-rays are incident through beryllium windows, those for the 0.5–3 Å detector being slightly thicker, and ionize the detector gas. The electrons produced by the ionization process are attracted by the positive voltage on an anode wire, and on their passage through the gas produce more ionizations and more electrons. The avalanche that results gives rise to a voltage pulse. The number of pulses recorded by the electronics of each ion chamber per unit time is a measure of the X-ray flux.

Proportional counters were widely used as X-ray detectors, e.g. by a spectrometer built by the Mullard Space Science Laboratory (University College London) on the NASA spacecraft *Orbiting Solar Observatory-4* (*OSO-4*: Culhane *et al.* (1969)). The different proportional counters were uncollimated, so that the spectrometer, housed in the rotating wheel section of *OSO-4*, moved past the Sun every two seconds, detecting the total radiation from the

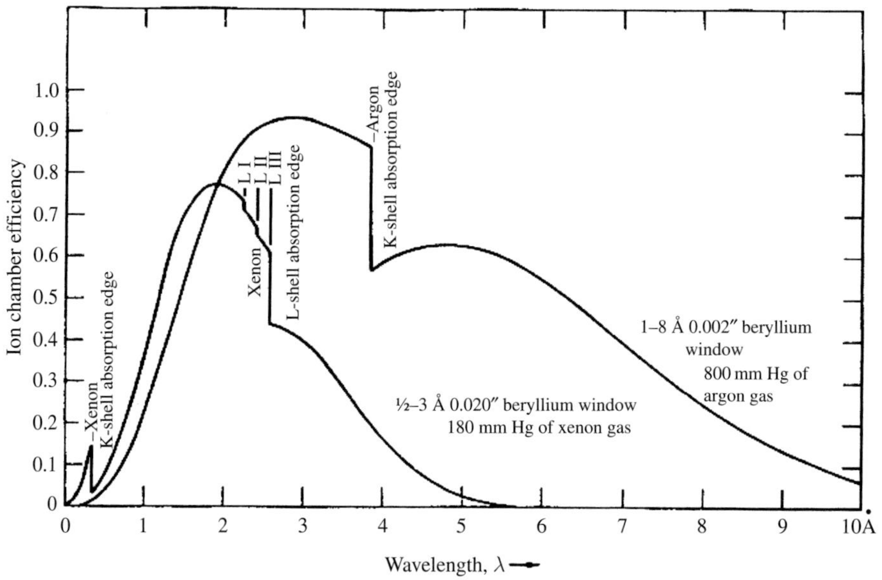

Fig. 7.10. Efficiencies of the 0.5–3 Å and 1–8 Å ion chambers of *GOES*. From Donnelly and Unzicker (1974).

Sun. Proportional counters, like ion chambers, detect X-ray photons on the principle of the ionization they produce in the gas of a detector, but with the height of the voltage pulse being proportional to the energy of the original X-ray photon. An energy (or wavelength) resolution of 15–20% can be achieved, i.e. at an energy of 6 keV (wavelength 2 Å), the energy (wavelength) resolution (FWHM) is $\sim$1 keV ($\sim$0.3 Å). This is adequate to resolve the Fe line feature at $\sim$1.9 Å as a bump on top of continuum emission – this line feature is emitted strongly during hot flares (see Section 10.2). Early proportional counters were often calibrated by in-flight systems, using small Fe^{55} radioactive sources producing 2.1 Å photons.

Scintillation detectors enable X-ray emission from solar flares to be observed, ranging from soft X-rays to much higher energies, up to many hundreds of keV. The scintillator material is often sodium iodide (NaI) crystal. X-ray photons enter the crystal, producing small flashes of optical light emission in the ionization process. The flashes are counted by photomultiplier detectors. Anti-coincidence techniques are used to filter out counts due to cosmic ray particles – a typical system is one in which the scintillation crystal is partly enclosed by a shell of similar material, with photomultipliers counting the optical flashes in both crystals (a cosmic ray particle is much more likely than an X-ray photon to pass through the shell and the crystal it encloses). With an energy resolution similar to proportional counters in the X-ray region, the Fe line complex at 6.7 keV (1.9 Å) can likewise just be resolved. Cryogenically cooled detectors, as with the germanium detectors making up the *RHESSI* spectrometer (Lin *et al.* (2002)), improve the energy resolution considerably at harder X-ray energies, where the emission during solar flare impulsive stages is dominated by non-thermal bremsstrahlung.

7.5.2 *Crystal spectrometers*

Much higher spectral resolution at X-ray wavelengths can be achieved with crystal spectrometers. Particular wavelengths λ can be selected by having a collimated X-ray source

Table 7.1. *Typical crystals used for observing solar X-rays.*

Crystal	2*d* (Å)	Typical wavelength (Å)	Typical values rocking curve (arcsec)	Peak reflectivity (per cent)
Potassium acid phthalate (KAP)	26.64	19.0	200	3
Acid diphosphate (ADP)	10.64	9.0	40	32
Quartz $(10\bar{1}0)^a$	8.51	6.0	44	10
Germanium $(220)^a$	4.00	3.17	50	47
Germanium $(422)^a$	2.31	1.85	10	50

[a] Numbers specify crystal diffracting planes.

incident on a flat crystal at a particular angle θ (measured from the crystal surface). Bragg diffraction occurs according to the formula

$$n\lambda = 2d \sin\theta, \tag{7.15}$$

where n is the Bragg diffraction order and d is the crystal plane spacing. Generally, the crystal reflectivity is highest for first-order ($n = 1$) diffraction. Collimation of X-rays incident from a source on the Sun can be achieved through two or more sets of identical grids consisting of fine wires that are opaque to the X-rays being observed. A proportional counter may be used to detect the diffracted X-ray photons. A complete spectrum over a chosen wavelength range can be built up by rotating the crystal about an axis, so that θ and therefore the wavelength λ of the detected X-rays change.

Monochromatic X-rays undergoing diffraction from a crystal produce a diffracted beam which, when scanned in angle by a detector, has a non-zero width $\Delta\theta$ called the rocking curve, the magnitude of which depends on the crystal material. An infinitely narrow spectral line is thus broadened into a line profile having a width $\Delta\lambda$ given by differentiating Equation (7.15):

$$\Delta\lambda = 2d \cos\theta \, \Delta\theta \tag{7.16}$$

for first-order diffraction. High-resolution spectroscopy therefore requires a choice of crystal with rocking curve as small as possible. In addition, the crystal peak reflectivity should be as high as possible. Table 7.1 gives parameters, including rocking curve widths and peak reflectivities, for Bragg-diffracting crystals that have been commonly used for observing solar X-rays. The crystal $2d$ values define the wavelength ranges available for each crystal – observed wavelengths must be less than $2d$.

Some large and sophisticated crystal spectrometers have included several crystals of different materials so that various wavelength ranges can be observed, as is desirable for low and high levels of solar activity. These include the SOLFLEX spectrometer on the *P78-1* spacecraft (Doschek (1983)) and the Flat Crystal Spectrometer on *SMM* (Acton *et al.* (1980)). The FCS had seven such crystals, all attached to a common, rotatable shaft, with a fine-wire collimator (square field of view, 14″ FWHM) to produce a nearly parallel beam of X-ray emission incident on the crystals. With the spectrometer fixed in space and pointed at the brightest element of an active region or flare, spectra could be observed by rotating the crystal shaft through a pre-defined range. The crystals were orientated such that, with the crystal shaft fixed, diffracted radiation from intense spectral lines in the wavelength range of each crystal were simultaneously incident on the detectors. Thus, by moving the whole spectrometer in a scanning motion, images of active regions or flares could be built up in these spectral lines. Figure 7.11(*a*) illustrates the principle of the FCS.

Curved crystals have been used in a number of solar spectrometers in the past 25 years. A range of wavelengths can be observed simultaneously by the fact that X-rays from a solar source, again incident via a collimator, have slightly different Bragg angles over the crystal face. Figure 7.11(*b*) shows the set-up of the *SMM* Bent Crystal Spectrometer. Diffraction occurs at angles θ that are slightly different (range θ_1 to θ_2) across the face of the bent crystal. The diffracted X-rays are detected by a proportional counter with position-sensitive anode. In a given time interval (typically a few seconds, so that rapidly developing flares can be observed), read-out of a complete spectrum over a range $\lambda_1 = 2d \sin \theta_1$ and $\lambda_2 = 2d \sin \theta_2$ can be accomplished. The wavelength range is related to the crystal radius of curvature R and its length L by

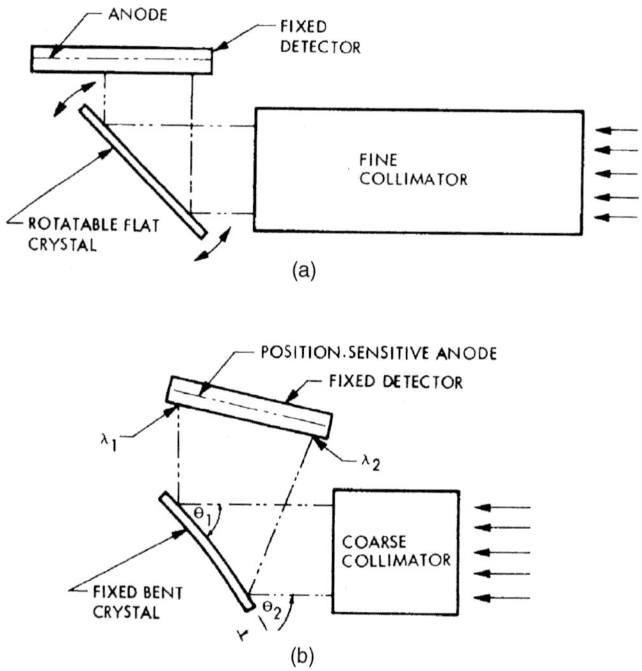

Fig. 7.11. Principle of the (*a*) Flat Crystal Spectrometer (FCS) and (*b*) Bent Crystal Spectrometer (BCS) on the *SMM* spacecraft. Solar X-rays in the direction indicated by arrows are incident on the crystals via collimators, undergo diffraction at each crystal of the FCS and BCS, and the diffracted X-rays are detected by proportional counters. For the BCS, diffraction occurs with different incident angles (from θ_1 to θ_2) across the bent crystal, and the diffracted X-rays are detected by a proportional counter having a position-sensitive anode. A complete spectrum, with wavelength range λ_1 to λ_2, is read out over short time intervals. From Acton *et al.* (1980).

$$R = \frac{L}{\theta_2 - \theta_1}. \tag{7.17}$$

Although the fabrication of bent crystals is far from easy (this is generally done by bending the crystal over a curved substrate), a bent crystal spectrometer has a number of advantages over those with flat scanning crystals. One is that it has no moving parts, with consequent reduced possibility of failure after launch. Another is the fact that a complete spectrum can be observed simultaneously, whereas a flat crystal spectrometer takes time to observe a particular range. This is a matter of some importance for observing fast-changing solar flares. Another advantage is the much larger effective crystal area compared with a flat crystal. The effective crystal area $A(\lambda)$ is given by (Rapley *et al.* (1977))

$$A(\lambda) = \frac{A_0}{\theta_2 - \theta_1} \frac{\lambda}{2d} R_{\text{int}}(\lambda), \tag{7.18}$$

where A_0 is the total crystal area and the integrated reflectivity is $R_{\text{int}} = \int R(\theta)\, d\theta$. Thus, the sensitivity of the S XV channel of the Bragg Crystal Spectrometer on *Yohkoh* was a factor of over 60 greater than that of the S XV channel of the *SMM* Flat Crystal Spectrometer (Culhane *et al.* (1991)).

A major problem with some crystal spectrometers, especially those working at short wavelengths, isthe background produced by fluorescence of the crystal material by solar X-rays. Thus, for quartz crystals, solar X-rays with wavelengths less than the K-edge of silicon (6.7 Å) produce Kα and other characteristic line radiation from the quartz. This radiates from the crystal in all directions and is incident, along with the diffracted solar X-ray emission, on the detector, forming a background. This radiation can be discriminated against in the detector electronics using the fact that the pulse height of the fluoresced radiation is generally different from the pulse heights of the solar X-rays. This was successfully done for two of the channels of the Polish Space Research Centre's RESIK (REntgenovsky Spektrometr s Izognutymi Kristalami) instrument on *CORONAS-F*, where pulse-height discriminator levels were carefully optimized over the lifetime of the instrument (Sylwester *et al.* (2005)). Fluorescence can also be reduced by the use of small collimators over the crystals themselves, so limiting the solid angle over which fluorescence is produced, as was done in some early solar crystal spectrometers built by the Lebedev group in Moscow, and by placing the crystal far from the detector.

For measuring line widths and shifts, which are of importance at the onset stages of flares, a crystal spectrometer's wavelength resolution is crucial. This depends, as has been mentioned, on the crystal rocking curve, but an additional factor is the broadening of a spectral line produced by the non-zero area of the emission region being observed. For finely collimated crystal spectrometers, this is not generally important, but for uncollimated crystal spectrometers such as the *P78-1* SOLFLEX spectrometer, the *Yohkoh* Bragg Crystal Spectrometer, and *CORONAS-F* RESIK instrument, 'spatial' broadening can be important. The spatial broadening in wavelength is given by Equation (7.16) with $\Delta\theta$ the spatial extent of the emitting region along the dispersion direction of the crystals. For the *P78-1* SOLFLEX spectrometer and the *Yohkoh* Bragg Crystal Spectrometer, a spatial extent in the north–south direction (this being the dispersion direction) of 1 arcmin (FWHM) amounted to a wavelength broadening of 1.2 mÅ in the S XV channel of the instrument, which in terms of a turbulent velocity is 43 km s^{-1}. Much larger turbulent velocities are measured during flare impulsive stages when the emitting region is comparatively small; but at late stages of flares with very long durations, which take the form of a slowly expanding X-ray arch, the spatial extent can be several arcminutes (Phillips *et al.* (2005a)), and can completely account for the apparent broadening of the S XV lines (see Section 10.7). Measuring flare turbulence velocities with uncollimated instruments therefore requires knowledge of the flare's spatial extent.

The observation of short-wavelength (blue-shifted) components in the X-ray spectra of flares in their impulsive stages is widely interpreted as the Doppler shift of material approaching the observer. Velocities of 150–300 km s^{-1} are commonly observed (Section 10.7), and sometimes much larger. The displacement in wavelength can also be produced by a secondary emission source spatially separated along the crystal dispersion axis. To distinguish a spatial from a velocity shift, a double crystal spectrometer is required, with the diffracting crystals orientated in directions facing each other – a so-called dopplerometer mode. This is the principle behind the Polish Diogeness instrument, a scanning flat crystal spectrometer on *CORONAS-F* (Sylwester *et al.* (2002)). Two of the four channels on Diogeness had identical quartz crystals ($10\bar{1}1$ diffraction plane), and observed Ca XIX line emission at 3.17 Å in dopplerometer mode. Spatially separated components in the flare resulted in a longer-wavelength shift for one channel, a shorter-wavelength shift for the other, but a Doppler shift due to approaching material always resulted in the short-wavelength direction for both channels.

7.5.3 *Microcalorimeters*

A new and highly sophisticated technology has been developed in recentyears to detect X-rays through the principle that X-ray photons incident on cryogenically cooled material produce tiny but measurable increases in temperature. The X-ray Spectrometer (XRS) instrument on the Japanese space mission *Suzaku*, designed to observe non-solar sources, was the first orbiting X-ray microcalorimeter, though shortly after the 2005 launch a failure of an interface between the spacecraft and dewar container led to the loss of liquid helium needed to maintain the temperature of the X-ray sensor to less than 0.1 K. Even so, measurements with internal calibration sources confirmed that the expected resolution of 7 eV (FWHM) at 6 keV (2mÅ at 2 Å) was achieved. The future *Constellation-X* mission, designed like *Suzaku* to observe non-solar X-ray sources, will carry the X-ray Microcalorimeter System (XMS), an array of 32 × 32 X-ray microcalorimeters, and will have a spectral resolution of ∼2.4 eV at 6 keV. At the wavelength of the Fe line complex at 6.7 keV, this resolution will be just adequate to observe the Fe XXIV satellite structure. The use of liquid helium for this newly developed technology will mean that the lifetimes of microcalorimeters in space will be limited to about two years.

7.6 Imaging instruments

An ideal imager employed in solar research needs to have high spatial resolution over a wide angular extent (∼1 degree) and a fast time cadence, as well as the ability to record images with a high degree of thermal purity. Since such instruments are beyond present-day technology, less challenging instruments have been constructed.

The VUV and EUV parts of the solar spectrum have numerous closely spaced lines, emitted by a multi-temperature plasma. Nevertheless, there are a few ranges where one, two, or three closely spaced lines, emitted by ions formed at nearly the same temperature, dominate the spectrum. Influenced by the properties of the solar spectrum, three classes of imagers – spectroheliograph, multilayer normal-incidence telescope, and telescope–spectrometer combination – have been designed. Spectroheliographs and multilayer normal-incidence telescopes are designed to record entire solar regions in a single exposure, while telescope–spectrometers raster across solar regions over periods of time.

7.6.1 The Skylab S082A spectroheliograph

The S082A slitless spectroheliograph on *Skylab* was based on a Wadsworth design (Tousey *et al.* (1977)). Its main optical element was a 3600 lines mm^{-1}, 400 cm radius concave grating, and its recording medium consisted of long strips of EUV-sensitive photographic film. A 1000 Å aluminium filter was used to suppress undesired long-wavelength radiation. The spectroheliograph covered the 173–333 Å and 320–480 Å ranges at 2 arcsec spatial resolution and the range between 480 and 630 Å with a reduced resolution. Its spectral dispersion was 1.39 Å mm^{-1} and the diameter of the solar image it produced was 18.3 mm, corresponding to a width of 26 Å in the dispersion direction.

Since the spectral line density in the solar EUV range is often higher than one per Å, it is not surprising that many of the S082A images, and in particular those produced by lines uniformly emitted by the entire disk, overlapped considerably. Nevertheless, there were cases where the images were not blended, so supporting many important studies. Such in particular were recordings associated with flares or small (1 arcmin) active regions that were significantly brighter than their surroundings. Even when no such regions were present, fairly unblended full disk images were obtained in the intense He II 304 Å, Ne VII 465 Å, Fe XV 284 Å, and Fe XVI 335 Å lines. A large collection of spectroheliograph images was published in a two-volume atlas (Feldman *et al.* (1987)). A copy of all the original spectroheliograph recordings is stored in the National Space Science Data Center.

7.6.2 Multilayer normal-incidence telescopes

The Extreme-ultraviolet Imaging Telescope (EIT) on *SOHO* (Delaboudinière *et al.* (1995)) and the *Transition Region and Coronal Explorer (TRACE)* (Handy *et al.* (1999)) are recent examples of EUV normal-incidence telescopes employing Mo–Si multilayer coatings. EIT and *TRACE* were designed to record images in narrow spectral ranges each dominated by bright lines emitted by similar temperature plasmas. The EIT instrument is a Ritchey–Chrétien telescope with an effective focal length of 165.2 cm, having a $45' \times 45'$ field of view and a spatial resolution of $2.6'' \times 2.6''$. *TRACE* is a Cassegrain telescope with an effective focal length of 866 cm, a $8.5' \times 8.5'$ field of view and a $0.5'' \times 0.5''$ spatial resolution. In each telescope the mirrors were divided into four quadrants, each with a multilayer coating peaking at different 6–10 Å wide bands. Common to both imagers are the 171 Å channel (nominal wavelength) dominated by Fe IX and Fe X lines, the 195 Å channel dominated by Fe XII, Fe XXIV, and Ca XVII lines, and the 284 Å channel reflecting a strong Fe XV line. The fourth quadrant of EIT is designed to reflect the 304 Å He II and Si XI lines and the *TRACE* fourth channel is designed to reflect four VUV wavelength bands near 1700 Å, 1600 Å, 1550 Å (C IV line), and 1216 Å (H I Ly-α line). During periods when there is low solar activity, the images provide good descriptions of quiet coronal plasmas. However, when flares occur, in addition to the Fe XXIV and Ca XVII emission in the 195 Å channel, detectable bremsstrahlung emission is also expected to be present in all channels (Feldman *et al.* (1999a), Phillips *et al.* (2005b)).

7.6.3 Telescope–spectrometer combinations

The traditional white-light spectroheliograph is the prototype of present-day telescope–spectrograph imagers used in space research, e.g. the CDS and SUMER instruments on *SOHO* and the EIS instrument on *Hinode*. Although the time required to record an image by such instruments is considerably longer than with a traditional imager, these instruments are able to produce images from a single line or a narrow part of the continuum in a crowded spectral range.

For the SUMER instrument, the telescope mirror is an off-axis paraboloid designed to image on the spectrometer slit any area within a $64' \times 64'$ field in two perpendicular directions, enough to cover the Sun and part of the corona off the solar limb. (The Sun's angular diameter from the distance of the *SOHO* spacecraft is 32 arcminutes.) The spectrometer optics in turn produce a stigmatic image of the entrance slit on the detector such that each pixel along a spectral line corresponds to a distinct part of the solar atmosphere approximately $1'' \times 1''$ in size. To record an image at a particular wavelength, the detector is set to the desired wavelength while the solar image is stepped in small increments across the slit in the east–west or west–east direction. Depending on the line intensity, integration times at each step are typically a few seconds to a few tens of seconds. As the solar image relative to the slit is moved from limb to limb in the east–west direction, a 300 or 120 arcsec slice of the solar image is imaged. When one such scan is completed, the solar image is raised or lowered by an appropriate distance in the north–south direction and the east–west scanning is repeated. Since the area of the solar disk as seen from *SOHO* corresponds to $\sim 3 \times 10^6$ arcsec^2, $\sim 10^4$ steps are needed to produce a full Sun image with the 1×300 arcsec^2 slit. Depending on the integration time at each step, it may take from several hours to more than a day to record a full-disk high-resolution image. A large collection of solar images recorded by SUMER were published in an atlas (Feldman *et al.* (2003b)).

7.6.4 Grazing-incidence imagers

In the X-ray range, spectral lines are largely ordered according to temperature, i.e. as the temperature of the plasma increases, the wavelengths of the X-ray lines produced tend to become progressively shorter. Grazing-incidence telescopes equipped with several filters, each transmitting a slightly different passband, are used in solar observations. Although such telescopes cannot single out a narrow temperature range, they can be used in conjunction with different filters to derive the emission measure distribution.

Skylab was the first major solar mission to carry two soft X-ray grazing-incidence telescopes (Underwood *et al.* (1977), Vaiana *et al.* (1977)). The next-generation Soft X-ray Telescope (SXT) was launched as part of the *Yohkoh* mission (Tsuneta *et al.* (1991)). SXT shared many of the features of the design of the *Skylab* soft X-ray grazing-incidence telescopes. It consisted of a 1.5 m focal length telescope utilizing two hyperboloid surfaces of revolution, two filter wheels, a mechanical shutter and a CCD detector optimized for X-rays. Pre-launch measurements showed that the approximate FWHM of a point source imaged by the telescope at any location on the solar surface was ~3 arcsec, and 50% of the image energy was enclosed in a 5 arcsec circle. The scattering wings of the telescope were much improved over those achieved with telescopes flown during the *Skylab* era. The CCD camera utilized a 1024×1024 virtual-phase technology with 18.3 μm pixel spacing (2.45 arcsec). Full Sun images could be obtained, and, during flares or other events, SXT could be commanded to obtain faster-cadence Partial Frame Images (PFI) of 64×64 pixels. In the PFI mode, a succession of five images could be recorded, one for each of the available filters, in 10 s.

The X-ray Telescope (XRT) on *Hinode* is similar in operation to the *Yohkoh* SXT but has improved spatial resolution (1 arcsec) and different filters, which allow the temperature structure of the corona to be examined over the range 0.5 MK to 10 MK, so including the quiet Sun X-ray corona as well as active regions and flares.

7.6.5 Modulation collimators as imagers

The *Reuven Ramaty High-Energy Solar Spectroscopic Imager* (*RHESSI*), launched in 2002, images solar flares in X-rays and gamma-rays over an energy range 4 keV to 17 MeV with a spatial resolution as high as 2.3 arcsec. In its spectrometer mode, the spectral resolution is $\sim$1 keV. Full descriptions of the mission, including the imaging capabilities, are given by Lin *et al.* (2002) and Hurford *et al.* (2002). Figure 7.12 is a schematic view of *RHESSI*. Unlike soft X-rays, focusing optics cannot be used for high-energy X-rays and gamma-rays, so imaging can in practical terms only be accomplished through rotation modulation collimator techniques. *RHESSI* has nine collimators consisting of a pair of grids separated by a distance of 1.55 m. Each grid is an array of equally spaced open slits and opaque slats, with the slats and slits of each grid pair parallel to one another. Behind the rear grids are nine cryogenically cooled pure germanium detectors, segmented such that the front segment is sensitive to X-rays up to $\sim$250 keV, the rear segment to higher-energy radiation, up to gamma-ray energies. The *RHESSI* spacecraft, normally pointed at Sun centre, rotates once every 4 s, so that X-rays from a solar source that is off-axis are modulated in a pattern that characterizes the offset angle and source direction. The angular resolution of the image so obtained depends on the pitch p of the grids. The smallest value of p is 34 μm, and succeeding values are a factor of $\sqrt{3}$ larger up to 2.75 mm. With the separation of the grid pairs $L = 1.55$ m, the angular resolution is given by $p/2L$. For the finest grids, the resolution is 2.3 arcsec, for the coarsest grids it is about 3 arcmin. In a half-rotation of *RHESSI*, or 2 s, some

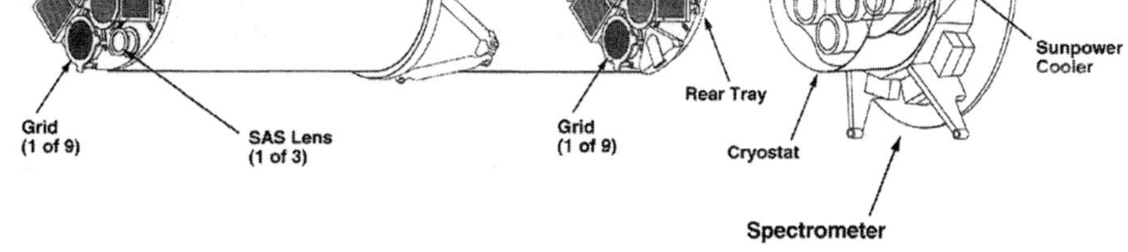

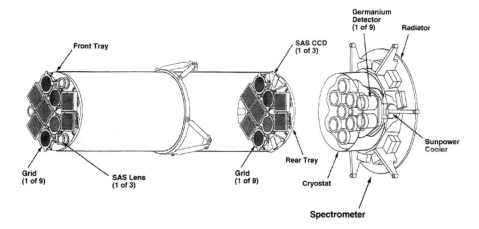

Fig. 7.12. Principle components of *RHESSI*. Two identical sets of nine grids are mounted on grid trays at the front and rear of the collimator, and a corresponding set of nine cooled germanium detectors is mounted behind the rear grids. The spacecraft, normally pointed at Sun centre, rotates once every 4 s. Courtesy the *RHESSI* team.

1100 Fourier components are measured by the rotation modulation collimators, allowing flare images to be resolved in much detail. This is a considerable improvement over the *Yohkoh* Hard X-ray Telescope (HXT) (Kosugi *et al.* (1991)) which operated on similar principles.

An aspect system on board the spacecraft enables *RHESSI* to be kept accurately pointed at the Sun. It consists of three lens–filter combinations mounted on the front grid that form visible-light images of the Sun at the rear grid where they are detected by linear diode arrays. A star scanner maintains the spacecraft roll angle.

Flare images are reconstructed from the modulation collimator data with elaborate analysis software. An initial estimate of the solar flare image is made with a back-projection method. This image is a convolution of the source and instrument response and so has side-lobes that are removed with a variety of techniques including, for example, a maximum-entropy method. Application of each of these methods involves a prediction of the source character, and the observed and predicted modulation patterns are compared. Iterations are then made until an acceptable agreement of predicted and observed patterns is achieved. A dynamic range, i.e. ratio of the brightest to the weakest source in the flare image, of up to about 100:1 can be obtained for the best conditions.

7.7 Instrument calibration

Establishing the physical properties of the solar atmosphere from UV and X-ray spectra requires the conversion of raw data in photon count or data number units from space-borne spectrometers and imagers into physical units – fluxes or intensities – for comparison with predictions from theoretical models and spectral codes. Such conversion is possible with an instrument flux calibration. The instrument calibration for narrow-band imagers consists of two steps: (i) conversion of observed photon counts into fluxes; (ii) conversion of fluxes into temperature and EM values. The determination of the temperature and EM response functions is usually done by convolving the spectrum predicted by a spectral code as a function of temperature with the instrumental wavelength response, while the true flux calibration is measured before launch with ground-based instrumentation. For spectrometers, the flux calibration consists of the conversion from observed photon counts to fluxes, and must be complemented with wavelength and line width calibration.

7.7.1 Wavelength calibration

Wavelength calibration consists of the relation between the position along the direction of dispersion of light in the detector and wavelength. Although this relation is determined by the optical design of the instrument and hence is known before launch, small movements in the optical components during flight and imperfections in their positioning can introduce systematic offsets of the wavelengths on the detector.

Such effects can be measured and corrected for by comparing profile centroids of all lines in the observed spectrum of a source whose physical parameters are well known with accurate wavelength measurements from laboratory plasmas.

7.7.2 Line width calibration

The width and shape of the profile of a spectral line are due both to the physical properties of the emitting source and to the instrument's characteristics. In order to determine the former, it is necessary to remove the contributions of the latter. Measurements of the instrumental broadening of a spectral line are usually done in the laboratory and yield wavelength-dependent results, but observed spectra can be used to check the accuracy of the removal of the instrumental component of the line width, and its stability with time. Following Equation (5.38), the FWHM $\Delta\lambda$ of a line with Gaussian profile, deconvolved from the instrumental broadening, can be written as

$$(\Delta\lambda)^2 = (\Delta\lambda)^2_{\text{meas}} - \left(\Delta\lambda\right)^2_{\text{instr}} = 4\ln 2\left(\frac{\lambda_0}{c}\right)^2\left(\frac{2k_{\text{B}}T_i}{M} + v^2_{\text{nth}}\right), \tag{7.19}$$

where $(\Delta\lambda)_{\text{instr}}$ is the instrumental broadening, $(\Delta\lambda)_{\text{meas}}$ is the observed FWHM, M and T_{i} are the ion mass and temperature, λ_0 is the rest wavelength of the line, and v_{nth} is the non-thermal velocity. From Equation (7.19), the quantity

$$\left(\frac{\Delta\lambda}{\lambda_0}\right)^2 = \frac{4\ln 2}{c^2}\left(\frac{2k_{\text{B}}T_{\text{i}}}{M} + v^2_{\text{nth}}\right) \tag{7.20}$$

is the same for all the lines of the same ion, since presumably these lines share the same value of v_{nth}. If the measured $\Delta\lambda/\lambda_0$ ratios do not have the same value, it is possible that the instrumental broadening was not properly removed. Therefore, if the wavelength range of the spectrometer includes several lines of the same ion at different wavelengths, the quantity $\Delta\lambda/\lambda_0$ can be measured to determine whether any systematic wavelength-dependent trend exists, and when this quantity is not constant, a wavelength-dependent correction factor can be measured.

This method requires several unblended lines of the same ion at very different wavelengths to be present in the instrumental wavelength range; also, it does not allow the detection of the presence of a constant bias in the $\Delta\lambda/\lambda_0$ quantity, and hence a constant offset of $(\Delta\lambda)_{\text{instr}}$, unless v_{nth} and T_{i} are known with accuracy.

7.7.3 *Flux calibration*

The flux calibration of a spectrometer can be checked using in-flight observations of spectra and applying diagnostic techniques. Here we describe the most important methods.

Line flux ratios

The use of flux ratios for in-flight intensity calibration has been known for many years (Griffin & McWhirter (1962), Héroux (1964)). It has more recently been applied to the *SOHO* CDS and SERTS spectrometers (Young *et al.* (1998), Brosius *et al.* (1998a, b), Del Zanna *et al.* (2001)). The technique consists of comparing predicted values of the flux ratios of lines having very different wavelengths with observations: from any systematic disagreement with wavelength it is possible to determine a correction factor to fluxes to bring the observed ratio in agreement with the predicted value. When many ratios at many different wavelengths are used, it is possible to build a curve of corrections as a function of wavelength relative to a reference line, constituting a correction to the adopted relative flux calibration. If no initial calibration is provided, this curve represents a complete relative flux calibration of the spectrometer. The flux ratios to be used in this technique must be carefully chosen, since they should be as independent as possible from the physical parameters of the emitting plasma to minimize uncertainties. Ratios of lines from the same ion should be used to remove uncertainties in the element and ion abundances, and minimize those in the plasma temperature.

Among ratios within the same ion, those involving lines from the same upper level are the best, since they only depend on the ratio of their Einstein coefficients for spontaneous radiation. When such ratios are unavailable in a spectrum, flux ratios of lines from different upper levels but independent of density can be used, although they are more sensitive to inaccuracies in the atomic physics and sometimes also to the electron temperature. Such ratios are rather frequent in the EUV and UV spectra below 2000 Å. This technique allows the measurement of the relative calibration of a spectrometer, or the correction from an initial one, for the wavelength range covered by the lines used in the flux ratios. It also provides the relative calibration of different channels of a spectrometer when the spectrum is observed by different detectors at different wavelengths, provided lines of the same ion are observed in all channels.

The main limitations to this technique are that it does not allow the absolute flux calibration to be measured, and also that when only few ratios from few ions are available, the calibration variation with wavelength is very uncertain. Also, uncertainties in atomic physics and the presence of blends in the lines used are issues to be carefully considered.

DEM/EM and L-function analysis

A DEM or an EM diagnostic technique can also be used to determine the relative intensity calibration when only a few line ratios are available. These methods, used for example by Landi *et al.* (1997), consist in calculating the DEM or EM of the emitting plasma with all the density-insensitive lines available in the dataset, and comparing the line fluxes calculated with the resulting DEM or EM with observations: any systematic difference can be used to measure a correction curve to the flux calibration. If the DEM of the emitting plasma is known, the L-function method (Section 5.2.2) can be applied to the density-insensitive lines of the same ion. The calibration correction function can be derived from the differences of the L-functions of lines at different wavelengths, since all L-functions of density-insensitive lines of the same ion should have the same value if the calibration were correct.

Continuum calibration

This method, first developed by Feldman *et al.* (2005b), combines free–free continuum radiation and line fluxes, and can be applied to isothermal plasmas whose temperature is not initially known. It is an iterative method, and capitalizes on the weak temperature dependence of the free–free radiation.

If the plasma is isothermal, a calibration correction function can be obtained by a four-step procedure: (i) the ratios R/P between the measured count rates or fluxes R of the free–free emission at several wavelengths and the predicted values P are calculated as a function of the electron temperature; their values at each wavelength are averaged over T and the standard deviation is taken as the uncertainty; (ii) the average R/P ratios are normalized relative to one of them, resulting in a first correction function to the flux calibration; (iii) line fluxes are corrected using the correction function, and the EM analysis is applied to measure the plasma temperature and EM; (iv) the values of P are calculated using the measured T and EM values, and used to derive an improved correction function by repeating the first three steps. The entire procedure is re-applied iteratively until convergence of the measured T_e, EM, and correction function. If the plasma is multithermal, the corrected fluxes will be used in step (iii) to measure the plasma DEM, to be used to calculate improved P values.

This method has the advantage of dispensing with knowledge of the plasma temperature structure, and as a by-product it provides measurements of the EM, T_e, or DEM. However, it

Table 7.2. *Significant spacecraft instrumentation for solar ultraviolet and X-ray spectral observations.*[a]

Spacecraft and instrument	Period of operations	Wavelength range (Å)[b]	Instrument description
Skylab	1973–4		
XUV Spectroheliograph, S082A		171–630	Slitless grating spectrograph
UV Spectrograph, S082B		970–1970, 1940–3940	Slit spectrograph
X-ray Spectrographic Telescope, S054		3.5–47	Grazing-incidence telescope
X-ray Telescope, S056		3–60	Grazing-incidence telescope
UV Spectrometer–Spectroheliometer		280–1350	Normal-incidence spectrograph
P78-1	1979–83		
SOLFLEX		1.82–1.97, 3.14–3.24	Crystal spectrometer
SOLEX		18.4–23.0	Crystal spectrometer
Solar Maximum Mission (*SMM*)	1980–9		
Flat Crystal Spectrometer		1.8–19.5	Crystal spectrometer
Bent Crystal Spectrometer		1.77–1.95, 3.17–3.23	Bent crystal spectrometer
Hinotori	1981–2		
SOX1, SOX2		1.72–1.95	Crystal spectrometer
Yohkoh	1991–2001		
Soft X-ray Telescope		3–45	Grazing-incidence telescope
Hard X-ray Telescope		15–100 keV	Fourier-synthesis imager
Bragg Crystal Spectrometer		1.76–1.89, 3.16–3.19, 5.01–5.11	Bent crystal spectrometer
Solar and Heliospheric Observatory (*SOHO*)	1995–		
UV Coronagraph Spectrometer (UVCS)		Selected sp. lines (Ly-α, O VI, etc.)	Occulter and grating spectrometer
Extreme-ultraviolet Imaging Telescope (EIT)		171, 195, 284, 304	Normal-incidence telescope with multilayer optics telescope
Coronal Diagnostic Spectrometer (CDS):			Grazing-incidence telescope
Normal Incidence Spectrometer (NIS)		308–381, 513–633	
Grazing Incidence Spectrometer (GIS)		151–221, 256–338, 393–493, 656–785	Grazing-incidence spectrometer

Solar Ultraviolet Measurements of Emitted Radiation (SUMER)		500–1610	Off-axis paraboloid telescope, Normal-incidence spectrometer
Transition Region and Coronal Explorer (*TRACE*)	1998–	171, 195, 304 Wide bands at 1700, 1600, 1550, 1216	Normal-incidence telescope with multilayer optics
CORONAS-F	2001–3		
RESIK		3.3–6.1	Bent crystal spectrometer
Diogeness		3.0–6.7	Scanning flat crystal spectrometer, two crystals in dopplerometer configuration
RHESSI	2002–	4 keV–17 MeV	Bi-grid rotation modulation collimators, spectrometer (Ge detectors) (Ge detectors)
Hinode	2006–		
Extreme-ultraviolet Imaging Spectrometer, EIS		170–211, 246–291	Normal-incidence spectrometer
X-ray Telescope, XRT		Soft X-rays	Grazing-incidence telescope
Solar EUV Rocket Telescope and Spectrograph (SERTS)	10 rocket flights (1983–2000)	Selected bands in the 280–420 range	Grazing-incidence telescope, stigmatic spectrograph
High Resolution Telescope Spectrometer (HRTS)	10 rocket flights (March 1, 1979– Sept. 30, 1997); flight on Spacelab-2	1200–1700	

[a] Since the time of *Skylab* (1973–4).
[b] Energy ranges (keV or MeV) are given for high-energy instruments.

requires that the spectral resolution and coverage be such as to include uncontaminated free–free emission. Also, free–free emission is rather weak in active regions at all wavelengths, so that this technique can only be applied to dense, hot regions or during the decay phase of flares.

7.8 Summary of spacecraft instruments

Ultraviolet and X-ray observations of the Sun have steadily improved over the space-flight era, which began in the late 1940s. The earliest observations gave us only crude information about the spectra and morphology of features in the solar atmosphere, but from the time of the *Skylab* observatory, manned by astronauts, a vast improvement in observational quality has been achieved through steady application of novel techniques and materials that have been described in this chapter.

A summary of some of the principal spacecraft and rocket-borne instrumentation is given in Table 7.2. It allows the reader to appreciate some of the major advances in the past few decades in the studies of ultraviolet and X-ray spectroscopy and imaging of the solar atmosphere.

8

Quiet Sun and coronal holes

8.1 Introduction

Studies of the properties of the solar atmosphere began with the onset of space-borne instruments working in X-rays and the ultraviolet. The first instruments had poor spatial resolution and conclusions regarding the structure were indirectly inferred from spectroscopic observations. Early ultraviolet spectrometers also had poor spectral resolution, from which the intensities of the stronger lines could give only a general idea of the distribution of emission measure with temperature (Pottasch (1964)). With assumptions of plane parallel geometry and hydrostatic equilibrium, and with neglect of any fine structure possibly present (i.e. filling factors of unity), the gross features of the solar atmosphere could be deduced. From this, various models (Athay (1976), Mariska (1992)) indicated the presence of a narrow transition region, height range less than 100 km, between the chromosphere and corona. Figure 1.1 shows the atmospheric structure according to theoretical models. A growing corpus of observations, particularly those starting from the *Skylab* mission, showed that the transition region had a much larger extent than was indicated in earlier models, leading to a revision of our ideas of its nature, which are discussed in this chapter.

Bray *et al.* (1991) state in their book that 'coronal loops are a phenomenon of active regions and there is growing evidence that they are the dominant structure in the higher levels (inner corona) of the Sun's atmosphere'. Indeed, the existence of large-scale coronal structures in quiet Sun regions was well known from white-light images during total solar eclipses for many years. They are observed to consist of large loops tapering to cusp-like apices beyond which are the coronal streamers. Coronagraph images show that during solar minimum the quiet Sun streamers extend out to several solar radii and can last for many days. Smaller-scale quiet Sun coronal structures were not observed with sufficient spatial resolution and contrast until the *Skylab* mission. Considerable improvements in the resolution of both spectroscopic and imaging observations in the ultraviolet and soft X-rays have given much insight into their nature. This is also discussed in this chapter.

The dynamic nature of the so-called quiet Sun has also been revealed in ever increasing clarity by spacecraft observations. Spectroscopic observations show the existence of non-thermal broadening of emission lines over all temperatures, with a persistent red shift in transition region lines. These observations also indicate the presence of waves, which have important implications for the heating of the solar atmosphere. Waves with longer periods have now been clearly seen propagating along structures such as polar plumes and coronal loops. Specific events, such as brightenings in monochromatic images of the quiet Sun or explosive phenomena revealed by temporarily broadened spectral lines, have been recognized in recent years. We describe them in this chapter together with an outline of their implications for the heating of the solar atmosphere.

8.2 Quiet Sun corona and coronal holes: structure

High-resolution (∼1 arcsec) images of the Sun from the *SOHO* SUMER instrument have allowed quiet Sun structures at temperatures below 8×10^5 K to be studied to greatest advantage. Figure 8.1 shows full Sun images in C III (977.02 Å) and S VI (933.38 Å) emission lines taken in 1996, when the Sun was near minimum activity. These lines have temperatures of formation equal to 5×10^4 K to 2.6×10^5 K respectively. There is little if any distinction between the near-equatorial regions and each of the polar regions where there are large coronal holes. Figure 8.2 shows a quiet Sun area near Sun centre in the C III 977.02 Å line. Clumps of emission coincide with the chromospheric network, consisting of densely packed, bright loop-like structures, lengths ∼10–20 arcsec and widths near the SUMER 1 arcsec spatial resolution. The interiors of the network cells are marked by fainter loop-like structures, randomly orientated, with a wider distribution of lengths, ∼10–50 arcsec.

The contrast between the network emission and that within the network cells is a function of temperature (Reeves (1976); Feldman *et al.* (1976a)): at temperatures of 40 000–100 000 K, the network structures are about five times brighter than the cell interiors. There is less contrast at lower (∼10 000 K) and higher ($\sim 8 \times 10^5$ K) temperatures.

Limb observations indicate that structures with temperatures of $\sim 2 \times 10^5$ K reach their maximum brightness at a height of ∼5 arcsec, or 3600 km (Doschek *et al.* (1976b); Feldman *et al.* (1976b)). Bohlin and Sheeley (1978) found that in coronal holes limb brightening in Ne VII lines ($\sim 5 \times 10^5$ K) extended to greater heights than in quiet regions, a behaviour most

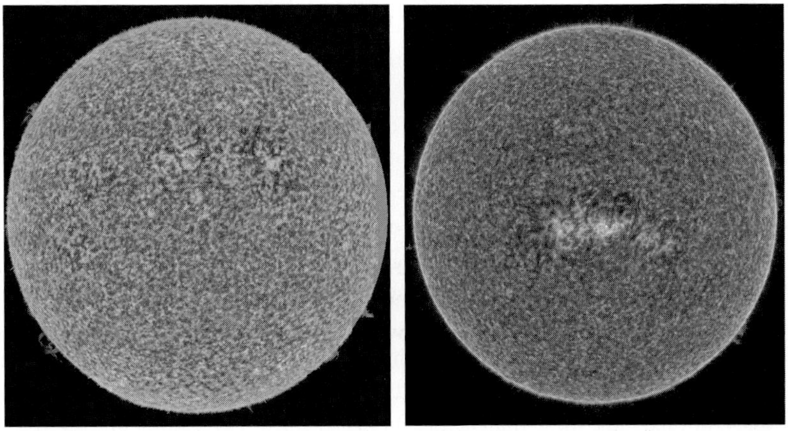

(a) (b)

Fig. 8.1. *SOHO* SUMER full Sun images taken near solar minimum. (*a*) C III (977.02 Å) image, taken over a period of 25 hours, from 22:40 UT on 1996 January 28 to 00:06 UT on 1996 January 30. (*b*) S VI (933.38 Å) image, taken over a period of 8 hours, from 23:02 UT on 1996 May 12 to 07:33 UT on 1996 May 13. Courtesy *SOHO* SUMER team.

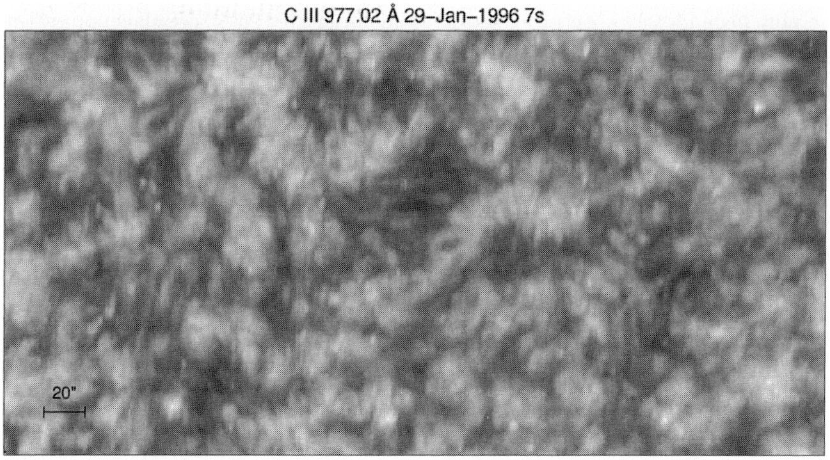

Fig. 8.2. A small region, measuring 373″ × 188″ arc at Sun centre observed by *SOHO* SUMER in the C III 977.02 Å line on 1996 January 29. This line is mainly formed at a temperature of ~40 000 K. Courtesy *SOHO* SUMER team.

likely influenced by the orientation of the coronal magnetic field (radial or super-radial in coronal holes, loop structures in the rest of the quiet Sun).

Coronal structures are evident in *TRACE* or *SOHO* EIT images made in ultraviolet lines with formation temperatures $\gtrsim 7 \times 10^5$ K. Two such images, made with *SOHO* EIT in its 171 Å and 195 Å filters, are shown in Figure 8.3, including the north polar region. The 171 Å filter includes emission from Fe IX (171.07 Å) and Fe X (174.53 Å) lines, emitted at temperatures $\sim 8 \times 10^5 - 1.0 \times 10^6$ K, and the 195 Å filter includes emission from the Fe XII (195.12 Å) line, emitted at $\sim 1.4 \times 10^6$ K. Within the coronal hole are bright points, revealed to be small bipolar regions when compared with photospheric magnetograms. Outside the hole boundary, the emission is in the form of densely packed loops with lengths ranging from a few arcseconds to several arcminutes. Their widths are close to the resolving power of the EIT instrument, ~2.6 arcsec.

Information about the heights of quiet Sun coronal loops can be obtained from limb brightening measurements in the extreme ultraviolet and soft X-ray spectral regions. In *SOHO* SUMER images, the extent of limb brightening at temperatures of 1 MK is ~12 arcsec, or 8700 km. Comparison of *TRACE* images at the limb in the 171 Å and 195 Å filters shows that the higher-temperature (1.4 MK) loop structures extend higher than those at lower temperatures (~1 MK), while below heights of 20 000 km (30 arcsec, or $R_\odot = 1.03$) the lower-temperature and higher-temperature loops are intermixed. The limb brightening in 171 Å images is more distinct than that in 195 Å images, indicating that the height distribution of the 1.4 MK material varies considerably. This is consistent with spectroscopic observations indicating that at heights of 1.03–1.5$R_\odot$ along the quiet equatorial regions the plasma is nearly isothermal with $T_e = 1.4$ MK. The fact that long-exposure *TRACE* images in the 284 Å filter (mostly Fe XV line emission) do not show the loop-like structures seen in the 171 Å and 195 Å filters indicates that quiet Sun loops have temperatures lower than ~2 MK, at which temperature the Fe^{+14} fractional ion abundance is very low.

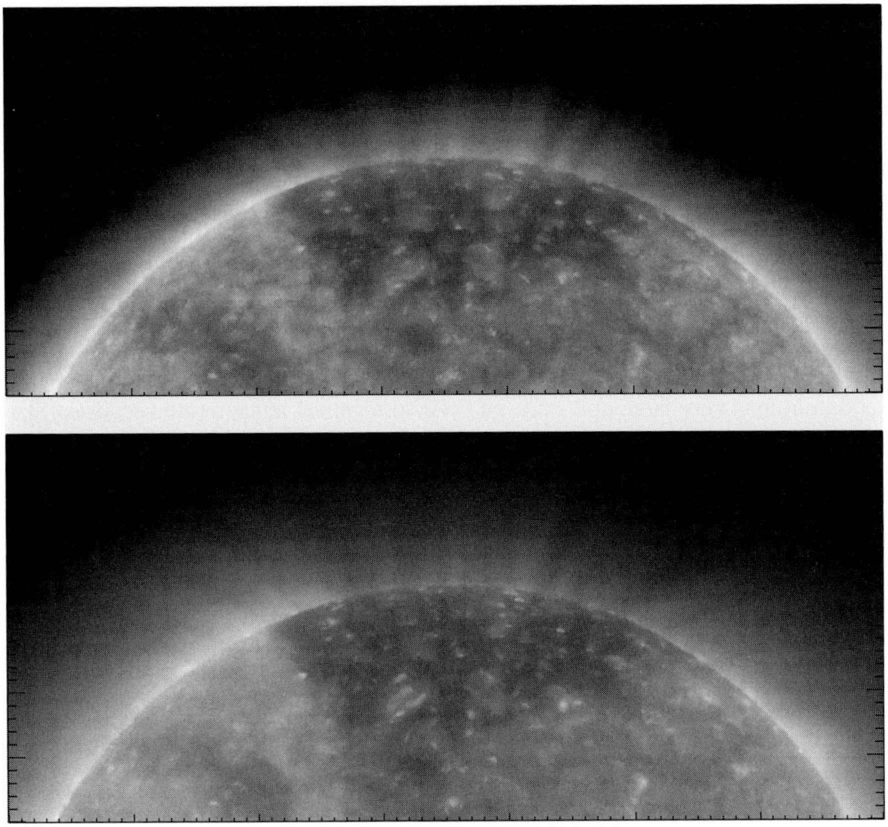

Fig. 8.3. The north polar region of the Sun taken in *TRACE* images (171 Å, *top*; 195 Å, *bottom*) on 1996 August 26, near solar minimum. An extensive polar hole is present, marked by decreased emission. Courtesy *SOHO* SUMER team.

A striking feature of polar regions, noted from white-light photographs during total solar eclipses (Koutchmy (1977)), is the presence of radiating structures called polar plumes. In extreme ultraviolet and soft X-ray solar images, they apparently radiate from small bright areas. They are evident in *SOHO* EIT images made in 171 Å and 195 Å wavelengths, and are just visible in 284 Å images. The temperature indicated is approximately 1.0–1.6 MK (DeForest *et al.* (1997)). Observations with SUMER (Hassler *et al.* (1997)) made in the O VI 1031.92 Å line with 1.5 arcsec resolution show the plumes out to at least $1.3 R_{\odot}$. Observations with the white-light coronagraph on the SPARTAN mission (Fisher & Guhathakurta (1995)) indicate that the widths (subtending 2.5° from the Sun's centre) of polar plumes were constant out to $5 R_{\odot}$, but *SOHO* LASCO observations indicate that there is a super-radial expansion at higher altitudes. Again, while the white-light SPARTAN observations indicated that polar plumes are very stable with time, EIT observations indicate changes over as little as 10 minutes for filamentary structures at the core of plumes (DeForest *et al.* (1997)). It was long presumed that polar plumes start from small bipolar structures in the low corona, but comparisons with magnetograms from the MDI instrument on *SOHO* show that the regions from which plumes radiate are in fact complex but basically unipolar regions coincident with chromospheric network structures.

8.3 Quiet Sun corona and coronal holes: plasma properties

8.3.1 Temperature

The solar ultraviolet spectrum in the 500–1500 Å range includes many lines emitted by ions having maximum fractional abundance over a range of coronal temperatures, and so their fluxes may be used to probe the temperature and emission measure of the corona in quiet Sun and coronal hole regions. A simple emission measure technique, described in Chapter 5 (Section 5.5.1) was applied to line fluxes recorded during a solar minimum period along equatorial regions. Contrary to earlier expectations of a continuous emission measure vs. temperature distribution in the corona (Figure 1.1), it has been found that the line-of-sight quiet Sun coronal plasma at heights of 1.05–1.5$R_\odot$ are isothermal at a temperature of 1.4 MK (Feldman *et al.* (1999b)). The fact that the number of structures along any line of sight above the quiet Sun limb is very large and since all have an identical temperature suggested that the finding is general and should apply to all other quiet Sun coronal plasma. Indeed, similar analysis of spectra recorded in other quiet solar regions during different parts of the solar cycle confirmed the earlier findings (Warren (1999), Widing *et al.* (2005), Landi *et al.* (2006b)). Figure 8.4, taken from a detailed analysis that includes tens of spectral lines, shows that at a height of $\sim$1.05 $R_\odot$ the temperature along the quiet Sun equatorial streamer line of sight is 1.35 MK and the emission measure is log EM $= 43.3\pm0.15$ (EM in cm^{-3}). The EM decreases nearly exponentially with height (one order of magnitude between 1.05$R_\odot$ and 1.25$R_\odot$).

Using a similar technique the emission measure above the polar coronal hole was also found to be isothermal at a temperature of 9×10^5 K (Feldman *et al.* (1998)). There does not appear to be isothermal plasma close to the solar surface with temperature 6×10^5 K.

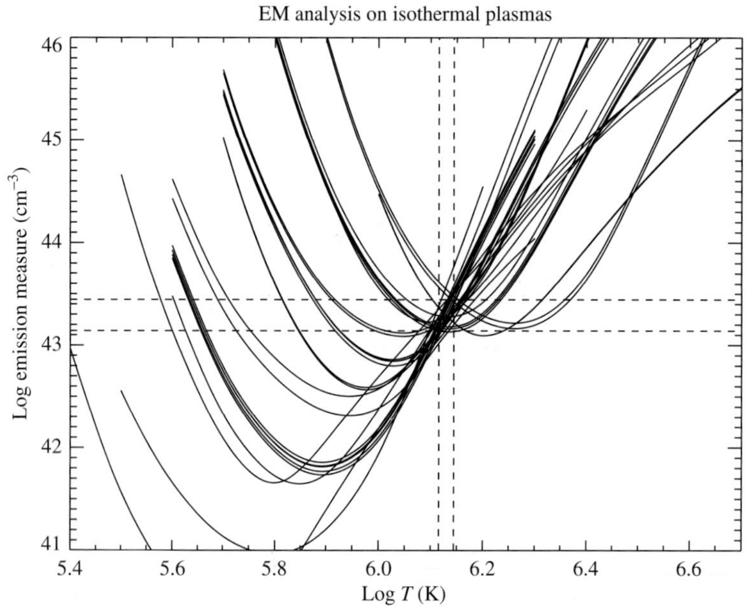

Fig. 8.4. Plot of the logarithm of emission measure against logarithm of temperature for a large number of intercombination (spin-forbidden) lines in the Be-, B-, C-, and Mg-like isoelectronic sequences emitted by a quiet Sun equatorial streamer. The derived temperature is 1.35 MK. Data from the SUMER instrument on *SOHO*. From Landi *et al.* (2002).

As strange as it may appear, no plasma at temperatures other than the three temperatures mentioned are found in quiescent coronal structures in the range $0.5\times10^6 < T_e < 2\times10^6$ K. For more on the isothermal nature of coronal plasmas, see Chapter 9 (Section 9.3.1).

8.3.2 *Variation of EUV line flux with height*

Studying the way in which the intensities of extreme ultraviolet lines decrease with radial distance off the solar limb is instructive and can give information about the physical properties of the solar atmosphere. The results of one such investigation, made by Feldman *et al.* (1998) using data from SUMER, are shown in Figure 8.5. Here, the intensities of strong lines are plotted with radial distance for a quiet equatorial region. The lines are due to Ne VIII, Mg X, Si XI, Si XII, Fe X, Fe XI, and Fe XII. The decrease with height in the atmosphere is approximately exponential, as would be expected for hydrostatic equilibrium in an isothermal atmosphere, temperature T:

$$P(r) = P(R_\odot)\exp(-h/H) \tag{8.1}$$

where H, the pressure scale height, is given by $kT/m_{\rm ion}g$ and $P(r)$ is the gas pressure at radial distance r (g is the solar gravity, $m_{\rm ion}$ the ion mass, and the solar surface is defined by $r = R_\odot$). The height range of the measured intensities is 1.05–1.5$R_\odot$ – the intensities of the Ne, Mg, and Si lines are a factor ~400 times weaker at the largest distances than near the limb.

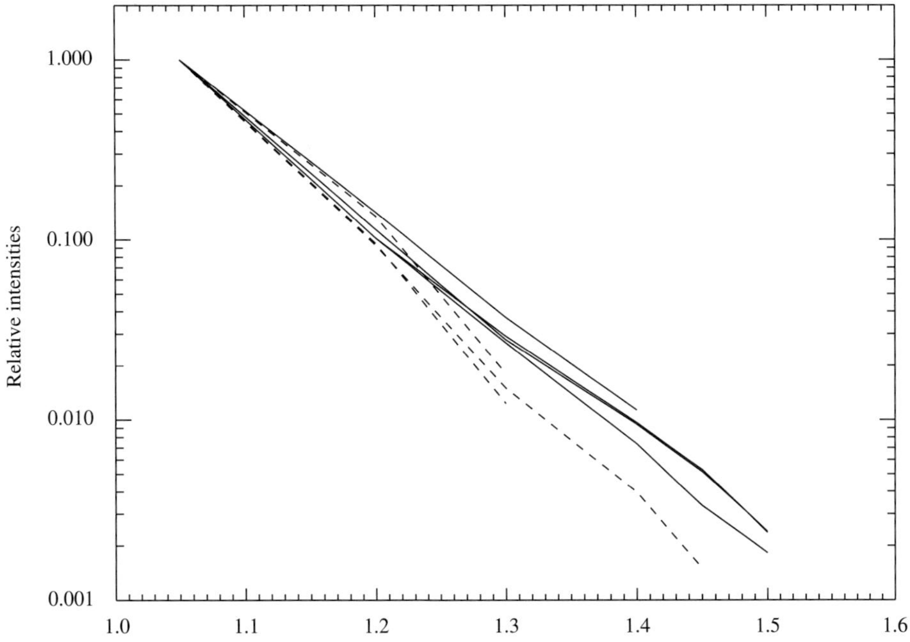

Fig. 8.5. Variation of EUV line intensity with height above the solar limb for a quiet equatorial region. The lines, observed with the *SOHO* SUMER instrument, are: Ne VIII 770.41 Å, Mg X 609.78 Å, Si XI 580.85 Å, Si XII 499.40 Å, Fe X 1028.02 Å, Fe XI 1467.06 Å, and Fe XII 1242.00 Å. The intensities are normalized to their values closest to the limb. The Fe line variations are plotted as dashed lines. From Feldman *et al.* (1998).

The Si XI and Si XII line intensities in Figure 8.5 decrease with height at the same rate as the Ne VIII and Mg X line intensities. However, the fractional abundances of the Si ions have a substantially different temperature dependence from those of either the Ne or Mg ions, so the similarity in the way the intensity decreases with height indicates that the temperature of the quiet Sun coronal plasma is unchanged from $1.05R_\odot$ out to at least $1.5R_\odot$. Since the quiet Sun region is nearly isothermal, the faster decrease of the Fe line intensities cannot be a temperature or emission measure effect. Most likely, it is due to gravitational settling. With hydrostatic equilibrium (Equation (8.1)), the scale height depends inversely on the ion mass, so it would be expected to be smaller for Fe ions than for lighter elements. This is discussed further in Chapter 11 (Section 11.3.4).

By contrast, coronal hole intensities behave slightly differently from quiet Sun equatorial regions. It is found that line intensity ratios vary with height above the limb in a way that implies a gradual increase in temperature with height. The temperature near the limb peaks at $\sim 8\times 10^5$ K, while at greater heights it reaches nearly 1×10^6 K. Possibly line-of-sight effects due to small contributions from 1.4 MK quiet Sun plasma to the measured coronal hole flux are responsible for the apparent temperature increase with height.

8.3.3 Electron densities

Density-sensitive line ratios provide the most reliable determinations of electron densities N_e in the solar atmosphere, but unfortunately the temperature range $2\times 10^4 \leq T_e \leq 3\times 10^5$ K includes only a few such ratios. The best known are those arising from transitions $2s^2 2p\ ^2P - 2s2p^2\ ^4P$, $2s2p\ ^3P - 2p^2\ ^3P$, and $3s3p\ ^3P - 3p^2\ ^3P$ in C, N, O, and Si ions, but they are only slightly sensitive to N_e in the range 10^9–10^{10} cm^{-3}. By comparing the flux of any of the above-mentioned lines with the flux of an allowed line emitted by an ion with similar temperature sensitivity, densities as high as 1×10^{13} cm^{-3} are derived (see Feldman and Doschek (1978b)).

Electron densities can be derived from the ratio of the flux of the spin-forbidden line of C III at 1908.73 Å, which is proportional to N_e, and the flux of the Si IV allowed line at 1402.77 Å which is proportional to N_e^2. Figure 8.6 (*a*) and (*b*) (from Feldman & Doschek (1978a)) shows the fluxes in these lines near the solar limb in a quiet Sun region and a coronal hole. Note that, between distances of 12 arcsec inside the limb and 4 arcsec off the limb, the C III line fluxes are all within a factor of 2. For distances greater than 4 arcsec above the limb, the C III line flux extends farther in the coronal hole. Figure 8.6(*c*) shows the Si IV 1402.77 Å/C III 1908.73 Å flux ratio in the same regions. The derived electron density in the quiet Sun region is about a factor 2 higher than in the coronal hole.

Electron densities can also be derived from the ratio of Si VIII lines at 1440.49 Å ($2s^2 2p^3\ ^4S_{3/2} - 2s^2 2p^3\ ^2D_{5/2}$) and 1445.76 Å ($2s^2 2p^3\ ^4S_{3/2} - 2s^2 2p^3\ ^2D_{3/2}$) for the quiet Sun corona and coronal holes. Banerjee *et al.* (1998) used this ratio to obtain the variation of electron density with height in the north and south polar coronal holes in 1996, with the Sun near solar minimum and the polar coronal holes at their maximum development. Values of N_e decreasing from 1.3×10^8 cm^{-3} at 27 arcsec above the limb ($1.03R_\odot$) to 1.6×10^7 cm^{-3} at 250 arcsecs above the limb ($1.25R_\odot$) were obtained for both north and south poles. A study by Feldman *et al.* (1999b) used the same Si VIII lines to measure electron densities in an equatorial quiet Sun coronal region. The ratio $I(1445.76)/I(1440.49)$ gave the following electron densities: $\sim 1.8\times 10^8$ cm^{-3} ($1.04R_\odot$), 3.8×10^7 cm^{-3} ($1.20R_\odot$), and 1.6×10^7 cm^{-3} ($1.30R_\odot$). Figure 8.7 shows the variation of the line ratio with distance from the solar limb. The two lines are too weak to extend measurements further out than radial

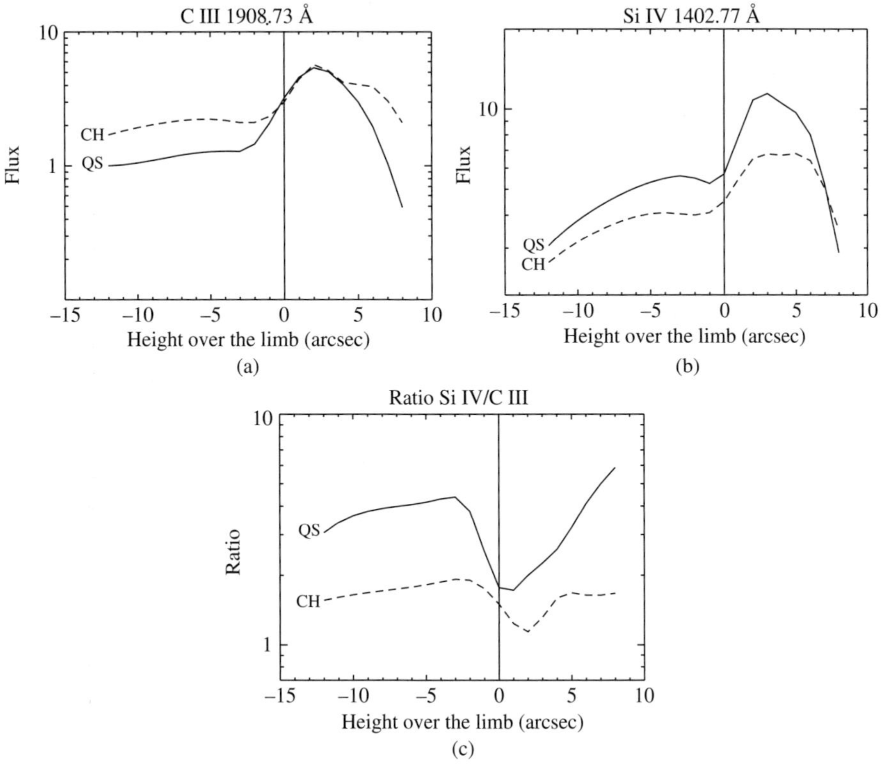

Fig. 8.6. (*a*) Flux of the C III 1908.73 Å line in the vicinity of the solar limb in a quiet Sun region and a coronal hole. (*b*) Similar plot for the flux of the Si IV 1402.77 Å line. (*c*) Ratio of the line fluxes.

distances of $1.3R_\odot$. On average, electron densities from these lines in coronal hole regions are lower by a factor 2 to 3.

A similar result was obtained by Gallagher *et al.* (1999b), from Si IX and Si X density-sensitive line pairs observed by the *SOHO* CDS instrument. The Si IX lines are at 349.90 Å ($2s^2\,2p^2\,{}^3P_2 - 2s\,2p^3\,{}^3D_3$) and 341.95 Å ($2s^2\,2p^2\,{}^3P_0 - 2s\,2p^3\,{}^3D_1$), the Si X lines at 356.05 Å ($2s^2\,2p\,{}^2P_{3/2} - 2s\,2p^2\,{}^2D_{5/2}$) and 347.42 Å ($2s^2 2p\,{}^2P_{1/2} - 2s2p^2\,{}^2D_{3/2}$). Measurements were taken over the south polar coronal hole and neighbourhood, with radial distances ranging from near the limb to $1.20R_\odot$. Again, coronal hole densities are approximately a factor 3 less than equatorial quiet Sun densities. The density data obtained offer a means of comparing with theoretical variations of density over a coronal hole. One such theory (Lima & Priest (1993)) involves hydrodynamic simulations without consideration of magnetic fields except that the basic geometry is one in which there is an equatorial region with confined magnetic structures and a polar region where the field lines are open and material flows unimpeded into interplanetary space. On the basis of this model, electron density is a function of radial distance R and position angle around the Sun θ, as well as parameters defining the ratio of equatorial and polar densities and the width of the density (and wind velocity) profile with θ. Figure 8.8 shows how remarkable the agreement is between measurements from each line ratio and this comparatively simple theory.

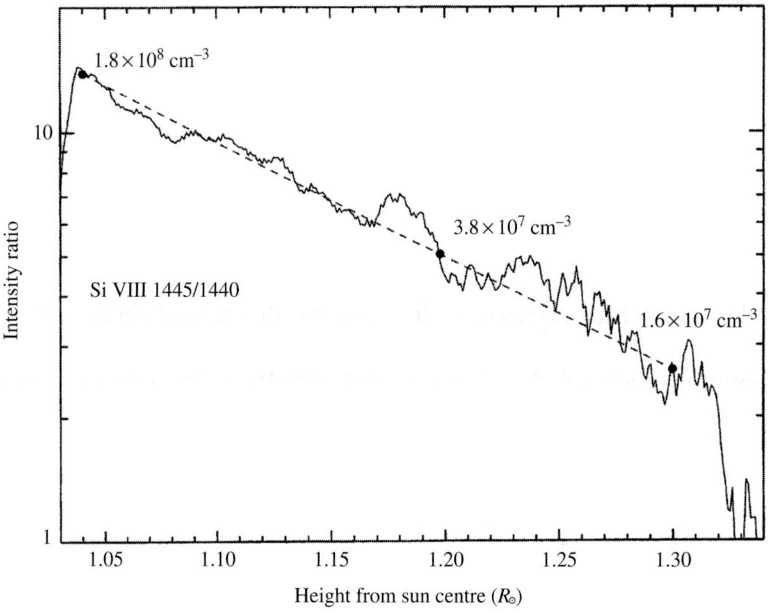

Fig. 8.7. Variation of electron density with distance from solar limb, using the Si VIII 1445 Å/1440 Å ratio. From Feldman *et al.* (1999b).

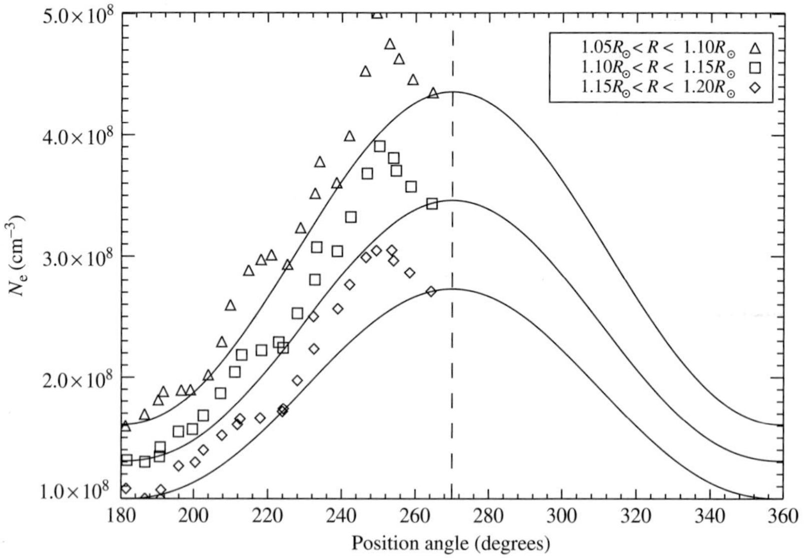

Fig. 8.8. Electron densities from Si IX and Si X line ratios, measured from *SOHO* CDS data, plotted against position angle θ over the south polar coronal hole, and models of Lima and Priest (1993). The two parameters of the theory, relating the ratio of equatorial and polar densities and the width of the density profile in θ, were evaluated by a χ^2 technique. Data from Gallagher *et al.* (1999b).

8.4 Quiet Sun transition region

8.4.1 *Line shifts*

A significant discovery about the nature of the transition region, made in the mid 1970s, is the persistent recessional velocity of material in the quiet Sun transition region. The observed velocities are only 5 to 15 km s^{-1}. No spacecraft instrument to date has a precisely defined absolute wavelength scale, so measured line positions must be made relative to reference lines such as those in the chromosphere or at the limb, each being presumed to be unshifted relative to the solar coordinate system. Different results for different lines have been obtained, so it is possible that the red shifts in lines are time dependent.

Measurements made with the slit spectrograph S082B on *Skylab* were made by Doschek *et al.* (1976a) using several lines, ranging from Si IV 1402.77 Å to O V at 1218.35 Å, with emitting temperatures between 70 000 K and 2.2×10^5 K. The spectral resolution was 0.06 Å for the range 1200–1565 Å, or a velocity resolution of 13 km s^{-1}, though most lines have profiles which have widths (FWHM) of about 20 km s^{-1}. However, much finer velocity resolution could be obtained by measuring line shifts relative to chromospheric lines, which included those from C I, S I, etc. A variety of quiet Sun features – network and network interiors, in and out of coronal holes – were observed. It was found that only some features near Sun centre showed red shifts, up to 15 km s^{-1} for the cooler (Si IV, C IV) lines, but more generally only 5–7 km s^{-1} shifts were measured for small ($\sim$2 arcsec) features in bright network structures. The line shifts were not necessarily zero near the solar limb as might be expected. There appeared to be hardly any line shifts for network cell interiors from the *Skylab* measurements, but this result is not certain because of the low emission from such areas.

Rather similar red shifts were found by Lites *et al.* (1976) using the Ultraviolet Spectrometer on *OSO-8* with more modest spatial resolution. Gebbie *et al.* (1981) found small red shifts for the lines of C II (1335.66, 1335.71 Å), Si IV (1393.76 Å), and C IV (1548.20 Å) observing with the UVSP instrument on *Solar Maximum Mission*, with highest values for the Si IV line, 4 km s^{-1}. Athay *et al.* (1983) also used UVSP results for the C IV line, obtaining slightly larger values. A correlation of red shift and line width was found.

The *Skylab* measurements of Doschek *et al.* (1976a) did not find red shifts for the higher-temperature O V line at 1218.35 Å, formed at 2.2×10^5 K, though the line in question is in the wings of the intense H Ly-α line at 1216 Å. Surprisingly, *SOHO* SUMER measurements by Brekke *et al.* (1997) found small red shifts for not only transition region ions such as O V, S V, and S VI, but also higher temperature ions such as Mg X (temperatures up to $\sim$1 MK), with red shifts of $\sim$5 km s^{-1}. More recent SUMER measurements (Doyle *et al.* (2002)) using lines of C III (977.02 Å), O VI (1031.91 Å), and Ne VIII (770.41 Å, 780.33 Å) indicated red shifts of these lines similar to those found by previous observers. Larger (13 km s^{-1}) red shifts were found for bright network structures compared with cell interiors for the Ne VIII line. Chae *et al.* (1998d) claimed to find a well-defined relationship of red shift amount with line temperature of formation, based on *SOHO* SUMER measurements, with red shifts of < 5 km s^{-1} for chromospheric and coronal temperatures but up to $\sim$12 km s^{-1} for transition region temperatures.

Whether transition region material is actually receding as implied by the observed red shifts is still not certain. Feldman (1983) points out that in cases where material is probably compressed as it descends, the emission (which varies as N_e^2) will naturally increase and possibly dominate over emission from rising volumes which have smaller densities. If the material really is descending, then one is forced to speculate on the nature of the mass balance, assuming one exists. Pneuman and Kopp (1978) find that the observed transition region red shifts imply mass flows that are of the same order as the upward mass flows in spicules, so that there is an approximate mass balance. Several other models exist to explain the red shifts, including mass flows as a result of asymmetries in low-lying loop structures (Spadaro *et al.* (1991)) or enhancements in density as the downflowing plasma radiatively cools (Reale *et al.* (1997)).

8.4.2 Nature of the transition region

As already indicated, solar atmospheric models, such as that due to Gabriel (1976) (Figure 1.7), imply the existence of a relatively cool ($\sim 10^4$ K) chromosphere with overlying hot ($\sim 10^6$ K) plasma confined by magnetic flux tubes, with a very thin transition region separating the chromosphere and corona. Model calculations indeed suggest that the transition is extremely thin, less than 100 km.

The earliest models (e.g. Pottasch (1964)) assumed spherically symmetric geometries for the solar atmosphere and a static nature. An energy balance in the transition region of the form

$$\nabla \cdot \mathbf{F}_{\text{cond}} = Q - L \tag{8.2}$$

was assumed, in which $\mathbf{F}_{\text{cond}}$ is the thermal conductive flux, Q the energy input, and L an energy loss term, commonly assumed to be in the form of radiative energy. Observed line fluxes gave values of temperature gradients. With the fluxes of several ultraviolet lines covering the expected temperature range of the transition region, and a model that relates temperature T_e and N_e (generally a constant pressure model, $N_e T_e$ = a constant), a relation between T_e and the temperature gradient dT_e/dh can be derived. From this, the thermal conductive flux can be derived. This was found to be nearly constant at around $1\text{–}2 \times 10^6$ erg cm^{-2} s^{-1} over the temperature range 10^5–10^6 K.

The model of Gabriel (1976), in which the conductive flux was channeled entirely into the magnetic field concentrations at the boundaries of the supergranules, alleviated some of the problems raised by the spherically symmetrical atmosphere models. The latter predicted unreasonably large values of F_{cond} and so large values of Q were needed to maintain the energy balance of Equation (8.2). With Gabriel (1976) model, the atmosphere was spatially resolved, and the value of F_{cond} was reduced, both over the network and in the centre of supergranular cells. From this, the thermal conductive flux was shown to be nearly constant over the temperature range of the 'upper' transition region, i.e. the temperature range 2×10^5–10^6 K (see Figure 8.9).

As can be seen from Figure 8.9(*b*), the constancy of conductive flux ceases to hold for the 'lower' transition region, where $T_e < 2 \times 10^5$ K – this plot is based on calculations by Mariska (1992), with emission measure distributions of Raymond and Doyle (1981) (see Figure 8.9(*a*)). This is equivalent to there being an excess of emission at these lower temperatures. Explanations for this included the transition region being a thin region of the atmosphere surrounding spicule-like structures observed in Hα spectroheliograms. The alternative model of Dere *et al.* (1987), based on HRTS observations from *Spacelab 2*, uses the fact that electron densities from O IV lines imply extremely small path lengths, of order

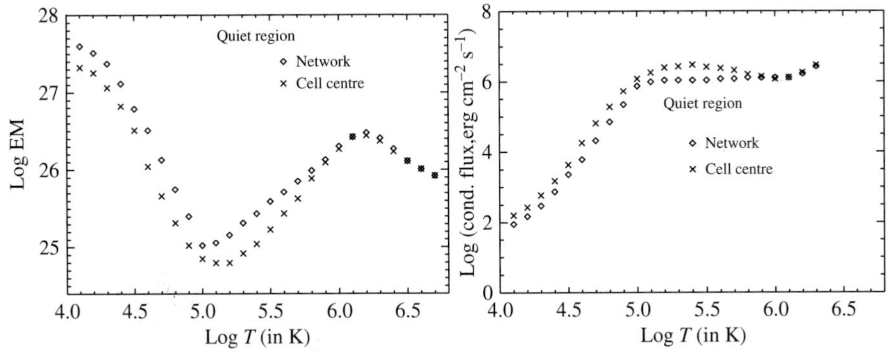

Fig. 8.9. (*Left*) Emission measure distribution for quiet Sun atmosphere, network cell centre and network, based on *Skylab* line intensities, analyzed by Raymond and Doyle (1981). (*Right*) Conductive flux for the transition region in the centre of a supergranular cell based on Raymond and Doyle's (1981) emission measure distribution and a constant gas pressure ($N_e T_e$ = constant), calculated by Mariska (1992).

a few km, yet the observed structures in the C IV lines (temperatures $\sim 10^5$ K) are more than 2000 km across. Their model requires the gross structures seen in C IV to be made up of fine filaments beyond the instrumental resolution of HRTS.

Feldman (1983) criticized early work that assumed a constant conductive flux in spherically symmetric approximations on the grounds that the emission measure distributions from transition region lines did not, as had been claimed, suggest a $T^{3/2}$ relation in the upper transition region, which would have implied a constant conductive flux. Instead, Feldman (1983, 1987) argued that much of the transition region ultraviolet line emission is from extremely fine structures – 'unresolved fine structures' – below the resolution of current space instruments, yet accounting for much of the transition region emission. Feldman (1983) speculated that the unresolved fine structures begin their lives as cool ($\sim$20 000 K) small loops but then heat and expand until disintegrating at a temperature of $\sim 5 \times 10^5$ K. The observed emission measure distributions mean that the loops spend more time as cool structures than hot. It is supposed that the unresolved fine structures have an upper temperature limit of at least 5×10^5 K, the temperature of observed Ne VIII lines, but not as high as 9×10^5 K, the temperature of the coronal Mg IX lines. At any time, the loops are in tight bundles containing both hot and cool loops, with all the loops being magnetically isolated from both the chromosphere and corona.

Wikstøl *et al.* (1997) have proposed a dynamic model for the transition region, in which a thin transition region is constantly responding to the shock waves produced by numerous tiny flares (nanoflares) which occur at magnetic loop tops. The transition region is then the response of the nanoflare shocks as they are reflected at the top of the chromosphere. The observed transition region emission at the limb is a consequence of many dynamic processes occurring within the resolution limits of any particular instrument.

8.5 Non-thermal broadening of ultraviolet emission lines

Measurements of ultraviolet line widths in the solar atmosphere give valuable information on the possible presence of non-thermal energy sources. In general, the widths of ultraviolet lines exceed thermal Doppler widths, assuming that the latter are determined by ion temperatures that are equal to electron temperatures. Although some recent work indicates that the ion temperatures do in fact often exceed electron temperatures in the corona, apparently dependent on phase of the solar cycle (Feldman *et al.* (2007)), the excess line width is generally expressed as a non-thermal turbulent velocity, either the root-mean-square value or the most probable velocity ξ. The spectrometer instrumental width must be sufficiently small to enable such measurements to be made. Such spectrometers were available even in the 1970s (e.g. Boland *et al.* (1973)), but much improved spectral resolution is now available with the *SOHO* SUMER and *Hinode* EIS instruments. In addition, good spatial resolution is a requirement, since it appears that non-thermal line broadening varies across the fine structures of the quiet Sun.

Line widths were obtained by Doschek *et al.* (1976b) from the *Skylab* S082B slit spectrograph with a resolution of 0.06 Å (equivalent to $\sim$10 km s^{-1}) and with a spatial resolution of approximately 2 arcsec. By averaging measurements of transition region lines inside and outside the solar limb within a polar coronal hole they obtained non-thermal velocities of between 16 and 27 km s^{-1}, with no appreciable correlation with temperature up to 2×10^5 K. Several works have used data from the NRL HRTS instrument, with its 1 arcsec spatial resolution and 0.05 Å spectral resolution. The long slit could be positioned perpendicular to the solar limb to get measurements at different heights. This procedure was used by Dere and Mason (1993) to get non-thermal velocities for a variety of solar features and emission lines with a range of emitting temperatures. The quiet Sun values for a wide range of ions (up to Fe XII, temperature 1.4×10^6 K) have non-thermal velocities from 4 to 27 km s^{-1} with no apparent correlation with temperature. Similar values ($\sim$16 km s^{-1}) were obtained for the C IV doublet at 1550 Å (Dere *et al.* (1987)).

Much improved line broadening measurements have been made with *SOHO* SUMER, with 1–2 km s^{-1} spectral resolution and $1'' \times 2''$ spatial resolution. Also, long, uninterrupted series of observations can be made, since *SOHO* (unlike Earth-orbiting spacecraft) does not suffer eclipses. The instrumental broadening still needs to be subtracted, but with pre-launch measurements available, this can be done in a straightforward way. Some earlier results were achieved by Seely *et al.* (1997) (coronal lines outside the limb) and by Chae *et al.* (1998a) (disk centre). The non-thermal velocities measured by Chae *et al.* (1998a) cover the temperature range 10^4 K (C I) to 1.4×10^6 K (Fe XII), with several lines in the intermediate, transition region range. A striking variation with temperature results from their data, with non-thermal velocities increasing from 3 km s^{-1} at chromospheric temperatures to $\sim$30 km s^{-1} at $\sim 10^5$ K, after which there is a decrease to $\sim$18 km s^{-1} at coronal temperatures. A correlation of non-thermal velocity with the flux of the feature is indicated.

The correlation of non-thermal velocities with the flux of the feature is particularly marked in SUMER measurements of some transition region lines by Landi *et al.* (2000) and Akiyama *et al.* (2005). The measurements of Landi *et al.* (2000) were carried out on spectra averaged over pixels whose line intensities were within pre-defined flux groups and whose line profiles were shifted to correct for pixel-to-pixel line shifts in order to increase the signal-to-noise ratio. Figure 8.10 shows their results for O V 629.73 Å and N V 1238.82 Å; a similar correlation was found in Si IV (1402.77 Å). A lack of correlation was indicated in the earlier result of Athay *et al.* (1983) for the C IV 1548 Å line using *SMM* data.

Possible variations of non-thermal broadening in coronal lines with height above the solar limb have been investigated with SUMER data also. This is of great interest since a variation

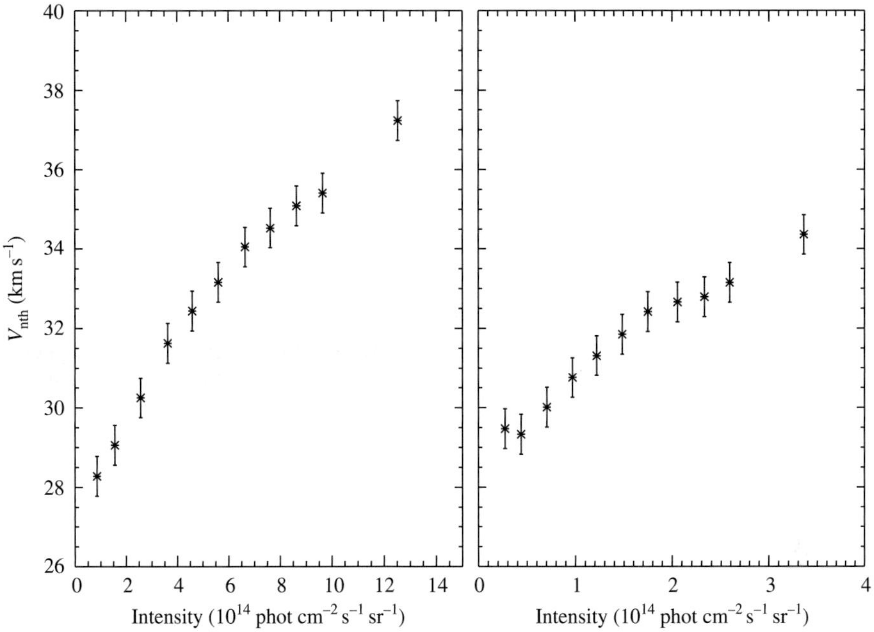

Fig. 8.10. Quiet Sun disk feature non-thermal velocity and flux for the O v 629.73 Å (*left*) and N v 1238.82 Å (*right*) lines. From Landi *et al.* (2000).

might suggest the existence of propagating waves, with obvious implications for coronal heating and solar wind acceleration. Banerjee *et al.* (1998) measured the width of the forbidden Si VIII line at 1445.75 Å at various heights above the limb up to 250 arcsec (180 000 km, or a radial distance of $1.25 R_\odot$). They obtained non-thermal velocities of about 27 km s^{-1} (27 arcsec) and 46 km s^{-1} (250 arcsec above the limb). These measurements were for interplume regions within a polar coronal hole. A dependence on height thus seems to be indicated by this work. In fact, with measured electron densities from the ratio of this Si VIII line with that at 1440.49 Å, an approximate relation exists between N_e and the fourth power of the indicated velocity, which Banerjee *et al.* (1998) attribute some significance to as this is the relation expected if the non-thermal velocities were a manifestation of propagating Alfvén waves. Indications that the interplume regions of coronal holes are the source of the solar wind rather than the plumes themselves are provided by O VI 1031.93 Å line width data by Hassler *et al.* (1997), who find that the line widths are $\leq 15\%$ narrower for plumes. Thus, Banerjee *et al.*'s (1998) result could imply that Alfvén waves accelerating the high-speed solar wind have an observational signature in the extreme ultraviolet range.

Doschek and Feldman (2000) used SUMER measurements of a coronal streamer, with the SUMER slit orientated in an east–west direction, to find that there was little or no correlation of line width with height. Table 8.1 gives the data in detail. The lines include those with formation temperatures up to 2 MK, and the height ranges are from $1.03 R_\odot$ to $1.45 R_\odot$. Non-thermal velocities between ~20 and 40 km s^{-1} are indicated by this study, independent of temperature and height range.

Table 8.1. *Measured line widths and velocities for ultraviolet lines observed by* SOHO *SUMER*.

Ion	Wvl. (Å)	FWHM (mÅ) and ξ (km s^{-1})[a]			
		1.03–1.13[b]	1.13–1.23	1.23–1.33	1.30–1.45
O VI	1031.93	288(34)	317(41)	320(42)	267(29)
Ne VIII	770.41	200(34)	207(36)	189(30)	213(38)
Mg X	609.79	137(27)	146(31)	150(33)	154(33)
Si VIII	1445.75	286(22)	285(22)	324(29)	
Si XII	499.40	96(21)	94(20)	108(27)	
Fe X	1028.04	229(35)	243(38)	220(33)	
Fe XII	1349.36	243(26)	256(28)	253(27)	296(34)

Data from Doschek & Feldman (2000).
[a] Non-thermal velocities ξ in parentheses.
[b] The height ranges are indicated by radial distances in units of $R_\odot$.

8.6 Quiet Sun dynamic phenomena

The term 'quiet Sun' is defined by the level of solar activity as measured by numbers of sunspots visible on the photosphere or amount of magnetic activity, both of which follow the well-known cycle of eleven years. However, on small length scales and short time intervals, the quiet Sun always displays activity manifested by dynamic phenomena. The photospheric convection and superconvection are obvious examples, as are the ever-changing chromospheric spicules visible on the limb in Hα light. At ultraviolet and X-ray wavelengths, the dynamic phenomena are more spectacular. An early and seminal study of ultraviolet radiation from the transition region was made by Brueckner and Bartoe (1983) using results from the HRTS instrument. The instrument, a Cassegrain telescope with a Wadsworth spectrograph, had a slit measuring $900'' \times 0.5''$ arc with the spectrograph covering the wavelength band 1175–1710 Å. In three flights of HRTS between 1975 and 1979 (on Black Brant rockets), a number of dynamic features were identified by broadened transition region lines, such as O IV and Si IV emission lines near 1400 Å, but most particularly the well-known C IV doublet at 1548, 1550 Å, emitted at 10^5 K. Brueckner & Bartoe (1983) recognized two broad categories, depending on the amount of line broadening: turbulent events (speeds < 250 km s^{-1}, spatial extents < 2 arcsec) and jets (much greater speeds, up to 400 km s^{-1}, with larger spatial extents). Figure 8.11 shows an example of a HRTS spectrogram with turbulent events and a jet. The more energetic jets are of much interest since the observed velocities exceeded the local sound speed ($\sim$150 km s^{-1}). If they were in the nature of expanding blobs of plasma, then shock waves would be likely to form, which might have implications for the heating of the solar atmosphere. Estimates of the numbers of events, extrapolated from those observed within the area of the HRTS slit to the whole Sun, averaged around 24 events per second. Simple estimates indicated that the energy and mass supplied by these events – 6×10^{27} erg s^{-1} and 2×10^{12} g s^{-1} – could explain the energy and mass requirements of the entire corona. Thus, some significance was attributed to these phenomena.

Further analysis of the HRTS turbulent events (Dere *et al.* (1989)) indicated that there was a slight spatial separation of the origin of the red shifts and blue shifts, so the broadened line profiles were not simply turbulence as their name would suggest: most literature now simply

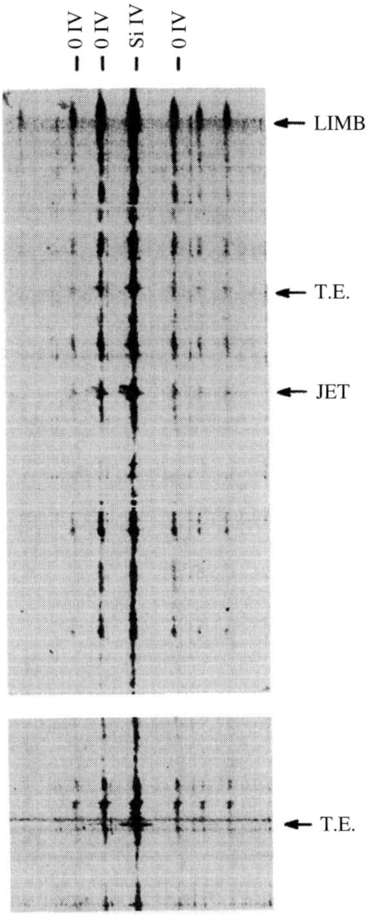

Fig. 8.11. HRTS spectrum from the first flight of the instrument, in 1975. This section shows the O IV (formation temperature 1.3×10^5 K) and Si IV (formation temperature 80 000 K) lines near 1400 Å. The 900″-long slit covers the solar disk out to and just beyond the limb. Turbulent events (T.E.) and a jet, shown as broadenings of the spectral lines, are indicated by arrows. From Brueckner & Bartoe (1983).

refers to them as explosive events. With *SOHO* SUMER spectra, in a number of spectral lines including the Si IV 1393.76 Å line, Innes *et al.* (1997a) found that there were half-hour-long bursts consisting of individual explosive events lasting from 1 to 6 minutes. Velocities of up to 180 km s^{-1} were measured. A bi-directional jet geometry was indicated (Innes *et al.* (1997b)), supporting a magnetic reconnection origin. Studies by Chae *et al.* (1998b,c) confirmed the existence of short bursts of events, with correlations in space and time with jets visible in the blue wing of the Hα line ('H$\alpha - 1$ Å' jets). Particularly revealing about the connection with the photospheric magnetic field was the observation that explosive events were correlated with cancellation events seen in fast-cadence ground-based magnetograms, in which small areas of positive and negative magnetic polarity, moving towards each other by supergranular convective flows, apparently merge.

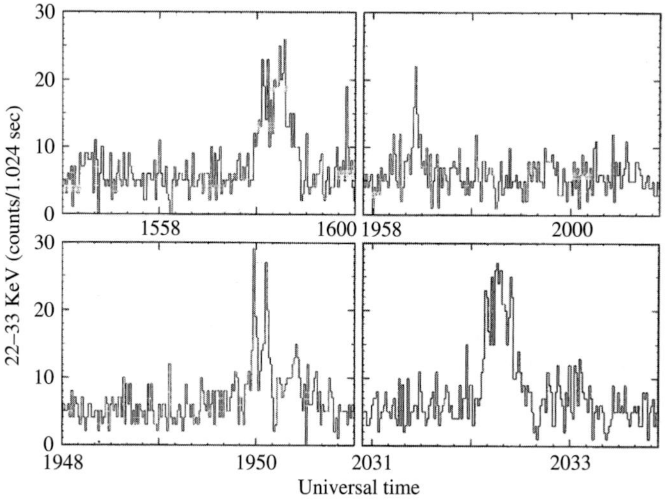

Fig. 8.12. Light curves for four of the largest 'microflares' seen by Lin *et al.* (1984) during 141 minutes of observation with a balloon-borne high-sensitivity hard X-ray spectrometer. The light curves are for the 22–33 keV channel: the total energy range of the instrument was 13–600 keV. Courtesy R. P. Lin.

Using a high-sensitivity balloon-borne X-ray spectrometer, Lin *et al.* (1984) found some 25 small hard X-ray flares during a 2 hour period when the Sun was near maximum activity, in June 1980. The peak fluxes of these flares at an energy of 20 keV averaged $\sim 10^{-2}$ photons cm^{-2} s^{-1} keV^{-1}, much smaller than previously observed hard X-ray bursts. Otherwise, these so-called microflares had very similar characteristics to larger hard X-ray flares in duration (seconds to tens of seconds), spectrum (power-law in photon energy), and being composed of smaller, elementary bursts. In the soft X-ray energies of *GOES*, these microflares were barely perceptible above the non-flaring CI background level. Figure 8.12 shows light curves of four of the microflares.

Further studies were later made with the high-sensitivity instruments on the *Yohkoh* spacecraft. With the *Yohkoh* SXT during solar minimum, Koutchmy *et al.* (1997) observed some tiny coronal flashes in a polar coronal hole during solar minimum with total radiated energies of only 10^{24} erg. A quiet Sun near-equatorial region was examined by Krucker *et al.* (1997) with the SXT for 91 minutes, and several small network flares, correlated with radio bursts, were observed having total radiative energies of between 10^{25} and 10^{26} erg. SXT images of a quiet Sun region at Sun centre over a three-day-long period, almost at the 1997 solar minimum, was analyzed by Preś and Phillips (1999). Six significant microflares, similar to the network flares of Krucker *et al.* (1997), were found, though in later unpublished work many more, fainter flares were identified. Extrapolating to the whole Sun, the events seen by Preś and Phillips (1999) represent some 2000 microflares per day.

As with the ultraviolet events noted by Brueckner and Bartoe (1983) and other observers, the possible energetic implications for many microflares occurring over the whole Sun's surface are worth examining. For X-ray flares particularly, the numbers of flares dN in an interval of the peak or total X-ray flux dE appear to follow a power-law dependence:

$$\frac{dN}{dE} = KE^{-\alpha}, \tag{8.3}$$

where K and α are constants. This has been found from soft X-ray flares (Drake (1971)) and for hard X-ray flares (Dennis (1985)). Of course, observational limits mean that there is always uncertainty whether the power-law dependence holds for all E, or whether different power-law indices α apply to different ranges of E. If the power-law dependence applies at very small flare total energies, such as are appropriate to microflares or even smaller events, one may examine the possibility that an accumulation of extremely small events – nanoflares or smaller – can account for the heating of the entire corona, either the quiet Sun or active corona. This is the basis of the ideas in which the corona is powered by tiny magnetic reconnection events (Levine (1974)) or more specifically nanoflares, each having a total energy below 10^{24} erg (Parker (1988)).

Hudson (1991) has noted that the integral of the energy contained in a range of flares with power-law frequency as in Equation (8.3),

$$P = \int_{E_0}^{E_1} \frac{dN}{dE} E\,dE = \frac{K}{2-\alpha}(E_1^{-\alpha+2} - E_0^{-\alpha+2}), \tag{8.4}$$

reveals whether tiny events have such a role in coronal heating. If α is less than 2, then very small flares as observed by present-day high-sensitivity instruments do not contribute to coronal heating unless the distribution of flare total energies steepens (or 'softens') below a certain value of E. In fact, the X-ray flare data of Drake (1971), Dennis (1985), and others indicate that α is marginally less than 2, around 1.8.

Studies of coronal brightenings in *TRACE* coronal images, those made in the 171 Å (Fe IX, Fe X) filter or 195 Å (Fe XII) filter, indicate somewhat different results from those obtained from X-ray studies. Parnell and Jupp (2000) obtained values of α a little above 2, though they found that strictly nanoflares, defined to be those events with energies in the range 10^{24}–10^{27} erg, are not adequate to explain the coronal heating requirements, as given by Withbroe and Noyes (1977). Parnell and Jupp (2000) state that extrapolation to picoflares (total energy down to 10^{21}–10^{24} erg) is needed for this.

The X-ray luminosity of the quiet Sun is of some importance in this respect, with values from *Skylab* data being widely quoted: Withbroe and Noyes (1977) give 2×10^{28} erg s^{-1}, and Vaiana and Rosner (1978) 5×10^{27} erg s^{-1}. However, measurements with *Yohkoh* (Acton, quoted by Haisch and Schmitt (1996)) and *GOES* indicate much smaller luminosities, of order 10^{26} erg s^{-1}. On this basis, the energies in the events observed by Parnell and Jupp (2000) are not far from those needed for the specifically quiet Sun corona.

In the extreme ultraviolet, accessible to the *SOHO* CDS instrument, a variety of dynamic phenomena can be recognized. Although line profiles cannot be easily measured with this instrument because of the relatively large instrumental profile, temporary brightenings, lasting a few minutes, have been seen in transition region lines (Harrison (1997), Gallagher *et al.* (1999a), Harra *et al.* (2000), Bewsher *et al.* (2002)). Brightenings ('blinkers') in the O V 629.73 Å line (Harrison (1997)) were observed to have durations in the range 7–26 minutes, occurring in bright portions of the quiet Sun network, with a birth rate of just over 1 per second over the whole Sun. The more detailed studies of Gallagher *et al.* (1999a), Harra *et al.* (2000), and Bewsher *et al.* (2002), in which automated means of recognizing blinkers were used, suggest an order-of-magnitude greater frequency than what was found by Harrison (1997). Some correlations with downward (red-shifted) velocities ($\gtrsim$ 20 km s^{-1}) have been found (Gallagher *et al.* (1999a), Madjarska & Doyle (2003)), and the occasional coincidence with explosive events seen with *SOHO* SUMER, though there is hardly any occasion when there is a coronal response (Bewsher *et al.* (2002)). The frequencies of brightenings in the network structures are much greater than those in the interiors of supergranular cells (Harra *et al.* (2000)). There is some indication that these brightenings are due to increases of density or possibly filling factor, though similar features in the C IV line at 1548.20 Å (Porter *et al.* (1987)) brighten as a result of impulsive heating. Studies with the *SOHO* EIT in the 304 Å filter (He II) show that blinkers are most likely to be identified with macrospicules (EUV equivalent of Hα spicules: Bohlin *et al.* (1975)) on the limb.

8.7 Coronal waves and oscillations

Since the quiet Sun corona is a fully ionized plasma, we expect to see a variety of MHD waves, though unlike the Earth's magnetosphere we must observe any wave phenomena remotely. Thus, we cannot directly observe pure Alfvén waves, for example, since (in the linear approximation at least) they are non-compressional and so there is no signature in line profiles. However, slow-mode MHD and sound waves can readily be detected through line widths. The observations have been discussed already (Section 8.5). In addition, there are many observations in the EUV and other wavelengths of oscillations which indicate the presence of waves in the corona. Full accounts of the theory of MHD waves and of the burgeoning new discipline of coronal seismology (the probing of magnetic field and other properties in the corona by observation of oscillations) are given by Aschwanden (2004). Here we summarize some of the quiet Sun observations of waves made in the EUV, although we omit large-scale wave phenomena (e.g. those known as EIT waves) associated with coronal mass ejections which are also discussed by Aschwanden (2004).

Observations of waves propagating along the polar plumes within coronal holes have been made with the *SOHO* EIT instrument. DeForest and Gurman (1998) took repeated images of the south polar coronal hole in 1996, near solar minimum, in the 171 Å (Fe IX, Fe X) passband of EIT. The cadence of the images is only 3 minutes. With a sequence of over 100 such images, they laid out images of a particular polar plume and identified features that they observed to propagate outwards along the length of the plume. From several such features, they derived a propagation speed of $\sim$100 km s^{-1}, which they identify with either the sound speed or the speed of a slow MHD wave. The features are observed to move through a height range from near the solar surface to $\sim$1.2 $R_\odot$. The flux of energy from such propagating waves is a few times 10^5 erg cm^{-2} s^{-1}, not far from the energy requirements of a coronal hole. Further out, UVCS observations in white light indicate similar wave phenomena (Ofman *et al.* (1997)). Wavelet analyses of *SOHO* CDS observations made in the O V 629.73 Å line indicate the presence of waves in a polar coronal hole with periods of about 10–25 s, occurring in bursts lasting $\sim$30 minutes. Again, there is evidence that such waves could supply the energy requirements of coronal hole regions.

Prominences or filaments have been observed in the EUV, and oscillations have been seen in them through Doppler-shifted material with *SOHO* SUMER. Bocchialini *et al.* (2001) have observed intensity and velocity oscillations with periods of 6–12 minutes. An interpretation of this and other SUMER observations of such waves is a fast-mode MHD wave.

Wave phenomena in the chromosphere have long been recognized through oscillations with three-minute periods seen in the Ca II H and K lines in the violet part of the visible-light spectrum (Carlsson & Stein (1997)). The emission inside supergranule cells, where magnetic fields are not dominant as at their edges, might be explained by a dynamic model in which there are vertical acoustic velocity perturbations, like a driving piston. However, this remains the subject of some controversy (Fossum & Carlsson (2005), Cuntz *et al.* (2007)). Observations with *SOHO* SUMER have established that oscillations also occur in the ultraviolet, most clearly in the ultraviolet continuum at wavelengths greater than 912 Å

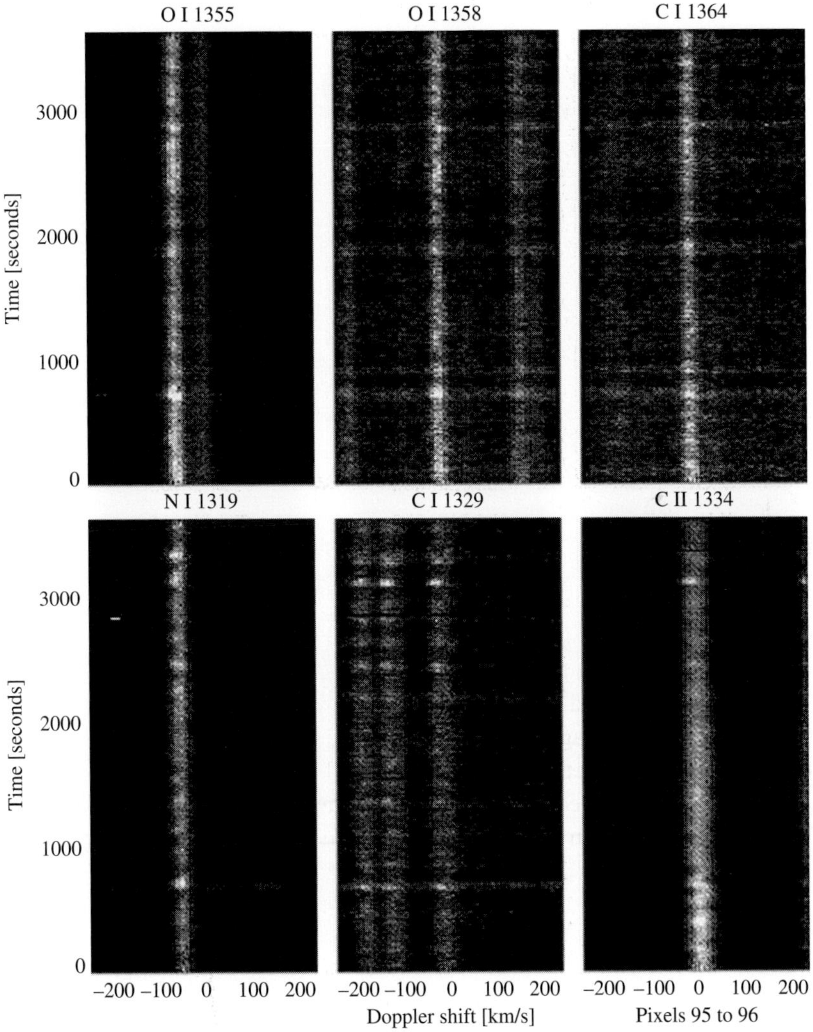

Fig. 8.13. Intensity profiles as a function of wavelength taken with the *SOHO* SUMER with its $1'' \times 120''$ slit for spectral lines near 1334 Å. Time is along the vertical axis (in s), and the horizontal wavelength scale is in km s^{-1}. Brightenings or 'grains' with an approximately 3 minute periodicity are evident, with small net blue shifts. From Carlsson *et al.* (1997).

(Carlsson *et al.* (1997)). They are also seen in chromospherically formed neutral carbon, nitrogen, oxygen, and helium ultraviolet lines in the SUMER range. As can be seen from Figure 8.13, the three-minute oscillations in intensity are very clear in lines of O I and C I. A blue shift in the lines, with velocity of $\sim$5 km s^{-1}, accompanies the intensity increases showing evidence for upward flows. It is clear, then, that the quiet Sun chromosphere, even in the non-magnetic supergranular cell interiors, is very dynamic.

9

Active regions

9.1 Introduction

As we have seen in Chapter 8, the so-called quiet Sun is in fact in a state of perpetual activity, with a huge number of localized changes – brightenings and material flows – occurring globally. Even within coronal holes there are bright points associated with bipolar magnetic regions which are subject to activity such as small flares and jets. There is in addition a large-scale activity that is associated with the familiar eleven-year solar cycle. In visible light, the most conspicuous examples of solar activity are sunspots, regions of high (up to 4000 G) magnetic fields, appearing as darker regions of the photosphere where the temperature is slightly depressed from the rest of the photosphere. Sunspots generally occur in pairs or groups, and it is found that in the chromosphere and corona there are counterparts to the photospheric sunspots in the form of chromospheric bright regions (known as plages) and loops stretching out into the corona, with temperatures up to 3 or 4 MK. These constitute *active regions*. They occur in latitude ranges that depend on the stage within the eleven-year activity cycle. At solar minimum, active regions are generally not completely absent but are seldom very large or well developed. They occur either side of the equator, in low latitudes, within $\sim 10^\circ$ north and south. The new activity cycle begins when active regions, sometimes accompanied by small sunspots, occur at high latitudes, latitudes 30° or more north and south. These small regions are short-lived, with any sunspots present lasting a day or so. However, the associated loops that are visible in the ultraviolet, extreme ultraviolet, and soft X-rays last for a few days. As the cycle proceeds, active regions become larger and last longer, and occur at lower latitudes. Maximum activity is attained when the sunspot number has its maximum. Then active regions are numerous, but still contained within a latitude band that is approximately 25° north and south, plus and minus a few degrees. Activity starts to decline when the sunspot number and the number and size of active regions decline, the latitudes continuing to approach the equator. The next cycle begins with the appearance of new active regions and spots at high latitudes, generally overlapping in time with the active regions of the old cycle near the equator.

9.2 Structure of active regions

9.2.1 Evolution of active regions

Active regions are associated with the magnetic field that emerges from subphotospheric regions, with sunspots arising from magnetic flux ropes that rise by magnetic buoyancy to the solar surface. A large active region emerges by stages. Its birth occurs with the brightening of quiet Sun network structures, observable in the visible-light Ca II H and K lines and ultraviolet lines with their origin in the chromosphere or transition region. The network cells become filled with bright plage areas, which brighten in the next day or so, becoming elongated in an east–west direction. Small spots start to appear, in the form of a pair or small group. They consist of leading and following spots – leading and following in the sense of solar rotation (east to west), with the leading spot slightly closer to the equator. The leading spot is generally larger and more well developed. The coronal loops associated with the active region are at this stage small and compact. The magnetic polarity of much of the leading part of the active region (including the leading sunspot) has one polarity, the following part has the opposite polarity. Active regions in the other hemisphere have this polarity pattern reversed. There may be fragments of opposite polarity near by, depending on the degree of complexity. In time, the leading and following sunspots become larger and separate, and are more parallel to the equator. More magnetic flux emerges, with continual small flaring occurring in the magnetically most complex part of the region. The newly emerged flux appears as small, bright loops in ultraviolet and as arch filament systems in Hα images. Their footpoints are marked by intensely bright regions with large flows marked by red and blue shifts in the Hα line (Ellerman bombs or moustaches).

Maximum development of the active region occurs when the chromospheric plage, as seen in Hα or in the ultraviolet He I and He II lines, reaches maximum brightness, the coronal loops are very bright and well developed, and the sunspots are largest and most complex. Flares are frequent and often very energetic, having very hot compact loops visible in X-rays and in ultraviolet lines. Separating areas of opposite magnetic polarities are so-called polarity inversion lines – when these are highly convoluted, flaring activity is very frequent. Along the less convoluted parts of the polarity inversion line, material gradually builds up to form a dark active region filament, visible as a prominence off the limb in Hα and the He I 584 Å and He II 304 Å EUV lines.

An active region decays by the gradual fading of the chromospheric plage and coronal loops, which expand, with the magnetic field simplifying from its earlier complex pattern. The sunspots become smaller, with the many small spots fading altogether. The last vestiges of the region are an area, approximately 100 000 km across, with a weak bipole structure, and the active region filament, which gradually migrates poleward to become a quiescent filament.

Figure 9.1 shows a well-developed active region observed by *SOHO* in December 1999, near the maximum of Cycle 23. The active region in white light, as recorded by the Michelson Doppler Imager (MDI), shows large sunspots with numerous small spots, the larger spots having dark central areas (umbrae) and lighter, filamentary outer areas (penumbrae). In the photospheric magnetic field recorded by MDI, the leading (west) parts of the active region are revealed as having positive polarity (white areas in the magnetogram), the following areas mostly negative polarity (dark areas), but with mixed polarity in the intermediate areas and to the south of the main spots. Bright plage occurs in the EIT 304 Å (He II) image, the emission being of chromospheric origin. The bright areas are well correlated with regions of strong magnetic field. In coronal emission, as recorded in the EIT 195 Å (Fe XII) image, many small loops are visible in a complex pattern near the spots as well as the area to the south. In addition, there are radiating sunspot plumes from the largest sunspots.

At times of activity maxima, many active regions dominate the ultraviolet and soft X-ray emission from the solar disk. Figure 9.2 (top) shows a *SOHO* EIT 195 Å solar image in June

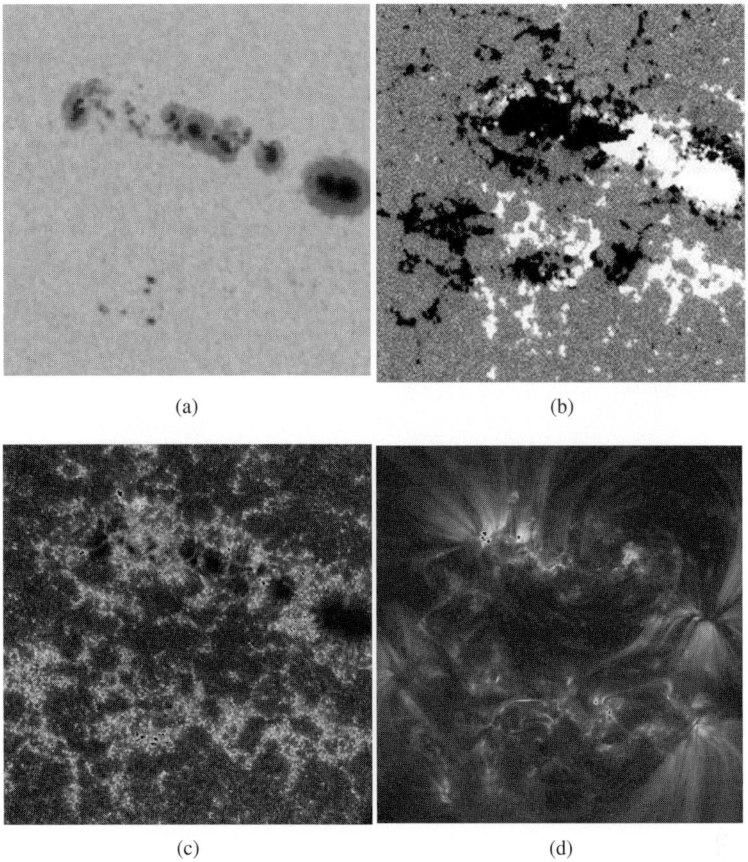

Fig. 9.1. Images of a well- developed active region on 1999 December 24 (06:24 UT). (*a*) MDI integrated intensity (white light) image showing sunspots. (*b*) MDI magnetogram showing photospheric magnetic field (white: positive polarity, black: negative). (*c*) *TRACE* 304 Å (He II) image. (*d*) *TRACE* 195 Å (Fe XII) image. Courtesy the MDI and *TRACE* teams.

1998, at a time of rising solar activity. The active regions appear in bands parallel to the equator, $\sim 25^\circ$ north and south latitude. Some are near their maximum development, with loops reaching $\sim 50\,000$ km above the photosphere, as can be seen from the active regions on the east and west limbs. At times of solar minimum, as in the EIT image in June 1996 (see Figure 9.2, bottom), only a few active regions at near-equatorial latitudes are present. One (slightly south and east of Sun centre) is an old, decaying region spread over a large area, with faint loops extending to high altitudes. Other regions, such as that slightly north and west of the Sun centre, are compact, with low-lying loops. At both poles there are conspicuous coronal holes, where the 195 Å emission is diminished. Within these holes, there are small bright points, visible in soft X-rays also, coincident with small bipolar regions in magnetograms.

9.2.2 Sunspots

The very earliest observers of sunspots recognized that there is a central, darker umbra surrounded by a lighter penumbra. High-resolution white-light images (such as those

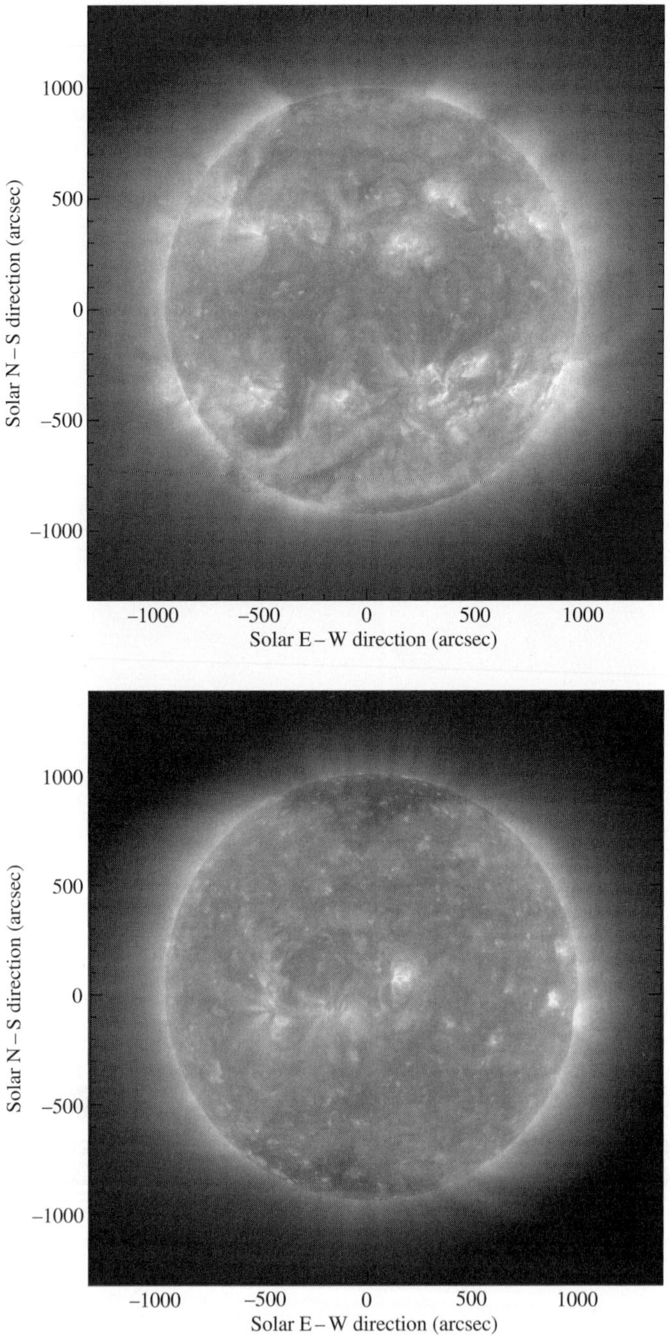

Fig. 9.2. Solar images obtained with *SOHO* EIT in its 195 Å filter. (*Top*) 1998 June 10, solar activity rising after the minimum two years before, with several active regions present on the Sun, mostly around 25° north and south latitude. (*Bottom*) 1996 June 7, near solar minimum, with few active regions on the Sun at near-equatorial latitudes and well-developed coronal holes at both poles. Courtesy *SOHO* EIT team.

from the Swedish Vacuum Tower Telescope in the Canary Islands and the SOT instrument on *Hinode*) have shown that the penumbra consists of elongated photospheric granules. Early extreme ultraviolet observations with the S055 instrument on *Skylab* (Foukal *et al.* (1974)) found that narrow plume-like structures radiated from well-developed sunspots, features much brighter than others in the active region observed. Further observations were made with the *SMM* UVSP instrument (Kingston *et al.* (1982)) at longer wavelengths and slightly lower temperatures; they appeared to differ from the original *Skylab* observations in their interpretation of the importance of the plumes imaged in transition region lines, temperatures between 10^5 K and 10^6 K.

A study by Maltby *et al.* (1998) made with the *SOHO* CDS instrument appears to resolve these differences. The existence of sunspot plumes is confirmed, with emission observed

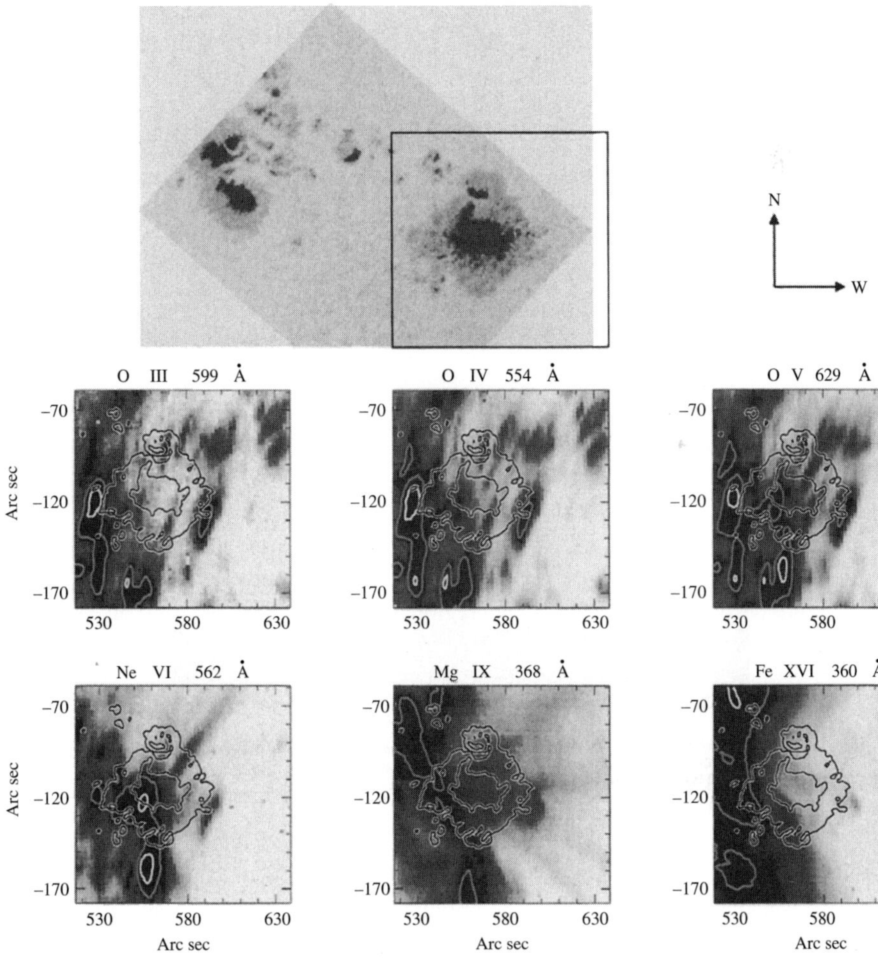

Fig. 9.3. *SOHO* CDS images of a sunspot on 1996 November 28 in the EUV lines indicated over each image. Dark areas show more intense regions. Note the sunspot plume in especially the Ne VI image. The sunspot was the largest one in active region NOAA 7999, photographed in white light with the National Solar Observatory's Vacuum Tower Telescope the previous day. Contours on the CDS images show the outline of this spot's umbra and penumbra. From Maltby *et al.* (1998).

in O IV 554.51 Å, O V 629.73 Å, and Ne VI 562.80 Å. These lines cover the temperature range 160 000 K to 400 000 K, a higher temperature range than in the observations made with *SMM*. Figure 9.3 shows images made in the various lines in this study, ranging from O III to coronal lines such as Mg IX 368.07 Å. A prominent sunspot plume radiates from the sunspot umbra towards the north-west in the O IV and O V images, and is particularly prominent in the Ne VI image. A variation in the plume width is apparent.

Dynamic phenomena associated with sunspots have been known since an outward or Evershed flow in the photosphere was identified, from a sunspot umbra out to its penumbra and a little beyond. In visible-light images, a moat consisting of outwardly moving bright points is present in large, often decaying spots (Sheeley (1969)). The visible-light Evershed flow is highest in the penumbra or just outside it, with greatest outflow velocities of 2 km s^{-1} being attained, but in chromospheric lines like Hα, the flow is inwards and occurs along the dark filaments making up the spot superpenumbra.

Ultraviolet and other observations of flows and waves give a rather unclear picture at present. Recent observations (Lites *et al.* (1998); Georgakilas *et al.* (2002)) suggest an inflow of material within the spot umbra but an outflow in the penumbra and beyond. The observations of Georgakilas *et al.* (2002) were made with the 1550 Å, 1600 Å, and 1700 Å filters of *TRACE*. In the 1700 Å filter particularly, striking bright and dark annular intensity fluctuations radiate from the sunspot, with velocities of 2 km s^{-1} at the spot penumbra, reducing to 1 km s^{-1} in the superpenumbra.

9.2.3 *Magnetic evolution of active regions*

With high-resolution magnetographs such as the MDI instrument on *SOHO*, we can follow the detailed evolution of active regions in the EUV and X-rays through changes in the photospheric magnetic field. Every active region follows a different evolution, so it is instructive to follow particular case studies. One example is the huge and highly flare-prolific active region NOAA 10314, which started life as a pair of magnetic bipoles within a transequatorial coronal hole (Morita & McIntosh (2005)). The two bipoles first became evident in MDI magnetograms on 2003 March 13, near Sun centre, then rapidly developed with westward motions of the major polarity centres, with the end result that a large single bipole structure developed just as the active region approached the west limb on March 19. A main polarity inversion line can be recognized from the start. (This is defined using criteria based on the observed magnetic field gradient and its derivative, as well as the constraint that the absolute value of the field changed by at least 300 G across the inversion line.) In the next few days, this polarity inversion line rotated, at a rate of approximately 20° per day. The flare productivity of the region meanwhile increased steadily, with flares of *GOES* importance C or sub-C. All occurred along the polarity inversion line. Eventually, the polarity inversion line reached an angle of 100 degrees relative to the initial orientation, and at this time major flaring commenced, with many M and some X flares. The rotation of the polarity inversion line then slackened, but with major flaring continuing (Figure 9.4). Seemingly, the angle of the polarity inversion line for this particular active region had a major impact on the flaring process and the evolution of the active region.

9.2.4 *X-ray bright points*

The small bright points in polar coronal holes and in quiet Sun areas offer a means of testing the relation of the photospheric magnetic field and their intensities in either ultraviolet

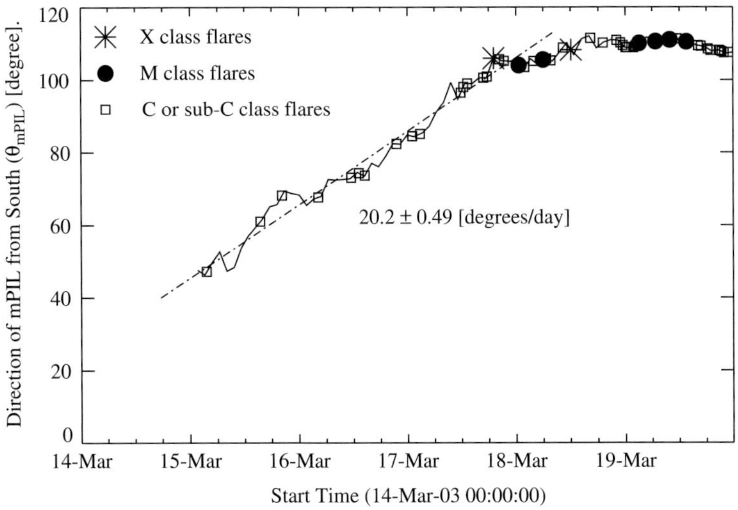

Fig. 9.4. The evolution of a major active region, NOAA 10314, during a six-day period in March 2003 is followed in this plot of the rotation angle, from a south direction ($\theta_{\rm mPIL}$), of the magnetic polarity inversion line (mPIL). The mPIL is defined by the magnetic field gradient exceeding a value of 200 G $\rm Mm^{-1}$ (1 Mm = 1000 km), and was determined from *SOHO* MDI magnetograms with 96 minute cadence. Courtesy S. Morita, from Morita & McIntosh (2005).

or X-ray emission. They are generally much simpler than the near-equatorial active regions, generally consisting of simple bipoles (Golub *et al.* (1977)). The magnetic field in regions away from bright points is in a continuous state of change, with field disappearing when centres of opposite polarity collide and with replenishment occurring from below (Schrijver *et al.* (1998)).

A study by Preś and Phillips (1999) of coronal bright points identified in *SOHO* EIT images made in the 195 Å (Fe XII) filter shows an intimate link between the evolution of the total flux of bright points and the magnetic flux Φ, or $\int B_l \, dS$, where B_l is the line-of-sight magnetic field strength as measured from photospheric magnetograms made with the *SOHO* MDI instrument, and the integral is over a surface area S covered by the bright point. A region near Sun centre was analyzed over a two-day period very close to solar minimum, and several bright points in the EIT images identified. A striking correlation of the magnetic flux Φ and the EIT 195 Å intensity results (Figure 9.5 shows one of the bright points investigated). Even fine features in the 195 Å light curve are reproduced to some extent in the magnetic flux curve, showing the intimate relation between the photospheric field and activity in the corona. As with most coronal bright points, the lifetime was about 1.5 days. The bright point in this case could be identified with the *Yohkoh* SXT instrument, and though the time cadence of these X-ray observations is not as short as with the *SOHO* observations, a clear correlation of the X-ray intensity with the EIT 195 Å flux and Φ is apparent. The flux of the EIT emission F_{195} is very nearly proportional to the magnetic flux Φ.

Table 9.1 gives measured energies for bright points in the survey of coronal bright points by Preś and Phillips (1999). The largest uncertainties in the estimated quantities are those in the size of the bright points. The volume V was taken to be a sphere or ellipsoid, depending

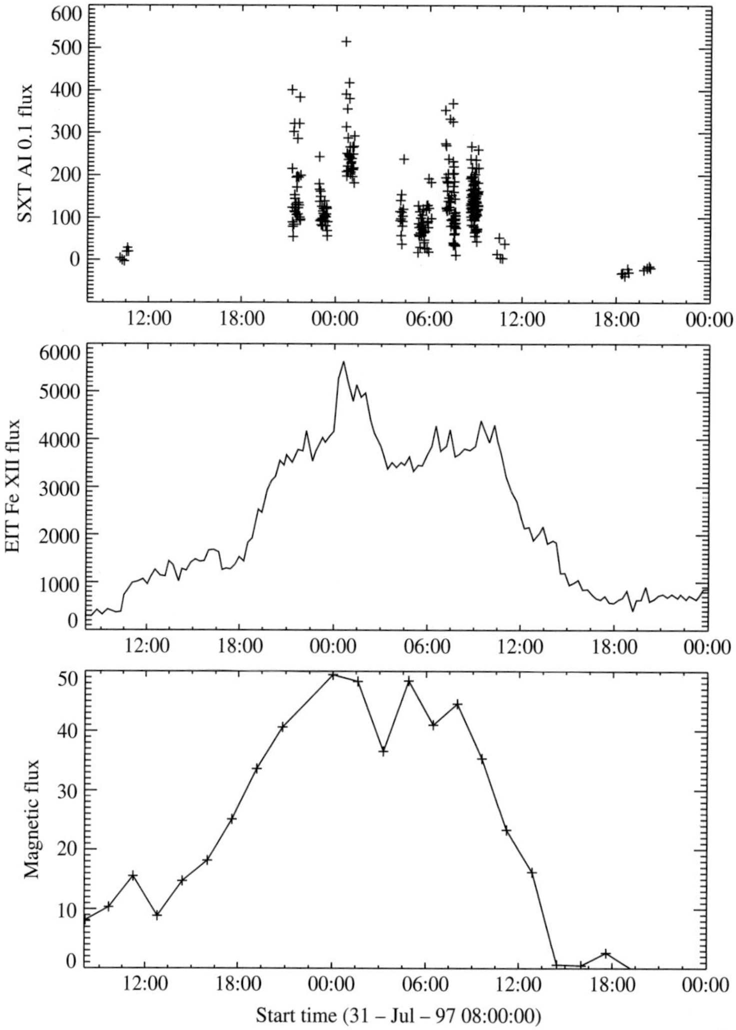

Fig. 9.5. Time history over a several-hour-long period of a coronal bright point near Sun centre (bright point A in Table 9.1). (*Top*) Variations in the soft X-ray flux recorded by the *Yohkoh* SXT (thin aluminium filter). (*Middle*) EUV flux recorded by the *SOHO* EIT (Fe XII, 195 Å filter). (*Bottom*) Magnetic flux estimated from photospheric magnetograms made by the *SOHO* MDI. From Preś & Phillips (1999).

on the appearance in the *SOHO* EIT images. For the radiative energy, the *Yohkoh* SXT gives some indication of the total X-ray flux in the 1–300 Å band through extrapolation of the SXT bands that have shorter wavelengths, but some estimates were also made from the EIT fluxes. These are less by a factor 10 or more than the conductive losses. The latter are calculated from the classical (Spitzer) expression for thermal conductivity (Equation (1.10)) using loop lengths from EIT images. The magnetic energies are from the *SOHO* MDI magnetograms. The conductive losses are, for the three bright points for which estimates were possible, within a factor 2 of the magnetic energies. This suggests that the magnetic energy of a bright point accounts for its energy losses.

Table 9.1. *Energies of coronal bright points observed by Preś and Phillips (1999).*

Bright point	E_{th} (erg) ($3N_e k_B T_e V$)	E_{rad} (1–300 Å, erg)	E_{cond} (erg)	E_{mag} (erg)
A	—	—	—	1×10^{29}
B	4×10^{26}	6×10^{27}	$> 4 \times 10^{28}$	1.4×10^{29}
C	9×10^{26}	1×10^{28}	1.4×10^{29}	8×10^{28}
D	—	—	—	$> 2 \times 10^{28}$
E	—	—	—	1.7×10^{29}
F	2×10^{27}	6×10^{27}	2.6×10^{29}	2.4×10^{29}

E_{th}: thermal energy, E_{rad}: radiative energy loss; E_{cond}: conductive energy loss; E_{mag}: magnetic energy.

9.3 Active region loops

The plasma confined in magnetic loops provides most of the emission in active regions so that loops can be thought of as the basic building blocks of active regions. An example of their importance is shown in Figure 9.6, showing an active region on the disk as seen by *TRACE* in its 171 Å and 195 Å filters. Differences between the images are due to temperature variations. Figure 9.7 shows monochromatic images of an active region on the limb taken by the *SOHO* CDS instrument with lines formed at very different temperatures, characteristic of chromosphere, transition region, and corona. Different loops show up the different temperatures.

The importance of active region loops has long been recognized, and substantial efforts have been made both to measure their physical properties and to develop theoretical models able to account for them. Loops can be roughly seen as magnetically confined and thermally

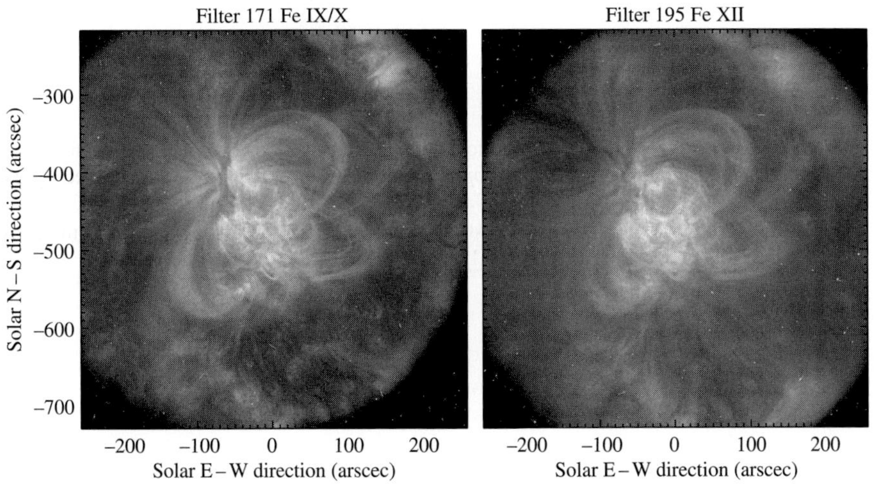

Fig. 9.6. An active region observed by *TRACE* in its 171 Å (*left*) and 195 Å (*right*) filters on 1998 October 13. Courtesy *TRACE* team.

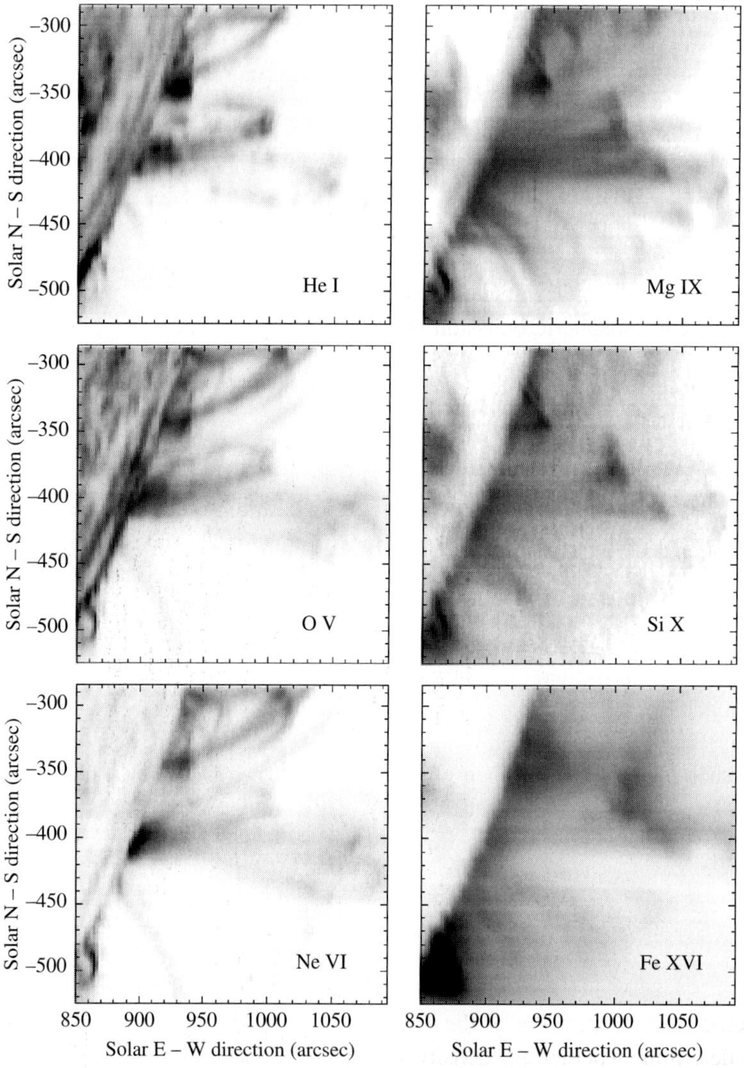

Fig. 9.7. Monochromatic images of an active region taken with the *SOHO* CDS spectrometer with different ions: He I (2×10^4 K), O V (2.4×10^5 K), Ne VI (4.3×10^5 K), Mg IX (9.8×10^5 K), Si X (1.3×10^6 K), and Fe XVI (3×10^6 K). Courtesy *SOHO* CDS team.

insulated pipes connecting two areas of the photosphere with different magnetic field polarity, called footpoints, since the plasma β (ratio of plasma and magnetic pressures) is very low and the conductivity of the plasma is very high.

9.3.1 Physical properties of active region loops

The shape of the loops varies according to the spatial resolution of the instrument observing them. Before the launch of *TRACE* and *SOHO*, coronal imagers and spectrometers seldom had a spatial resolution better than a few arcseconds, so that loops appeared as uniform toroidal structures with approximately constant cross-section. Only with the advent of high spatial resolution (1 arcsec) imagers were the apparently uniform loops revealed to be in the form of filamentary loops with 1 arcsec diameters or less. These strands follow the main loop structure from footpoint to footpoint, and possibly have different physical properties. Unfortunately, no conclusive evidence has been found to support the claim that elementary loop strands within large loops have been completely resolved with current instrumentation. The cross-section of large-scale loops (comprising several elementary strands) is currently an open problem in loop physics. Images of active region loops in the corona suggest that the loop cross-section is approximately constant, yet the lack of clear connection between the coronal loop emission (e.g. in Ne VIII lines) with transition region or chromospheric emission (e.g. C IV and H I lines) pointed out by Landi and Feldman (2004) points towards a sharp reduction of the cross-section at chromospheric temperatures.

Simple theoretical considerations predict that the plasma temperature along the loop axis should vary smoothly from chromospheric values at the footpoints to a maximum in the corona, as a result of heating processes and cooling by radiation and conduction. However, imagers and spectrometers on board *TRACE* and *SOHO* have shown that loops are almost isothermal along their entire length with temperatures in the range $1–3 \times 10^6$ K, and some decrease in temperature only at the footpoints where also emission from ions formed in cooler conditions is present (Neupert *et al.* (1998), Lenz *et al.* (1999), Landi & Landini (2004)). Several authors attempted to measure the temperature structure across the loop either with an emission measure or a differential emission measure (DEM) analysis, in an effort to determine whether the loop plasma could be considered locally isothermal at a particular position along the loop, or multi-thermal as the presence of multi-temperature strands requires. However, differing results have been obtained, with some authors (Brković *et al.* (2002), Del Zanna & Mason (2003)) finding loops to be isothermal and others (Schmelz *et al.* (2005)) finding them to be multi-thermal. Unfortunately, the spatial resolution of current spectrometers does not allow a clear separation between individual loop structures and the surrounding plasma, and so a definitive measurement of the loop thermal structure across its axis cannot yet be made.

Background subtraction and the relative weakness of spectral lines that are density sensitive have prevented the measurement of the density profile along the loop axis with sufficient accuracy to determine whether any density variation occurs along the loop axis. In general, constant densities along the loop axis in the $8.5 \leq \log N_e \leq 9.5$ range are compatible with measurement uncertainties, although some indication of a decrease of the density at the loop top seems to be present (Del Zanna & Mason (2003), Ugarte-Urra *et al.* (2005)).

Dynamics in loops can be detected directly by high-cadence series of images as motions of the entire structure or as intensity changes due to waves or plasma condensations travelling along the loop. Otherwise, Doppler shifts in line wavelengths can be readily measured by spectrometers if the direction of the velocity is not perpendicular to the line of sight. For example, Winebarger *et al.* (2002) measured velocities as high as 40 km s^{-1} in an active region loop system, concluding that plasma dynamics should play an important role in loop modelling. Robbrecht *et al.* (2001) observed propagating slow magneto-acoustic waves as disturbances originating from small-scale brightenings at the loop footpoints, while de Moortel *et al.* (2002) observed longitudinal oscillations at footpoints, damped within 12 000 km. Damped longitudinal oscillations were also observed by Wang *et al.* (2002)

following small and large flare activity, originating together with enhancements of Fe XIX emission at the beginning of flares.

9.3.2 *Loop models*

Theoretical loop models were developed as an attempt to reproduce the morphology and the observed physical properties of active region loops. In particular, they aim at understanding the physical mechanism responsible for loop heating. Loop models fall into two broad categories: magnetohydrodynamic (MHD) or hydrodynamic models. The difference between the two lies in the treatment of magnetic field.

In MHD models, the equations of magnetohydrodynamics are explicitly taken into account and solved to determine the magnetic configuration of an entire active region, using as boundary conditions the magnetic and velocity fields measured in the photosphere; also, these models describe the plasma heating using magnetic reconnection theory. An example of such models is the one developed by Gudiksen and Nordlund (2005). Hydrodynamic models, on the other hand, focus on individual loops, and the magnetic field is merely assumed to confine the plasma within a loop, like a pipe; also, the heating is assumed to be a function of one or more loop parameters, e.g. exponential heating concentrated at the footpoints, or simply uniform heating. Hydrodynamic models solve the equations of conservation of mass, momentum, and energy, and are able to predict the temperature, density, pressure and velocity profiles as a function of time of individual loops, and are better suited for detailed comparison with diagnostic results from spectrometers. Examples of such models are the one-dimensional hydrodynamic models of Landini and Monsignori Fossi (1975) and Rosner *et al.* (1978), and their extension to dynamic régimes by Landini and Monsignori Fossi (1981) and Serio *et al.* (1981).

9.3.3 *Open problems in loop physics*

There are many open problems in loop physics, and some of the most basic questions are still unanswered. The foremost problem in loop physics is to understand what are the physical processes that heat the loop plasma and maintain its temperature stable up to a few million degrees. MHD models rely on several different approximations to describe magnetic reconnection as the main mechanism for heating the plasma, but so far have failed to provide a widely accepted picture; also, their predictions are not easy to compare with observations. Hydrodynamic models on the contrary attempted to determine the shape of the heating function as a function of the loop physical parameters (usually temperature or position along the loop) by comparing the observed temperature and density profiles with those predicted adopting a specific heating function since, as shown by Priest *et al.* (2000), the heating function determines the temperature profile along the loop. Two different kinds of heating have been studied: steady-state and time-dependent heating; the former as a time-independent function of temperature or position along the loop, the latter as a distribution of nanoflare events occurring randomly in the loop with a user-defined frequency and energy spectrum. However, the modelling of nanoflares has been hindered by the need to calculate explicitly the ionization state of all the elements in a plasma as a function of time, since the ion abundances in a nanoflare picture can be very different from those in equilibrium (e.g. Bradshaw & Mason (2003)). Attempts at determining the functional form of steady-state heating have also failed, owing to the problems of background subtraction (Reale (2002)), and such comparisons have only allowed uniform and exponential steady-state heating to be ruled out.

The temperature structure of loops remains a mystery, mostly because the spatial resolution in spectrometers is presently inadequate to allow a single loop structure to be resolved from its surroundings and its physical properties measured. In fact, no consensus has been reached on whether the loop plasma is isothermal or multi-thermal across its section, despite the importance of the issue: multi-thermal loop structure would be a direct indication of the presence of sub-resolution loop strands at different temperatures, and a strong constraint on heating mechanisms, whereas an isothermal plasma would imply that sub-resolution strands, if present, are all at the same temperature, again providing very strong constraints on heating models.

The temperature and density profiles along the loop length are still a puzzle despite numerous efforts to reproduce them. Measurements obtained with the *SOHO* spectrometers and imagers and with *TRACE* have unambiguously shown that the loop plasma is denser than that predicted by theoretical models, and much more nearly isothermal. An example is shown in Figure 9.8, where the measured and predicted density and temperature profiles are shown as a function of position along the loop length (from Landi & Landini (2004)): predicted profiles were calculated using an array of plasma velocity values and heating functions, but none showed agreement with observations.

The loop cross-section is often assumed to be constant along the entire loop, and several *TRACE* images indicate that this is the case in the coronal portion of the loop, encompassing most of the loop length. However, little is known about the cross-section at lower temperatures. Recently, Landi and Feldman (2004) showed that the loop cross-section must decrease greatly at temperatures of 10^5 K or lower, since the emission predicted by uniform-cross-section models at such temperatures was far too high no matter what heating function was used. They also showed that a variable cross-section would not only greatly reduce the loop emission in the chromosphere and transition region, but also causes the loop temperature profile to vary far less than in the uniform-cross-section case. An example is shown in Figure 9.9, where the temperature profile and the emission measure–temperature plot are shown for two loop models whose only difference lies in the cross-section shape.

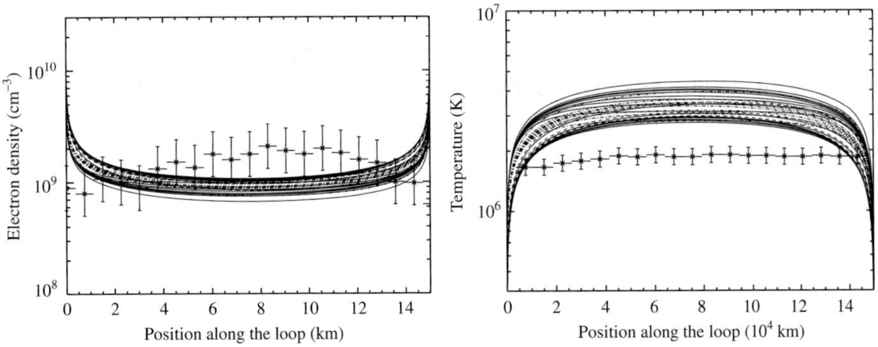

Fig. 9.8. Predicted and measured (points) values of electron density (*left*) and temperature (*right*) in an active region loop. From Landi & Landini (2004).

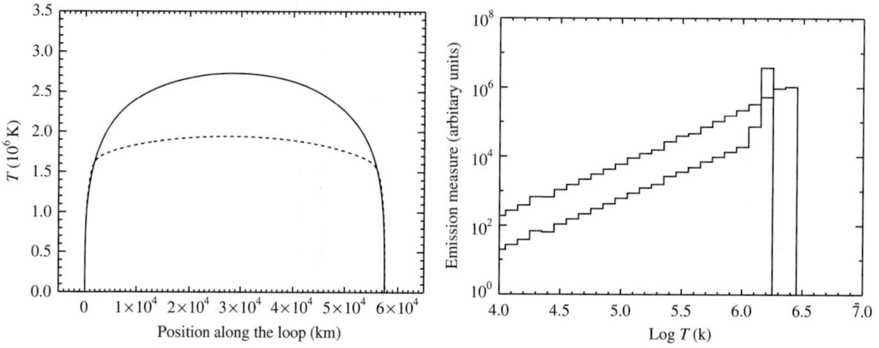

Fig. 9.9. Temperature profile along a loop (*left*) and emission measure versus temperature (*right*) predicted using a constant-cross-section loop model (full line) and a variable-cross-section loop model (dashed line). From Landi & Feldman (2004).

Another outstanding problem is the relation of active region loops with widely differing temperatures. As with quiet Sun loops, there appears to be no direct connection between loops having transition-region temperatures ($\sim 10^5$ K) and those with temperatures $\gtrsim 1$ MK, as shown by Figure 9.7. This is apparent from a comparison of *Yohkoh* SXT images with those from the *SOHO* CDS instrument made by Matthews and Harra-Murnion (1997). It was concluded from this study that, while there were some $\sim 10^5$ K loops that may have cooled from coronal temperatures, many loops had certainly not. Feldman and Laming (1994) similarly found, from *Skylab* data, that the flux of Ne VII and Mg IX line emission in off-limb loops was uncorrelated, suggesting that the lower-temperature structures were not thermally connected to those at coronal temperatures.

9.3.4 Loop oscillations

Spatial oscillations of coronal loops were first observed in *TRACE* 171 Å images during a flare on 1998 July 14 by Aschwanden *et al.* (1999). Five separate loops were identified following the flare, with lengths around 130 000 km, transverse oscillation amplitudes of about 4000 km, and periods of 256 s. The periods and phases for the five loops are similar, so indicating a common trigger for the oscillations. This is undoubtedly the flare itself, the oscillations being consistent with a radially propagating disturbance of about 700 km s^{-1}.

Full details of the analysis procedure are given by Aschwanden (2004) (Chapter 7), so only a brief discussion will be given here. The oscillations have been identified with magneto-acoustic oscillations of the fast kink mode. This gives us valuable information about the coronal magnetic field strength B, since the kink-mode period $P_{\rm kink}$ is related to B through

$$B = \frac{L}{P_{\rm kink}}\sqrt{8\pi\rho_{\rm i}(1+\rho_{\rm e}/\rho_{\rm i})}, \qquad (9.1)$$

where the external and internal mass densities are $\rho_{\rm e}$ and $\rho_{\rm i}$ respectively and the loop length is L. Nakariakov and Ofman (2001) used observations of the 1998 July 14 oscillating loops, including a density $N_{\rm e} = 2 \times 10^9$ cm^{-3} (uncertainty about a factor of 2) and density contrast $\rho_{\rm e}/\rho_{\rm i} = 0.1$, to get a field strength of 13 ± 9 G and Alfvén speed 756 ± 100 km s^{-1}. One of the largest uncertainties is in the density estimates. Future investigations of such oscillations might profitably use density-dependent line ratios of coronal lines to improve estimates.

Slow-mode oscillations in high-temperature (Fe XIX and Fe XXI) loops have been observed as Doppler shifts with the *SOHO* SUMER instrument as well as spatial shifts with *TRACE* and the EIT instrument on *SOHO* (Wang *et al.* (2002)). The oscillations are most likely triggered by a small disturbance near one of the footpoints of the loops. The oscillations, with periods of several minutes (up to 31 minutes), are damped, probably through the cooling of the loops by thermal conduction. The phase speed of the oscillations, determined from the observed periods and lengths of the loops, are approximately the local sound speed, and this together with the observed lag of intensity oscillations compared with those in velocity is consistent with slow-mode oscillations.

There is clearly much to be capitalized on these observations. With kink-mode oscillations, for instance, there is the possibility of getting valuable information about the coronal magnetic field in active region and quiet Sun loops. A new branch of solar physics is emerging called MHD coronal seismology, based on ultraviolet observations of coronal loops, which is likely to gain much importance in the next few years.

9.4 Active region densities

A number of studies have been made of electron densities in active regions using the diagnostic methods outlined in Chapter 5 (Section 5.2). Here we outline some of these.

Three active regions near the solar limb were observed with the *Skylab* S082B spectrometer in an automatic limb scan mode (Feldman & Doschek (1977, 1978a)). The $2'' \times 60''$ slit of the instrument was placed at a tangent to the limb, spectra were taken at $-12''$, $-4''$, $-2''$, 0, $2''$, $4''$, $6''$, $8''$, $12''$, and $20''$ relative to the limb (negative numbers indicate positions inside the limb). For the chromosphere, the profiles of high members of the H I Balmer series depend on electron density through the Stark effect, non-thermal mass motions, and temperature (see Section 5.2.4). In analyzing the profiles of Balmer lines with transitions from levels with upper quantum numbers of 9 to 31 (wavelengths between 3645 Å and 3835 Å), it was found that, although the Balmer lines at $2''$ above the limb were 2–4 times brighter in the active regions than quiet Sun and coronal hole regions, the electron densities in all regions were nearly the same, namely, 2×10^{11} cm^{-3}, to an accuracy of better than 30%. For an assumed temperature of 7800 K, the non-thermal mass motions in the regions were between 0 and 14 km s^{-1}.

As indicated in Chapter 8 (Section 8.3.3), electron densities can be derived from the ratio of the C III spin-forbidden line at 1908.73 Å and the Si IV line at 1402.77 Å for quiet Sun regions and coronal holes. This line ratio has also been examined for active regions by Feldman and Doschek (1977). In contrast to the quiet Sun results, the active region line fluxes in some limb locations are not only an order of magnitude larger but markedly irregular. The striking result is that, in spite of the irregular changes in the Si IV line flux at heights of 0 to $+6''$, a volume from which most of the transition-region emission occurs, the fluxes of the C III 1908.73 Å line emission are the same to within a factor ≤ 2, implying that within the same volumes the number of C^{+2} ions is quite similar. Since the composition in the transition region is nearly constant (see Chapter 11) this implies that at heights $\leq 6''$ the total number of particles with a temperature in the range 60 000–70 000 K in these active regions is similar. Clearly, the large flux differences are due to the large variation in electron densities. Note that, owing to the fact that the fractional ion abundances of C^{+2} and Si^{+3} peak at slightly different temperatures, the actual electron density ratios in the 60 000 K regions may differ somewhat from those shown in Figure 9.10(*c*) and Table 9.2.

Table 9.2. *Electron densities in two active regions above the limb derived from the Si* IV *1402.77 Å and C* III *1908.73 Å line ratios.*

Region	Electron densities (cm^{-3})						
	−12″	−4″	−2″	0	2″	4″	6″
B	3.2(11)	1.4(11)	1.0(11)	6.0(10)	2.8(10)	2.0(10)	2.0(10)
C	5.6(10)	1.2(11)	1.8(11)	8.0(10)	1.1(11)	1.0(11)	1.1(11)

Numbers in parentheses indicate powers of 10. Regions B and C are labelled in Figure 9.10. Positions relative to the limb are indicated by angular measures (negative numbers: inside the limb; positive numbers: off the limb).

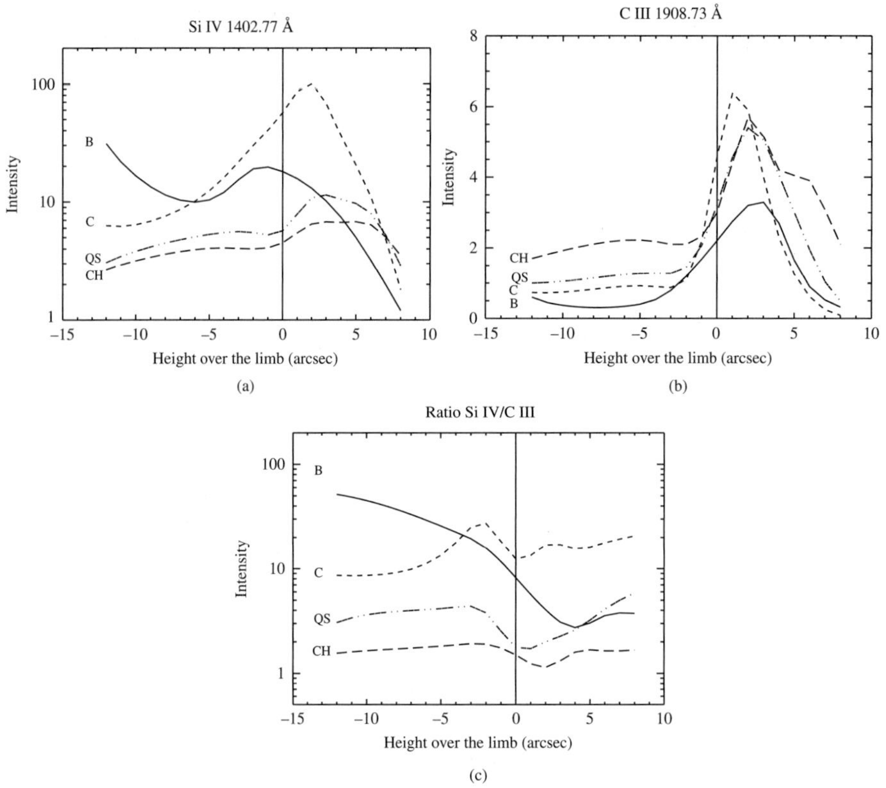

Fig. 9.10. (*a*) Si IV 1402.77 Å line flux across the solar limb for two active regions studied by Feldman and Doschek (1978a). (*b*) As for (*a*), but for the spin-forbidden C III 1908.73 Å line. (*c*) Flux ratio of Si IV 1402.77 Å to C III 1908.73 Å across the solar limb. From Feldman & Doschek (1978a).

In principle, if the line-of-sight relative fluxes and electron densities as a function of height near the solar limb are known, together with the absolute flux of the region when it passed the Sun's central meridian, the filling factor of the emitting region can be derived. Using the fact that electron densities in the quiet Sun transition region are nearly constant, Feldman *et al.*

(1979) used Abel-inversion of the observations near the limb to derive the height distribution of the plasma. By combining the height distribution and observed electron densities with the absolute flux measurement near Sun centre, they derived a quiet Sun transition region filling factor of ~ 1 %. Since the amount of plasma in quiet Sun and active regions is nearly the same while the electron densities in active regions are often an order of magnitude higher, they concluded that in parts of some active regions the filling factors are often an order of magnitude smaller, or ~ 0.1 %.

9.5 Active region transient phenomena

Various transient phenomena have been observed in active regions which may be important in energy requirements of active regions. We will deal with flares, which are the most obvious of them, in Chapter 10. Many of these transient phenomena can be observed with visible-light instruments, and have been familiar to solar physicists for many years using Hα spectroheliographs or telescopes with Hα interference filters. Among these are surges and sprays which can be seen to best advantage on the limb. Both are ejections of material, those in sprays often exceeding the solar escape velocity. Observations in soft X-rays with *Yohkoh* and in the EUV with *TRACE* and the CDS and SUMER instruments on *SOHO* have established that these transient phenomena are very energetic and the temperatures involved are very high, and contain large amounts of energy.

9.5.1 *X-ray transient brightenings*

An extensive series of studies made by Shimizu (Shimizu *et al.* (1992, 1994), (Shimizu (1995)) has shown that active regions often have transient brightenings in the form of loops, either single or multiple, or sometimes small point-like features. Several are shown in Figure 9.11, a series of partial frame images in the thin aluminium (Al.1) and aluminium/magnesium/manganese sandwich (Al/Mg/Mn) filters of the *Yohkoh* SXT. In this six-hour period of this highly dynamic active region (NOAA 6891) in October 1991, near solar maximum and shortly after the launch of *Yohkoh*, many transient brightenings were observed. Loops with lengths between a few thousand km and 50 000 km were observed to brighten over short periods of time in various locations within the active region. On average, for NOAA 6891, a brightening was seen every 2.5 minutes. More than half were multiple-loop brightenings. For active regions less developed than NOAA 6891, transient brightenings occur about once per hour. The brighter examples are often detectable as small increases in the whole-Sun emission as observed by *GOES*.

Studies made by Shimizu (1995) of the frequency of transient brightenings as a function of energy lost in the form of radiation and conduction, or total thermal energy contained by the plasma, show that smaller transient brightenings tend to be the most numerous. In fact, like quiet Sun brightenings in the form of small flares over the whole Sun (Section 8.6), there seems to be a power-law dependence,

$$\frac{dN}{dE} = E^{-\alpha}, \tag{9.2}$$

where E is either the energy lost or the thermal energy. The power-law index α varies somewhat according to how the transient brightenings are recognized in the samples taken by Shimizu (1995), but all appear to have $\alpha < 2$ when E is the thermal energy, given by $3N_e k_B T_e V$ (V estimated volume of the transient loop brightening). A filling factor of one

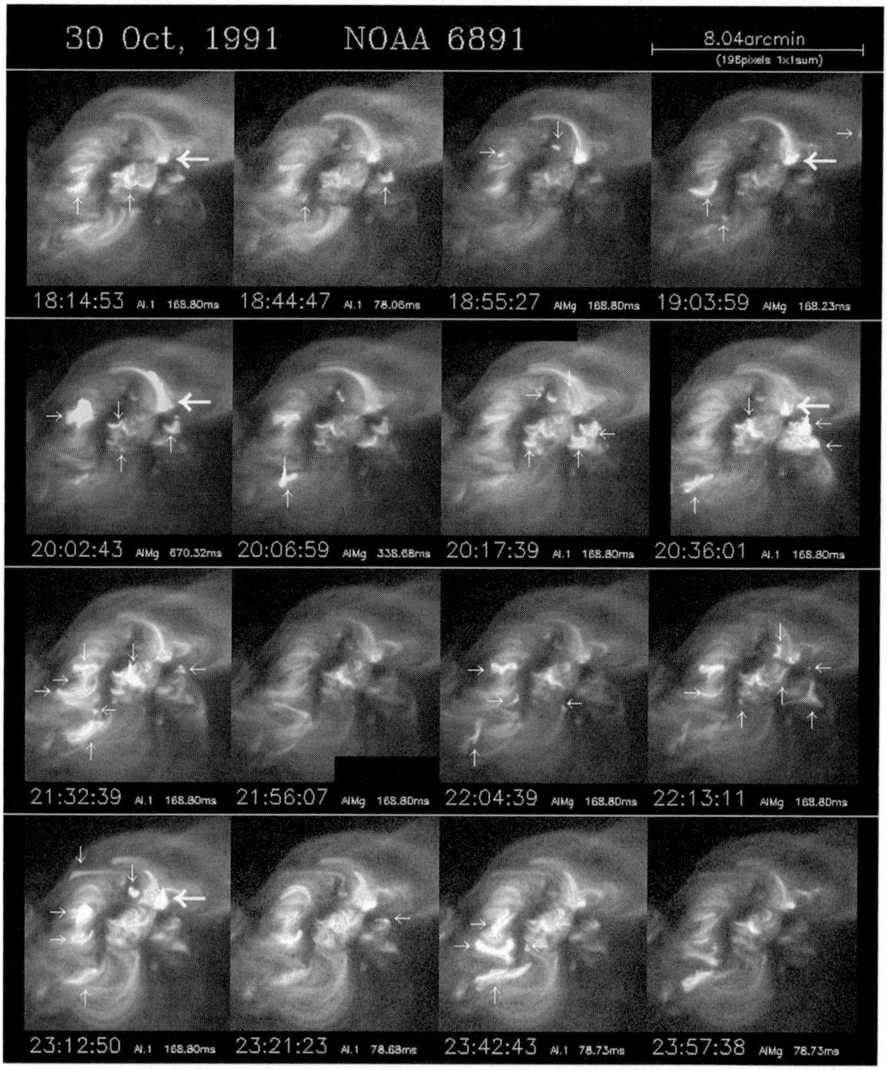

Fig. 9.11. Active region NOAA 6891 observed in the thin Al and Al/Mg/Mn filters of the SXT on *Yohkoh* during a six-hour period on 1991 October 30. Many transient brightenings in loops are observed, shown by arrows (arrow length indicates importance of brightening). Courtesy, T. Shimizu, from Shimizu *et al.* (1992).

is assumed. Thus, for a range $10^{27} < E < 10^{29}$ (E in erg), the value of α is 1.5 to 1.6, with deviations at either end of this range owing to selection effects. This would apparently mean that transient brightenings with energies extrapolated down to very small values cannot power the active regions analyzed by Shimizu (1995). Taking the transient brightenings in the energy ranges observed by the SXT, for one particular active region (NOAA 7260 in August 1992), the total of energy lost by radiation and conduction and thermal energy is 3×10^{27} erg s^{-1}, a factor 5 less than that required to heat the active region, $\sim 1.5 \times 10^{28}$ erg s^{-1}. However, some assumptions are made about the energies which should be noted. The ratios of fluxes in two different SXT filters were measured in order to give electron temperature, which can only give rather crude estimates. More importantly, electron densities are estimated from a combination of emission measures and volumes measured from SXT images, with values of 10^9–10^{10} cm^{-3} resulting. These are almost certainly too low, probably by up to a factor 10, because of small filling factors. This is likely to have a profound effect on conclusions regarding the energetics of transient brightenings and their role in active region heating. From these studies, one can conclude that the question is still quite open. Shimizu (1995) points out that the role of ultraviolet brightenings and jets is uncertain and may provide any 'missing' energy from those estimated from the SXT transient brightenings in active region heating.

9.5.2 Jets and sprays

A significant discovery made from early *Yohkoh* observations was that of X-ray jets, transient X-ray phenomena showing a collimated flow of material (Shibata *et al.* (1992, 1994)). There appear to be several classes of jets, depending on their association. Some are associated with X-ray bright points, some with active regions, and some with flares. Figure 9.12 shows the evolution of one such jet and the associated active region (NOAA 6918) over a period of a few hours. The jet is most prominent between 11:28 UT and 11:36 UT, but there are changes in the active region, particularly the bright point from which the jet emerges, up till 13:08 UT when a void in the bright point is apparent. Flows of approximately 100 km s^{-1} are observed, though this velocity is a lower limit as the direction of the jet cannot be determined. The total length of the jet is some 200 000 km. The width of

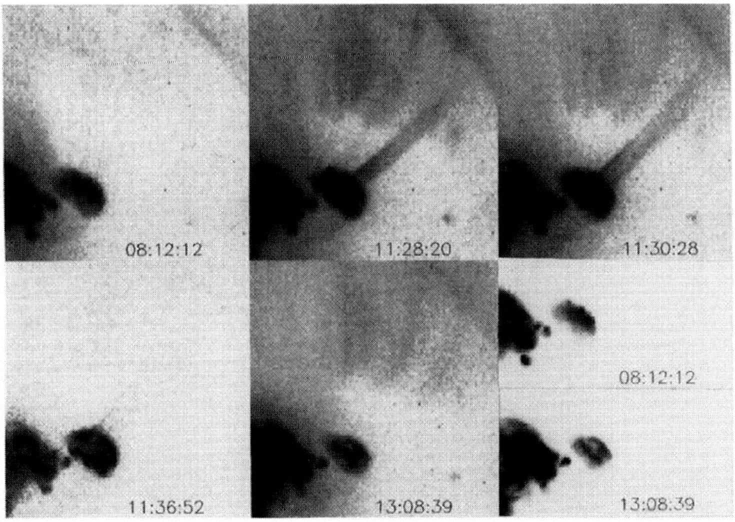

Fig. 9.12. X-ray jet observed with the *Yohkoh* SXT on 1991 November 12 from active region NOAA 6918. Times (UT) are indicated on each frame. The bottom right frame shows images taken from the first and last frames to indicate the difference between them. The images were taken from full-frame images (FFIs) in the Al.1 and Al/Mg/Mn filters. Courtesy K. Shibata, from Shibata *et al.* (1992).

this particular jet is constant as far as can be determined from the SXT images, so the motion is very nearly collimated. Examples of converging or diverging flows have also been found.

The physical conditions of jets in a sample taken by Shibata *et al.* (1992) have been found from the ratio of emission in different SXT filters. Typical temperatures of approximately 3×10^6 K and column emission measures of 10^{26} cm^{-5} are indicated. With a unit filling factor assumed, and a jet width of 20 000 km, a density of $N_e = 2 \times 10^9$ cm^{-3} results for the jet footpoint, and 4×10^8 cm^{-3} for the midpoint of the jet. Like the density estimates for active region transient brightenings, these are likely to be lower limits to the actual densities. The typical mass of a jet is 2×10^{13} g, practically as large as that of the active region. With a lower limit on the velocity of 100 km s^{-1}, the kinetic energy is $\sim 10^{27}$ erg. This is small compared with the thermal energy of the jet, $E_{\rm th} = 3 N_e k_B T_e V$, which is $\sim 3 \times 10^{28}$ erg.

Not all jets take straight trajectories. Some meander or undulate, suggesting that the magnetic field lines guiding their motion are themselves helical or twisted. This occurred for a jet on 1992 November 8, where the total length was as much as 400 000 km. This particular jet is interesting in that the site of its releasc seems to have been the site for several other jets: there seem to be locations where jets are liable to occur repeatedly.

The resemblance of X-ray jets with Hα jets is striking, and has led to searches for coincidences. One particular occurrence is mentioned by Shibata *et al.* (1992) which shows striking contrasts between the Hα and X-ray structures. The X-ray jet evolved from brightenings at its footpoint, as did the Hα surge. Though the two were nearly co-spatial, the X-ray jet extended up to only 5000 km, with velocity ~ 100 km s^{-1}, while the Hα surge extended to much greater heights, $\sim 45\,000$ km, but with smaller velocity (~ 30 km s^{-1}). If the jet density was about 10^{10} cm^{-3}, as is suggested by its width, temperature, and emission measure, its mass (10^{12} g) and kinetic energy (3×10^{25} g) were likely to be much less than those of the Hα surge. The Hα surge density was probably at least 10^{11} cm^{-3}, suggesting a mass of 10^{14} g and kinetic energy of 10^{28} erg.

Jets associated with large active regions or emerging flux regions in near-equatorial latitudes are described by Shibata *et al.* (1994) as two-sided-loop structures, but those from small bipolar regions in coronal holes leave behind anemone-like structures in which small loops connect the central bright point of one magnetic polarity to the surrounding unipolar field area with opposite polarity.

Jets observed from large, flare-productive active regions generally occur on the western edge of the active region, just ahead of the leading sunspot. It is here that there are small satellite spots associated with Hα surge activity, and also the greatest likelihood of seeing associations of X-ray jets and Hα surges.

10

Solar flares

10.1 Introduction

Flares are the most energetic eruptions in the solar atmosphere. Some 10^{32} erg may be released in the largest flares, with electromagnetic radiation extending from gamma-rays to long-wavelength radio waves. Much of the energy appears in the form of soft and hard X-ray emission. Flares occur in regions of magnetic complexity on the Sun, and are invariably associated with active regions. The development of flares is marked by a very rapid impulsive phase near the onset associated with bursts of hard X-ray emission, some ultraviolet line emission, and microwave radio emission; individual bursts of emission within this phase last only fractions of a second, with groups of bursts lasting a few minutes. Hard X-ray emission is in the form of non-thermal bremsstrahlung with flux decreasing as a power-law with photon energy ($(h\nu)^{-\alpha}$, α = constant) extending from around 10 keV (about 1 Å) to over 100 keV (about 0.1 Å). This emission is thought to arise from energetic, most likely beamed, electrons accelerated from some region in the corona and directed towards the upper chromosphere. Often, tens of seconds before the impulsive bursts begin, there is a slow rise of soft ($\lambda \gtrsim 1$ Å) X-ray emission that continues after the impulsive burst has ended, reaching a maximum usually within minutes after onset. The decline of the soft X-ray maximum occurs over time scales varying from a few minutes to many hours, sometimes more than a day. The emission is due to hot plasma contained in magnetic flux loops which are low-lying in the initial flare development but rise with time. Here, we will describe how results from spectrometers and imaging instruments observing in the ultraviolet and soft X-ray parts of the spectrum have enabled the physical conditions of flares to be deduced.

10.2 X-ray flare spectra

For the past forty years or so, spacecraft spectrometers have been used to obtain details of soft X-ray spectra. The earliest instruments were rocket-borne, but a common problem was the appropriate timing of the rocket launch – usually, by the time the spectrometers were ready to be exposed, the very hot, early phase of the flare was already over. Later, small spacecraft observatories such as the NASA *Orbiting Solar Observatories* (*OSO*), the *P78-1* spacecraft, and the Soviet *Intercosmos* satellites, then larger, dedicated missions like the NASA *Solar Maximum Mission* and *Reuven Ramaty High Energy Solar Spectroscopic Imager* (*RHESSI*), the Japanese *Yohkoh* spacecraft, and the Russian *CORONAS-F* mission provided stable platforms for spectrometers.

Rapidly scanning crystal spectrometers on spacecraft enabled all stages including the high-temperature initial development of flares to be observed. A breakthrough was achieved in the early 1970s with the resolving of the complex of lines near 1.9 Å by Neupert and Swartz (1970) and Grineva *et al.* (1973) due to He-like Fe, and the H-like and He-like Ca lines near 3.0 Å and 3.2 Å by Doschek *et al.* (1979) with such spectrometers. The advent of bent crystal spectrometers improved the time resolution still further, notable instruments being the Bent Crystal Spectrometer on *Solar Maximum Mission* (Acton *et al.* (1980)) and the Bragg Crystal Spectrometer on *Yohkoh* (Culhane *et al.* (1991)). These instruments also had the advantage of much larger sensitive areas than equivalent flat, scanning crystal spectrometers, but the disadvantage that only limited wavelength ranges could be covered. Thus, scanning crystal spectrometers are still needed for looking at large wavelength ranges, e.g. for studying the flare-excited lines in the 10–20 Å range, which includes many strong Fe lines having transitions between n=2 and n=3, 4, etc. levels. The flat crystal spectrometers SOLEX (on the *P78-1* spacecraft: McKenzie *et al.* (1980b)) and Flat Crystal Spectrometer (FCS, on the *SMM* spacecraft: Phillips *et al.* (1982)) observed several flares in this range, including those of diagnostically useful lines. More recently, the RESIK instrument on the *CORONAS-F* spacecraft (Sylwester *et al.* (2005)) has observed the 3–6 Å spectrum.

Of the broad-band X-ray instruments observing flares, the most familiar are the ion chamber detectors on *GOES*, which routinely monitor solar X-rays in the 0.5–4 Å and 1–8 Å bands. Both channels are sensitive to the many weak flares and minor activity in active regions that occur near solar maximum, so care must be taken in deducing temperatures and emission measures from flares themselves. Many authors analyzing *GOES* data subtract a pre-flare emission level from both channels. Early observations (Culhane & Phillips (1970)) with proportional counters had energy resolution that was high enough to observe the Fe line complex at 1.9 Å, and the slope of the free–free and free–bound continua in the 1–4 Å region enabled the flare electron temperature T_e and volume emission measure $\int N_e^2\, dV$ to be deduced. The *RHESSI* mission, although primarily intended to observe the hard X-ray and gamma-ray spectra of flares, observes the X-ray continuum in this range as well as the Fe line complex and the higher-energy group of lines at ~1.55 Å (8.0 keV), the Fe/Ni line feature, which is due to $1s^2 - 1snp$ $(n > 2)$ lines of Fe XXV and associated satellites and He-like Ni lines. Only one early crystal spectrometer covered this region (Neupert *et al.* (1967)), so although the ~0.3 Å (or 1 keV in energy units) resolution of *RHESSI* does not allow the resolution of individual lines, it is making useful observations of a range hitherto unexplored with higher-resolution spectrometers. *RHESSI* also observes the 1–3 Å solar continuum with no instrumental contamination, assuming proper background subtraction, and so line-to-continuum observations can be made from which Fe/H abundance ratios can be deduced.

Forming images of flares over limited wavelength regions gives us information about the location of the hottest parts of the flare emission. X-ray telescopes with wavelength-discriminating filters such as those on *Skylab* have provided useful data, but the Soft X-ray Telescope (SXT: Tsuneta *et al.* (1991)) on *Yohkoh* (operational 1991–2001), with a spatial resolution of 2 arcsec, has given us a much improved view of the flaring plasma, particularly flare loop structures on the solar limb. Of the six filters, the Be 119 μm (~3–7 Å) and Al 11.6 μm (thick Al: ~8–11 Å) filters were particularly useful for observing hot flare structures. Comparison of flare images from SXT with hard X-ray images from the Hard X-ray Telescope (HXT) (Kosugi *et al.* (1991), spatial resolution ~6 arcsec) indicated the location of hard X-ray thermal emission above soft X-ray loops (Masuda *et al.* (1994)). More recently, the relative locations of hard and soft X-ray flare emission have been investigated using the imaging spectroscopy capabilities of the *RHESSI* instrument (Linhui *et al.* (2002)), with improved spatial resolution.

Using plasma diagnostic techniques outlined in Chapter 5, we can find considerable information about the flaring plasma's properties – electron temperatures T_e and densities N_e from spectral line ratios, ion temperatures or plasma turbulent velocities from spectral line widths, and directed motions from spectral line displacements. With complementary data from X-ray imaging instruments, volumes may be estimated and compared with estimates of N_e and emission measures EM; from such comparisons, filling factors can be established. X-ray imaging instruments show that the soft X-ray emission of flares is in the form of loops, which appear to climb as the flare develops, with long-duration flares showing features 100 000 km or more above the solar surface towards the end of their slow decay. Flare intensity depends on the degree of magnetic complexity of the flare location, as deduced from photospheric field measurements, with regions of strong magnetic gradient giving rise to the most intense flares. Soft X-ray line and continuum emission shows that electron temperatures may reach 30 MK, with emission measures up to $\sim 10^{50}$ cm^{-3}. There appears to be a continuous sequence of flare total emission down to microflares and perhaps smaller, where the total energy released is $\sim 10^{26}$ erg (Lin *et al.* (1984), Koutchmy *et al.* (1997)).

There is a rather weak correlation of X-ray flare importance, measured by the *GOES* 1–8 Å flux (Chapter 1, Section 1.6), with Hα flare brightness and area. The background or non-flaring level of 1–8 Å emission is a good indicator of general solar activity. At times of solar minimum (as in 1996 and 2007), the background level frequently falls below the minimum detectable (A1) level. On the other hand, exceptionally strong flares may have maximum 1–8 Å intensities greatly exceeding X1, resulting in the saturation of the *GOES* detectors, as happened with the flare of 2003 November 4, the greatest of cycle 23 (importance deduced by extrapolation to be about X28).

Some of the strongest lines in the 1–22 Å range of flare X-ray spectra are those due to He-like ions of abundant elements that have a high fractional abundance over a wide range of temperatures. This is due to the stable nature of the ground configuration of these ions – two $1s$ electrons forming a closed shell. The resonance lines of He-like ions and their satellites are not only prominent features of X-ray flare spectra but are important for diagnostic purposes. Early observations (Neupert & Swartz (1970), Grineva *et al.* (1973)) showed that the Fe XXV lines accounted for much of the emission of the complex at $\sim$1.9 Å, with the resonance or w line at 1.850 Å (Table 10.1). Other lines on the long-wavelength side of the w line (in order of wavelength) are due to Fe XXV intercombination (x, y) and forbidden (z) lines, and to Fe XXIV satellites formed by dielectronic recombination and inner-shell excitation. Later spectrometers enabled this crowded spectral region to be further resolved, and with the advent of bent crystal spectrometers spectra with both high spectral resolution and high time resolution could be obtained, so that the changing ratios of the satellites to the resonance line could be evaluated during the developing stages of soft X-ray flares. The SOX2 scanning spectrometer on *Hinotori* (Tanaka *et al.* (1982)) had a spectral resolution of 0.15 mÅ, one of the best ever achieved with a spaceborne crystal spectrometer, and produced many excellent flare spectra. The spectra shown in Figure 10.1 are at the peak and during the decay stages of three large flares in 1979 taken with the *P78-1* SOLFLEX scanning spectrometer. The dielectronically formed Fe XXIV satellites j and k (labelled on each spectrum) are important as the j/w flux ratio depends on T_e, while the inner-shell Fe XXIV satellite q enables the

Table 10.1. *Helium-like ion line and principal satellite notation and wavelengths.*

Ion	Transition and notation[a]	Fe $Z = 26$	Ca $Z = 20$	Ar $Z = 18$	S $Z = 16$	Si $Z = 14$
He	$1s^2\,{}^1S_0 - 1s2p\,{}^1P_1$ $(w)^a$	1.8499[b]	3.1769	3.9488	5.0385	6.6477
He	$1s^2\,{}^1S_0 - 1s2p\,{}^3P_2$ (x)	1.8552	3.1889	3.9656	5.0629	6.6848
He	$1s^2\,{}^1S_0 - 1s2p\,{}^3P_1$ (y)	1.8595	3.1925	3.9691	5.0662	6.6879
He	$1s^2\,{}^1S_0 - 1s2s\,{}^3S_1$ (z)	1.8680	3.2111	3.994	5.101	6.740
Li	$1s^2\,2p\,{}^2P_{3/2} - 1s2p^2\,{}^2D_{5/2}$ (j)	1.8660	3.2097	3.992	5.1010	6.742
Li	$1s^2\,2p\,{}^2P_{1/2} - 1s2p^2\,{}^2D_{3/2}$ (k)	1.8631	3.2058	3.989	5.0964	6.738
Li	$1s^2\,2s\,{}^2S_{1/2} - 1s2s2p\,({}^1P)\,{}^2P_{3/2}$ (q)	1.8610	3.2003	3.981	5.0861	6.718
Li	$1s^2\,2s\,{}^2S_{1/2} - 1s2s2p\,({}^1P)\,{}^2P_{1/2}$ (r)	1.8643	3.2015	3.983	5.087	6.720

[a] Line letter notation is that of Gabriel (1972).
[b] Wavelengths are from line list in Appendix 2, Kelly (1987), Lemen *et al.* (1984).

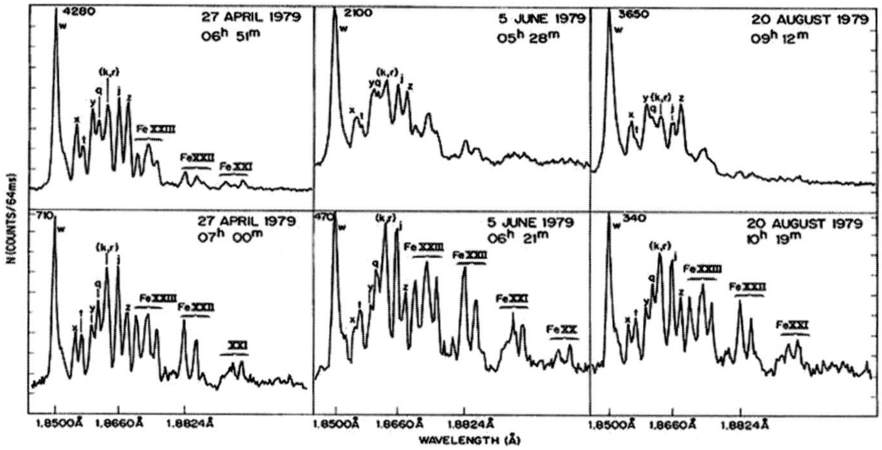

Fig. 10.1. X-ray spectra during three large flares in 1979, taken with the high-resolution *P78-1* SOLFLEX spectrometer. The spectra were taken by scanning over the 1.84–1.94 Å range, and include the Fe XXV resonance line and associated satellites. Those in the upper panels were taken near the peak of each flare, those in the lower panels were taken during the flare decay, and show how the satellites/resonance line ratio inversely depends on the temperature. The vertical scale is photon counts, with the peak photon counts given in the top left of each panel. From Doschek *et al.* (1980).

concentration of Fe^{+23} ions relative to Fe^{+24} ions to be determined through the q/w flux ratio. Very similar spectra have been obtained from tokamak plasma devices, e.g. the Princeton Large Torus (PLT) (Bitter *et al.* (1979)), for which the temperature is similar to flares but the densities a little higher ($T_e \approx 20$ MK, $N_e \approx 10^{13}$–10^{14} cm^{-3}).

Helium-like ion spectra from lower-Z ions are also prominent in solar flare spectra, especially those of abundant elements. Because of an approximately Z^4 dependence of the satellite line intensities relative to the He-like ion w line, the satellites are much less evident in the spectra of Ca XIX and S XV than in Fe XXV spectra (see Table 10.1 for wavelengths). The

Ca lines were seen with the flat crystal spectrometer on *P78-1*, the bent crystal spectrometers on *SMM* and *Yohkoh*, and the double-flat crystal spectrometer Diogeness on the *CORONAS-F* spacecraft. For Ca and S, the diagnostically useful satellite j is blended with the He-like ion z line, but (depending on the spectral resolution of the instrument) satellite k is distinguishable from line z. Helium-like Mg (Mg XI) and Si (Si XIII) lines and the accompanying satellites are intense line groups in flare spectra. Helium-like Ne (Ne IX) lines at $\sim$13.4 Å are present in X-ray flare spectra but are seriously blended with several lines of Fe ions, notably a multiplet of Fe XIX (Bhatia *et al.* (1989); Landi & Phillips (2005)). Lines of less abundant, odd-Z elements – Cl and K – are features of spectra from the RESIK instrument on *CORONAS-F*.

The Lyman lines of H-like ions are also prominent in flare spectra. Those from H-like Fe (Fe XXVI) are only present for the early stages of very powerful flares, where the separate components of the Ly-α line are present together with dielectronic satellites on the long-wavelength side due to He-like Fe. They were observed to great advantage by the high-resolution spectrometer on *Hinotori* (Tanaka (1986)). The Ly-α lines of Ca XX were observed by the SOLFLEX instrument on *P78-1*. X-ray flare images in the Ly-α lines of Mg XII were made with the RES-C spectropolarimeter on *CORONAS-F* (Sylwester *et al.* (2005)).

As well as lines due to H-like and He-like ions, there are numerous multiplets of lines due to various Fe ions extending over the 8–18 Å range. The strongest of them, and indeed in the entire X-ray spectrum below 20 Å, are the $2s^2 2p^6 - 2s^2 2p^5\, 3l$ ($l = s, d$) lines of Ne-like Fe (Fe XVII). As with the $1s^2$ electrons in He-like ions, the Ne-like ion stage is relatively stable and exists over a wide range of temperatures, from active regions to flares, by virtue of the close outer shell of electrons ($2p^6$). The resonance line, $2s^2 2p^6\,{}^1S_0 - 2s^2 2p^5\, 3d\,{}^1P_1$, at 15.013 Å, is the most intense of the group and the anomalously high oscillator strength of the transition may result, under some circumstances, in resonant scattering in dense flare or active region plasmas (Phillips *et al.* (1997)). Other ions with spectral lines well represented in common flare spectra include those of Fe XVIII and Fe XIX. During the decay stages of flares observed by the FCS on *SMM* and SOLEX on *P78-1*, the observed lines involve transitions from the $n = 2$ to $n = 3$ shells of Fe ions. Unfortunately, in the case of the *SMM* FCS, undue attention was paid in the early stages of the spacecraft mission to obtaining images rather than spectra and no spectral scans during most of the most intense flares of 1980 were obtained. After a gap in observations owing to the failure of the *SMM* attitude control unit, mission operations were resumed in 1984, but by then solar activity was greatly reduced. More spectral scans could be undertaken in greater numbers, though of generally less intense flares than in 1980. A fortunate circumstance occurred during a flare in 1985 when the FCS performed a scan over the peak of a class M4.5 flare. Many strong lines were observed in the region between 8 and 10 Å that were unfamiliar at the time, but which later analysis (Fawcett *et al.* (1987), Landi & Phillips (2005)) revealed to be due to $n = 2$ to $n = 4$ and $n = 5$ transitions in F-like to Li-like Fe ions. The Fe XXIV $1s^2\,2s\,{}^2S_{1/2} - 1s^2\,4p\,{}^2P_{3/2}$, $1s^2\,2s\,{}^2S_{1/2} - 1s^2\,4p\,{}^2P_{1/2}$ doublet (7.983 Å and 7.993 Å) are the hottest lines of the group, being emitted at temperatures of more than 10 MK.

The *RHESSI* mission has observed many thousands of X-ray flares. Commonly, the spectrum in the 4–12 keV range is characterized by a thermal continuum and line features at 6.7 keV ($\sim$1.9 Å, Fe line feature) and 8.0 keV ($\sim$1.55 Å, Fe/Ni line feature). These are discussed in Chapter 5 (Section 5.7.2). In an analysis of several flares during which the continuum and both line features were well observed, equivalent widths were obtained

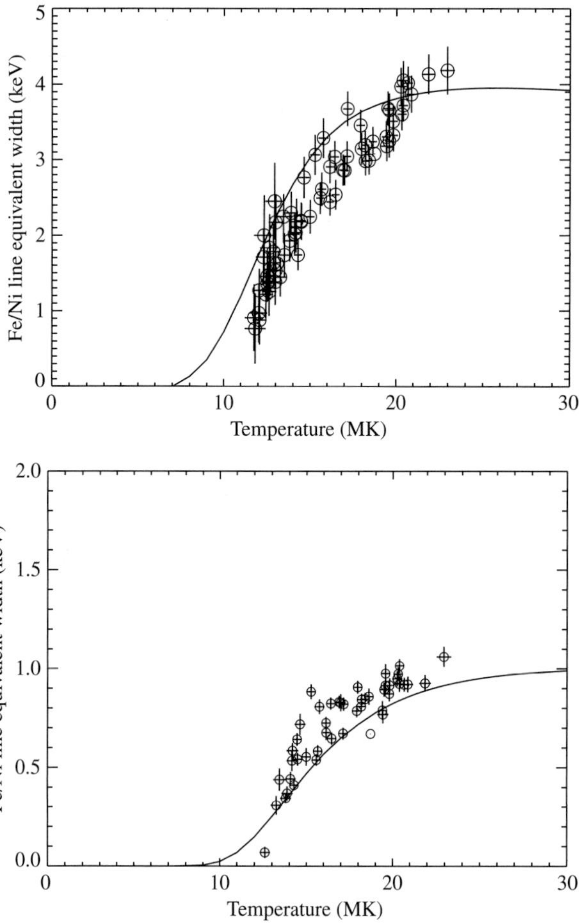

Fig. 10.2. Observed equivalent widths (in keV, points with error bars) of the Fe line feature at 6.7 keV (*top*) and Fe/Ni line feature at 8.0 keV (*bottom*) plotted against temperature T_e as determined from the slope of the continuum in the ~4–12 keV range using *RHESSI* spectra during the 2002 May 31 M2 flare. The theoretical variation of each equivalent width is shown by the solid lines, based on calculations of Phillips (2004). Courtesy C. Chifor.

for a range of observed temperatures from the continuum slope with energy. These were compared with those theoretically derived (Phillips *et al.* (2006a)). Figure 10.2 shows the observed equivalent widths and theoretical variation with T_e for spectra through the course of one of the flares analyzed. The agreement in shape appears to confirm the atomic parameters for the lines making up both line features, and the absolute magnitude confirms the assumed coronal value for the Fe/H abundance, taken from Feldman and Laming (2000). There is a slight displacement of points from the theoretical curve in T_e that might be related to errors in the ion fractions of Mazzotta *et al.* (1998).

10.3 Ultraviolet flare spectra

There are several ultraviolet lines of ions present in high-temperature flare plasmas that have been observed and used as diagnostics of flare plasmas. In the range of the

NRL S082B slit spectrometer on the *Skylab* mission, lines of Fe XVIII (974.86 Å), Fe XIX (1118.10 Å), and Fe XXI (1354.08 Å) are present, all due to forbidden transitions within the ground configuration; e.g. the Fe XXI line is due to $1s^2 2s^2 2p^2\,{}^3P_0 - 1s^2 2s^2 2p^2\,{}^3P_1$. This line was also observed by the UVSP instrument on *SMM*. Although useful for observing flares, its close proximity to a C I line sometimes makes interpretation difficult. Flare spectra with the much more sensitive *SOHO* SUMER and CDS spectrometers in the 500–1500 Å range include lines of Fe XVII–Fe XXIII as well as highly ionized stages of the less abundant elements Cr, Mn, Co, and Ni. At still shorter wavelengths, there are strong lines due to Fe XXIII at 263.74 Å due to the transition $1s^2\,2s^2\,{}^1S_0 - 1s^2\,2s\,2p\,{}^3P_1$ and Fe XXIV at 192.03 Å (transition $1s^2\,2s\,{}^2S_{1/2} - 1s^2\,2p\,{}^2P_{3/2}$) and 255.11 Å ($1s^2\,2s\,{}^2S_{1/2} - 1s^2\,2p\,{}^2P_{1/2}$). Images of flares with the S082A slitless spectrograph on *Skylab* in the 255.11 Å line and those in other lines (Widing & Cheng (1974)) showed the location of the hot flare plasma – often a very confined region, as was later seen by the *Yohkoh* SXT – relative to cooler plasma. Flare images obtained with the EIT on *SOHO* and with *TRACE* in their 195 Å filters show emission structures that are evident in the 171 Å filters, and so can be identified with temperatures characteristic of coronal ions such as Fe XII (195 Å) or Fe IX (171 Å). Emission that is strong in the 195 Å filter but absent in the 171 Å filter can be attributed to the 192.03 Å Fe XXIV line which has a much larger temperature. Thus, in Figure 1.10, the spine structure in the *TRACE* 195 Å disk image of a large flare is due to Fe XXIV line emission. Two Fe XIX lines, at 592.24 Å (due to $2s^2 2p^4\,{}^3P_2 - 2s^2 2p^4\,{}^1D_2$) and 1118.07 Å (due to $2s^2\,2p^4\,{}^3P_2 - 2s^2\,2p^4\,{}^3P_1$) are prominent in the wavelength range of the CDS and SUMER instruments on *SOHO*, and have been used for flare studies (Brosius & Phillips (2004); Innes *et al.* (2001)).

Figure 10.3 shows an image of an intense disk flare observed by the *Skylab* slitless spectrograph (S082A) in an old active region, together with a ground-based observatory magnetogram. The flare, which is discussed by Widing and Cheng (1974) and Widing and

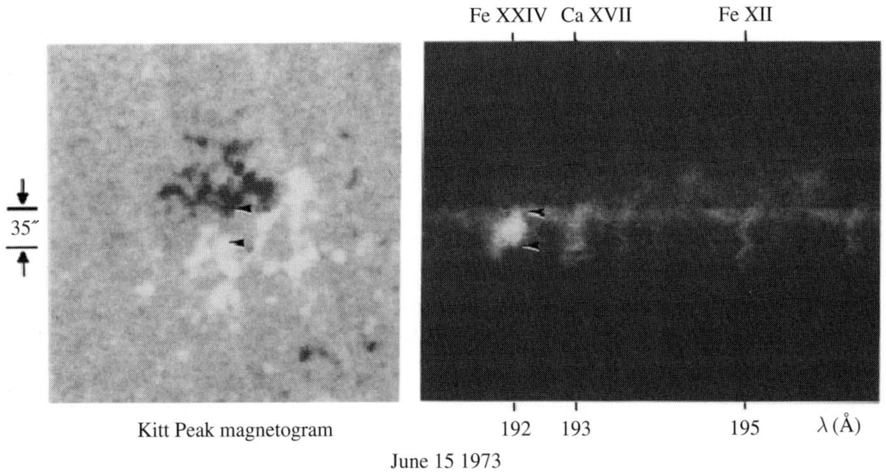

Fig. 10.3. Kitt Peak National Observatory magnetogram (*left*) and *Skylab* S082A slitless spectrograph images of the flare of 1973 June 15 (*right*) showing how the Fe XXIV 192.02 Å emission fits into the gap between the positive and negative polarity areas of the magnetogram (marked by short arrows). From Widing & Cheng (1974).

Dere (1977), had a well-developed double-ribbon structure but with a small high-temperature structure visible in Fe XXIII and Fe XXIV images. The chromospheric He II line (256.32 Å) emission occurs at the loop footpoints (marked by arrows in Figure 10.3). A bright loop-top region is visible in the Fe XXIV lines at 192.03 Å and 255.11 Å lines (formed at 14–18 MK) during the rise, the peak, and the beginning of the decay phase. This confined region neatly fits in the gap between the two areas of He II line emission, over the magnetic polarity inversion line. The footpoints are also visible in the Ca XVII 192.88 Å line formed at 6 MK.

10.4 Flare temperatures and emission measures

Of the several ways of determining the flare electron temperature T_e, by far the most reliable makes use of the temperature sensitivity of the dielectronic satellite lines near resonance lines of the He-like or H-like ions of particular elements (Chapter 5, Section 5.3.1). It has been extensively used in analyzing X-ray spectra from spectrometers on the *P78-1* (SOLFLEX), *SMM* (BCS), *Hinotori*, *Yohkoh* (BCS), and *CORONAS-F* (RESIK) spacecraft observing Fe, Ca, S, and other ion line emission.

The flux ratio of lines emitted by adjacent ions of the same element can also give a useful indication of T_e. The temperature sensitivity of the 1–3 Å continuum slope has been used for analysis of broad-band X-ray spectra such as those from the *RHESSI* spectrometer. Cruder estimates may also be obtained by taking the ratio of two spectral bands, such as the two channels of *GOES* or the emission in two filters in imaging instruments (e.g. *Yohkoh* SXT or *TRACE*). A review of some of these methods is given by Phillips (1991). Here we give details of and results from the method using dielectronic satellites.

Temperature estimates from dielectronic satellite lines in X-ray spectra, based on the theory developed by Gabriel (1972), are the most direct, and involve no assumptions about the emitting plasma. The atomic data necessary in the derivation of T_e are of great importance. Gabriel (1972) used approximate Hartree–Fock calculations or published work to derive theoretical satellite-to-line w flux ratios, but the later calculations of Bely-Dubau *et al.* (1979, 1982a,b) used the University College London package of programs including the SUPERSTRUCTURE and AUTOLSJ codes to get satellite line wavelengths and intensity factors for Fe and Ca spectra. For the equivalent lines of other, generally lower-Z, ions, also enhanced during flares, satellite line intensity factors have been estimated from the UCL programs (see Dubau *et al.* (1994) for S XIV satellites), the Hartree–Fock procedure given by Cowan (1981), or the calculations of the Lebedev group in Moscow (Vainshtein & Safronova (1978, 1985)). A comparison by Harra-Murnion *et al.* (1996) of satellite intensity factors given by different methods for the S XIV lines shows differences of only 1–2% for the strong satellites j and k but factors of up to 2 for weaker satellites; such differences make negligible differences to the satellite spectra or to temperatures derived from the k/w ratio.

Bely-Dubau *et al.* (1982a,b) used the UCL distorted wave (DW) code in intermediate coupling to calculate the excitation rates of the He-like ion lines w, x, y, and z, while other authors have used this or other procedures, e.g. Pradhan *et al.* (1981) (DW), Sampson & Clark (1980), Zhang & Sampson (1987) (Coulomb–Born approximation). These latter authors have included the effects of autoionizing resonances in the collision strengths that are important for lighter (smaller-Z) ions such as He-like Mg, particularly in the $1s^2\,^1S_0 - 1s2s\,^3S_1$ excitation rate. More recent work has used collision strengths calculated with the close-coupling R-matrix code of the Queen's University Belfast group (Burke & Robb (1975), Berrington *et al.* (1995)). The autoionizing resonances are more exactly calculated with this code, so R-matrix results ought to give an improvement over those obtained by other methods. The atomic data for He-like ion excitation rates of interest to solar flares were reviewed by Dubau (1994).

For Fe spectra, temperatures T_e are estimated from the flux ratio j/w or k/w. For corresponding spectra of lighter elements Ca (observed by *SMM* and *Yohkoh*) and S (observed by *Yohkoh*), line j is blended with line z, so temperatures must be estimated from k/w or (in the case of *Yohkoh* spectra) a line feature that is a blend of two $n = 3$ satellites called $d13$ and $d15$ (notation of Bely-Dubau *et al.* (1979)) near line w. The Ca and S satellites have much reduced intensity relative to line w because of the smaller atomic number. For the *SMM* Ca spectra, where all the principal $n = 2$ satellites were observed, estimates of the fraction of Li-like Ca to He-like Ca are complicated by the blend of the Ar XVII ($1s^2$–$1s4p$) line with the Ca XVIII q line (Doschek & Feldman (1981)).

Apart from times when a superhot component is present, the Fe XXV w line and Fe XXIV satellites near 1.85 Å measure the hottest part of the plasma, so these are of greatest interest for flares. Values of T_e as high as 25 MK have been found from measurements with the *P78-1*, *SMM*, *Hinotori*, and *Yohkoh* spectrometers, with temperatures roughly correlating with *GOES* flare class. A typical behaviour for flares of M or X *GOES* class is the attainment of a temperature of about 20 MK shortly before the peak in the Fe XXV line w flux, followed by a gradual cooling (e.g. Sterling *et al.* (1994, 1997)). Lower-temperature ions such as Ca XIX and S XV remain visible for longer periods for most flares. At late stages of long-duration flares, when the emitting region appears as a confined blob high up in the solar atmosphere, at the apex of a large loop, Ca XIX and S XV spectra obtained by the *Yohkoh* BCS indicated a practically isothermal plasma, with both S and Ca temperatures equal to ~6 MK (Phillips *et al.* (2005a)). Figure 10.4 illustrates this for a *GOES* X9 class flare on 1992 November 2/3, the decay of which lasted 19 hours. At the onset of large flares, when Fe XXV, Ca XIX, and S XV spectra appear simultaneously, the Fe XXV temperature is generally several MK higher than those from Ca XIX and S XV spectra. It appears that in the flare rise phase, the flaring plasma has a huge range of temperatures, with hot ion spectra (such as Fe XXV) appearing simultaneously with cooler ion spectra (Ca XIX, S XV, and much cooler ions still); but in the late stages of long-duration flares the main X-ray-emitting plasma is much more nearly isothermal. This is broadly confirmed by the appearance of flares in images from *TRACE*: at early stages, the flare emission typically consists of many loops, both low temperature (Fe XII) and only a few with high temperature (Fe XXIV). At later stages, the emitting structure is more simplified in that the many loops are low temperature only.

It has frequently been found that, assuming ionization equilibrium holds in flare plasmas at the temperature T_e derived from j/w or k/w ratios, Fe XXIV lines q and r are overintense relative to their theoretical values, based on calculations of Bely-Dubau *et al.* (1982a). This has also been found from X-ray spectra from the Princeton Large Torus (Bitter *et al.* (1979)), so this is not a problem peculiar to solar conditions. Such results might suggest significant departures from ionization equilibrium, but for the decay stages of flares this seems unlikely unless the electron densities are very low, $N_e \sim 3 \times 10^{10}\,\mathrm{cm}^{-3}$ (Phillips *et al.* (1974); Phillips (2004)). More likely the excitation cross-sections for these lines or the ionization fractions, based on ionization equilibrium calculations, are erroneous. Ca XVIII satellites q and r have also been observed to be over-intense as with Fe spectra, though the blend of the Ar XVII line with line q makes this less certain. Assuming that these anomalies are due to incorrect ionization fractions, Antonucci *et al.* (1984) estimated the ratio of the Li-like to He-like

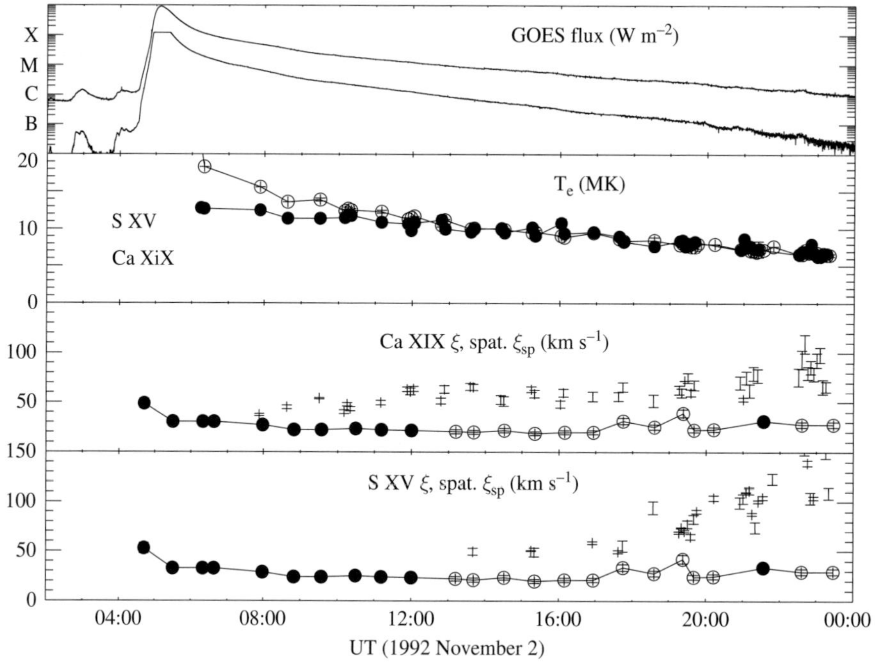

Fig. 10.4. Data derived from the *Yohkoh* BCS for a *GOES* class X9 flare on 1992 November 2/3. The *GOES* fluxes are given in the top panel, estimates of T_e from Ca XIX (filled circles) and S XV (open circles) (*second panel*). The bottom two panels show measured values of the most probable turbulent velocity ξ from Ca and S spectra (points with small error bars), with values of ξ that would be obtained from the spatial extent of the flare plasma from *Yohkoh* SXT images (filled and open circles). Based on work by Phillips *et al.* (2005a).

stage of Ca using *SMM* BCS spectra, finding that the observed ratio was a factor 2 higher at relatively high temperatures than most ionization equilibrium calculations available at that time.

At very early stages in the flare, around the hard X-ray impulsive stage, emission from H-like Fe (Fe XXVI) has been observed with spectrometers that includes the Fe XXVI Ly-α doublet (1.778, 1.784 Å) and Fe XXV satellites which fall on the long-wavelength side of the Fe XXVI lines. A particularly intense satellite feature known as J occurs at 1.792 Å, made up of three relatively strong satellites with transitions $1s2l - 2l2p, l = s, p$ (Dubau *et al.* (1981)). As with the Fe XXIV satellites near the Fe XXV line w, the intensity ratio of J to the Fe XXVI lines varies inversely with T_e as given by the calculations of Dubau (Dubau *et al.* (1981), Pike *et al.* (1996)). Using the calculations of Dubau *et al.* (1981), Tanaka (1986) using *Hinotori* data found that the temperatures from the $J/(\mathrm{Ly}\alpha_1 + \mathrm{Ly}\alpha_2)$ were frequently higher than those from the j/w ratio from the Fe XXIV satellites near the Fe XXV line w by several MK. Tanaka (1986) referred to flares with Fe XXVI temperatures higher by more than ~10 MK as having a superhot component. Indeed, support for a superhot component in some very powerful flares had been supplied by observations of the X-ray continuum in the 17–40 keV energy range (Lin *et al.* (1981)). The more sensitive (but lower-resolution) BCS detector on *Yohkoh* found that the Fe XXVI temperatures were generally higher than those from Fe XXV spectra: Figure 10.5 shows the *Hinotori* and *Yohkoh* Fe XXV and Fe XXVI temperatures plotted against

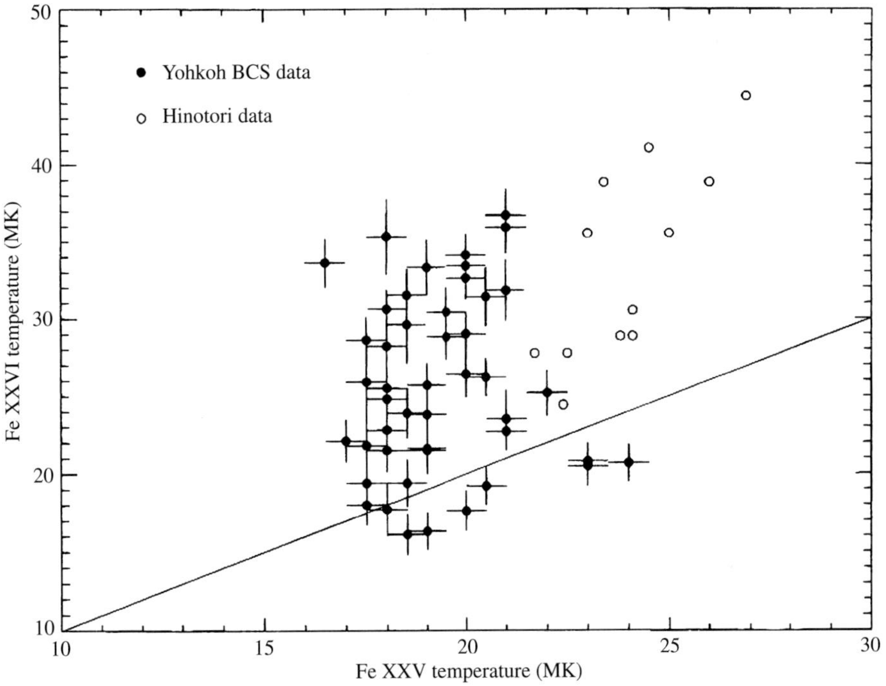

Fig. 10.5. Temperatures deduced from Fe XXVI spectra plotted against those from Fe XXV spectra to show the existence of a superhot component in some intense flares. From Pike *et al.* (1996).

each other for flares of medium to high intensity. The observation of huge numbers of flares by *RHESSI* has since confirmed that the superhot component is more developed for larger flares, less for weaker flares. It appears that the *Hinotori* observations were of very large flares where the superhot component was very pronounced.

X-ray flare spectra due to elements lighter than S have the disadvantage that the satellites are much weaker. Also, as with Ca and S, line j of Li-like Ar, Si, Al, and Mg is blended with line z of the He-like stage, so T_e must be estimated with the $d13$, $d15$ blend or satellite k. For such spectra, measurement of T_e using the so-called G flux ratio, $(x+y+z)/w$ (Gabriel and Jordan (1969)), is possible (e.g. Keenan & McCann (1987) for Al XII lines), though account must still be taken of the blend of the j line with line z and unresolved satellites with line w. In any event, comparison of observed spectra of He-like ions with low atomic number offers a means of checking the excitation rate calculations for the principal lines $w-z$. In the case of Mg XI spectra, for example, good agreement was found between the R-matrix calculations of Keenan *et al.* (1992) and line flux ratios during flares observed by the *SMM* FCS. This also applies to calculated Ar XVII line spectra, collision strengths of the principal lines being calculated by Keenan and McCann (1987), and solar (FCS) spectra (Phillips *et al.* (1993)), but more particularly for spectra from the Alcator C tokamak at the MIT Plasma Fusion Center, where the predicted density dependence in the $R=z/(x+y)$ ratio is confirmed using electron densities measured independently by laser interferometer measurements (Phillips *et al.* (1994b)). For Fe, Ca, and S spectra, the differences in the $w-z$ line fluxes between

R-matrix and calculations of Zhang and Sampson (1987) and those from the UCL codes appear to be $\lesssim$ 10% for the w line but up to 25% for the z line (Phillips *et al.* (2005c)). When compared with observed flare spectra (e.g. from the *SMM* BCS), there are significant discrepancies between the observed and calculated y and z line fluxes, the observed being too high, for all calculations; the discrepancy is slightly reduced with the R-matrix calculations. This has been noted several times in the past with spectra from tokamaks such as those from the Princeton Large Torus (Bitter *et al.* (1979)), with no satisfactory explanation available. Spectra from Electron Beam Ion Trap (EBIT) devices, with the electron beam energy adjusted to values just above the excitation energies of the $w-z$ lines, would be ideal for investigating this discrepancy, but EBIT Fe XXV spectra available to Phillips *et al.* (2005c) do not quite resolve the question.

Yohkoh soft X-ray images of flares when observed above the limb appear as well-defined loop-like structures that are much more intense at the top than at the footpoints, as can be seen in *Yohkoh* SXT images of the long-duration flare of 1992 November 2 (Figure 10.6).

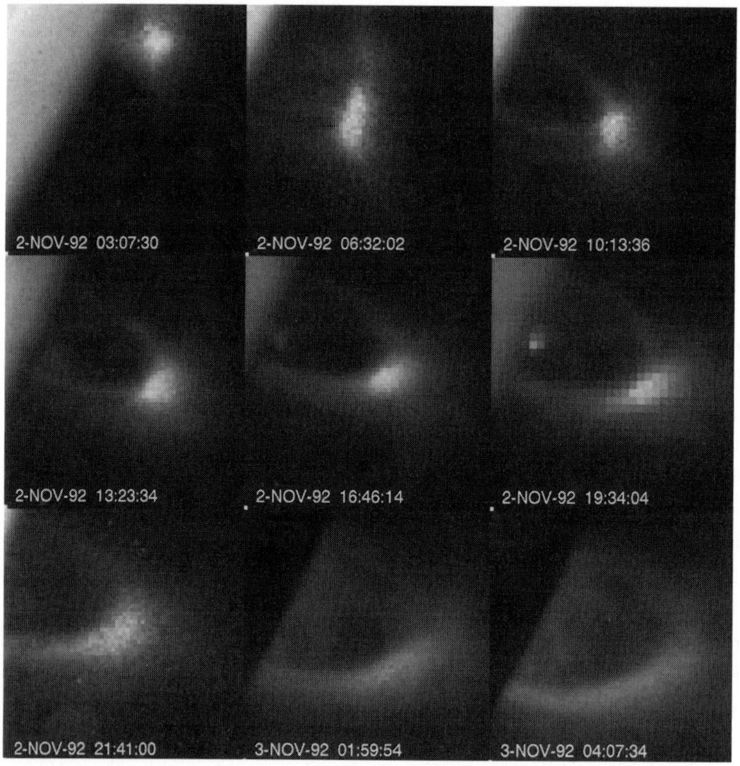

Fig. 10.6. Images of the 1992 November 2/3 flare obtained over a period from 03:07:50 UT (November 2) to 04:06:30 UT (November 3) using the *Yohkoh* SXT instrument. The limb is evident in particularly the last two images. The images were taken in a variety of SXT filters (e.g. the Be filter for the first three images), but the appearance is approximately the same regardless of the filter. The fluxes are scaled to the brightest pixel in each image, so the fluxes of one image should not be compared with those in any other. The size of each image is 2.61×2.61 arcmin^2. From Feldman *et al.* (1995).

Obtaining spatial distributions of T_e in flaring plasmas has only been possible using broad-band techniques. In principle a finely collimated crystal spectrometer observing Fe, Ca, or other lines could obtain more precise temperature maps of flares, but in practice this is precluded because of the relatively low sensitivity of such instruments. Temperature maps from the *Yohkoh* SXT show an unexpectedly non-uniform temperature distribution in, for example, the soft X-ray loop-top source, with a high-temperature ridge nearest the location of the hard X-ray loop-top source which *Yohkoh* HXT images show as overlying the soft X-ray loop. This suggested to Tsuneta *et al.* (1997) that the soft X-ray plasma is somehow heated by the overlying hard X-ray source. However, the ratio of emission in different SXT bands at each location to derive temperature maps relies on precise co-alignment of the images, and work by Siarkowski *et al.* (1996) suggests that the often-observed high-temperature ridge may be an artifact produced by the movement of the SXT filter wheel which induces slight spacecraft mis-pointing. What is evident from *Yohkoh* SXT temperature maps is that there is a high degree of non-uniformity, with the highest temperatures present in the bright soft X-ray loop-top source, but with apparently lower temperatures in the loop legs (e.g. Feldman *et al.* (1996)).

10.4.1 Flare differential emission measure

Flare temperatures estimated from the ratio of emission in the *GOES* channels are generally much lower than those from line ratios of ions such as Fe XXV and Ca XIX, so lower-temperature and higher-temperature material must co-exist within a flare (as mentioned, this does not appear to apply to the late stages of long-duration flares). Much effort, observational and theoretical, has been made to find the temperature distribution of material, or rather the DEM (Chapter 5, Section 5.6) of flare plasmas over the years. In principle, the DEM for a flare plasma may give important information about the cooling mechanisms occurring. Unfortunately, the inherent uncertainties in deriving the DEM from the integral equation (5.46) with kernel functions $G(T)$ covering wide temperature ranges are so large that it is arguably doubtful whether any significant physical information can be obtained. However, some authors have found fairly consistent forms for the DEM using spectral line and broad-band fluxes, the results of McTiernan *et al.* (1999) being a case in point, where they use a combination of *Yohkoh* BCS spectra and SXT filter fluxes at different stages in a flare's development. They obtained DEMs that are bimodal in T, with peaks at temperatures of $\sim$5 MK and $\sim$20 MK, for the flare rise and maximum stages, but which have a monotonic decline with T for late stages of the flare. This is confirmed by other workers (Kępa *et al.* (2004)). As indicated earlier, *TRACE* images of disk flares often show loops that emit in Fe XII radiation as well as Fe XXIV line emission from the loop tops. This would indicate the presence of both a cool component (a few MK) and a hotter component (10–20 MK) at the maximum stages of flare development, so agreeing qualitatively with the DEM curves.

Single-loop flare models (Emslie (1985)) do not, however, give such functional forms. The model loops in the work of Emslie (1985) and others cool by radiation and by conduction via the loop footpoints which are in the cool chromospheric regions; it is the conduction cooling, giving rise to large temperature gradients ∇T near the footpoints, that leads to emission measures nearly inversely proportional to ∇T (since the electron density is almost constant over most of the loop's length). Thus there are very large values of DEM at the loop top, in the form of an upward cusp. This is hardly consistent with the observations. The explanation for this apparent discrepancy must be the observational fact that at least major flares consist of a multiplicity of loops, most of them at rather low temperatures (~1–5 MK), with a few or even only one at very high (~10–20 MK) temperatures, as is seen in *TRACE* images. The hot loop or loops would give rise to an upward cusp in the DEM on a generally declining functional form that the cool loops give rise to. This DEM is then the function appropriate to the flare. In practice, this function is unfolded from observations consisting of line, continuum, or broad-band fluxes, all of which have wide $G(T)$ (or equivalent) functions that are in a sense the temperature 'resolution' of the observations. The cusp form is then smoothed out to give the forms that McTiernan *et al.* (1999) and others derive, namely a hump feature with maximum at 10–20 MK on a declining background.

After flare maximum, the hump at 10–20 MK appears to be insignificant and the DEM is a monotonically declining function, with power-law dependence, according to McTiernan *et al.* (1999). If the DEM really were power-law in temperature, much simplification could be allowed in various analyses such as those deriving element abundances from spectra. With RESIK spectral line observations during the decay stages of four long-duration flares, Phillips *et al.* (2003) derived relative abundances of K, Ar, and S using functional forms like DEM $\propto T^{\alpha}$ (power-law) and DEM $\propto \exp(-T/T_0)$ (exponential), with α and T_0 constants for each time in the flare decay. A suitable way of determining either α or T_0 is to use the flux ratio of two lines emitted by different ions (Si XIII and Si XIV lines were chosen in the analysis of Phillips *et al.* (2003)). If no high-resolution spectra are available, some progress in determining α or T_0 may be made using the fluxes in the *GOES* 0.5–4 Å and 1–8 Å bands. More realistically, the true flare DEM function does not, as is implied by these functional forms, extend to indefinitely high temperatures; rather there must be some cut-off temperature, but finding this extra parameter would then need more than simply the ratio of two lines or the fluxes in the two *GOES* channels.

10.5 Flare densities

Widely differing values of electron densities have been measured in solar flare plasmas which has given rise to some confusion in the past. The issue of flare densities is important in several different contexts, e.g. in evaluating the plasma β (ratio of gas pressure to magnetic pressure), or in deciding the relative importance of radiative and conductive cooling of a flare plasma. In the X-ray and ultraviolet ranges, the only direct means of obtaining densities is from the ratio of line fluxes. We will outline some of those commonly used for solar flares. There are other indirect ways of deriving densities which we also mention.

Flux ratios of the He-like ion forbidden (z) line to the intercombination (y) line are density sensitive (Gabriel and Jordan (1969); see Chapter 5, Section 5.2). Under solar flare conditions the $R = y/z$ ratio in O VII, Ne IX and Mg XI are expected be reliable electron density indicators. Ratios in He-like ions from heavier elements cease to be density sensitive for $N_e \leq 10^{13}$ cm^{-3}, a value higher than most estimates in flares. The O VII lines become sensitive at densities of 3×10^{10} cm^{-3} while Mg XI becomes sensitive at densities of 1×10^{12} cm^{-3}. The *P78-1* SOLEX spectrometer (McKenzie *et al.* (1980a); Doschek *et al.* (1981)) observed the O VII line ratios during a large flare over a 50 minute interval. The electron densities were observed to increase during the flare rise phase, reaching $\sim1\times10^{12}$ cm^{-3} at the time of peak flare intensity. During the decay phase the electron densities decreased to their pre-flare values. Figure 10.7 shows the changes in the O VII lines at various stages during the flare.

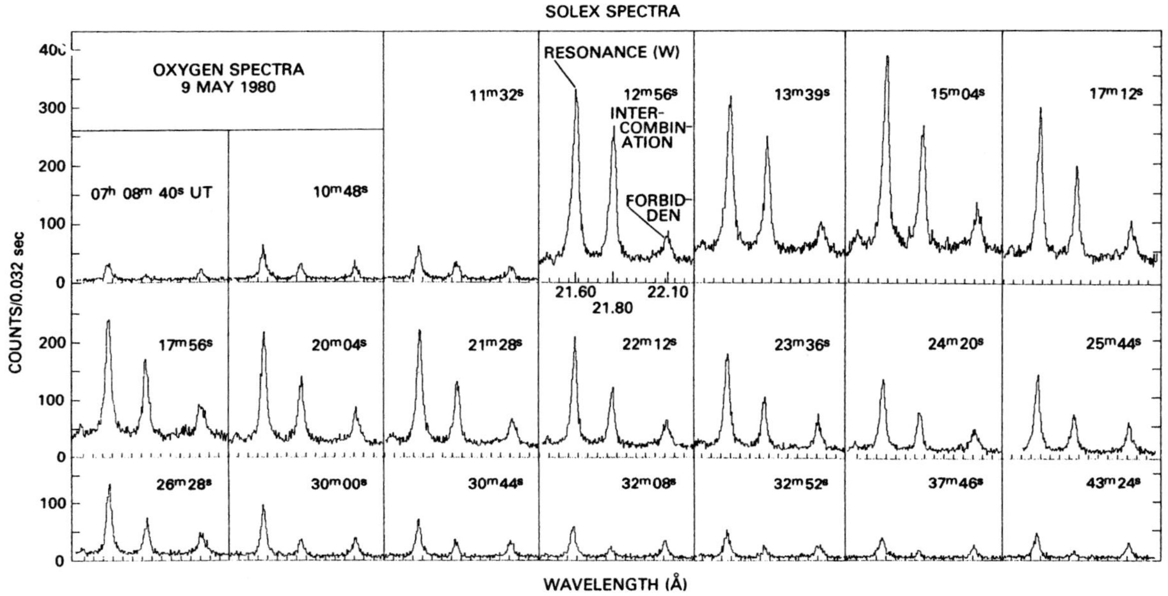

Fig. 10.7. Changing relative fluxes of the O VII X-ray lines during a large flare on 1980 May 9 observed by the *P78-1* SOLEX instrument. The relative flux of line y (21.804 Å) increases and that of line z (22.101 Å) decreases during the rise phase, the y/z ratio reaching a maximum at the flare peak (approximately 03:04 UT). The density N_e, derived from y/z, is estimated to exceed 10^{12} cm^{-3} at peak. On the decay, the relative flux of y decreases and z increases, indicating decreasing N_e. The resonance line w is at 21.602 Å. From Doschek *et al.* (1981).

A similar result was found from *SMM* FCS observations of the Ne IX lines, though the blending from Fe XIX lines (some of which are themselves density sensitive) makes this less certain (Bhatia *et al.*, 1989). A peak value of $N_e \approx 10^{12}$ cm^{-3} was obtained, again at the flare peak. *SMM* FCS data for the Mg XI lines were also used to measure N_e during a compact flare: at flare peak, $N_e \approx 5 \times 10^{12}$ cm^{-3} was obtained (Linford & Wolfson, 1988). For relatively low temperatures (i.e. during the initial rise or final stages), the Mg VIII satellite j blends with the z line, but this is unlikely to affect the result of Linford and Wolfson (1988), since the j line should be weak compared with z at flare peak.

Unfortunately, the O VII, Ne IX, and Mg XI lines provide no information regarding the high temperature ($T_e \gtrsim 10$ MK) flare plasmas (the contribution functions of the O VII, Ne IX, and Mg XI X-ray lines maximize at 2 MK, 3.5 MK, and 6 MK). Electron densities in the $T_e \approx 10$–20 MK range can in principle be obtained from various Fe (and also Ca and Ni) ions emitting lines with transitions in the $2s^2 2p^k - 2s2p^{k+1}$ and $2s^2 2p^k - 2s^2 2p^{k-1} nd$ ($n = 3,4$) arrays. Lines of the $2s^2 2p^k - 2s2p^{k+1}$ array appear near 100 Å for Fe, while lines of the $2s^2 2p^k - 2s^2 2p^{k-1} nd$ array appear in the 8 – 16 Å range. Unfortunately, there are to date no high-quality solar flare spectra available in the 100 Å range. However, flare spectra in the 8 – 16 Å range have been obtained by the *SMM* FCS and the *P78-1* SOLEX instruments. Spectral scans over a region near 12 Å during the decay of an M1.5 flare on 1980 August 25 include density-sensitive Fe XXI lines (Phillips *et al.* (1982); Landi & Phillips (2005)). Two of them, at 12.282 Å ($2p^2\,^3P_0 - 2p3d\,^3D_1$) and 12.395 Å ($2p^2\,^3P_1 - 2p3d\,^3D_1$), should be evident at low densities and indeed are present with correct relative intensities, but a third, with transition $2p^2\,^3P_2 - 2p3d\,^3D_3$ at 12.32 Å, is absent from this spectrum. The 12.32 Å line is expected to increase strongly at $N_e \gtrsim 10^{12}$ cm^{-3} (Mason *et al.* (1979)), so its absence indicates an upper limit to N_e of this amount. A more interesting result was found from spectral scans of lines in the 7–10 Å range on the rise phase and near the peak of an M4.5 flare on 1985 July 2 (Phillips *et al.* (1996)). The FCS was pointed at the brightest part of this flare. Two Fe XXI lines, due to $2p^2\,^3P_1 - 2p4d\,^3D_1$ (9.492 Å) and $2p^2\,^3P_1 - 2p4d\,^1D_2$ (9.497 Å), were observed only 6 minutes after the flare peak; the 9.497 Å line increases strongly with N_e, and the observed ratio implies $N_e = 10^{13}$ cm^{-3}. A similar ratio is obtained from two Fe XXII lines. This extraordinarily high density exceeds even that estimated from the Mg XI lines by Linford and Wolfson (1988). It implies very short cooling times by radiation: a time for the flare plasma to cool by 1 MK at this time (when $T_e \approx 17$ MK) is $\sim$1.4 s. Although such a high value of N_e seems improbable, there seems to be little doubt about either the observation or the accuracy of the calculations. It would appear that the FCS scanned these lines at an opportune time and location within the flare. Possibly a few minutes later a much smaller density would have been measured. The short radiation cooling time clearly shows that there was a continuing energy input into the flare plasma.

Because of the density sensitivity of the populations in the ground configuration levels of the Fe^{+21} and Fe^{+22} ions, the dielectronic lines are density sensitive (Chapter 5, Section 5.2.1). Phillips *et al.* (1983) give extensive calculations of the intensities of dielectronic lines (with contributions made by inner-shell excitation) of Fe XIX to Fe XXII between 1.877 Å and 1.918 Å, i.e. over limited wavelength ranges accessible to instruments like *SMM* BCS and *P78-1* SOLFLEX. Lines of Fe XX at 1.901–1.909 Å in particular are sensitive to N_e when $N_e \gtrsim 10^{12}$ cm^{-3}. Using the Cowan Hartree–Fock code, the density sensitivity and synthetic spectra were calculated, then compared with several *SMM* BCS spectra during flares in 1980. All the spectra were found to be in the low-density limit, $N_e \lesssim 10^{12}$ cm^{-3}. The density

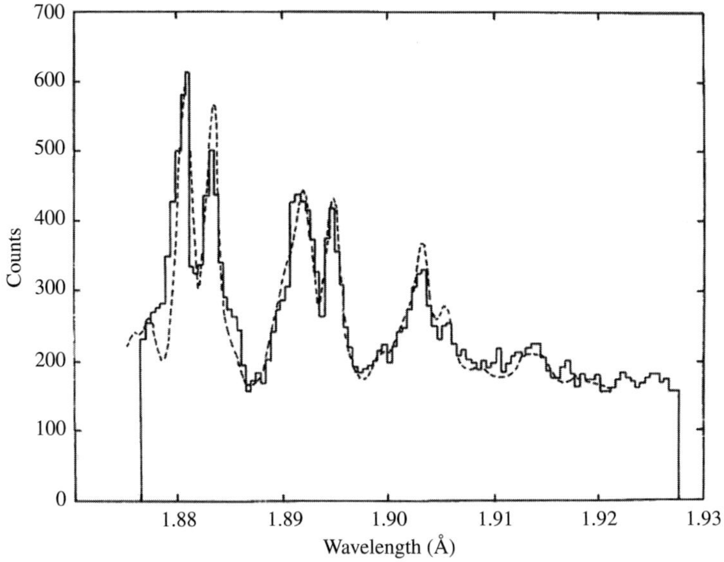

Fig. 10.8. Part of the X-ray spectrum from the Princeton Large Torus (Hill *et al.* (1979)) in the region of dielectronic satellites of Fe XXII (1.88 Å), Fe XXI (1.89 Å), and Fe XX (1.90 Å) (histogram) compared with a calculated spectrum of Phillips *et al.* (1983), to illustrate the density dependence of the Fe XX lines. The observed and calculated spectra have $N_e = 2\times10^{14}$ cm^{-3}. In the low-density limit ($N_e < 10^{12}$ cm^{-3}), there are two Fe XX line features at 1.903 and 1.906 Å.

sensitivity was confirmed for a spectrum in this same region obtained from the Princeton Large Torus (PLT: Hill *et al.* (1979)). This is illustrated in Figure 10.8. At the PLT density, $\sim10^{14}$cm^{-3}, the Fe XX line group has a maximum at 1.903 Å whereas solar flare spectra in the low-density limit show two line features at 1.903 and 1.906 Å.

Subsequent to the work of Phillips *et al.* (1983), BCS spectra were examined for the 1985 July 2 flare when high values of N_e were indicated from Fe XXI and Fe XXII observed by the FCS. However, the Fe XX dielectronic lines were found to be in the low-density limit, i.e. apparently inconsistent with the result from the FCS spectra at the same time. However, this is almost certainly because of the much larger field of view of the BCS (6 arcmin FWHM) than that of the FCS (14 arcsec FWHM). The emission observed by the BCS is averaged over a correspondingly larger area, and most likely includes parts of the flare that have much lower densities than those observed at the flare's brightest point by the FCS.

Estimates of N_e can also be made by the indirect method of combining volume emission measure and volume from flare images, V, as described in Chapter 5 (Section 5.2.3). This was extensively used for *Skylab* observations with the S054 X-ray telescope (Pallavicini *et al.* (1977)) and of images in high-temperature lines recorded by the S082A spectroheliograph (Widing & Dere (1977)). An isothermal plasma was generally assumed. Estimates were given by Pallavicini *et al.* (1977) for a survey of limb flares using volumes from the *Skylab* X-ray telescope and emission measures from the *SOLRAD* X-ray spacecraft; they are between 7×10^9 and 2×10^{11} cm^{-3}. Subsequent spacecraft imaging instruments with improved spatial resolution, culminating in the *TRACE* mission, show that the estimated volumes were probably overestimated because of the telescope's substantial point spread function, resulting in underestimated filling factors. Most probably, electron densities for these flares were well above 10^{11} cm^{-3}. For a compact flare observed with the *Skylab* S082 spectroheliograph and imaged in several extreme ultraviolet lines, N_e was estimated to be 3×10^{11} cm^{-3} for line emission having a characteristic temperature of 7 MK, and somewhat lower at higher temperatures. Similar analyses were done by Doschek *et al.* (1995) for loop-top sources using *Yohkoh* SXT and BCS data, and by Aschwanden and Benz (1997) for peak emission of flares observed with the SXT, with $N_e = 2\times10^{10}$–2.5×10^{11} cm^{-3}.

Densities can be estimated from the cooling times of very impulsive, compact flares for which radiation is the principal cooling mechanism. We deal with this in the next section. The near-ionization equilibrium conditions for many flares impose a lower limit on N_e, as is discussed in Section 10.10.

10.6 Flare morphology

X-ray and extreme ultraviolet telescopes have enabled flare images to be obtained with spatial resolution as high as ~1 arcsec (725 km). Early rocket-borne telescopes and those on board the *Skylab* manned observatory (operating 1973–4) (Underwood *et al.* (1977), Vaiana *et al.* (1977), Tousey *et al.* (1977)) showed that the basic flare geometry consisted of loop-like structures, with the hottest part (temperature ~20 MK) visible as an unresolved core (Widing & Cheng (1974), Kahler *et al.* (1975), Petrasso *et al.* (1975), Widing & Dere (1977)). The *Skylab* mission flew near the time of the solar minimum preceding Cycle 21, and so only a handful of flares was observed. Also, although the spatial resolution of ~2 arcsec was better than any previous instrument, the detailed time development of flares could not be recorded owing to the use of photographic film to record images. Some imaging instruments on *Solar Maximum Mission* and *Hinotori* obtained further results for flares in Cycle 21, but the high-resolution Soft X-ray Telescope and Hard X-ray Telescope on *Yohkoh* (operating 1991–2001) enabled much improved insight of flare structures in the X-ray region. The use of CCD cameras and high-resolution optics enabled the SXT to record emitting regions with a spatial resolution of 2.4 arcsec and a time resolution of 2 s. The developing stages of even short-lived flares with confined spatial extent could be followed. As with nearly all structures in the corona emitting in soft X-rays or extreme ultraviolet radiation, the emission is optically thin, so that a single line of sight may contain many individual emitting regions which cannot be distinguished. A flare's loop morphology, for instance, may not be very clearly delineated when the flare is observed near the centre of the solar disk, but on the other hand views of flares on the solar limb do not clearly show the footpoints of the loop structures.

Early SXT results showed unexpected images of flares with very short durations. A study of 38 such flares was made by Feldman *et al.* (1994), with *GOES* classifications of up to M5 and occurring mostly on the solar disk. SXT images were made with the filter having the highest-energy response, the 119 μm Be filter, with FWHM response approximately 3–7 Å. They show that the emitting plasma was often in the form of a bright point, not more than 1 pixel in extent at flare maximum, implying that even these high-resolution images were not properly resolving the flare plasma. Occasionally a faint loop structure was visible, with a tiny loop-top source much brighter than the rest of the loop. Temperatures, measured from *Yohkoh* BCS Fe XXV spectra, using the satellite-to-resonance line ratio j/w, show a rapid cooling, a typical value for a 1 MK decrease being 13 s. The absence of a complete loop structure suggests that conductive cooling of these flares did not occur, rather the hot spots or loop-top sources were conductively isolated and so must have cooled by radiation. If $\Delta t_{\rm rad}$ is the radiative cooling time for this plasma to cool by an electron temperature $\Delta T_{\rm e}$, the electron density is

$$N_{\rm e} = \frac{3k_{\rm B}\Delta T_{\rm e}}{\Delta t_{\rm rad} R(T_{\rm e})}, \tag{10.1}$$

where $R(T_{\rm e})$ is the radiative power loss function (see Figure 1.11). For the 38 flares in this study the mean value of $\Delta t_{\rm rad}$, the time for the Fe XXV temperature to decrease by 1 MK, was 14 s, giving $N_{\rm e} \approx 10^{12}$ cm^{-3}. The average emission measure for the peak intensity of the flares was 3×10^{48} cm^{-3}. A value $N_{\rm e} \approx 10^{12}$ cm^{-3} implies volumes of 3×10^{24} cm^3, and if assumed to be spherical, a diameter of 1790 km, or 1 SXT pixel, as observed.

How is the hot flare plasma to be conductively isolated if it is part of a magnetic loop with footpoints in the chromosphere? The field lines must be tangled in a highly complex way. Jakimiec *et al.* (1997) speculated that the field lines fold on themselves in such a way that there are many small-scale reconnections keeping the flare plasma energized. Alternatively, the increase then decrease of $N_{\rm e}$ as measured from the O VII X-ray line ratios (McKenzie *et al.* (1980a)) and the confined nature of the plasma might suggest the existence of a plasma pinch, as observed in some high-energy laboratory plasmas (Lee (1974)). Possibly a slowly increasing current exists that eventually breaks down leading to the plasma pinch, with high density. It would appear that the origin of impulsive X-ray flares is open to a number of options rather than the current sheet hypothesis that is conventionally applied to all flares (e.g. Forbes & Acton (1996)).

Several long-duration flares, i.e. those with decay times of many hours or even more than a day, were observed by the SXT on *Yohkoh*. Unfortunately, the complete development through the flare decay was often interrupted by observing sequences for flares occurring elsewhere on the Sun. One flare of great interest, discussed previously in this chapter (see also Feldman *et al.* (1995); Phillips *et al.* (2005a)), occurred on 1992 November 2, with peak soft X-ray flux at around 03:10 UT. The rise time was about 40 minutes and the decay time was many hours – the *GOES* X-ray emission smoothly decays and eventually fades to the pre-flare background at about 09:00 UT on November 3, some 30 hours after the flare first became evident. The flare loop footpoints were some 2400 km over the south-west limb of the Sun at the flare start, and so invisible from the Earth. Solar rotation carried them some 47 300 km over the limb 30 hours later. Figure 10.4 (top panel) shows the X-ray flux in both channels of *GOES*. *Yohkoh* BCS observed most of the flare, but near the peak time the BCS detectors were saturated. Line ratios in the BCS channels viewing the Ca XIX and S XV lines indicate $T_{\rm e} \approx$12–15 MK at 06:00 UT on November 2 and only 6 MK by 00:00 UT on November 3, a drop in temperature of 1 MK in 2–3 hours. If no energy were added and the cooling were by radiation only, an electron density of $1-2\times10^{9}$ cm^{-3} is implied from Equation (10.1). This is much too low for such an intense flare, and indicates instead that energy is continuously supplied during the flare decay phase. The SXT recorded flare images up to about 12:00 UT with its Be 119 μm filter. These show two or three bright loops above the south-west limb up till approximately 12:00 on November 2, after which a single loop is apparent. The most striking feature, however, is a bright source, extent about 9000 km, located near or at the apex of the loop (see Figure 10.6). This feature, with flux some ten times the emission in the loop legs, climbs with time, from 25 000 km above the limb at about 04:00 UT to 66 000 km at 22:00 UT. When corrected for solar rotation, the heights above the photosphere are approximately 27 000 km to 90 000 km, a rate of increase of 0.5–1 km s^{-1}. Eventually, emission extends along the whole of the southern loop leg. Thus, a fairly confined loop-top source is observed over a period of several hours.

Thus, while the soft X-ray emission in flares has a basic loop geometry, the emission is far from uniform along the length of the loop, both for short-lived and for long-duration flares. This has been commented on in the literature several times, although there remains the possibility of geometrical factors combined with the fact that the optical thinness of the plasma sometimes gives rise to a loop-top brightening that is only apparent, not real (Forbes & Acton (1996)). The reality of a high-temperature spine structure in the disk flare illustrated in Figure 1.10, identifiable with hot loop-top sources in a loop arcade, is however undeniable, and strongly suggests that non-uniformity of X-ray emission along a loop is real and requires an explanation that is not forthcoming from the assumption of a plasma loop cooling by classical thermal conductivity.

10.7 Mass motions in flares

Flare spectral line widths and shifts provide information on line-of-sight motions, while flare images give information about motions perpendicular to the line-of-sight. Spectral line profile observations with slit spectrometers or uncollimated spectrometers in the EUV and VUV have been used to derive line-of-sight motions. The FWHM of a spectral line observed by an uncollimated spectrometer depends on the instrumental width $\Delta\lambda_{\rm I}$, source dimensions $\Delta\lambda_{\rm sp}$, $T_{\rm ion}$, and turbulent or non-thermal mass motions as follows:

$$\mathrm{FWHM_{obs}} = \left[(\Delta\lambda_{\rm I})^2 + (\Delta\lambda_{\rm sp})^2 + 4\ln 2 \left(\frac{\lambda}{c}\right)^2 \left(\frac{2k_{\rm B}T_{\rm ion}}{m_{\rm ion}} + \xi^2\right)\right]^{1/2}, \quad (10.2)$$

where ξ is the most probable non-thermal velocity of the random plasma motion, $T_{\rm ion}$ is the ion temperature, and $m_{\rm ion}$ is the atomic mass of the ion producing the spectral line. This equation assumes that all the broadening factors making up the line – instrumental, spatial, thermal Doppler, and turbulent mass motions – have Gaussian components. This is not quite the case, though frequently this is a reasonable approximation. Also, it is generally assumed that $T_{\rm e} = T_{\rm ion}$ which, unless there is a heating mechanism by which ions are preferentially heated, is justified for flare plasmas where equilibrium times between electrons and protons are measured in seconds. Observations indicate that the profiles of flare lines emitted by ~10 MK plasmas are practically always wider than those from thermal motions alone.

The first high-quality Fe XXV and Ca XIX X-ray flare spectra from which line shapes and shifts could be reliably determined were recorded by the *P78-1* SOLFLEX instrument and by the *SMM* Bent Crystal Spectrometer (BCS). Their instrumental widths $\Delta\lambda_{\rm I}$ and the flare temperatures $T_{\rm e}$ (from spectral line ratios) could be determined with accuracy. Thus, when analyzing SOLFLEX or BCS spectral data, the only unknowns in Equation (10.2) are $\Delta\lambda_{\rm sp}$ and ξ. If $\Delta\lambda_{\rm sp}$ is small, ξ can be obtained. Various analyses have found that for short- and medium-duration events non-thermal velocities ξ are greatest at or close to flare onset. From thereon it decreases monotonically until reaching some minimum value that remains fairly constant for the rest of the event. In long-duration flares, measured widths of Ca XIX and S XV resonance lines with the *Yohkoh* BCS appear to indicate that late in the flare development ξ starts to rise again. However, when allowance is made for the considerable spatial extent of the flare at temperatures like those characterizing Ca XIX and S XV (Phillips *et al.* (2005a)), the value of ξ does in fact continue to decline, with an asymptotic value of $\sim$40 km s^{-1}. Figure 10.4 (bottom two panels) illustrates this for the long-duration flare on 1992 November 2/3. This is very similar to those obtained with the much higher resolution SUMER spectrometer. For M and X flares observed by the SOLFLEX instrument, typical values of ξ at onset are $\sim$150 km s^{-1} (Doschek *et al.* (1979,1980)), declining monotonically to $\sim$70 km s^{-1}.

Early in the rise phase of disk flares, a short-wavelength (or blue-shifted) component is present, nearly always less intense than the unshifted component, generally disappearing well before flare maximum. Interpreting the blue shift as a Doppler motion, the velocity of approach implies averages about 150–300 km s^{-1}, but in extreme cases is as much as 800 km s^{-1} (Feldman *et al.* (1980b)). Examples of two M3 flares, with unusually intense blue-shifted components with large displacements from the stationary component, are shown Figure 10.9. The blue-shifted component appears as an asymmetrical extension to the blue wing of the resonance line. Shortly after flare onset the flux in the blue-shifted component was about 50% of the the stationary component's flux, but disappeared within a few minutes.

Blue-shifted components in a large number of *Yohkoh* BCS spectra of fairly intense flares were studied in detail by Mariska *et al.* (1993) and Mariska (1994). Estimates of the velocity and relative flux of the blue-shifted component depend on knowing the spectral location of lines emitted by the stationary component (the spectrometer being uncollimated). This was assumed to be the spectrum emitted by the flare in its decay phase, when the blue-shifted component had disappeared. For each flare, rise phase spectra (when the Fe XXV line flux reached 25% of its maximum) were compared with spectra emitted late in the decay phase. The average centroids of Fe XXV, Ca XIX, and S XV resonance lines in the flare rise phase were shifted to shorter wavelengths by amounts that are (in terms of velocities) 76, 63, and 58 km s^{-1} respectively. No correlation was found between the centroid shift and the peak flux, rise time or total flare duration, or any centre-to-limb variation. However, a good correlation was found between shifts in the Fe XXV, Ca XIX, and S XV lines.

Flare mass motions in the EUV or VUV lines have been observed, the first and most convincing observations being of the 1354.08 Å Fe XXI forbidden line recorded by the *Skylab* S082B normal-incidence spectrometer. Mass motions during the decay phases of several flares, with typical velocities of 20–50 km s^{-1}, were observed (Cheng *et al.* (1979)). The observations indicated little or no bulk line-of-sight motion. The *SOHO* SUMER instrument has recorded a number of flare spectra, some during flare onsets (Innes *et al.* (2001), Landi *et al.* (2003), Wang *et al.* (2003), Feldman *et al.* (2004)). Generally, SUMER spectra show non-thermal mass motions from line widths of only 30–40 km s^{-1} late in flare decays. At flare onset, SUMER observations also indicate that the entire high-temperature plasma exhibits damped oscillations lasting for a few cycles, with amplitudes initially $\sim$100 km s^{-1} and periods $\sim$10 minutes. There do not appear to be any other bulk plasma motions larger than a few km s^{-1}.

Brosius and Phillips (2004) report on coordinated observations of an M2 disk flare on 2001 April 24 with CDS on *SOHO*, the BCS and HXT instruments on *Yohkoh*, the EIT instrument on *SOHO*, and *TRACE*. The impulsive stage of the flare was observed, with blue-shifted components evident in BCS Ca XIX and S XV spectra. By a fortunate circumstance, the CDS instrument was performing a spatial scan of the active region in which the flare occurred, and one of the pixels of the $4'' \times 4'$ slit coincided with a bright point, interpreted by Brosius and Phillips (2004) as a loop footpoint, visible in the 195 Å *TRACE* image of the active

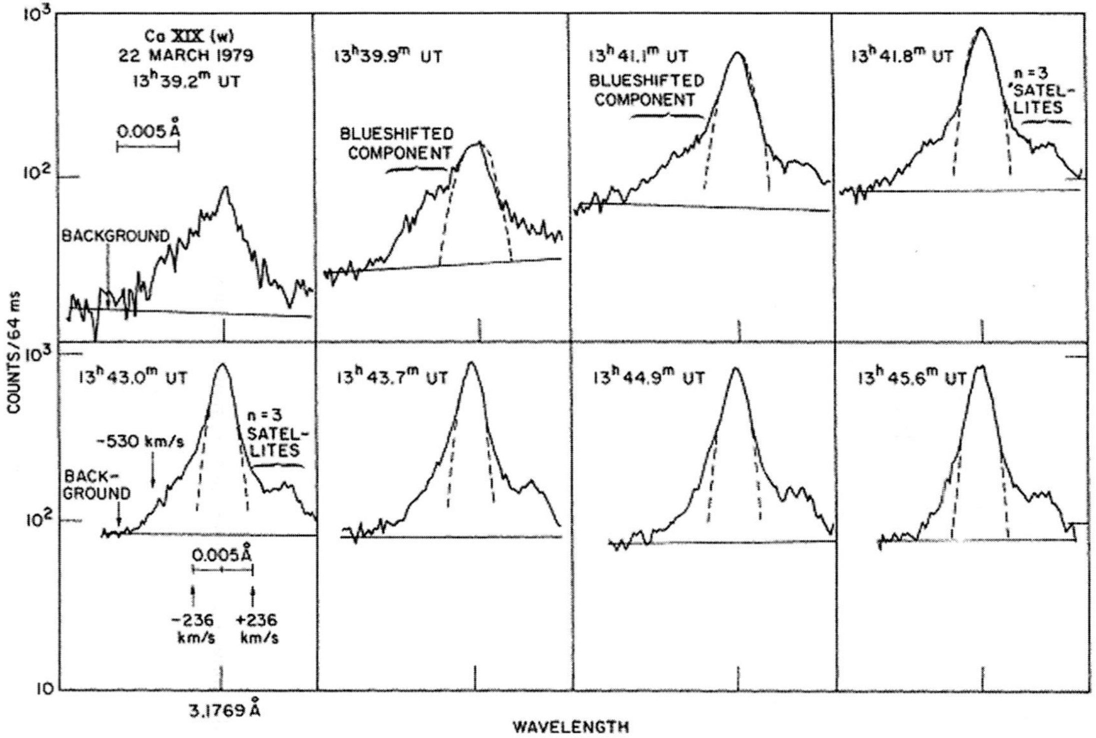

Fig. 10.9. Time series of the profile of the Ca XIX w line during the rise phase of a flare on 1979 March 22 (UT indicated in each plot, logarithmic flux scale). The observations were made with the *P78-1* SOLFLEX instrument. The partly resolved Ca XVIII n = 3 dielectronic satellites are visible on the long-wavelength side of the line w. Approach velocities of up to ~530 km s^{-1} are indicated by the blue-shifted component. From Feldman *et al.* (1980b).

region. One of the lines in the CDS study included the Fe XIX 592.2 Å line, emitted at 8 MK. During the flare impulsive stage, the $4'' \times 20''$ CDS pixel corresponding to the *TRACE* bright point shows Fe XIX line profiles having a blue-shifted component with upward velocity of 64 km s^{-1} at the same time (to within a few s) that the *Yohkoh* BCS recorded 78 km s^{-1} upward velocities in the Ca and S lines. This remains one of the best established illustrations of the existence of upward velocities at a loop footpoint.

10.8 Non-thermal electrons

Although commonly believed, the existence of beams of non-thermal electrons at the flare impulsive stage is hard to prove beyond doubt. A possible diagnostic exists in X-ray spectra (Chapter 5, Section 5.9.3, see Gabriel & Phillips (1979)), relying on the fact that, in Fe XXV spectra (as the chief example), the w line is excited by electrons with energies above 6.70 keV (the excitation energy of w) but dielectronic satellite j is excited by electrons with energy 4.69 keV and the blended dielectronic satellite feature $d13$ and $d15$ is excited by electrons with energy 5.82 keV. When the *P78-1* SOLFLEX and *SMM* BCS flare spectra first became available, however, it was immediately clear that the diagnostic suffered from the problem that the line widths were very large at precisely the time that non-thermal electrons were present, i.e. the flare impulsive stage, and so large that the $d13$, $d15$ feature could never be distinguished from line w, even for limb flares when blue shifts are almost non-existent. This appears to be the case for spectra from lower-Z ions, e.g. Ca XIX, also. However, confirmation of the diagnostic has been made by Beiersdorfer *et al.* (1993) for PLT tokamak plasmas which have substantially higher electron densities than in solar flares. There are indications from RESIK Ar XVII spectra that the k and j lines, which are blended with the Ar XVII z line, are sometimes anomalously intense. It is speculated that occasionally there are non-thermal electrons with energies of about 2.2 keV that give rise to excitation of the j and k satellites but not the Ar XVII w line – in a sense, this is the reverse of the situation discussed by Gabriel and Phillips (1979): the satellites are enhanced by non-thermal electrons but not line w. If confirmed, it is analogous to excitation of the satellites in an Electron Beam Ion Trap (EBIT) device, where beams of precisely calibrated mono-energetic electrons are directed at ions of a particular species.

10.9 X-ray fluorescence lines in flares

Kα inner-shell Fe lines on the long-wavelength side of the Fe XXV w line at 1.850 Å are often present in high-resolution flare spectra taken with the *P78-1* SOLFLEX spectrometer and the *SMM* BCS (Figure 10.10). Observations of the Fe Kα line flux with the BCS (Parmar *et al.* (1984)) showed that these lines are mostly excited by fluorescence (Chapter 4, Section 4.9.2). This is illustrated by the agreement of ratios of the Kα line flux to the total continuum flux above an energy of 7.11 keV with the theoretical curves (Figure 10.11). The best agreement is for flare heights of between zero and ~30 000 km. There is in general no Kα line emission for limb flares.

The centre-to-limb variation of the Fe Kβ lines (Section 4.9.2) is confirmed by *Yohkoh* BCS observations (Phillips *et al.* (1994a)). Because of the fact that the Kα lines are formed by photospheric Fe and Fe XXV lines and nearby satellites are formed by collisional excitation of Fe in the corona, Phillips *et al.* (1994a) point out that the ratio of the fluorescence lines to the Fe XXV lines should be a measure of the photospheric-to-coronal abundance of Fe.

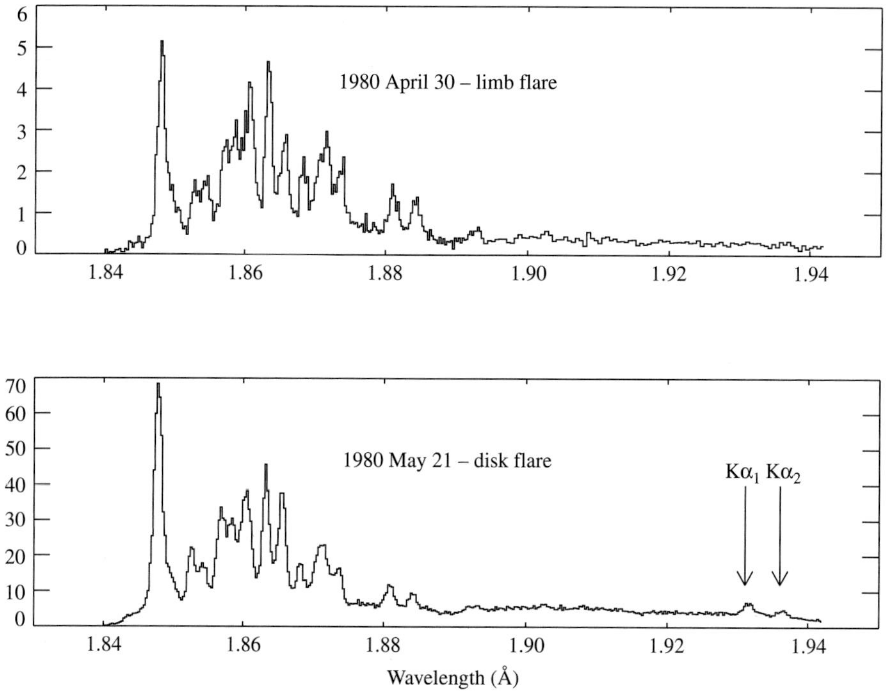

Fig. 10.10. *SMM* BCS spectra in the 1.84–1.94 Å range showing the Fe fluorescence lines Kα_1 (1.936Å) and Kα_2 (1.940Å) for a disk flare (*bottom*) and their absence in a limb flare (*top*).

Not all iron atoms in the photosphere are in neutral form, for the temperature there ($\sim$6000 K) is high enough to produce once-ionized Fe. However, the Fe III Kα doublet is spectroscopically indistinguishable from those of Fe II or indeed of other ions up to those formed at coronal temperatures, e.g. Fe XVIII. This is because the screening of the $[1s]-[2p]$ hole transition by the outer 24 electrons in the case of Fe II is almost the same as for the outer 8 electrons of Fe XVIII. This ceases to be true of higher ions of Fe, the observed lines with $1s - 2p$ transitions being due to dielectronic recombination and inner-shell excitation. The formation of satellite lines near the Kα lines, seen in the *SMM* BCS spectra (Parmar *et al.* (1984)), may be due to target Fe atoms that are doubly ionized, with a $1s$ and $2s$ electron removed (Veigele (1970)).

Fluorescence may create an L-shell vacancy in Fe atoms rather than one in the K shell, with the vacancy filled by an electron in the M shell. Several components of the resulting Lα line are thereby formed, with the most intense at 17.622Å. Although tentatively identified in a flare spectral scan with the *SMM* FCS (Phillips *et al.* (1982)), the line feature at this wavelength is almost certainly due to an intense Fe XVIII line with unusual transition ($2s\,2p^6\,{}^2S_{1/2} - 2s^2\,2p^4\,({}^1D)\,3p\;{}^2P_{3/2}$: see Cornille *et al.* (1994)).

There remains the possibility of some excitation of the Fe Kα lines by nonthermal electrons during the impulsive stages of flares (Emslie *et al.* (1986)). The collisional excitation calculations by Phillips and Neupert (1973), appropriate for a thin-target model (Fe Kα line excitation along the length of a flare loop by non-thermal electrons that lose only small

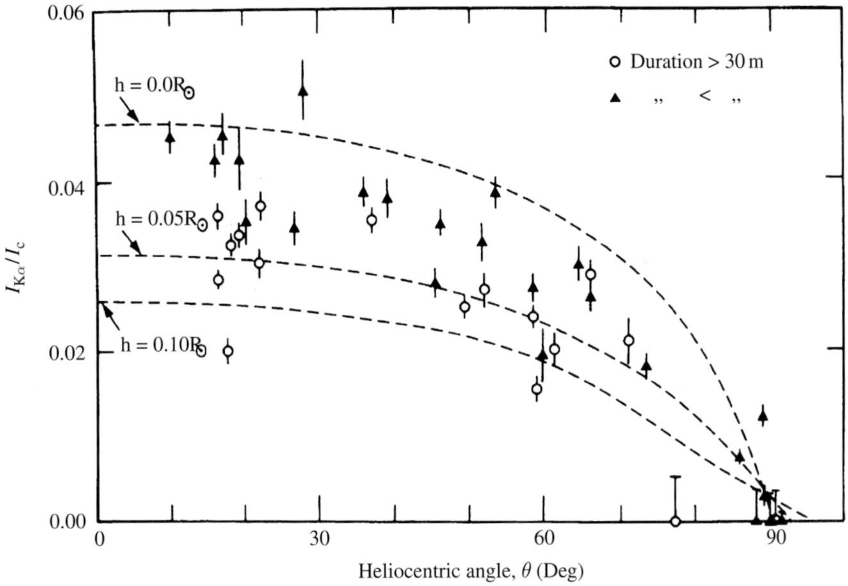

Fig. 10.11. Ratio of the Fe Kα line flux to the flux in the continuum at energies > 7.11 keV plotted against heliocentric angle θ for flares observed by the BCS on *SMM*: flares are categorized by duration (less or greater than 30 minutes). The dashed curves, based on calculations of Bai (1979), are for heights h indicated ($1\,R_\odot = 695\,000$ km). A photospheric Fe abundance $A_{\rm Fe}/A_{\rm H} = 5.5\times10^{-5}$ is assumed. From Parmar *et al.* (1984).

amounts of energy by collisions), were re-done for a thick target (non-thermal electrons excited at a flare loop top and travelling down to the loop footpoints, and there losing all their energy). The calculated emission was then compared with that observed during a relatively modest flare on 1980 March 29 but for which there is a hard X-ray burst at the impulsive stage with a very hard spectrum. Much of the Fe Kα line emission, observed by the *SMM* BCS, can be explained by fluorescence, as with practically all *SMM* flares, but there remains a 2σ peak of emission at the time of the hard X-ray impulse. The amount of emission agrees with that calculated in the thick-target approximation by Emslie *et al.* (1986). Thus, in particular cases, non-thermal electron excitation of the Fe Kα lines may not be negligible.

Kα lines have only been observed in solar flare spectra in the case of Fe, and remain the only lines in the X-ray range excited by photon excitation. The strong dependence on Z of the fluorescence yield means that even in elements like Ca ($Z = 20$) the Kα lines are weak. For Ni ($Z = 28$), the Kα lines would be relatively somewhat stronger than for Fe, but the much lower abundance of Ni has so far prevented the complex of lines at 1.53–1.60 Å (due to He-like Ni and dielectronic satellite lines and the Kα lines) being observed in solar flare spectra at crystal spectrometer resolution.

10.10 Ionization equilibrium in flares

The assumption of ionization equilibrium in such a dynamic plasma as a solar flare might seem to be highly questionable. In fact, even the lower limits on electron densities in flare plasmas, estimated from measured volumes in X-ray images and volume emission measures (in the range $7\times10^9 - 2\times10^{11}$ cm^{-3} for *Skylab* X-ray limb flares (Pallavicini *et al.* (1977)), are generally high enough to ensure a near approximation to ionization equilibrium. This may be shown for the hot plasma emitting Fe XXV lines by comparing the maximum rate of change of T_e (from $I(j)/I(w)$ ratios), during either its heating phase (positive dT_e/dt) or cooling phase (negative dT_e/dt), with respectively the ionization or recombination time. The ionization time is $\tau_{ion} = Q_{ion}N_e$ where Q_{ion} is the rate coefficient (cm^3 s^{-1}) for ionization of Fe^{+23} to Fe^{+24} ions, while the recombination time is $\tau_{rec} = \alpha_{rec}N_e$ where α_{rec} is the rate coefficient for recombination of Fe^{+25} to Fe^{+24} ions. Figure 10.12 shows measured values of $(dT_e/dt)/T_e$, i.e. inverse characteristic times for temperature changes, during a flare observed by the Fe XXV channel of the *Yohkoh* BCS instrument, and inverse ionization and recombination times, assuming $N_e = 10^{10}$ cm^{-3}. During both the heating and cooling phase of this flare, the absolute values of the inverse characteristic heating or cooling times are always much less than the inverse ionization or recombination times respectively. For densities larger than 10^{10} cm^{-3}, the inverse ionization and recombination times are correspondingly larger. So for high ionization stages of Fe, ionization equilibrium is a good approximation for this typical flare. This is likely to be true of Ca, S, and many other ions also since the ionization and recombination coefficients are of the same order.

A rather more sophisticated treatment is to consider the time-dependent ionization equilibrium equation for a flare plasma, with a whole set of ions $i-1$, i, $i+1\ldots$ of an element having number densities N_{i-1}, N_i, $N_{i+1}\ldots$ that are a function of time t as a flare develops

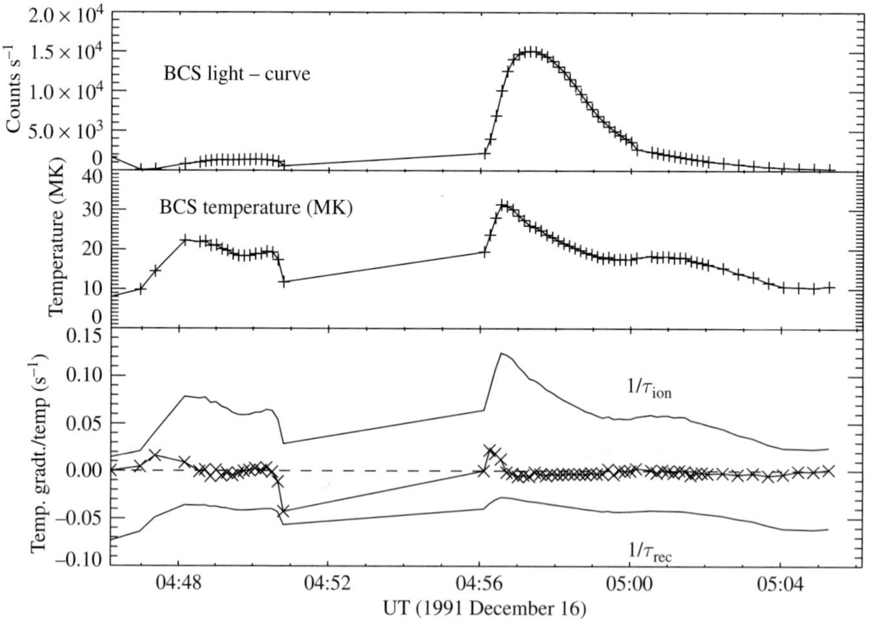

Fig. 10.12. Light curve in the *Yohkoh* BCS Fe XXV channel and T_e measured from j/w ratios (*top two panels*) with values of $(dT_e/dt)/T_e$, i.e. inverse characteristic times for temperature changes (curve with crosses) and inverse ionization (positive) and recombination (negative) times (*bottom*). From Phillips (2004).

or decays. With Q and α the ionization and recombination coefficients, the time dependence of ion stage i is given by

$$\frac{dN_i}{dt} = \frac{dN_i}{dT_e} \times \frac{dT_e}{dt} = N_e[Q_{i-1}N_{i-1} - (\alpha_i + Q_i)N_i + \alpha_{i+1}N_{i+1}]. \tag{10.3}$$

Taking various assumed values of N_e and with the necessary data for ionization and recombination coefficients, it is possible to solve for the various N_i with initial values of temperature $T_e(t=0)$ and N_i (these being the steady-state ionization equilibrium values) and measured values of dT_e/dt as in Figure 10.12. For high-temperature laboratory plasmas, Burton and Wilson (1961) found analytic solutions, but for flare plasmas numerical analysis based on Runge–Kutta techniques is generally necessary to invert the tridiagonal matrix of α and Q values on the right-hand side of Equation (10.3). This was done for various flares in which the Fe XXV 1.9 Å lines were observed by the Goddard Space Flight Center crystal spectrometer on *OSO-3* (Phillips *et al.* (1974)). It was found that the line flux per unit emission measure was hardly distinguishable from the steady-state ionization equilibrium. This placed a lower limit on the electron density of about 3×10^{10} cm^{-3}, since for lower values of N_e the time-dependent values of $N(Fe^{+24})$ departed significantly from the steady-state values. This probe of ionization equilibrium in flare plasmas is also a diagnostic of electron density and could be applied to much higher-quality data now available.

10.11 *GOES* observations of flares

10.11.1 *Characteristics of GOES light curves*

There is a huge (over five orders of magnitude) variation in peak flare flux in soft X-ray wavelengths, not counting the likely existence of microflares. *GOES* light curves show a basic pattern of exponential rise and exponential decay, sometimes with fluctuations, often unrelated to the flare itself. If a flare light curve shows complexity, it may often be resolved into two or more rise-and-decay curves. There is no evidence of extremely rapid (few seconds) fluctuations as occurs at high energies ($h\nu \gtrsim 30$ keV) during the flare impulsive stage.

The soft X-ray flux of a flare during its rise $[F_r(t)]$ and decay $[F_d(t)]$ phases can be approximately expressed by exponentials:

$$F_r(t) = F_r(t_0)\exp[(t-t_0)/\tau_r] \tag{10.4}$$

and

$$F_d(t) = F_d(t_0)\exp[-(t-t_0)/\tau_d], \tag{10.5}$$

with τ_r and τ_d the characteristic *e*-folding rise and fall times. Values of τ_d vary widely, from a lower limit of 7 s to many hundreds of seconds. Values of τ_r may be as short as only a few seconds (Feldman *et al.* (1984), (1994)). In a study of some 50 flares observed by the *Yohkoh* BCS Ca XIX channel, all of *GOES* C1 importance or greater and with light curves all completely observed within a *Yohkoh* daytime period (approximately one hour), it was found that τ_d/τ_r lay between 1 and 6 (up to 30 for flares with durations greater than an hour) with no clear relationship between τ_d/τ_r and either τ_r, τ_d, or the flare peak flux.

The importance of an X-ray flare may be expressed by the maximum *GOES* 1–8 Å flux or the maximum emission measure or possibly the maximum *GOES* electron temperature, though these do not in general occur simultaneously. They are nevertheless related, as has been found from an extensive survey of 868 flares ranging in *GOES* importance from A2 to

X2, all observed with the *Yohkoh* BCS (Phillips & Feldman (1995), Feldman *et al.* (1996)). The temperatures and the times of peak emission were determined for each flare from BCS using ratios of dielectronic satellites to resonance lines in Fe XXV, Ca XIX, and S XV. The Fe XXV lines were used in determining peak emission and temperatures larger than 14 MK, the Ca XIX lines were used to determine peak emission in the 8.5–14 MK range, and the S XV lines were used in the remaining cooler flares. In the most intense flares, the *Yohkoh* BCS Fe XXV detector was saturated, so X-ray fluxes, temperatures, and EMs were determined from the *GOES* 0.5–4 Å and 1–8 Å flux ratios. Figure 10.13 (left panel) is a plot of the logarithm of the *GOES* 1–8 Å peak flux against the peak temperature, showing that the two are fairly closely related. A best-fit solution is

$$F(T_e) = 3.5 \times 10^{0.185 T_e - 9.0}, \tag{10.6}$$

where $F(T_e)$ is the flux measured by the *GOES* 1–8 Å flux in W m^{-2} and T_e is in MK. All flares in the survey, from class A2 to X2, are included. Thus, no flare with X-ray importance greater than M1 has peak T_e lower than 15 MK, and no class with X-ray importance less than A9 has $T_e > 10$ MK. The peak temperatures of A2 flares are approximately 5–7 MK.

The emission measure and T_e are similarly related, as indicated by the sample of 868 flares as well as all the major flares between 1977 and 1991 (Cycles 21 and 22: Garcia & McIntosh (1992)). Figure 10.13 (right panel) shows the emission measure plotted against peak temperatures for flares in both samples. The emission measures vary over three orders (5×10^{46} to 1×10^{50} cm^{-3}), T_e from 5 MK to 35 MK. The best-fit functional form is

$$\mathrm{EM}(T_e) = 1.7 \times 10^{0.13 T_e + 46.0}, \tag{10.7}$$

where EM(T_e) is given in units of cm^{-3} and T_e in units of MK. The slope of the best-fit line is smaller than for the X-ray flux–T_e relation.

There appears to be no systematic study involving flux or emission measure as a function of temperature at times during the rise phase of flares. In a limited study of M and X flares, it has been found that early in the rise phase temperatures were less than or equal to 14 MK.

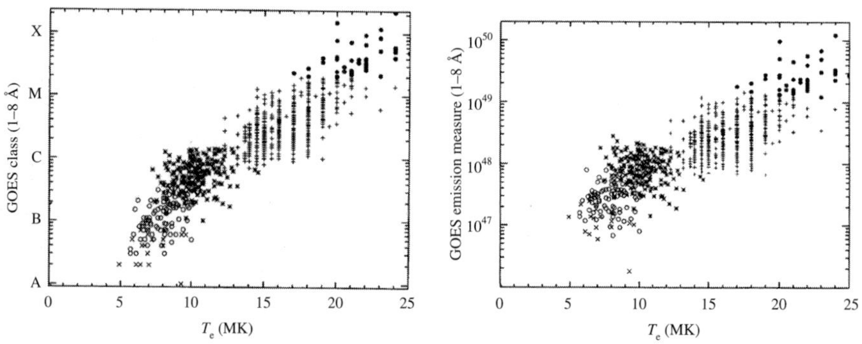

Fig. 10.13. (*Left*) *GOES* 1–8 Å peak flux for 868 flares plotted against electron temperature T_e determined from the *Yohkoh* BCS using spectra in the Fe XXV, Ca XIX, and S XV channels as appropriate. (*Right*) GOES 1–8 Å emission measure plotted against T_e from *Yohkoh* BCS spectra. The *GOES* class range for the flares in this analysis is from A2 to X2. Based on Feldman *et al.* (1996).

However, at about the time when the flux reached a C5 level, the temperature, even for major events, had already reached or exceeded 80% of its maximum value. An example of temperature as a function of time during the flare rise phase is given by Sterling *et al.* (1994). Since electron temperatures during rise phases are related to maximum temperatures and thus to the maximum fluxes, they can be used to predict the maximum size of the eruption.

10.11.2 Flare occurrence rates as a function of solar activity

The frequency of flares depends on solar activity but the distribution of flare X-ray importance does not. The flare occurrence rate was first studied in detail by Drake (1971) over a period of high activity. During this period, the non-flare background measured by *SOLRAD* was approximately C1, so only flares of class C1 or greater were studied. A study by Feldman *et al.* (1997a) of the flare occurrence rate included much fainter flares, down to importance A2, using *GOES* data. The following is based on this work.

During solar maximum periods, flares of small and moderate size erupt at a high frequency. The X-ray emission from these flares create a fairly uniform background at C1 class levels or higher. As a result, flares of class A and B blend with the background and become undetectable. Figure 10.14 illustrates the nature of the X-ray background during periods of high and low solar activity. While on 1991 October 23 the background level corresponded to a C2 level, on 1994 June 22 it was equivalent only to an A1 level, a factor of 200 lower.

X-ray background levels are critically important in determining the expected occurrence number per hour of flares in a particular X-ray class, particularly for class A and B flares. Flares must be counted within given flux levels using some graduation system that depends on the background level. In the study of Feldman *et al.* (1997a), flares were partitioned into six domains, according to the X-ray background (L) as follows: (1) $L \leq$ A1; (2) A1 $< L \leq$ A4; (3) A4 $< L \leq$ B1 ; (4) B1 $< L \leq$ B4; (5) B4 $< L \leq$ C1; (6) C1 $< L \leq$ C4. Flares were

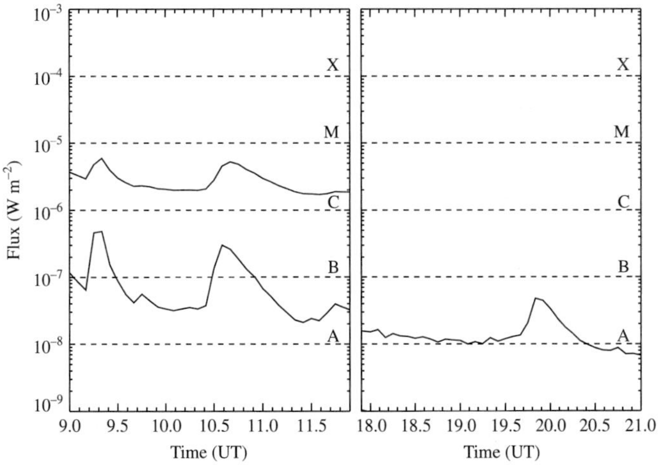

Fig. 10.14. *GOES* light curves for two small flares on 1991 October 23 (*left*) and 1994 June 22 (*right*). The upper curve in the left panel and the single curve in the right are the 1–8 Å light curves (the lower curve in the left panel is the 0.5–4 Å light curve). From Feldman (1996).

placed in particular domains only if their 1–8 Å peak fluxes (before background subtraction) were at least twice as large as the domain's maximum background.

Histograms of the number of events in a flux interval as a function of their background-subtracted peak X-ray flux for each of the six domains are shown in Figure 10.15. Using a linear least-squares fit distribution, values defining the number of events per class were derived. The number of counted events in each domain was converted into flare occurrence rates per hour by dividing them by the total hours of observations in the domain. Figure 10.15 shows that there is a linear dependence between the logarithm of the number (N) of flares observed within a specified flux interval and the logarithm of the flare peak flux (F), i.e.

$$\frac{dN}{dF} = C\,F^{-(\alpha+1)}, \tag{10.8}$$

where C is a constant that depends on the background level and the slope is defined such that it is consistent with the values in Figure 10.15. The data in the figure represent the integration of Equation (10.8) over flux intervals. The intervals were arbitrarily defined as flux changes by a factor 1.5, i.e. the number of events in a flux interval from F_i to $1.5 \times F_i$ were counted. In terms of Equation (10.8), the number of events N within the interval from F_i to $1.5 \times F_i$ is

$$\Delta N = \left(\frac{1.5^{-\alpha} - 1}{-\alpha}\right) \times C\,F_i^{-\alpha}. \tag{10.9}$$

Some of the variations in slope in Figure 10.15 may be statistical. The mean value of α is 0.88. The number of events per flux interval, dN/dF, is thus $F^{-(\alpha+1)} = F^{-1.88}$, in agreement with previous work (e.g. Drake (1971)). This indicates that the total number of

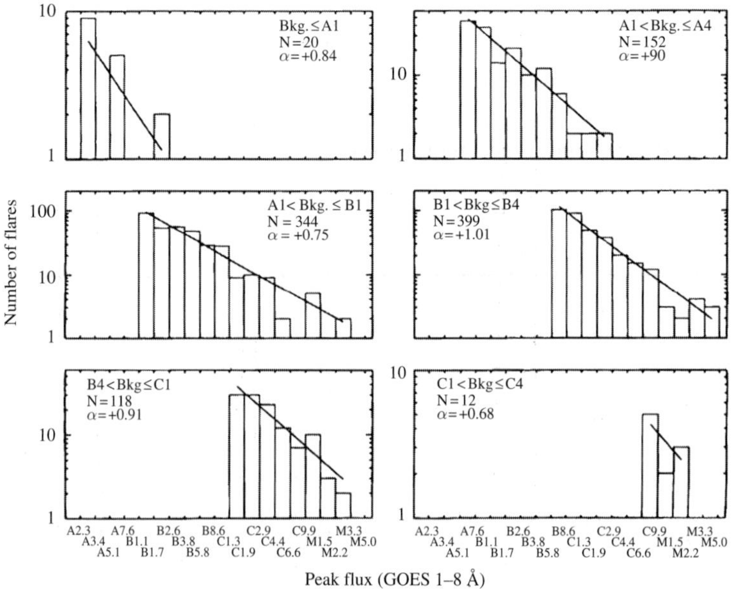

Fig. 10.15. Histograms of the number of events in the six flux intervals described in the text as a function of their FWHM. From Feldman *et al.* (1997a).

Table 10.2. *Number of flares per hour with flux higher than a class F as a function of GOES 1–8 Å flux.*

Background level	F=A1	F=B1	F=C1	F=M1	F=X1
C1 < background ≤ C4	79	10	1.4	1.8(−1)[a]	2.4(−2)
B4 < background ≤ C1	15	2.0	2.6(−1)	3.5(−2)	4.6(−3)
B1 < background ≤ B4	3.0	4.0(−1)	5.2(−2)	6.8(−3)	9.1(−4)
A4 < background ≤ B1	7.4(−1)	9.8(−2)	1.3(−2)	1.7(−3)	2.2(−4)
A1 < background ≤ A4	2.2(−1)	2.9(−2)	3.9(−3)	5.1(−4)	6.8(−5)
background ≤ A1	9.0(−2)	1.2(−2)	1.6(3)	2.1(−4)	2.7(−5)

[a] Powers of ten are indicated by numbers in parentheses.

flares per hour, regardless of activity level, increases as flare peak flux decreases; however, the ratio of the number of flares with different flux values is nearly constant.

An estimate of the number of flares within a given flux interval that occur per hour for each of the background domains can be calculated by assuming an average slope $-(\alpha+1) = -1.88$. Table 10.2 lists the total number of flares per hour brighter than a given X-ray class as a function of the X-ray background. The calculations are particularly interesting for the faintest flares because they cannot be observed directly with uncollimated instrumentation when the X-ray backgrounds are high.

The above result shows that the number of flares per hour regardless of the activity level increases as the peak flux decreases. How far this relationship extends as the peak flare flux decreases is as yet unclear. As mentioned in Section 10.11.1, peak electron temperatures of flares decrease with decreasing peak X-ray flux. For example, for flares that are an order of magnitude fainter than class A1 the peak temperature will most likely be much lower than 5×10^6 K and may approach the temperature of the quiescent active region that the flare occurs in. Since for such temperatures, the concept of a flare becomes ill-defined, the background emission corresponding to class A1 is a combination of both quiescent active region emission as well as emission from many very faint flares.

10.11.3 Duration vs. peak brightness in solar flares

In the work of Feldman *et al.* (1997a), the 'reduced lifetime' of flares was defined to be the full width (in time) at half maximum *GOES* 1–8 Å flux (FWHM). In most cases, and in particular when flares erupt at a sufficiently slow rate that two flares never appear at the same time, the FWHM could be determined to an accuracy of 10%. In cases where events overlap in time, the FWHM determinations may be uncertain to within 20–30%. Lifetimes of soft X-ray flares range from about one minute for impulsive events to many hours for long-duration events, and the magnitudes of their fluxes may change by over five orders of magnitude. In spite of this, the values of the flare reduced lifetimes in each of the six domains selected according to background (see Section 10.11.2) are quite similar, as is evident from the two flares shown in Figure 10.14. The distribution of FWHM values peaks at around 6–8 minutes, with a tail extending toward longer lifetimes. It should be noted that, although the maximum reduced lifetime shown on the graph is 32 minutes, long-duration events have not been excluded from this study.

11

Element abundances

11.1 Introduction

Since the early days of solar spectroscopy, the Solar System composition has been considered to be among the most fundamental set of parameters in astrophysics. Russell (1929) in his pioneering research succeeded in deriving the abundances of the 56 elements whose signatures he recognized in the photospheric spectrum of the Sun. Since then a significant body of research regarding the abundances of all elements in the solar photosphere has been accumulated. Much of the research on the composition of the solar photosphere is summarized in a number of review articles, e.g. Cameron (1970), Anders & Grevesse (1989), Grevesse & Sauval (1998), Grevesse *et al.* (2005).

The Sun, and the nebula out of which it was formed, is composed (by numbers of atoms or ions) of $\sim$90% hydrogen, $\sim$10% helium, and slightly over 0.1% of heavier elements. Oxygen (atomic number $Z = 8$), the most abundant element after H and He, has an abundance that is over three orders of magnitude lower than H. Ne (10), Mg (12), Si (14), and Fe (26) have abundances over four orders of magnitude lower than the abundance of H. The abundances of Na (11), Al (13), Ca (20), and Ni (28) are five to six orders of magnitude lower than that of H, and K (19) is almost seven orders lower. Copper (29) and other heavier elements are more than seven orders of magnitude less abundant than H. Spectral lines emitted by ions whose atomic numbers are larger than $Z = 28$ are thus generally not detectable in spectra from the solar atmosphere at temperatures larger than 10 000 K, and so our discussion will be limited to elements with atomic numbers $Z \leq 28$, and with abundance ratios relative to hydrogen of at least 1×10^{-7}.

The twenty most abundant elements in the solar photosphere, which are relevant to solar and stellar upper atmosphere studies, are given in Table 11.1. The elements in the table can be divided into two distinct groups. The first group includes non-volatile elements such as Na, Mg, Al, Si, P, S, K, Ca, Cr, Mn, Fe, and Ni. The second group, which consists of volatile elements, can further be divided into two subgroups. The first subgroup includes the volatile elements H, N, C, O, and Cl, and the second includes the noble gases He, Ne, and Ar. Abundances of all but the noble gases can in principle be derived from solar photospheric spectra. The abundances of the non-volatile elements can also be obtained from the analysis of a particular class of meteorites. The noble gas composition cannot be measured using the above two methods and some other means of determination is required.

Table 11.1. *First ionization potential, photospheric and solar upper atmosphere abundances above quiet Sun regions.*

Z	Element	FIP (eV)	Photospheric abund.[a]	Quiet Sun coronal abund.[a]
1	H	13.6	12.00[b]	12.00
2	He	24.6	10.899 ± 0.01[b]	10.9
6	C	11.3	8.58 ± 0.08[d]	8.58
7	N	14.5	8.02 ± 0.08[d]	8.02
8	O	13.6	8.88 ± 0.08[c]	8.88
10	Ne	21.6	8.10 ± 0.08[c]	8.10
11	Na	5.1	6.30 ± 0.03[b]	6.90–7.10
12	Mg	7.6	7.55 ± 0.02[b]	8.15
13	Al	6.0	6.46 ± 0.02[b]	7.06–7.26
14	Si	8.2	7.54 ± 0.02[b]	8.14
15	P	10.5	5.46 ± 0.04[b]	5.46
16	S	10.4	7.34 ± 0.08[e]	7.34
17	Cl	13.0	5.26 ± 0.06[b]	5.26
18	Ar	15.8	6.45 ± 0.04[e]	6.45
19	K	4.3	5.11 ± 0.05[b]	5.71–5.91
20	Ca	6.1	6.34 ± 0.03[b]	6.94–7.14
24	Cr	6.8	5.65 ± 0.05[b]	6.25–6.45
25	Mn	7.4	5.50 ± 0.03[b]	6.10
26	Fe	7.9	7.47 ± 0.03[b]	8.07
28	Ni	7.6	6.22 ± 0.03[b]	6.82–7.02

[a] Abundances are expressed logarithmically, with H = 12.
[b] Lodders (2003).
[c] Feldman & Widing (2007).
[d] Lodders (2003), increased by a factor 0.19 following Feldman & Widing (2007).
[e] Phillips *et al.* (2003).

Meteorites arrive in a variety of forms. At one extreme are the iron-type meteorites and at the opposite extreme are the meteorites called the CI carbonaceous chondrites. The chondrites most likely resulted from the break-up of asteroids that were not subject to the kind of differentiation affecting the major planets. The presence of water and organic compounds in CI carbonaceous chondrites is a testimony to the fact that they have not at any time been subject to high-temperature processes. Because of their unique properties, Goldschmidt (1937) suggested that the composition of the non-volatile elements in the CI carbonaceous chondrite meteorites and in the solar photosphere are the same. At present it is accepted that this is indeed the case.

In early studies, significant differences existed between the non-volatile elemental abundances derived from meteorites and from the photospheric spectra. However, with improved calculations of oscillator strength values, elemental abundances derived from the solar photosphere became almost identical to those derived from meteoritic studies. Recently, Lodders (2003) published a revised list of laboratory-derived abundances of the CI carbonaceous chondrites. The photospheric abundances for the non-volatile elements in Table 11.1 are adopted from that list.

Volatile elements are not completely retained in meteoritic material; thus, their photospheric abundances must be obtained by other means. For the volatile elements H, C, N, O, and Cl, which are easily excited under photospheric conditions, the photosphere is the best source of abundance values. Noble gases are not represented at all in spectra emitted by the solar photosphere, and so their abundances in Table 11.1 have been taken from measurements of spectra emitted by the solar upper atmosphere known to have photospheric abundances (Feldman & Widing (2003, 2007)).

It was not until the 1960s that the first spectroscopic observations of the upper solar atmosphere (in the $\lambda < 2000$ Å range) were obtained from space. The spectral resolution of the early observations was fairly good, though the spatial resolution was poor. Pottasch (1963, 1964) was the first to present evidence, from early space ultraviolet solar spectra, of composition differences between the solar upper atmosphere and the photosphere. In scaling the ultraviolet abundance to H he found that Mg, Al, and Si were three times more abundant in the solar upper atmosphere compared to the then standard photospheric abundances. Unfortunately, these results did not attract much attention. The first conclusive findings of differences between solar upper atmosphere and photospheric composition resulted from *in situ* solar wind (SW) and solar energetic particle (SEP) observations. The information obtained showed that the composition of the solar wind is not always the same as the composition of the photosphere. While the composition in the fast solar wind resembles to a large extent the photospheric composition, in the slow wind the abundances of the non-volatile elements are increased by a factor of 4–5.

Figure 11.1 (left) is a schematic representation of the SW and SEP findings. The solar wind and solar energetic particles are believed to originate from the solar atmosphere and in particular from the corona, thus retaining a coronal composition. It was discovered that elements can be divided into two groups, according to their first ionization potential (FIP): low-FIP (i.e. those with FIP < 10 eV) and high-FIP (FIP > 10 eV) elements. At first, sulphur (FIP=10.3 eV) was considered to be an intermediate element, but there is some evidence that the division in FIP should be between 9 and 10 eV, with S belonging to the high-FIP group. Solar wind and solar energetic particle measurements established that, on average, the ratio of abundances between the low-FIP and high-FIP groups in the quiet corona is a factor of 3–4 higher than the corresponding photospheric ratios. Figure 11.1 (right) displays the

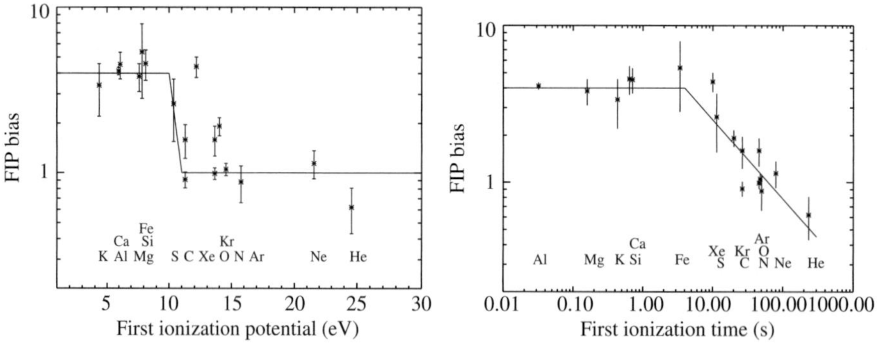

Fig. 11.1. FIP bias as a function of first ionization potential (FIP, *left*) and first ionization time (FIT, *right*) in the solar wind and in the solar energetic particles, as reported by von Steiger *et al.* (1997).

dependence of the ratio between coronal and photospheric abundances as a function of the first ionization time.

Meyer (1985) reviewed spectroscopic measurements of upper solar atmosphere abundances already published in the scientific literature and showed that these measurements generally supported the SW and SEP findings. Relative to H, it was not at first clear whether elements with low FIP were enhanced or those with high FIP depleted in the corona. However, it was later shown that the modifications were the result of increases in the abundance of low-FIP elements and not the decreases in the high-FIP ones. Notice that all the high-FIP elements, with the exception of S, Ar, and the very low abundance elements P and Cl, have atomic numbers smaller than or equal to 10, while atomic numbers of all low-FIP elements are larger than 10. Inspired by SW and SEP findings, and by Meyer's (1985) work, many studies were conducted over the past two decades. For details, see review articles by Feldman (1992), Saba (1995), Fludra *et al.* (1999), Feldman & Laming (2000), Feldman & Widing (2003), and Raymond (2004).

11.2 Solar wind abundances

Properties of the solar wind have been investigated for over three decades by arrays of *in situ* analyzers installed on a variety of satellites. Over the years, as the sophistication of the analyzing devices and their sensitivity improved, measurements of the charge, mass, and velocity of the intercepted particles were progressively refined. A fairly detailed description of solar wind properties, such as the degree of ionization, speed, and composition, has now emerged.

During solar minimum periods, the solar wind consists of two main components. The first component, apparently originating from solar quiet regions, is called the slow-speed solar wind, and has a velocity of $\sim$400 km s^{-1}. A second component originates in coronal holes and is called the fast-speed solar wind, with velocity $\sim$750 km s^{-1}. During periods of high activity there is an additional solar wind component that is associated with transient phenomena and is made up of what are referred to as solar energetic particles. We will not refer to solar energetic particle events here any further: information regarding the chemical composition in these events can be found from Reames (1994,1998) and references therein.

At first most solar wind remote-sensing observations were obtained from instruments aboard satellites launched in either low-Earth orbits or orbits round the Sun, always in low heliographic latitudes. As a result the measurements reflected the solar wind component emerging from solar regions located at low latitudes, i.e. from non-polar coronal hole regions. It was only when some polar coronal holes extended into sufficiently low solar latitudes that the wind originating from such regions was detected. A graphic display of the solar wind properties came from the *Ulysses* observatory. *Ulysses* was launched jointly by ESA and NASA in October 1990 in a solar orbit passing almost over each solar pole. *Ulysses* reached the solar south pole region (maximum heliographic latitude of 80.2°) in June–November of 1994 and the north pole a year later (June–September 1995). *Ulysses* was launched shortly after solar maximum (which occurred in 1990) and completed its first orbit in early 1998, well before the solar maximum in 2001. Neugebauer (1999) summarizes the findings during its first solar orbit. In its second orbit, *Ulysses* passed over the south pole in late 2000 and over the north pole a year later, a time period that included solar maximum activity.

Figure 11.2 shows the speed of the solar wind as measured by *Ulysses* in its first orbit around the Sun. When *Ulysses* was close to the ecliptic plane, the solar wind velocity was

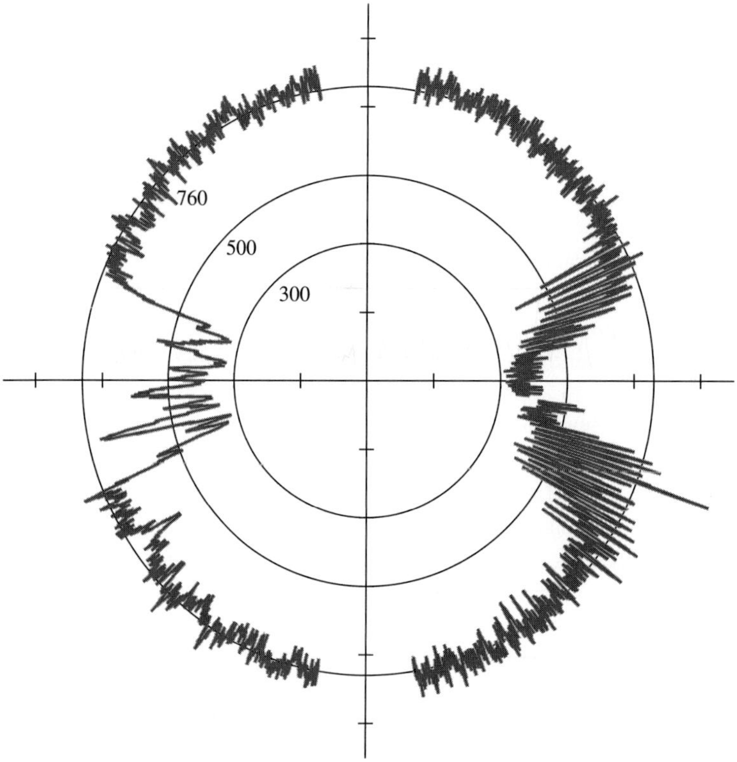

Fig. 11.2. Polar plot of solar wind speed (km s^{-1}) as a function of heliographic latitude observed by *Ulysses* in its out-of-ecliptic phase between 1994 and 1995. From McComas *et al.* (1998).

close to 400 km s^{-1}, but at higher southern latitudes *Ulysses* encountered recurrent wind streams from equatorial extensions of a polar coronal hole. The recurrent high-speed wind indicated a rotation period close to the sidereal mid-latitude surface rotation period of ~26 days. Later, with *Ulysses* now at higher latitudes, the spacecraft became completely immersed in the polar coronal hole wind, with solar wind speeds constant at 750 km s^{-1}.

The slow-speed solar wind, originating in quiet Sun regions, and the fast-speed solar wind, originating in coronal hole regions, are different not only in their speed but also in other plasma properties including densities, temperatures, and elemental abundances. The wind velocities are more than an order of magnitude larger than velocities measured in either quiet Sun or coronal hole regions close to the solar surface. The large solar wind velocities are thus undoubtedly the result of acceleration processes occurring far away from the solar surface. In contrast, the plasma temperature and composition in the solar wind and near the solar surface are quite similar. They are in large part the result of processes occurring close to the solar surface.

Figure 11.3 shows the velocity, temperature, and relative elemental abundance ratios C/O and Mg/O as measured by *Ulysses* during a 200 day period in 1992. The times during which the wind speed was high were associated with coronal hole regions that reached low solar

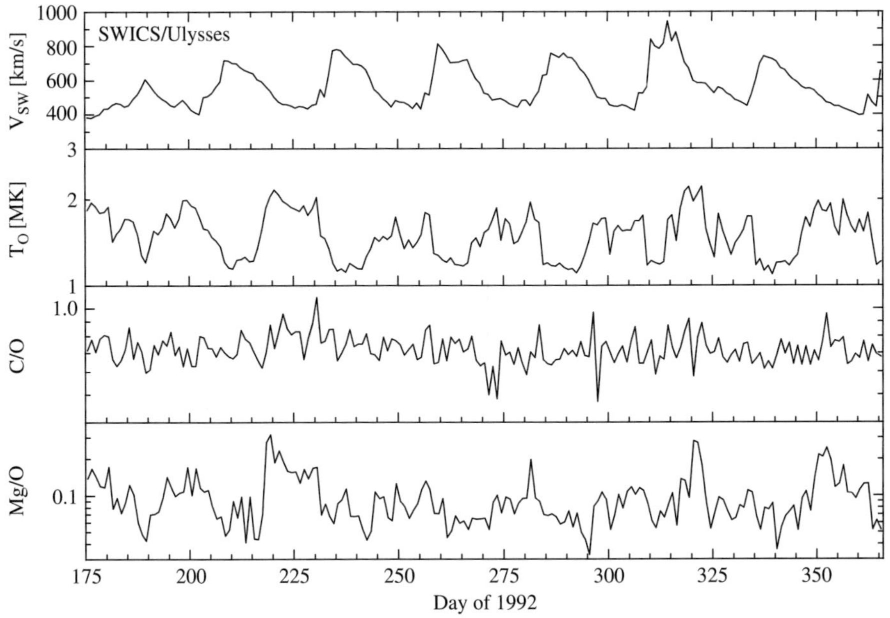

Fig. 11.3. Solar wind He^{++} ion velocity, oxygen freezing-in temperature, C/O and Mg/O abundance ratios (top to bottom panels) from the SWICS instrument on *Ulysses* during the second half of 1992. A distinct stream structure developed, with a high-speed stream appearing once per solar rotation (~26 day sidereal rotation period). The heliocentric distance of *Ulysses* was between 5.4 AU and 5.0 AU and heliographic latitude between $-12°$ and $-23°$. From von Steiger *et al.* (1997).

latitudes. Similarly, the slow-speed wind was associated with times when quiet regions covered the low-latitude solar areas. There is an apparent anti-correlation between the wind speed and plasma temperature as well as between wind speed and the Mg/O abundance ratio. When the wind speed is at a maximum, the temperature approaches a minimum ($T_e \approx 10^6$ K); when the wind speed approaches a minimum, the temperature approaches a maximum ($T_e = 1.5 - 2 \times 10^6$ K). These distinctions are not surprising, since to a large extent they reflect the coronal temperature above the wind-emerging regions. The coronal temperature above quiet regions – $T_e \approx 1.4 \times 10^6$ K – is reflected in the slow-speed wind temperature, while the coronal temperature above polar coronal hole regions – $T_e \sim 9 \times 10^5$ K – is reflected in the fast-speed wind temperature.

11.3 Abundances in quiet Sun and coronal hole regions

To help characterize the composition of the solar upper atmosphere and to quantify composition changes, we define the quantity *FIP bias* as the ratio of the elemental abundance in the solar upper atmosphere to its value in the photosphere, i.e.

$$\mathrm{FIP_{bias}} = A_{\mathrm{El}}^{\mathrm{SUA}} / A_{\mathrm{El}}^{\mathrm{Phot}}, \tag{11.1}$$

where $A_{\mathrm{El}}^{\mathrm{SUA}}$ and $A_{\mathrm{El}}^{\mathrm{Phot}}$ are the abundances of the element El in the solar upper atmosphere and in the photosphere respectively.

11.3.1 The average FIP bias in coronal hole and quiet Sun regions

When viewed through narrow-temperature-band filters in the $3 \times 10^4 \leq T_e \leq 7 \times 10^5$ K range, polar coronal hole plasmas and equatorial quiet Sun plasmas look fairly similar (Figure 8.1). However, at $T_e > 8 \times 10^5$ K, the quiet Sun equatorial region's morphology becomes distinctly different from that of polar coronal regions (Figure 8.3). The $3 \times 10^4 \leq T_e \leq 7 \times 10^5$ K plasmas in both quiet Sun and coronal hole regions appear to be confined in small loop-like structures. In the hotter quiet Sun regions, the plasma is confined in loops also, but these structures are much longer than in coronal holes and may reach lengths of $> 5 \times 10^5$ km. The morphology of polar coronal holes for which $T_e \approx 9 \times 10^5$ K is much simpler and, other than sporadic polar plumes present in the region, the plasma is not confined in closed magnetic structures.

The number of loop-like structures per unit area in quiet Sun and coronal hole regions is too large to enable present-day instruments to resolve individual structures and follow their time development. Thus, in quiet Sun and coronal hole regions, information is available only on average abundance values close to the solar surface ($h \leq 1.5\ R_\odot$), on the composition changes with temperature, and on the composition changes with height.

11.3.2 Transition region in coronal hole and quiet Sun during solar minimum

Using Mg VI/Ne VI EUV line intensity ratios, Feldman and Widing (1993) derived a FIP bias of 1.5–2 for transition region ($3 \times 10^4 \leq T_e \leq 7 \times 10^5$ K) structures in quiet Sun and coronal hole regions. Young and Mason (1998) derived a FIP bias of $\sim$2 in transition region cell interiors and cell boundaries. More recently, Young (2005) derived a FIP bias of 1.25 for the network and 1.66 for the cell centres. Laming *et al.* (1995) analyzed the transition region spectra of the unresolved Sun (Sun viewed as a star) and found a FIP bias of slightly larger than 1. These results indicate that the $3 \times 10^4 \leq T_e \leq 7 \times 10^5$ K quiet Sun and coronal hole plasmas have a FIP bias of 1.5 ± 0.5.

11.3.3 Solar coronal plasmas

Coronal quiet Sun plasmas during solar minimum are nearly isothermal. Using theoretical emissivities of lines from high-FIP and low-FIP elements, the FIP bias vs. temperature can be obtained (Feldman *et al.* (1998)). Both high-FIP and low-FIP lines intersect at a temperature of $\log T_e \approx 6.13$ ($T_e = 1.35 \times 10^6$ K). High-FIP lines show a FIP bias of $\sim$1 while the low-FIP lines indicate a FIP bias of 3–4 (see Figure 11.4). Landi *et al.* (2006b) studied the coronal composition early in the rise of the activity cycle (in 1998) in over 36 locations distributed in a rectangular grid bounded in the east–west direction by $1R_\odot$ to $1.5\ R_\odot$ and in the north–south directions by $-0.5R_\odot$ and $+0.5R_\odot$. With the exception of the extreme south-west corner, which had coronal-hole-like plasma properties, the region exhibited typical quiet Sun properties, i.e. the derived FIP bias at heights $h \leq 1.4R_\odot$ was $\sim$3.5 with a tendency for somewhat increased values at greater heights. As indicated in Table 11.1, on average it is assumed that in the quiet Sun corona the low-FIP elements with FIP values between 7.6 and 8.2 eV (i.e. Mg, Si, Fe, and Ni) are enriched by a factor of 4 while ones having a lower FIP are enriched by at least that factor. For more on the second group, see Section 11.6.

Studies of the $T_e = 8 \times 10^5$ K polar coronal hole plasmas indicate that in such regions the FIP bias is very close to 1 (Figure 11.5). As is the case in the solar wind, coronal hole plasmas undergo little if any composition modification once emerging from beneath the photosphere.

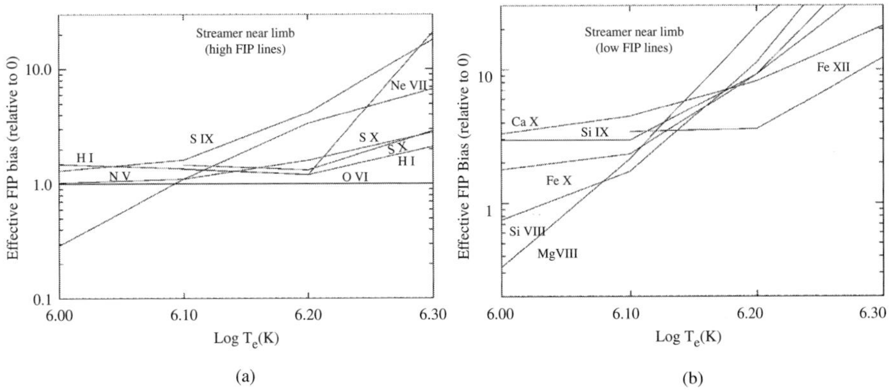

Fig. 11.4. Effective FIP bias as a function of temperature for the corona above a quiet Sun equatorial region: (*a*) high-FIP lines; (*b*) low-FIP lines. Lines from Li-like ions (Ne VIII, Na IX, Mg X) are not shown. From Feldman *et al.* (1998).

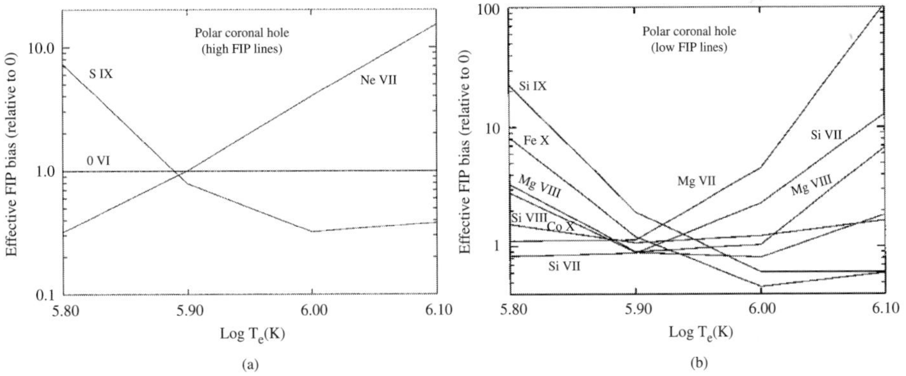

Fig. 11.5. The effective FIP bias as a function of temperature for the 9×10^5 K coronal hole plasma: (*a*) high-FIP lines, (*b*) low-FIP lines. From Feldman *et al.* (1998).

11.3.4 Elemental settling in the quiet corona

Quiet Sun coronal loops, which may exceed heights of a solar radius, survive for days. Since in such loops the electron density varies widely between the loop footpoints and loop tops, and the plasma turbulence is small, elemental settling may occur. This results in a FIP bias modification as a function of height. This is illustrated in Figure 8.5 (Chapter 8) where the normalized fluxes of Ne VIII ($Z = 10$), Mg X ($Z = 12$), Si XI, and Si XII ($Z = 14$) were plotted against height in the solar atmosphere in the western hemisphere. All these elements have similar atomic weights. Since their fluxes decrease with height at practically the same rate, we can conclude that the temperature does not vary appreciably with height. More importantly, the fact that the flux ratios of the high-FIP to low-FIP lines do not vary with height shows that the FIP bias also does not vary with height, at least for elements with similar atomic weights. However, the fluxes of Fe X, Fe XI, and Fe XII lines, also shown in

Figure 8.5, decrease much more rapidly in the 1.05–1.45$R_{\odot}$ range; this can be accounted for by element settling. Another example of element settling is most likely indicated by the observation of Raymond *et al.* (1997), who found, using *SOHO* UVCS instrument data, that the O/H abundance ratio in an equatorial streamer was a factor 3–10 times lower than in the photosphere.

11.4 FIP bias in distinct solar structures

Plasmas confined in loop-like structures at the instant of emergence from beneath the photosphere can be expected to have photospheric composition. However, once above the photosphere the composition becomes modified. The amount of modification has been found to depend on the age of the structure. We give some examples of this.

11.4.1 An eruptive prominence

Widing *et al.* (1986) studied properties of a plasma volume some 10 minutes after it was violently ejected from the photosphere at a speed of ~400 km s^{-1}, reaching temperatures of 10^6 K. Typically, coronal O/Mg abundance ratios are ~6 (Meyer (1985)) but in the ejected plasma it was 17.8, quite similar to the photospheric ratio. The eruptive prominence is an indication that ejected plasmas while undergoing intense heating maintain their photospheric composition for at least the first ten minutes.

11.4.2 Active regions

Shortly after emerging from under the photosphere, plasma trapped in rising active region magnetic fields is expected to have photospheric composition, as indeed is the case (Sheeley (1995, 1996), Widing (1997)). The FIP bias of hot ($2 - 5 \times 10^6$ K) active region plasmas that are more established was studied in the soft X-rays using flux ratios of O VIII and Ne IX lines (O and Ne are high-FIP elements) and Mg XI and Fe XVII lines (Mg and Fe are low-FIP elements) (Strong *et al.* (1988), McKenzie and Feldman (1992)). While low-FIP flux ratios of Fe XVII to Mg XI show small variations and are in close agreement with CI chondrites values, the abundance ratio of the O VIII and Ne IX lines shows a factor of ~2 variations. It is not clear whether this variability is due to inaccuracies in the atomic physics, to differential emission measure effects, or to actual composition modification.

McKenzie and Feldman (1992) found that the flux ratios of Fe XVII to Ne IX in different active regions vary by factors of 1–4. With the same line ratios, Strong *et al.* (1991) and Saba and Strong (1992) found variations in Fe/Ne abundance ratios as high as a factor 7.

Widing and Feldman (1995) studied the Mg/Ne abundance ratios in a sample of active regions using EUV lines. For three of the active regions, the FIP bias derived from the ratio of Mg VI to Ne VI line fluxes reached values of 4.8, 5.4, and 5.9, showing that for active regions observed at high spatial resolution the measured FIP bias could exceed 4 by significant amounts. In some old active regions, the FIP bias values can be significantly higher (8–16) (Widing & Feldman (1992), Feldman (1992), Young & Mason (1997), Dwivedi *et al.* (1999), Feldman *et al.* (2004)).

11.4.3 Flares

Flares are transient phenomena lasting from tens of minutes to tens of hours. Very impulsive flares probably do not last long enough for their chemical composition to be changed from the photosphere if the material in flare loops originated in or beneath the photosphere. Thus, the *Skylab* 1973 December 2 flare erupted in or very near the photosphere and as expected its composition as derived from EUV lines was nearly photospheric (Feldman & Widing (1990)). Sylwester *et al.* (1990), Fludra *et al.* (1990), Sterling *et al.* (1993), and others studied a large number of flares from their 1.8–3.2 Å X-ray spectra. In doing so they found FIP biases of 1 to 4. Similarly, Strong *et al.* (1991) and McKenzie and Feldman (1992) studied large populations of flares from their soft X-ray spectra and also found FIP biases of 1–4. Feldman *et al.* (2004) used the *SOHO* SUMER high-spatial-resolution observations to derive the composition at a height of $\sim 1.1 R_\odot$ in a series of recurring flares. By comparing the flux of the free–free continuum with the fluxes of highly ionized Ca and Fe lines, they obtained for the low-FIP elements a FIP bias of 8–10, a value significantly larger than the earlier mostly X-ray measurements.

11.4.4 High-temperature quiescent prominences

Prominences are long lasting, mostly relatively cool structures, though some have temperatures of up to $\sim 5\times 10^5$ K. Spicer *et al.* (1998) in surveying the *Skylab* database analyzed seven such structures, finding that two had FIP biases of ~ 1 and the remaining five had values intermediate between 1 and 4.

11.5 FIP bias rate of change in the solar atmosphere

At birth, plasmas in the solar upper atmosphere have an FIP bias of 1, and within the following few hours the FIP bias is practically unchanged. However, in the average quiet Sun corona and in the slow-speed solar wind, FIP biases of 4–5 are detected. Furthermore, in some old active regions and in some flaring plasmas the FIP bias values can be significantly higher. Thus, the most interesting issues associated with elemental abundances in the solar atmosphere are the FIP rate of change and the maximum FIP biases a plasma volume can have.

The composition of active region plasmas, born near the east limb, can be monitored from birth to decay, or their disappearance beyond the west limb – a time period of about a week to ten days. In searching the *Skylab* spectroheliograph records, Widing and Feldman (2001) identified six fairly bright active regions that were born near the east limb and lasted for at least three days. The composition of each of the regions was determined for as long as the active region features were visible. Figure 11.6 shows, for four of the regions, FIP bias values plotted against time since emergence. As can be seen, shortly after emergence the composition of the active-region plasma was photospheric, but within a few hours it began to change. After 2–3 days, FIP bias values of 4–5 occurred, and after 4.6 and 5.5 days the FIP bias values were 8 and 9 respectively. The trend of increasing bias was similar for all active regions.

In order to establish the rate of FIP bias change in various types of solar atmosphere structures, one needs to identify well-defined regions immediately after birth and follow them for a long time. Unfortunately, this has only ever been accomplished using *Skylab* spectroheliograph data. Assuming that the rate of change derived for most magnetically confined solar atmosphere plasmas is similar to the one derived for the *Skylab* small active region, we can conclude the following: since for quiet Sun coronal plasmas, average FIP bias values are ~ 4, such structures last on average 2–3 days. The cooler quiet Sun and coronal hole transition region structures having a FIP bias of ~ 1.5 last on average only a few hours.

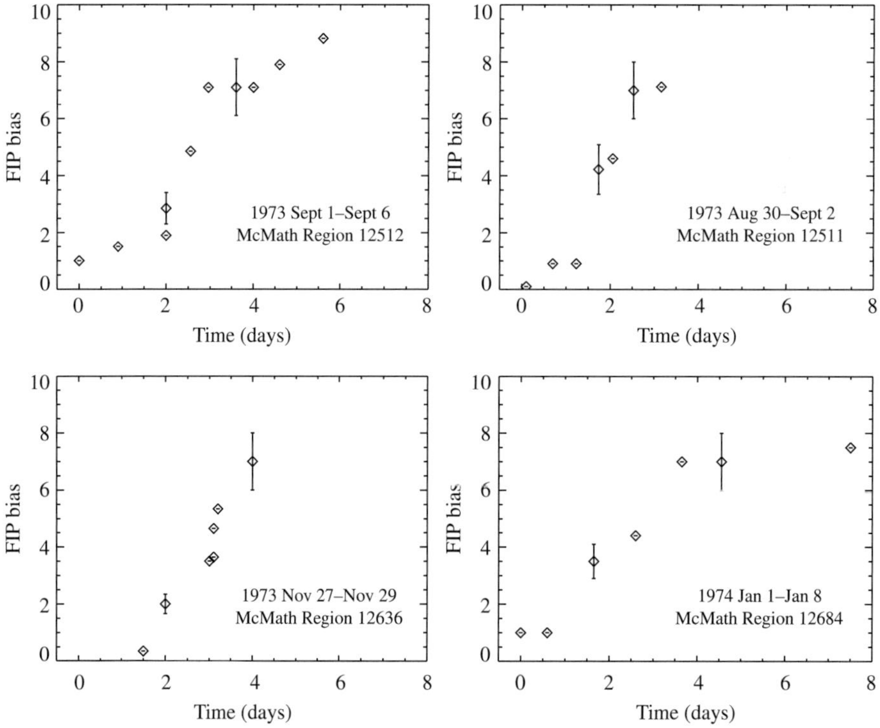

Fig. 11.6. FIP bias values plotted against the time interval from the emergence of new active regions, based on *Skylab* data. From Widing & Feldman ((2001)).

The fact that a FIP bias of ~15 is often found for some active regions implies that such plasmas have lasted for weeks.

11.6 Is the FIP bias magnitude FIP dependent?

The most abundant low-FIP elements in the solar photosphere can be divided into three groups according to their relative abundances. The first group of elements consists of Mg (FIP = 7.6 eV), Si (8.2 eV), and Fe (7.9 eV), with relative photospheric abundances of about $3-4 \times 10^{-5}$ times the abundance of H. The second group of elements consists of Na (FIP = 5.1 eV), Al (6.0 eV), Ca (6.1 eV), and Ni (7.6 eV), with photospheric abundances of $2-3 \times 10^{-6}$ the abundance of H. The third group of elements consists of K (FIP = 4.3 eV) alone, which has a photospheric abundance of 1.3×10^{-7} times that of H. With the exception of Ni, the FIP of elements of the second group are also lower by 2–3 eV than those of the first group. Similarly, relative to elements in the first group, the abundance of K is reduced by more than two orders of magnitude and its FIP by 3.3–3.9 eV. The cumulative number of nuclei relative to H and He combined as a function of the FIP is shown in Figure 11.7.

In the photosphere, the abundance ratio Ca/Fe is 0.074 and Ca/Mg 0.062 (Table 11.1). Doschek *et al.* (1985) used He-like and H-like flare lines and their satellites ($T_e \approx 1 \times 10^7$ K) recorded by the *P78-1* SOLFLEX instrument to derive the Ca abundance in solar flares. According to them, the abundance ratio of Ca/Fe = 0.093 is a factor of 1.3 higher than the

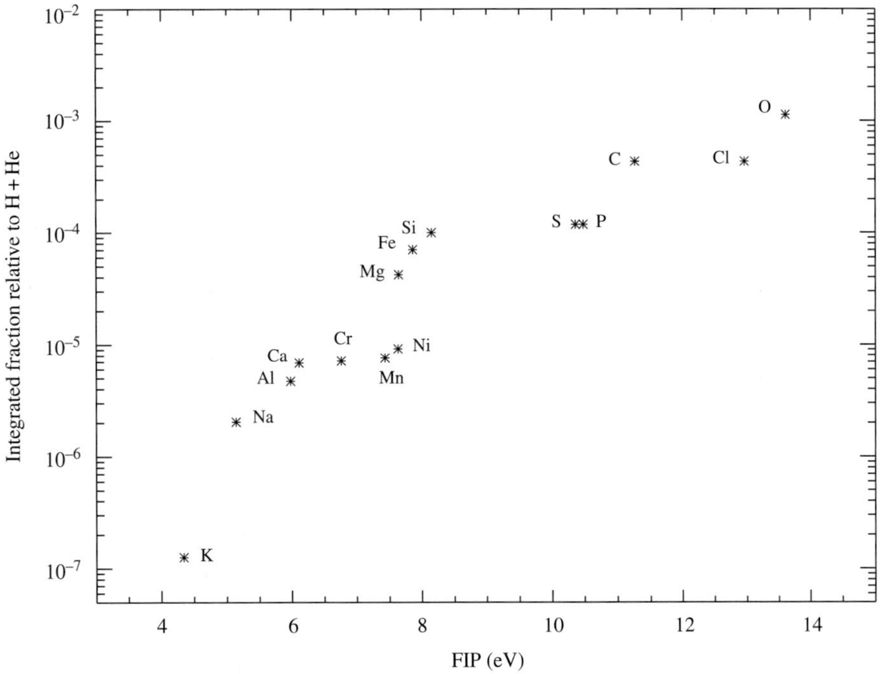

Fig. 11.7. Cumulative number of particles relative to the sum of H and He as a function of the first ionization potential.

photospheric ratio. Using elemental abundances from seven flares derived from the Bragg Crystal Spectrometer on *Yohkoh*, Fludra *et al.* (1993) derived an average abundance ratio of Ca/Fe = 0.092, or 1.3 times the photospheric value. The Ca/Fe abundance ratio was derived for a number of flares recorded by spectrometers on *SMM* and on *Yohkoh* (Phillips & Feldman (1991, 1997)). The average abundance ratio in the flares they measured was Ca/Fe = 0.156, a factor of 2.2 larger than the photospheric value. Schmelz and Fludra (1993) used a flare decay phase spectrum, recorded by the *SMM* BCS, to derive an abundance ratio of Ca/Fe = 0.086. In analyzing EUV lines from an active region, Monsignori Fossi *et al.* (1994) derived a ratio of Ca/Fe = 0.13, or 2.1 times the photospheric ratio. As indicated in the above quoted measurements, the Ca/Fe abundance ratios are not the same in all solar regions; nevertheless, the derived ratios are always larger than the accepted photospheric ratio.

In the photosphere the Al/Mg abundance ratio is 0.078 (Table 11.1). McKenzie and Feldman (1992) derived the Al/Mg abundance ratio for six flares. For one flare, they derived an abundance ratio of 0.071, while for the remainder the ratios varied between 0.098 and 0.134, with uncertainties of 0.029 or less. Monsignori Fossi *et al.* (1994) from their EUV line study derived an abundance ratio of Al/Mg = 0.17. As was the case with Ca, the Al abundance ratio relative to Mg is enhanced relative to the photosphere by as much as a factor of 2.

Potassium (K) has the lowest FIP (4.34 eV) among the more abundant elements in the solar upper atmosphere. In the photosphere, K/Ca = 0.059 (Table 11.1). Using SOLFLEX

Table 11.2. *Abundance of elements with FIP bias lower than a maximum value FIP_m relative to H and He combined.*

FIP_m (eV)	$N(FIP < FIP_m)/N(H + He)$
10	$\sim 1 \times 10^{-4}$
7	$\sim 7 \times 10^{-6}$
5	$\sim 1 \times 10^{-7}$
4.2	$\sim 3 \times 10^{-10}$

X-ray flare spectra, Doschek *et al.* (1985) measured an abundance ratio of K/Ca = 0.10, a factor 1.7 higher than in the photosphere.

Owing to the significantly lower concentrations of elements from the second and third groups, uncertainties associated with their abundances are often quite large, thus making it difficult to decide conclusively whether their FIP bias is different from the FIP bias of elements from the first and second groups. Although the effect of the FIP on the FIP bias needs further study, there are indications that such a dependence may exist. As listed in Table 11.1, as a result of the above discussion the FIP bias of elements with FIP $\leq$ 7 eV in the quiet Sun corona is expected to be on average 4–6.

11.7 Number of free electrons in the photosphere

Knowledge of the element abundances of solar plasmas and of the dependence on temperature allows the determination of the amount of free electrons and therefore of the conductivity of these plasmas. At all temperatures larger than 10^5 K, H and He are completely ionized so that even if the heavier elements retain many or even most of their electrons, the number of free electrons is comparable to that of protons and heavy nuclei. However, close to the temperature minimum the temperature is sufficiently low to cause H and He to be mostly in their neutral state, so that the number of free electrons is strongly dependent on the abundances of all the other elements whose first ionization potential is sufficiently low to allow them to be ionized even at photospheric temperatures. The number of free electrons is very important because it determines the conductivity of the solar plasma, usually assumed to be very high.

The ratio of the total number of atoms with a FIP less than a specific value to H and He combined is a very sensitive function of the FIP value, as shown in Table 11.2. In the FIP interval between C (FIP = 11.5 eV) and Na (5.2 eV), the abundance ratio relative to H is reduced by around three orders of magnitude. When the temperature decreases to values for which only the lowest-FIP elements are ionized, the total number of free electrons decreases greatly. This means that the conductivity of such cool plasmas will be decreased by orders of magnitude from the conductivity of the corona (Feldman (1993)).

In the photosphere and temperature-minimum region, the temperature is low enough that only a fraction of even low-FIP elements are ionized. Models for these low-temperature regions are shown in Figure 11.8. In the VAL model (Vernazza *et al.* (1981)) the value of N_e/N_H is determined by using measured intensities from optically thin and thick lines, continuum radiation in the 1600 Å region, and photospheric elemental abundances. These

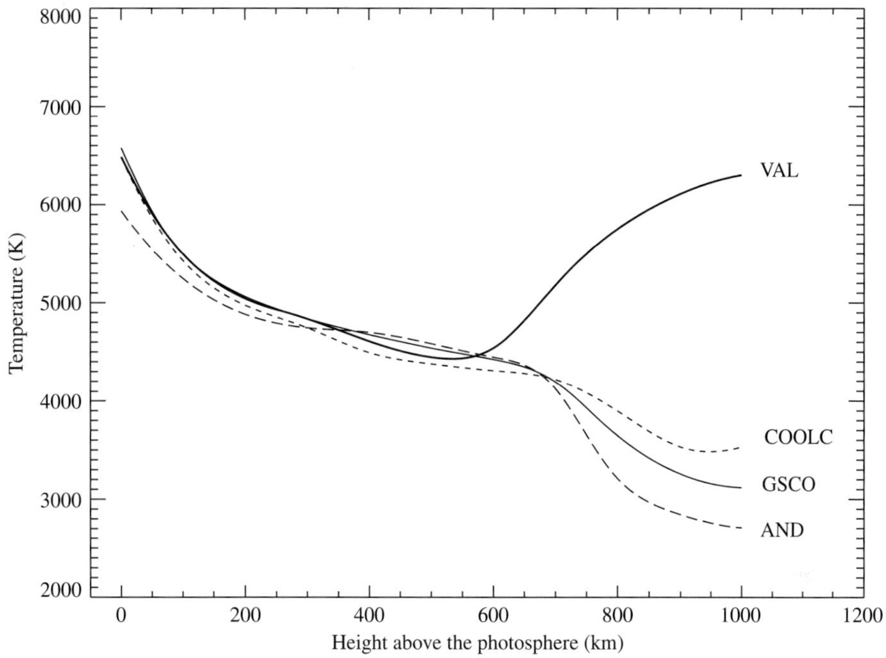

Fig. 11.8. Models of the lower solar atmosphere: temperature dependence on the height above the solar photosphere. VAL: Vernazza *et al.* (1981); COOLC: Ayres *et al.* (1986); GSCO: Grevesse & Sauval (1991); AND: Anderson (1989).

models predict that the minimum temperatures in the lower solar atmosphere are in excess of 4000 K, so that Mg, Si, and Fe are substantially ionized while the higher-FIP abundant elements are mostly neutral. If this is so, the number of free electrons is $\sim 10^{-4}$ times the number of nuclei. Ayres (1981, 1990) showed that the empirical one-component models of the solar outer layers (Maltby *et al.* (1986), VAL, and Holweger & Müller (1974)) fail to explain the very optically thick lines, namely, H Ly-α, Ca II H and K, as well as the infrared CO molecular lines at the centre of the disk and their centre-to-limb variations. In order to resolve the discrepancy, Ayres (1981) proposed a thermal bifurcation model for the solar atmosphere composed of two thermally distinct zones of hot ($T \geq 5000$ K) and cool ($T \leq 4000$ K) plasma, which co-exist at the same altitude. Ayres *et al.* (1986) found that the filling factor of the surface coverage of the hot regions should be less than 10% at the level of formation of the CO lines, while the cool plasma with CO occupies more than 90% of the solar surface. Grevesse and Sauval (1991) proposed a new model (GSCO, Figure 11.8) based on high-resolution solar spectra of molecular lines in the infrared, predicting temperatures significantly lower than those predicted by Ayres *et al.* (1986) (COOLC). A theoretical non-LTE model in radiative equilibrium by Anderson (1989) (AND) looks similar in shape to the COOLC and GSCO semi-empirical models; however, it predicts even lower temperatures ($T_e \approx 2700$ K) in the cooler parts of the lower solar atmosphere.

Solar atmospheric regions for which $T_e \approx 2700$ K will contain a significantly lower fraction of free electrons than regions predicted by earlier models like the VAL. In fact, if there are cool enough regions where the majority of ions with FIP > 5 eV are neutral, the ratio of electrons to H atoms can be only $N_e/N_H \approx 1 \times 10^{-7}$. The electrical conductivity of such a region is reduced by orders of magnitude from those predicted by earlier models. If the temperature in some solar regions is so low that even K is mostly neutral, the fraction of free electrons relative to neutral atoms may be extremely small.

11.8 Summary

Although the source of solar atmosphere plasmas is the photosphere, the chemical compositions of such plasmas are not identical as would be expected. In coronal holes, where solar upper atmosphere plasmas are not confined by the local magnetic field, and in hot plasma shortly after emerging from beneath the photosphere, the composition is photospheric; however, in other regions this is clearly not the case. Measurements indicate that in regions confined by the magnetic field, the FIP bias is time dependent. In freshly emerging plasma volumes, the FIP bias is nearly photospheric, but as an active region ages the FIP bias increases on a time scale of hours or days, and for old regions it can reach values as high as 15. From a study of several small active regions it appears that the FIP bias increases linearly in time. At small heights ($h \leq 1.2R_{\odot}$) the enrichment values of the various low-FIP elements are independent of their atomic weight; at greater heights, elemental settling may occur.

The plasma composition variations can be used as tracers for processes occurring in the solar atmosphere. They can be used to identify the sources from which plasmas high in the solar atmosphere emerged, e.g. flares, mass ejections, the various solar wind components and solar energetic particles. Most important, the understanding of the apparently linear FIP bias increase with time (periods of days) in magnetically confined regions may lead to an understanding of the coronal heating process.

Appendix I Units

Appendix 1 Units

The units in this book are mostly based on the c.g.s. system (centimetre, gram, second) which is still widely used in solar physics literature. The unit for wavelength of radiation used frequently here is the angstrom, abbreviated Å, and named after A. J. Ångström, the Swedish physicist. The angstrom was originally defined in terms of the wavelength of a spectral line of cadmium, but since 1960 it has been defined as 10^{-10} of a metre, where a metre, the fundamental unit of SI units, is defined as the length of the path travelled by light in a vacuum in 1/299 792 458 s.

The following table allows c.g.s. units (second column) of common physical quantities referred to in this book to be converted to SI units (third column) which are standard in most other branches of physics. In addition, the table gives the SI unit corresponding to the electronvolt as a unit of energy and temperature in high-energy solar physics.

Physical quantity	To convert from	To	Multiply by
length	centimetre (cm)	metre (m)	10^{-2}
area	cm^2	m^2	10^{-4}
volume	cm^3	m^3	10^{-6}
mass	gram (g)	kilogram (kg)	10^{-3}
velocity	cm s^{-1}	m s^{-1}	10^{-2}
mass density	g cm^{-3}	kg m^{-3}	1000
number density	cm^{-3}	m^{-3}	10^6
force	dyne	N	10^{-5}
energy	erg	J	10^{-7}
photon energy	electronvolt (eV)	J	1.602×10^{-19}
pressure	dyne cm^{-2}	Pa	10
power	erg s^{-1}	W	10^{-7}
wavelength	Å	nm	0.1
frequency	Hz	Hz	1
temperature	eV	K	11 604.8

Appendix 2 Line lists

Principal spectral lines in solar disk spectrum

Ion	Wvl. (Å)	Wvl.[L] (Å)	Transition	Ref.
Ni XXVIII	1.544	1.536	$1s\ ^2S_{1/2}-2p\ ^2P_{3/2}$	11
Ni XXVIII		1.536	$1s\ ^2S_{1/2}-2p\ ^2P_{1/2}$	
Fe XXV	1.567	1.573	$1s^2\ ^1S_0-1s\ 3p\ ^1P_1$	11
Ni XXVII	1.587	1.589	$1s^2\ ^1S_0-1s\ 2p\ ^1P_1$	11
Fe K$\beta_{1,3}$	1.757		[1s]–[3p]	
Fe XXVI	1.780	1.778	$1s\ ^2S_{1/2}-2p\ ^2P_{3/2}$	11
Fe XXVI		1.783	$1s\ ^2S_{1/2}-2p\ ^2P_{1/2}$	
Fe XXV(w)	1.8499	1.850	$1s^2\ ^1S_0-1s\ 2p\ ^1P_1$	15
Fe XXV(x)	1.8552	1.855	$1s^2\ ^1S_0-1s\ 2p\ ^3P_2$	15
Fe XXIV	1.8568	1.856	$1s^2\ 2s\ ^2S_{1/2}-1s\ 2s\ 2p\ ^2P_{3/2}$	15
Fe XXV(y)	1.8595	1.860	$1s^2\ ^1S_0-1s\ 2p\ ^3P_1$	15
Fe XXIV (q)	1.8610	1.861	$1s^2\ 2s\ ^2S_{1/2}-1s\ 2s\ 2p\ ^2P_{3/2}$	15
Fe XXIV	1.8631	1.863	$1s^2\ 2p\ ^2P_{1/2}-1s\ 2p^2\ ^2D_{3/2}$	15
Fe XXIV (j)	1.8660	1.866	$1s^2\ 2p\ ^2P_{3/2}-1s\ 2p^2\ ^2D_{5/2}$	15
Fe XXV(z)	1.8680	1.868	$1s^2\ ^1S_0-1s\ 2s\ ^3S_1$	15
Fe XXIII	1.8704	1.870	$1s^2\ 2s^2\ ^1S_0-1s\ 2s^2\ 2p\ ^1P_1$	15
Fe XXIII	1.8754	1.876	$1s^2\ 2s\ 2p\ ^3P_2-1s\ 2s\ 2p^2\ ^3D_3$	15
Fe XXII	1.8824	1.882	$1s^2\ 2s^2\ 2p\ ^2P_{1/2}-1s\ 2s^2\ 2p^2\ ^2D_{3/2}$	15
Fe XXI	1.8966	1.897	$1s^2\ 2s^2\ 2p^2\ ^3P_2-1s\ 2s^2\ 2p^3\ ^3D_3$	15
Fe XX	1.9051	1.906	$1s^2\ 2s^2\ 2p^3\ ^2D_{3/2}-1s\ 2s^2\ 2p^4\ ^2D_{3/2}$	15
Fe XX	1.9075	1.908	$1s^2\ 2s^2\ 2p^3\ ^4S_{3/2}-1s\ 2s^2\ 2p^4\ ^4P_{5/2}$	15
Fe Kα_1	1.9360		$[1s_{1/2}]-[2p_{3/2}]$	15
Fe Kα_2	1.9400		$[1s_{1/2}]-[2p_{1/2}]$	15
Ca XX	3.0185	3.109	$1s\ ^2S_{1/2}-2p\ ^2P_{3/2}$	15
Ca XX	3.0239	3.024	$1s\ ^2S_{1/2}-2p\ ^2P_{1/2}$	15
Ca XIX(w)	3.1769	3.177	$1s^2\ ^1S_0-1s\ 2p\ ^1P_1$	15
Ca XVIII	3.1822	3.183	$1s^2\ 3p\ ^2P_{3/2}-1s\ 2p\ 3p\ ^2D_{5/2}$	15
Ca XIX(x)	3.1889	3.189	$1s^2\ ^1S_0-1s\ 2p\ ^3P_2$	15
Ca XIX(y)	3.1925	3.193	$1s^2\ ^1S_0-1s\ 2p\ ^3P_1$	15
Ca XVIII (q)	3.2003	3.201	$1s^2\ 2s\ ^2S_{1/2}-1s\ 2s\ 2p\ ^2P_{3/2}$	15
Ca XVIII (k)	3.2066	3.207	$1s^2\ 2p\ ^2P_{1/2}-1s\ 2p\ 2p\ ^2D_{3/2}$	15
Ca XIX(z)	3.2111	3.211	$1s^2\ ^1S_0-1s\ 2s\ ^3S_1$	15

Ion	Wvl. (Å)	Wvl.[L] (Å)	Transition	Ref.
Ca XVIII	3.225	3.224	$1s^2\ 2p\ ^2P_{3/2}-1s\ 2p\ 2p\ ^4P_{5/2}$	11
K XVIII	3.532	3.532	$1s^2\ ^1S_0-1s\ 2p\ ^1P_1$	29
K XVIII	3.548	3.546	$1s^2\ ^1S_0-1s\ 2p\ ^3P_2$	29
K XVIII		3.549	$1s^2\ ^1S_0-1s\ 2p\ ^3P_1$	
K XVIII	3.571	3.571	$1s^2\ ^1S_0-1s\ 2s\ ^3S_1$	29
Ar XVIII	3.733	3.731	$1s\ ^2S_{1/2}-2p\ ^2P_{3/2}$	29
Ar XVIII		3.737	$1s\ ^2S_{1/2}-2p\ ^2P_{1/2}$	
Ar XVII	3.949	3.949	$1s^2\ ^1S_0-1s\ 2p\ ^1P_1$	29
Ar XVII	3.967	3.966	$1s^2\ ^1S_0-1s\ 2p\ ^3P_2$	29
Ar XVII		3.969	$1s^2\ ^1S_0-1s\ 2p\ ^3P_1$	
Ar XVII	3.994	3.994	$1s^2\ ^1S_0-1s\ 2s\ ^3S_1$	29
Cl XVII	4.186	4.185	$1s\ ^2S_{1/2}-2p\ ^2P_{3/2}$	29
Cl XVII		4.191	$1s\ ^2S_{1/2}-2p\ ^2P_{1/2}$	
Cl XVI	4.444	4.444	$1s^2\ ^1S_0-1s\ 2p\ ^1P_1$	29
Cl XVI	4.466	4.464	$1s^2\ ^1S_0-1s\ 2p\ ^3P_2$	29
Cl XVI		4.468	$1s^2\ ^1S_0-1s\ 2p\ ^3P_1$	
Cl XVI	4.496	4.497	$1s^2\ ^1S_0-1s\ 2s\ ^3S_1$	29
S XVI	4.729	4.727	$1s\ ^2S_{1/2}-2p\ ^2P_{1/2,3/2}$	29
S XV	5.039	5.039	$1s^2\ ^1S_0-1s\ 2p\ ^1P_1$	29
S XV	5.066	5.063	$1s^2\ ^1S_0-1s\ 2p\ ^3P_2$	33
S XV		5.067	$1s^2\ ^1S_0-1s\ 2p\ ^3P_1$	
S XV	5.102	5.102	$1s^2\ ^1S_0-1s\ 2s\ ^3S_1$	29
Si XIV	6.179	6.180	$1s\ ^2S_{1/2}-2p\ ^2P_{3/2}$	26
Si XIV	6.186	6.186	$1s\ ^2S_{1/2}-2p\ ^2P_{1/2}$	26
Si XIII	6.647	6.648	$1s^2\ ^1S_0-1s\ 2p\ ^1P_1$	26
Si XIII	6.685	6.685	$1s^2\ ^1S_0-1s\ 2p\ ^3P_2$	26
Si XIII	6.688	6.688	$1s^2\ ^1S_0-1s\ 2p\ ^3P_1$	26
Si XIII	6.740	6.740	$1s^2\ ^1S_0-1s\ 2s\ ^3S_1$	26
Mg XII	7.102	7.106	$1s\ ^2S_{1/2}-3p\ ^2P_{3/2}$	26
Mg XII	7.105	7.107	$1s\ ^2S_{1/2}-3p\ ^2P_{1/2}$	26
Al XIII	7.170	7.176	$1s\ ^2S_{1/2}-2p\ ^2P_{3/2}$	26

continued.

Ion	Wvl. (Å)	Wvl.^{L} (Å)	Transition	Ref.
Al XIII		7.176	$1s\ ^2S_{1/2}–2p\ ^2P_{1/2}$	
Fe XXIV	7.368	7.370	$1s^2\ 2p\ ^2P_{1/2}–1s^2\ 5d\ ^2D_{3/2}$	18
Fe XXIV	7.454	7.460	$1s^2\ 2p\ ^2P_{3/2}–1s^2\ 5s\ ^2S_{1/2}$	18
Mg XI	7.473	7.473	$1s^2\ ^1S_0–1s\ 4p\ ^1P_1$	18
Al XII	7.757	7.757	$1s^2\ ^1S_0–1s\ 2p\ ^1P_1$	18
Al XII	7.807	7.804	$1s^2\ ^1S_0–1s\ 2p\ ^3P_2$	18
Mg XI	7.850	7.851	$1s^2\ ^1S_0–1s\ 3p\ ^1P_1$	18
Mg XI	7.864	7.863	$1s^2\ ^1S_0–1s\ 3p\ ^3P_1$	33
Al XII	7.871	7.872	$1s^2\ ^1S_0–1s\ 2s\ ^3S_1$	18
Fe XXIII	7.934	7.934	$2s\ 2p\ ^1P_1–2s\ 5s\ ^1S_0$	18
Fe XXIV	7.986	7.984	$1s^2\ 2s\ ^2S_{1/2}–1s^2\ 4p\ ^2P_{3/2}$	18
Fe XXIV	7.996	7.993	$1s^2\ 2s\ ^2S_{1/2}–1s^2\ 4p\ ^2P_{1/2}$	18
Na XI	8.019	8.021	$1s\ ^2S_{1/2}–4p\ ^2P_{3/2}$	18
Mg XI	8.069	8.070	$1s^2\ 2p\ ^2P_{1/2}–1s\ 2p\ 3p\ ^2D_{3/2}$	18
Fe XXII	8.090	8.090	$2s^2\ 2p\ ^2P_{1/2}–2s^2\ 5d\ ^2D_{3/2}$	18
Fe XXIV	8.233	8.233	$1s^2\ 2p\ ^2P_{1/2}–1s^2\ 4d\ ^2D_{3/2}$	18
Fe XXIII	8.304	8.303	$2s^2\ ^1S_0–2s\ 4p\ ^1P_1$	18
Fe XXIV	8.317	8.316	$1s^2\ 2p\ ^2P_{3/2}–1s^2\ 4d\ ^2D_{5/2}$	18
Fe XXIII		8.315	$2s^2\ ^1S_0–2s\ 4p\ ^3P_1$	
Mg XII	8.419	8.419	$1s\ ^2S_{1/2}–2p\ ^2P_{3/2}$	18
Mg XII	8.424	8.425	$1s\ ^2S_{1/2}–2p\ ^2P_{1/2}$	18
Na XI	8.454	8.459	$1s\ ^2S_{1/2}–3p\ ^2P_{3/2}$	33
Na XI		8.460	$1s\ ^2S_{1/2}–3p\ ^2P_{1/2}$	
Mg XI	8.523	8.521	$1s\ 2s\ ^3S_1–2s\ 2p\ ^3P_2$	18
Fe XXIII	8.549	8.550	$2s\ 2p\ ^3P_1–2s\ 4d\ ^3D_2$	18
Fe XXI	8.573	8.573	$2s^2\ 2p^2\ ^3P_0–2s^2\ 2p\ 5d\ ^3D_1$	18
Fe XXIII	8.615	8.614	$2s\ 2p\ ^3P_2–2s\ 4d\ ^3D_3$	18
Fe XXI	8.644	8.643	$2s^2\ 2p^2\ ^3P_2–2s^2\ 2p\ 5d\ ^3F_3$	18
Fe XXI		8.644	$2s^2\ 2p^2\ ^3P_2–2s^2\ 2p\ 5d\ ^3F_2$	
Fe XXII	8.715	8.715	$2s^2\ 2p\ ^2P_{1/2}–2s\ 2p\ 4p\ ^2D_{3/2}$	18
Fe XXII	8.722	8.722	$2s^2\ 2p\ ^2P_{1/2}–2s\ 2p\ 4p\ ^2P_{1/2}$	18
Fe XXII	8.753	8.753	$2s^2\ 2p\ ^2P_{1/2}–2s\ 2p\ 4p\ ^4D_{1/2}$	18
Fe XXIII	8.815	8.814	$2s\ 2p\ ^1P_1–2s\ 4d\ ^1D_2$	18
Fe XXIII	8.907	8.906	$2s\ 2p\ ^1P_1–2s\ 4s\ ^1S_0$	18
Fe XXII	8.976	8.977	$2s^2\ 2p\ ^2P_{1/2}–2s^2\ 4d\ ^2D_{3/2}$	18
Fe XX	9.069	9.069	$2s^2\ 2p^3\ ^4S_{3/2}–2s^2\ 2p^2\ 5d\ ^4P_{3/2}$	18
Fe XX		9.069	$2s^2\ 2p^3\ ^4S_{3/2}–2s^2\ 2p^2\ 5d\ ^4P_{5/2}$	
Fe XXII	9.075	9.074	$2s^2\ 2p\ ^2P_{3/2}–2s^2\ 4d\ ^2D_{3/2}$	18
Fe XXII		9.075	$2s^2\ 2p\ ^2P_{3/2}–2s^2\ 4d\ ^2D_{5/2}$	
Mg XI	9.169	9.169	$1s^2\ ^1S_0–1s\ 2p\ ^1P_1$	23
Mg XI	9.225	9.228	$1s^2\ ^1S_0–1s\ 2p\ ^3P_2$	18
Mg XI	9.233	9.231	$1s^2\ ^1S_0–1s\ 2p\ ^3P_1$	18
Fe XXII	9.253	9.260	$2s\ 2p^2\ ^2D_{3/2}–2s\ 2p\ 4d\ ^2D_{3/2}$	18
Mg XI	9.316	9.314	$1s^2\ ^1S_0–1s\ 2s\ ^3S_1$	18
Fe XXII	9.362	9.362	$2s\ 2p^2\ ^2P_{1/2}–2s\ 2p\ 4d\ ^2D_{3/2}$	18
Fe XXII	9.392	9.392	$2s\ 2p^2\ ^2D_{3/2}–2s\ 2p\ 4s\ ^2P_{1/2}$	18
Fe XXI	9.476	9.475	$2s^2\ 2p^2\ ^3P_0–2s^2\ 2p\ 4d\ ^3D_1$	18
Ne X	9.481	9.481	$1s\ ^2S_{1/2}–5p\ ^2P_{3/2}$	18
Ne X		9.481	$1s\ ^2S_{1/2}–5p\ ^2P_{1/2}$	
Fe XXI	9.547	9.542	$2s^2\ 2p^2\ ^3P_1–2s^2\ 2p\ 4d\ ^3D_1$	18
Fe XXI		9.548	$2s^2\ 2p^2\ ^3P_1–2s^2\ 2p\ 4d\ ^3P_2$	
Fe XIX	9.691	9.691	$2s^2\ 2p^4\ ^3P_2–2s^2\ 2p^3\ 5d\ ^3S_1$	18
Fe XIX		9.691	$2s^2\ 2p^4\ ^3P_2–2s^2\ 2p^3\ 5d\ ^3D_3$	
Fe XXI		9.690	$2s^2\ 2p^2\ ^1S_0–2s^2\ 2p\ 4d\ ^1P_1$	
Fe XXI		9.690	$2s\ 2p^3\ ^3S_1–2s\ 2p^2\ 4d\ ^3D_2$	
Ne X	9.713	9.708	$1s\ ^2S_{1/2}–4p\ ^2P_{3/2}$	18
Ne X		9.708	$1s\ ^2S_{1/2}–4p\ ^2P_{1/2}$	
Fe XX	9.998	9.998	$2s^2\ 2p^3\ ^4S_{3/2}–2s^2\ 2p^2\ 4d\ ^4P_{5/2}$	18
Fe XX		9.998	$2s^2\ 2p^3\ ^4S_{3/2}–2s^2\ 2p^2\ 4d\ ^4P_{3/2}$	
Fe XX		9.998	$2s^2\ 2p^3\ ^4S_{3/2}–2s^2\ 2p^2\ 4d\ ^4P_{1/2}$	
Na XI	10.025	10.023	$1s\ ^2S_{1/2}–2p\ ^2P_{3/2}$	22
Na XI		10.029	$1s\ ^2S_{1/2}–2p\ ^2P_{1/2}$	

Ne X	10.245	10.238	$1s\ ^2S_{1/2}–3p\ ^2P_{3/2}$	23
Ne X		10.240	$1s\ ^2S_{1/2}–3p\ ^2P_{1/2}$	
Fe XIX	10.632	10.632	$2s^2\ 2p^4\ ^3P_0–2s^2\ 2p^3\ 4d\ ^3D_1$	18
Fe XIX		10.632	$2s^2\ 2p^4\ ^3P_2–2s^2\ 2p^3\ 4d\ ^3S_1$	
Fe XIX	10.641	10.641	$2s^2\ 2p^4\ ^3P_2–2s^2\ 2p^3\ 4d\ ^3P_2$	18
Fe XVIII		10.641	$2s^2\ 2p^5\ ^2P_{1/2}–2s^2\ 2p^4\ 5d\ ^2D_{3/2}$	
Fe XIX	10.655	10.655	$2s^2\ 2p^4\ ^3P_2–2s^2\ 2p^3\ 4d\ ^3D_3$	18
Fe XIX	10.684	10.684	$2s^2\ 2p^4\ ^3P_2–2s^2\ 2p^3\ 4d\ ^3F_3$	18
Fe XVII	10.769	10.770	$2s^2\ 2p^6\ ^1S_0–2s^2\ 2p^5\ 6d\ ^1P_1$	18
Fe XIX	10.816	10.816	$2s^2\ 2p^4\ ^3P_2–2s^2\ 2p^3\ 4d\ ^3D_3$	18
Fe XXIII	10.981	10.980	$2s^2\ ^1S_0–2s\ 3p\ ^1P_1$	18
Ne IX	11.000	11.000	$1s^2\ ^1S_0–1s\ 4p\ ^1P_1$	18
Na X		11.003	$1s^2\ ^1S_0–1s\ 2p\ ^1P_1$	
Fe XXIII	11.014	11.018	$2s^2\ ^1S_0–2s\ 3p\ ^3P_1$	18
Fe XVII	11.025	11.023	$2s^2\ 2p^6\ ^1S_0–2s\ 2p^6\ 4p\ ^1P_1$	18
Fe XVII	11.129	11.133	$2s^2\ 2p^6\ ^1S_0–2s^2\ 2p^5\ 5d\ ^1P_1$	18
Fe XVII	11.250	11.253	$2s^2\ 2p^6\ ^1S_0–2s^2\ 2p^5\ 5d\ ^3D_1$	18
Fe XVIII	11.324	11.324	$2s^2\ 2p^5\ ^2P_{3/2}–2s^2\ 2p^4\ 4d\ ^2S_{1/2}$	18
Fe XVIII		11.324	$2s^2\ 2p^5\ ^2P_{3/2}–2s^2\ 2p^4\ 4d\ ^2F_{5/2}$	
Fe XVIII	11.420	11.420	$2s^2\ 2p^5\ ^2P_{3/2}–2s^2\ 2p^4\ 4d\ ^2F_{5/2}$	18
Fe XVII		11.420	$2s^2\ 2p^6\ ^1S_0–2s^2\ 2p^5\ 5s\ ^1P_1$	
Fe XVIII	11.525	11.525	$2s^2\ 2p^5\ ^2P_{3/2}–2s^2\ 2p^4\ 4d\ ^2D_{5/2}$	18
Fe XVIII		11.525	$2s^2\ 2p^5\ ^2P_{3/2}–2s^2\ 2p^4\ 4d\ ^2D_{3/2}$	
Ni XIX	11.539	11.539	$2p^6\ ^1S_0–2s\ 2p^6\ 3p\ ^1P_1$	18
Ne IX	11.545	11.547	$1s^2\ ^1S_0–1s\ 3p\ ^1P_1$	18
Fe XX	11.739	11.739	$2s^2\ 2p^3\ ^4S_{3/2}–2s\ 2p^3\ 3p\ ^4P_{5/2}$	18
Fe XXIII		11.737	$2s\ 2p\ ^1P_1–2s\ 3d\ ^1D_2$	
Fe XXII	11.770	11.797	$2s^2\ 2p\ ^2P_{1/2}–2s^2\ 3d\ ^2D_{3/2}$	18
Fe XVII	12.125		$2s^2\ 2p^6\ ^1S_0–2s^2\ 2p^5\ 4d\ ^1P_1$	18
Ne X	12.134	12.132	$1s\ ^2S_{1/2}–2p\ ^2P_{3/2}$	18
Ne X		12.137	$1s\ ^2S_{1/2}–2p\ ^2P_{1/2}$	
Fe XVII	12.263	12.264	$2s^2\ 2p^6\ ^1S_0–2s^2\ 2p^5\ 4d\ ^3D_1$	18
Fe XXI	12.285	12.282	$2s^2\ 2p^2\ ^3P_0–2s^2\ 2p\ 3d\ ^3D_1$	18
Fe XXI	12.398	12.395	$2s^2\ 2p^2\ ^3P_1–2s^2\ 2p\ 3d\ ^3D_1$	18
Ni XIX	12.429	12.425	$2p^6\ ^1S_0–2p^5\ 3d\ ^1P_1$	18
Fe XXI		12.422	$2s^2\ 2p^2\ ^3P_1–2s^2\ 2p\ 3d\ ^3D_2$	
Ni XIX	12.652	12.656	$2p^6\ ^1S_0–2p^5\ 3d\ ^3D_1$	18
Fe XX	12.812	12.812	$2s^2\ 2p^3\ ^4S_{3/2}–2s^2\ 2p^2\ 3d\ ^4P_{1/2}$	18
Fe XVIII		12.812	$2s^2\ 2p^5\ ^2P_{3/2}–2s\ 2p^5\ 3p\ ^2D_{5/2}$	
Fe XX	12.827	12.827	$2s^2\ 2p^3\ ^4S_{3/2}–2s^2\ 2p^2\ 3d\ ^4P_{3/2}$	18
Fe XX	12.846	12.845	$2s^2\ 2p^3\ ^4S_{3/2}–2s^2\ 2p^2\ 3d\ ^4P_{5/2}$	18
Fe XX	12.951	12.951	$2s^2\ 2p^3\ ^4S_{3/2}–2s^2\ 2p^2\ 3d\ ^4D_{5/2}$	18
Fe XX	13.052	13.052	$2s^2\ 2p^3\ ^4S_{3/2}–2s^2\ 2p^2\ 3d\ ^4F_{5/2}$	18
Fe XX		13.045	$2s^2\ 2p^3\ ^2D_{5/2}–2s^2\ 2p^2\ 3d\ ^2D_{5/2}$	
Fe XX	13.267	13.267	$2s^2\ 2p^3\ ^2D_{5/2}–2s^2\ 2p^2\ 3d\ ^4F_{7/2}$	18
Fe XX		13.267	$2s^2\ 2p^3\ ^2P_{1/2}–2s^2\ 2p^2\ 3d\ ^2P_{1/2}$	
Fe XX		13.269	$2s^2\ 2p^3\ ^2D_{5/2}–2s^2\ 2p^2\ 3d\ ^2P_{3/2}$	
Fe XIX	13.429	13.430	$2s^2\ 2p^4\ ^3P_2–2s^2\ 2p^3\ 3d\ ^1F_3$	18
Ne IX	13.448	13.447	$1s^2\ ^1S_0–1s\ 2p\ ^1P_1$	18
Fe XVIII		13.448	$2s^2\ 2p^5\ ^2P_{3/2}–2s\ 2p^5\ 3p\ ^4D_{5/2}$	
Fe XIX	13.466	13.456	$2s^2\ 2p^4\ ^3P_2–2s^2\ 2p^3\ 3d\ ^3S_1$	18
Fe XIX		13.462	$2s^2\ 2p^4\ ^3P_0–2s^2\ 2p^3\ 3d\ ^3P_1$	
Fe XIX	13.505	13.506	$2s^2\ 2p^4\ ^3P_2–2s^2\ 2p^3\ 3d\ ^3D_2$	18
Fe XX		13.505	$2s\ 2p^4\ ^4P_{3/2}–2s\ 2p^3\ 3d\ ^4D_{3/2}$	
Fe XXI		13.507	$2s\ 2p^3\ ^3D_1–2s\ 2p^2\ 3s\ ^3P_0$	
Fe XIX	13.522	13.525	$2s^2\ 2p^4\ ^3P_2–2s^2\ 2p^3\ 3d\ ^3D_3$	18
Ne IX	13.555	13.553	$1s^2\ ^1S_0–1s\ 2p\ ^3P_1$	18
Fe XIX		13.554	$2s^2\ 2p^4\ ^3P_2–2s^2\ 2p^3\ 3d\ ^1P_1$	
Fe XIX		13.555	$2s^2\ 2p^4\ ^3P_2–2s^2\ 2p^3\ 3d\ ^3P_2$	
Ne IX	13.700	13.699	$1s^2\ ^1S_0–1s\ 2s\ ^3S_1$	18
Ni XIX	13.780	13.779	$2p^6\ ^1S_0–2p^5\ 3s\ ^1P_1$	18
Fe XX		13.781	$2s^2\ 2p^3\ ^4S_{3/2}–2s^2\ 2p^2\ 3s\ ^4P_{5/2}$	

continued.

Ion	Wvl. (Å)	Wvl.^{L} (Å)	Transition	Ref.
Fe XIX	13.799	13.799	$2s^2\,2p^4\,{}^3P_2$–$2s^2\,2p^3\,3d\,{}^3D_3$	18
Fe XVII	13.826	13.823	$2s^2\,2p^6\,{}^1S_0$–$2s\,2p^6\,3p\,{}^1P_1$	18
Ni XIX	14.042	14.043	$2p^6\,{}^1S_0$–$2p^5\,3s\,{}^3P_1$	18
Fe XIX		14.042	$2s\,2p^5\,{}^3P_2$–$2s\,2p^4\,3d\,{}^3F_3,\,{}^3D_2$	
Ni XIX	14.078	14.077	$2p^6\,{}^1S_0$–$2p^5\,3s\,{}^3P_2$	18
Fe XVIII	14.206	14.206	$2s^2\,2p^5\,{}^2P_{3/2}$–$2s^2\,2p^4\,3d\,{}^2D_{5/2}$	18
Fe XVIII	14.206	14.208	$2s^2\,2p^5\,{}^2P_{3/2}$–$2s^2\,2p^4\,3d\,{}^2P_{3/2}$	
Fe XVIII	14.259	14.258	$2s^2\,2p^5\,{}^2P_{3/2}$–$2s^2\,2p^4\,3d\,{}^2S_{1/2}$	18
Fe XVIII		14.259	$2s^2\,2p^5\,{}^2P_{3/2}$–$2s^2\,2p^4\,3d\,{}^2F_{5/2}$	
Fe XVIII	14.362	14.363	$2s^2\,2p^5\,{}^2P_{1/2}$–$2s^2\,2p^4\,3d\,{}^2D_{3/2}$	18
Fe XVIII	14.376	14.376	$2s^2\,2p^5\,{}^2P_{3/2}$–$2s^2\,2p^4\,3d\,{}^2D_{5/2}$	18
Fe XVIII	14.421	14.419	$2s^2\,2p^5\,{}^2P_{1/2}$–$2s^2\,2p^4\,3d\,{}^2P_{3/2}$	18
Fe XVIII		14.421	$2s^2\,2p^5\,{}^2P_{3/2}$–$2s^2\,2p^4\,3d\,{}^4P_{5/2}$	
Fe XVIII	14.536	14.537	$2s^2\,2p^5\,{}^2P_{3/2}$–$2s^2\,2p^4\,3d\,{}^2F_{5/2}$	18
Fe XVIII	14.555	14.549	$2s^2\,2p^5\,{}^2P_{3/2}$–$2s^2\,2p^4\,3d\,{}^4P_{3/2}$	18
Fe XVIII	14.584	14.584	$2s^2\,2p^5\,{}^2P_{3/2}$–$2s^2\,2p^4\,3d\,{}^4P_{1/2}$	18
Fe XIX	14.669	14.669	$2s^2\,2p^4\,{}^3P_2$–$2s^2\,2p^3\,3s\,{}^3D_3$	18
Fe XVIII		14.670	$2s^2\,2p^5\,{}^2P_{1/2}$–$2s^2\,2p^4\,3d\,{}^2D_{3/2}$	
Fe XIX	14.992	14.992	$2s^2\,2p^4\,{}^3P_2$–$2s^2\,2p^3\,3s\,{}^3S_1$	18
Fe XIX		14.992	$2s^2\,2p^4\,{}^1D_2$–$2s^2\,2p^3\,3s\,{}^1D_2$	
Fe XVII	15.014	15.015	$2s^2\,2p^6\,{}^1S_0$–$2s^2\,2p^5\,3d\,{}^1P_1$	18
Fe XX	15.060	15.060	$2s\,2p^4\,{}^4P_{3/2}$–$2s^2\,2p^2\,3p\,{}^4D_{5/2}$	18
Fe XX		15.063	$2s\,2p^4\,{}^4P_{5/2}$–$2s^2\,2p^2\,3p\,{}^4D_{3/2}$	
Fe XIX	15.081	15.081	$2s^2\,2p^4\,{}^3P_2$–$2s^2\,2p^3\,3s\,{}^5S_2$	18
O VIII	15.180	15.176	$1s\,{}^2S_{1/2}$–$4p\,{}^2P_{3/2}$	18
O VIII		15.177	$1s\,{}^2S_{1/2}$–$4p\,{}^2P_{1/2}$	
Fe XIX		15.163	$2s^2\,2p^4\,{}^3P_0$–$2s^2\,2p^3\,3s\,{}^3S_1$	
Fe XIX		15.180	$2s\,2p^5\,{}^3P_1$–$2s\,2p^4\,3s\,{}^3P_0$	
Fe XIX	15.208	15.208	$2s\,2p^5\,{}^3P_2$–$2s\,2p^4\,3s\,{}^3P_2$	18
Fe XVII	15.263	15.262	$2s^2\,2p^6\,{}^1S_0$–$2s^2\,2p^5\,3d\,{}^3D_1$	18
Fe XVII	15.454	15.450	$2s^2\,2p^6\,{}^1S_0$–$2s^2\,2p^5\,3d\,{}^3P_1$	18
Fe XVIII	15.627	15.627	$2s^2\,2p^5\,{}^2P_{3/2}$–$2s^2\,2p^4\,3s\,{}^2D_{5/2}$	18
Fe XVIII	15.830	15.830	$2s^2\,2p^5\,{}^2P_{3/2}$–$2s^2\,2p^4\,3s\,{}^4P_{3/2}$	18
Fe XVIII	15.873	15.873	$2s^2\,2p^5\,{}^2P_{1/2}$–$2s^2\,2p^4\,3s\,{}^2D_{3/2}$	18
Fe XVIII		15.873	$2s^2\,2p^5\,{}^2P_{3/2}$–$2s^2\,2p^4\,3s\,{}^4P_{1/2}$	18
O VIII	16.007	16.006	$1s\,{}^2S_{1/2}$–$3p\,{}^2P_{3/2}$	18
O VIII		16.007	$1s\,{}^2S_{1/2}$–$3p\,{}^2P_{1/2}$	
Fe XVII		16.005	$2s^2\,2p^6\,{}^1S_0$–$2s^2\,2p^5\,3p\,{}^1D_2$	
Fe XVIII		16.007	$2s^2\,2p^5\,{}^2P_{3/2}$–$2s^2\,2p^4\,3s\,{}^2P_{3/2}$	
Fe XVIII	16.034	16.035	$2s^2\,2p^5\,{}^2P_{1/2}$–$2s^2\,2p^4\,3s\,{}^2P_{1/2}$	18
Fe XIX		16.027	$2s\,2p^5\,{}^3P_1$–$2s^2\,2p^3\,3p\,{}^3P_0$	
Fe XVIII	16.076	16.076	$2s^2\,2p^5\,{}^2P_{3/2}$–$2s^2\,2p^4\,3s\,{}^4P_{5/2}$	18
Fe XIX	16.109	16.110	$2s\,2p^5\,{}^3P_2$–$2s^2\,2p^3\,3p\,{}^3P_2$	18
Fe XVIII	16.168	16.141	$2s\,2p^6\,{}^2S_{1/2}$–$2s\,2p^5\,3s\,{}^2P_{3/2}$	18
Fe XVII	16.778	16.777	$2s^2\,2p^6\,{}^1S_0$–$2s^2\,2p^5\,3s\,{}^3P_1$	18
Fe XVII	17.054	17.050	$2s^2\,2p^6\,{}^1S_0$–$2s^2\,2p^5\,3s\,{}^1P_1$	18
Fe XVII	17.099	17.097	$2s^2\,2p^6\,{}^1S_0$–$2s^2\,2p^5\,3s\,{}^3P_2$	18
Fe XVIII	17.624	17.591	$2s\,2p^6\,{}^2S_{1/2}$–$2s^2\,2p^4\,3p\,{}^2P_{3/2}$	18
O VII	18.628	18.627	$1s^2\,{}^1S_0$–$1s\,3p\,{}^1P_1$	18
O VIII	18.972	18.972	$1s\,{}^2S_{1/2}$–$2p\,{}^2P_{3/2}$	18
O VIII		18.973	$1s\,{}^2S_{1/2}$–$2p\,{}^2P_{1/2}$	
N VII	20.913	20.910	$1s\,{}^2S_{1/2}$–$3p\,{}^2P_{3/2}$	21
N VII		20.911	$1s\,{}^2S_{1/2}$–$3p\,{}^2P_{1/2}$	
O VII	21.601	21.602	$1s^2\,{}^1S_0$–$1s\,2p\,{}^1P_1$	18
O VII	21.807	21.804	$1s^2\,{}^1S_0$–$1s\,2p\,{}^3P_2$	18
O VII		21.807	$1s^2\,{}^1S_0$–$1s\,2p\,{}^3P_1$	18
O VII	22.100	22.101	$1s^2\,{}^1S_0$–$1s\,2s\,{}^3S_1$	18
N VII	24.78	24.779	$1s\,{}^2S_{1/2}$–$2p\,{}^2P_{3/2}$	24
N VII		24.785	$1s\,{}^2S_{1/2}$–$2p\,{}^2P_{1/2}$	
Ca XIII	26.03	26.033	$2s^2\,2p^4\,{}^3P_2$–$2s^2\,2p^3\,3d\,{}^3D_3$	1
C VI	28.46	28.465	$1s\,{}^2S_{1/2}$–$3p\,{}^2P_{3/2}$	1
C VI		28.466	$1s\,{}^2S_{1/2}$–$3p\,{}^2P_{1/2}$	

N VI	28.78	28.787	$1s^2\ ^1S_0$–$1s\ 2p\ ^1P_1$	1	Fe XVI	54.72	54.710	$3p\ ^2P_{3/2}$–$4d\ ^2D_{5/2}$	1
N VI	29.07	29.084	$1s^2\ ^1S_0$–$1s\ 2p\ ^3P_1$	1	Mg X	57.88	57.876	$1s^2\ 2s\ ^2S_{1/2}$–$1s^2\ 3p\ ^2P_{3/2}$	1
N VI	29.53	29.534	$1s^2\ ^1S_0$–$1s\ 2s\ ^3S_1$	1	Fe XV	59.40	59.405	$3s\ 3p\ ^1P_1$–$3s\ 4d\ ^1D_2$	1
Ca XI	30.45	30.448	$2p^6\ ^1S_0$–$2p^5\ 3d\ ^1P_1$	1	Fe XVI	62.88	62.872	$3p\ ^2P_{1/2}$–$4s\ ^2S_{1/2}$	1
S XIV		30.469	$1s^2\ 2s\ ^2S_{1/2}$–$1s^2\ 3p\ ^2P_{1/2}$		Mg X	63.30	63.295	$1s^2\ 2p\ ^2P_{3/2}$–$1s^2\ 3d\ ^2D_{3/2,5/2}$	1
Si XII	31.01	31.012	$1s^2\ 2s\ ^2S_{1/2}$–$1s^2\ 4p\ ^2P_{3/2}$	1	Fe XVI	63.72	63.711	$3p\ ^2P_{3/2}$–$4s\ ^2S_{1/2}$	1
Si XII		31.023	$1s^2\ 2s\ ^2S_{1/2}$–$1s^2\ 4p\ ^2P_{1/2}$		Fe XVI	66.25	66.249	$3d\ ^2D_{3/2}$–$4f\ ^2F_{5/2}$	1
S XIV	32.55	32.560	$1s^2\ 2p\ ^2P_{3/2}$–$1s^2\ 3d\ ^2D_{5/2}$	1	Fe XVI	66.36	66.357	$3d\ ^2D_{5/2}$–$4f\ ^2F_{7/2}$	1
S XIV		32.575	$1s^2\ 2p\ ^2P_{3/2}$–$1s^2\ 3d\ ^2D_{3/2}$		Si VIII	69.65	69.632	$2p^3\ ^4S_{3/2}$–$2s^2\ 2p^2\ 3s\ ^4P_{5/2}$	2
Ca XII	32.66	32.657	$2s^2\ 2p^5\ ^2P_{3/2}$–$2s^2\ 2p^4\ 3s\ ^4P_{5/2}$	1	Fe XV	70.05	70.054	$3s\ 3d\ ^3D_3$–$3s\ 4f\ ^3F_{3,4}$	1
Fe XVI		32.652	$3p\ ^2P_{3/2}$–$7d\ ^2D_{5/2}$		Mg IX	72.30	72.312	$2s\ 2p\ ^1P_1$–$2s\ 3d\ ^1D_2$	1
C VI	33.73	33.734	$1s\ ^2S_{1/2}$–$2p\ ^2P_{3/2}$	1	Fe XV	73.47	73.472	$3d\ ^1D_2$–$3s\ 4f\ ^1F_3$	1
C VI		33.740	$1s\ ^2S_{1/2}$–$2p\ ^2P_{1/2}$		Mg VIII	75.03	75.034	$2s^2\ 2p\ ^2P_{3/2}$–$2s^2\ 3d\ ^2D_{5/2}$	1
S XIII	35.67	35.667	$2s\ 2p\ ^1P_1$–$2s\ 3d\ ^1D_2$	1	Fe XVI	76.50	76.497	$3d\ ^2D_{5/2}$–$4p\ ^2P_{3/2}$	1
Fe XVI	36.75	36.749	$3s\ ^2S_{1/2}$–$5p\ ^2P_{3/2}$	1	Fe IX	83.446	83.457	$3s^2\ 3p^6\ ^1S_0$–$3s^2\ 3p^5\ 4d\ ^3P_1$	3
Fe XVI	36.80	36.803	$3s\ ^2S_{1/2}$–$5p\ ^2P_{1/2}$	1	Mg VII	84.024	84.026	$2s^2\ 2p^2\ ^3P_2$–$2s^2\ 2p\ 3d\ ^3D_3$	3
S XII	37.60	37.602	$2s\ 2p^2\ ^2D_{5/2}$–$2s\ 2p\ 3d\ ^2F_{7/2}$	1	Fe XI	86.765	86.772	$3s^2\ 3p^4\ ^3P_2$–$3s^2\ 3p^3\ 4s\ ^3D_3$	3
Fe XVI	40.14	40.153	$3p\ ^2P_{3/2}$–$5d\ ^2D_{5/2}$	1	Fe XI	87.018	87.025	$3s^2\ 3p^4\ ^3P_2$–$3s^2\ 3p^3\ 4s\ ^3D_2$	3
C V	40.27	40.268	$1s^2\ ^1S_0$–$1s\ 2p\ ^1P_1$	1	Ne VIII	88.082	88.082	$1s^2\ 2s\ ^2S_{1/2}$–$1s^2\ 3p\ ^2P_{3/2}$	3
C V	40.72	40.731	$1s^2\ ^1S_0$–$1s\ 2p\ ^3P_1$	1	Ni X	91.59	91.527	$3s^2\ 3p^6\ 3d\ ^2D_{3/2}$–$3s^2\ 3p^6\ 4f\ ^2F_{5/2}$	20
Si XII	40.91	40.911	$1s^2\ 2s\ ^2S_{1/2}$–$1s^2\ 3p\ ^2P_{3/2}$	1	Fe XVIII	93.84	93.923	$2s^2\ 2p^5\ ^2P_{3/2}$–$2s\ 2p^6\ ^2S_{1/2}$	1
Si XI	43.75	43.763	$2s^2\ ^1S_0$–$2s\ 3p\ ^1P_1$	1	Fe X	94.016	94.012	$3s^2\ 3p^5\ ^2P_{3/2}$–$3s^2\ 3p^4\ 4s\ ^2D_{5/2}$	3
Si XII	44.02	44.019	$1s^2\ 2p\ ^2P_{1/2}$–$1s^2\ 3d\ ^2D_{3/2}$	1	Fe X	95.370	95.374	$3s^2\ 3p^5\ ^2P_{1/2}$–$3s^2\ 3p^4\ 4s\ ^2D_{3/2}$	3
Si XII	44.16	44.165	$1s^2\ 2p\ ^2P_{3/2}$–$1s^2\ 3d\ ^2D_{5/2}$	1	Fe X	96.119	96.121	$3s^2\ 3p^5\ ^2P_{3/2}$–$3s^2\ 3p^4\ 4s\ ^2P_{3/2}$	3
Si XII	45.68	45.691	$1s^2\ 2p\ ^2P_{3/2}$–$1s^2\ 3s\ ^2S_{1/2}$	1	Ne VIII	98.263	98.260	$1s^2\ 2p\ ^2P_{3/2}$–$1s^2\ 3d\ ^2D_{5/2}$	3
Fe XVI	46.72	46.718	$3d\ ^2D_{5/2}$–$5f\ ^2F_{7/2}$	1	Fe XIX	101.55	101.550	$2s^2\ 2p^4\ ^3P_2$–$2s\ 2p^5\ ^3P_1$	17
Si XI	49.22	49.222	$2s\ 2p\ ^1P_1$–$2s\ 3d\ ^1D_2$	1	Fe XXI	102.2172	102.217	$2s^2\ 2p^2\ ^3P_2$–$2s\ 2p^3\ ^3S_1$	17
Fe XVI	50.35	50.361	$3s\ ^2S_{1/2}$–$4p\ ^2P_{3/2}$	1	Fe IX	103.564	103.566	$3s^2\ 3p^6\ ^1S_0$–$3s^2\ 3p^5\ 4s\ ^1P_1$	3
Fe XVI	50.56	50.565	$3s\ ^2S_{1/2}$–$4p\ ^2P_{1/2}$	1	Fe XVIII	103.95	103.973	$2s^2\ 2p^5\ ^2P_{1/2}$–$2s\ 2p^6\ ^2S_{1/2}$	19
Si XI	52.30	52.298	$2s\ 2p\ ^1P_1$–$2s\ 3s\ ^1S_0$	1	Fe IX	105.209	105.208	$3s^2\ 3p^6\ ^1S_0$–$3s^2\ 3p^5\ 4s\ ^3P_1$	3
Fe XVI	54.13	54.127	$3p\ ^2P_{1/2}$–$4d\ ^2D_{3/2}$	1	Fe XIX	108.37	108.355	$2s^2\ 2p^4\ ^3P_2$–$2s\ 2p^5\ ^3P_2$	17

continued.

Ion	Wvl. (Å)	Wvl.^{L} (Å)	Transition	Ref.
Fe XIX	109.9519	109.952	$2s^2\,2p^4\,^3P_0$–$2s\,2p^5\,^3P_1$	17
Fe XIX	111.80	111.695	$2s^2\,2p^4\,^3P_1$–$2s\,2p^5\,^3P_1$	17
Fe XX	113.45	113.349	$2s^2\,2p^3\,^2D_{5/2}$–$2s\,2p^4\,^2D_{5/2}$	17
Fe XXII	114.4101	114.410	$2s^2\,2p\,^2P_{3/2}$–$2s\,2p^2\,^2P_{3/2}$	17
Fe XXI	117.18	117.500	$2s^2\,2p^2\,^3P_1$–$2s\,2p^3\,^3P_1$	17
Fe XX	118.70	118.680	$2s^2\,2p^3\,^4S_{3/2}$–$2s\,2p^4\,^4P_{1/2}$	17
Fe XIX	119.92	119.984	$2s^2\,2p^4\,^3P_1$–$2s\,2p^5\,^3P_2$	17
Fe XXI	121.2129	121.213	$2s^2\,2p^2\,^3P_2$–$2s\,2p^3\,^3P_2$	17
Fe VIII	130.943	130.941	$3s^2\,3p^6\,3d\,^2D_{3/2}$–$3p^6\,4f\,^2F_{5/2}$	3
Fe VIII	131.247	131.240	$3s^2\,3p^6\,3d\,^2D_{5/2}$–$3p^6\,4f\,^2F_{7/2}$	3
Fe XXIII	132.83	132.906	$2s^2\,^1S_0$–$2s\,2p\,^1P_1$	17
Fe XX	132.8405	132.840	$2s^2\,2p^3\,^4S_{3/2}$–$2s\,2p^4\,^4P_{5/2}$	17
Fe XXII	135.73	135.791	$2s^2\,2p\,^2P_{1/2}$–$2s\,2p^2\,^2D_{3/2}$	17
Fe XXI	142.1436	142.144	$2s^2\,2p^2\,^3P_1$–$2s\,2p^3\,^3D_2$	17
Ni X	144.212	144.216	$3s^2\,3p^6\,3d\,^2D_{3/2}$–$3s^2\,3p^5\,3d^2\,^2D_{3/2}$	3
Ni X	144.986	144.988	$3s^2\,3p^6\,3d\,^2D_{5/2}$–$3s^2\,3p^5\,3d^2\,^2D_{5/2}$	3
Ca XII	147.274	147.282	$2s^2\,2p^5\,^2P_{1/2}$–$2s\,2p^6\,^2S_{1/2}$	3
Ni XI	148.374	148.377	$3s^2\,3p^6\,^1S_0$–$3s^2\,3p^5\,3d\,^1P_1$	3
O VI	150.089	150.090	$1s^2\,2s\,^2S_{1/2}$–$1s^2\,3p\,^2P_{1/2,3/2}$	3
Ni XII	152.154	152.154	$3s^2\,3p^5\,^2P_{3/2}$–$3s^2\,3p^4\,3d\,^2D_{5/2}$	3
Ni XII	154.179	154.162	$3s^2\,3p^5\,^2P_{3/2}$–$3s^2\,3p^4\,3d\,^2D_{5/2}$	3
Fe XXII	156.0193	156.019	$2s^2\,2p\,^2P_{3/2}$–$2s\,2p^2\,^2D_{5/2}$	17
Ni XIII	157.730	157.729	$3s^2\,3p^4\,^3P_2$–$3s^2\,3p^3\,3d\,^3D_3$	3
Ni XIV	164.146	164.146	$3s^2\,3p^3\,^2D_{5/2}$–$3s^2\,3p^2\,3d\,^2F_{7/2}$	3
Fe VIII	167.495	167.486	$3p^6\,3d\,^2D_{3/2}$–$3p^5\,3d^2\,^2D_{3/2}$	3
Fe VIII	168.176	168.173	$3p^6\,3d\,^2D_{5/2}$–$3p^5\,3d^2\,^2D_{5/2}$	3
Fe VIII	168.548	168.545	$3p^6\,3d\,^2D_{5/2}$–$3p^5\,3d^2\,^2P_{3/2}$	3
Fe VIII	168.933	168.930	$3p^6\,3d\,^2D_{3/2}$–$3p^5\,3d^2\,^2P_{1/2}$	3
Fe IX	171.074	171.073	$3s^2\,3p^6\,^1S_0$–$3s^2\,3p^5\,3d\,^1P_1$	6
O VI	172.934	172.936	$1s^2\,2p\,^2P_{1/2}$–$1s^2\,3d\,^2D_{3/2}$	4
O VI	173.080	173.080	$1s^2\,2p\,^2P_{3/2}$–$1s^2\,3d\,^2D_{5/2}$	4
Fe X	174.531	174.531	$3s^2\,3p^5\,^2P_{3/2}$–$3s^2\,3p^4\,3d\,^2D_{5/2}$	4
Fe X	175.263	175.263	$3s^2\,3p^5\,^2P_{1/2}$–$3s^2\,3p^4\,3d\,^2D_{3/2}$	4
Fe X	177.239	177.240	$3s^2\,3p^5\,^2P_{3/2}$–$3s^2\,3p^4\,3d\,^2P_{3/2}$	4
Fe XI	180.401	180.408	$3s^2\,3p^4\,^3P_2$–$3s^2\,3p^3\,3d\,^3D_3$	4
Fe X		180.441	$3s^2\,3p^5\,^2P_{1/2}$–$3s^2\,3p^4\,3d\,^2P_{1/2}$	
Fe XI	181.131	181.137	$3s^2\,3p^4\,^3P_0$–$3s^2\,3p^3\,3d\,^3D_1$	4
Fe XI	182.167	182.169	$3s^2\,3p^4\,^3P_1$–$3s^2\,3p^3\,3d\,^3D_2$	4
O VI	183.944	183.937	$1s^2\,2p\,^2P_{1/2}$–$1s^2\,3s\,^2S_{1/2}$	4
O VI	184.113	184.117	$1s^2\,2p\,^2P_{3/2}$–$1s^2\,3s\,^2S_{1/2}$	4
Fe X	184.536	184.537	$3s^2\,3p^5\,^2P_{3/2}$–$3s^2\,3p^4\,3d\,^2S_{1/2}$	4
Fe VIII	185.216	185.213	$3p^6\,3d\,^2D_{5/2}$–$3p^5\,3d^2\,^2F_{7/2}$	4
Fe XII	186.880	186.854	$3s^2\,3p^3\,^2D_{3/2}$–$3s^2\,3p^2\,3d\,^2F_{5/2}$	4
Fe XII		186.877	$3s^2\,3p^3\,^2D_{5/2}$–$3s^2\,3p^2\,3d\,^2F_{7/2}$	
Fe XI	188.216	188.232	$3s^2\,3p^4\,^3P_2$–$3s^2\,3p^3\,3d\,^3P_2$	4
Fe XI	188.299	188.299	$3s^2\,3p^4\,^3P_2$–$3s^2\,3p^3\,3d\,^1P_1$	30
Fe XXIV	192.03	192.029	$1s^2\,2s\,^2S_{1/2}$–$1s^2\,2p\,^2P_{3/2}$	27
Fe XII	192.394	192.394	$3s^2\,3p^3\,^4S_{3/2}$–$3s^2\,3p^2\,3d\,^4P_{1/2}$	4
Ca XVII	192.883	192.892	$2s^2\,^1S_0$–$2s\,2p\,^1P_1$	6
Fe XII	193.509	193.509	$3s^2\,3p^3\,^4S_{3/2}$–$3s^2\,3p^2\,3d\,^4P_{3/2}$	4
Fe XII	195.117	195.119	$3s^2\,3p^3\,^4S_{3/2}$–$3s^2\,3p^2\,3d\,^4P_{5/2}$	6
Fe XII		195.179	$3s^2\,3p^3\,^2D_{3/2}$–$3s^2\,3p^2\,3d\,^2D_{3/2}$	
Fe XIII	196.525	196.540	$3s^2\,3p^2\,^1D_2$–$3s^2\,3p\,3d\,^1F_3$	4
Fe XII	196.640	196.640	$3s^2\,3p^3\,^2D_{5/2}$–$3s^2\,3p^2\,3d\,^2D_{5/2}$	4
Fe XIII	200.021	200.022	$3s^2\,3p^2\,^3P_1$–$3s^2\,3p\,3d\,^3D_2$	4
Fe XIII	201.121	201.128	$3s^2\,3p^2\,^3P_1$–$3s^2\,3p\,3d\,^3D_1$	4
Fe XIII	202.044	202.044	$3s^2\,3p^2\,^3P_0$–$3s^2\,3p\,3d\,^3P_1$	4
Fe XIII	203.164	203.164	$3s^2\,3p^2\,^3P_1$–$3s^2\,3p\,3d\,^3P_0$	6
Fe XIII	203.826	203.797	$3s^2\,3p^2\,^3P_2$–$3s^2\,3p\,3d\,^3D_2$	4
Fe XIII		203.828	$3s^2\,3p^2\,^3P_2$–$3s^2\,3p\,3d\,^3D_3$	
Fe XIII	204.263	204.263	$3s^2\,3p^2\,^3P_1$–$3s^2\,3p\,3d\,^1D_2$	4
Fe XVII	204.665	204.665	$2s^2\,2p^5\,3s\,^1P_1$–$2s^2\,2p^5\,3p\,^1S_0$	6

Fe XIII	204.942	204.945	$3s^2\,3p^2\,{}^3P_2-3s^2\,3p\,3d\,{}^3D_1$	4
K XVI	206.258	206.253	$2s^2\,{}^1S_0-2s\,2p\,{}^1P_1$	6
Fe XIII	209.617	209.621	$3s^2\,3p^2\,{}^3P_1-3s^2\,3p\,3d\,{}^3P_2$	4
Fe XIII	209.916	209.919	$3s^2\,3p^2\,{}^3P_2-3s^2\,3p\,3d\,{}^3P_1$	4
Fe XIV	211.316	211.318	$3s^2\,3p\,{}^2P_{1/2}-3s^2\,3d\,{}^2D_{3/2}$	4
Fe XIII	213.770	213.771	$3s^2\,3p^2\,{}^3P_2-3s^2\,3p\,3d\,{}^3P_2$	4
Si VIII	214.757	214.759	$2s^2\,2p^3\,{}^2D_{3/2}-2s\,2p^4\,{}^2P_{1/2}$	6
Si VIII		216.922	$2s^2\,2p^3\,{}^2D_{5/2}-2s\,2p^4\,{}^2P_{3/2}$	
Fe IX	217.100	217.101	$3s^2\,3p^6\,{}^1S_0-3s^2\,3p^5\,3d\,{}^3D_1$	4
S XII	218.179	218.201	$2s^2\,2p\,{}^2P_{3/2}-2s\,2p^2\,{}^2P_{3/2}$	4
Fe IX	218.943	218.937	$3s^2\,3p^6\,{}^1S_0-3s^2\,3p^5\,3d\,{}^1D_2$	6
Fe XIV	219.123	219.131	$3s^2\,3p\,{}^2P_{3/2}-3s^2\,3d\,{}^2D_{5/2}$	4
Fe XIV	220.082	220.085	$3s^2\,3p\,{}^2P_{3/2}-3s^2\,3d\,{}^2D_{3/2}$	4
Fe XIII	221.822	221.827	$3s^2\,3p^2\,{}^1D_2-3s^2\,3p\,3d\,{}^1D_2$	4
Si IX	223.744	223.744	$2s^2\,2p^2\,{}^3P_0-2s\,2p^3\,{}^3S_1$	4
S IX	224.736	224.726	$2s^2\,2p^4\,{}^3P_2-2s\,2p^5\,{}^3P_2$	4
Si IX	225.021	225.025	$2s^2\,2p^2\,{}^3P_1-2s\,2p^3\,{}^3S_1$	4
Si IX	226.998	227.002	$2s^2\,2p^2\,{}^3P_2-2s\,2p^3\,{}^3S_1$	4
Si IX	227.35	227.362	$2s^2\,2p^2\,{}^1D_2-2s\,2p^3\,{}^1P_1$	10
S XII	227.47	227.500	$2s^2\,2p\,{}^2P_{1/2}-2s\,2p^2\,{}^2S_{1/2}$	10
Fe XIII	228.167	228.159	$2s^2\,2p^3\,{}^2D_{3/2}-2s\,2p^4\,{}^2D_{3/2}$	4
S X		228.166	$2s^2\,2p^3\,{}^2D_{3/2}-2s\,2p^4\,{}^2D_{3/2}$	
S X	228.70	228.694	$2s^2\,2p^3\,{}^2D_{5/2}-2s\,2p^4\,{}^2D_{5/2}$	10
He II	231.444	231.454	$1s\,{}^2S_{1/2}-8p\,{}^2P_{1/2,3/2}$	4
He II	232.57	232.584	$1s\,{}^2S_{1/2}-7p\,{}^2P_{1/2,3/2}$	10
He II	234.356	234.347	$1s\,{}^2S_{1/2}-6p\,{}^2P_{1/2,3/2}$	4
Ni XVIII	236.335	236.337	$3p\,{}^2P_{3/2}-3d\,{}^2D_{3/2}$	30
He II	237.333	237.331	$1s\,{}^2S_{1/2}-5p\,{}^2P_{3/2}$	4
O IV	238.624	238.570	$2s^2\,2p\,{}^2P_{3/2}-2s^2\,3d\,{}^2D_{5/2}$	30
O IV		238.579	$2s^2\,2p\,{}^2P_{3/2}-2s^2\,3d\,{}^2D_{3/2}$	
Ni XVI	239.488	239.550	$3s^2\,3p\,{}^2P_{1/2}-3s\,3p^2\,{}^2S_{1/2}$	30
S XI	239.834	239.817	$2s^2\,2p^2\,{}^3P_0-2s\,2p^3\,{}^3P_1$	30
Fe XIII	240.713	240.696	$3s^2\,3p^2\,{}^3P_0-3s\,3p^3\,{}^3S_1$	4
Fe IX	241.739	241.739	$3s^2\,3p^6\,{}^1S_0-3s^2\,3p^5\,3d\,{}^3P_2$	4
He II	243.026	243.027	$1s\,{}^2S_{1/2}-4p\,{}^2P_{3/2}$	4
He II		243.027	$1s\,{}^2S_{1/2}-4p\,{}^2P_{1/2}$	3
Fe XV	243.771	243.794	$3s\,3p\,{}^1P_1-3s\,3d\,{}^1D_2$	6
Fe IX	244.911	244.909	$3s^2\,3p^6\,{}^1S_0-3s^2\,3p^5\,3d\,{}^3P_1$	4
Si VI	246.01	246.005	$2s^2\,2p^5\,{}^2P_{3/2}-2s\,2p^6\,{}^2S_{1/2}$	10
Fe XIII	246.208	246.211	$3s^2\,3p^2\,{}^3P_1-3s\,3p^3\,{}^3S_1$	4
S XI	246.90	246.895	$2s^2\,2p^2\,{}^3P_2-2s\,2p^3\,{}^3P_2$	10
Ni XVII	249.177	249.178	$3s^2\,{}^1S_0-3s\,3p\,{}^1P_1$	4
Fe XVI	251.074	251.063	$3p\,{}^2P_{1/2}-3d\,{}^2D_{3/2}$	4
Fe XIII	251.953	251.956	$3s^2\,3p^2\,{}^3P_2-3s\,3p^3\,{}^3S_1$	4
Fe XIV	252.197	252.201	$3s^2\,3p\,{}^2P_{1/2}-3s\,3p^2\,{}^2P_{3/2}$	4
Fe XXII	253.17	253.170	$2s^2\,2p\,{}^2P_{3/2}-2s\,2p^2\,{}^4P_{5/2}$	10
Si X	253.795	253.788	$2s^2\,2p\,{}^2P_{1/2}-2s\,2p^2\,{}^2P_{3/2}$	4
Fe XXIV	255.11	255.114	$1s^2\,2s\,{}^2S_{1/2}-1s^2\,2p\,{}^2P_{1/2}$	10
He II	256.320	256.317	$1s\,{}^2S_{1/2}-3p\,{}^2P_{3/2}$	4
He II		256.318	$1s\,{}^2S_{1/2}-3p\,{}^2P_{1/2}$	
Si X	256.351	256.366	$2s^2\,2p\,{}^2P_{1/2}-2s\,2p^2\,{}^2P_{1/2}$	4
S XIII	256.686	256.685	$2s^2\,{}^1S_0-2s\,2p\,{}^1P_1$	4
S X	257.136	257.147	$2s^2\,2p^3\,{}^4S_{3/2}-2s\,2p^4\,{}^4P_{1/2}$	4
Fe X	257.246	257.259	$3s^2\,3p^5\,{}^2P_{3/2}-3s^2\,3p^4\,3d\,{}^4D_{5/2}$	6
Fe XIV	257.370	257.395	$3s^2\,3p\,{}^2P_{1/2}-3s\,3p^2\,{}^2P_{1/2}$	6
Si X	258.373	258.371	$2s^2\,2p\,{}^2P_{3/2}-2s\,2p^2\,{}^2P_{3/2}$	4
S X	259.494	259.497	$2s^2\,2p^3\,{}^4S_{3/2}-2s\,2p^4\,{}^4P_{3/2}$	4
Si X	261.056	261.044	$2s^2\,2p\,{}^2P_{3/2}-2s\,2p^2\,{}^2P_{1/2}$	4
Fe XVI	262.984	262.976	$3p\,{}^2P_{3/2}-3d\,{}^2D_{5/2}$	4
Fe XXIII	263.74	263.766	$2s^2\,{}^1S_0-2s\,2p\,{}^3P_1$	27
S X	264.233	264.231	$2s^2\,2p^3\,{}^4S_{3/2}-2s\,2p^4\,{}^4P_{5/2}$	4
Fe XIV	264.787	264.790	$3s^2\,3p\,{}^2P_{3/2}-3s\,3p^2\,{}^2P_{3/2}$	4

continued.

Ion	Wvl. (Å)	Wvl.^{L} (Å)	Transition	Ref.
Fe XIV	270.524	270.522	$3s^2\ 3p\ ^2P_{3/2}-3s\ 3p^2\ ^2P_{1/2}$	4
Si X	271.992	272.006	$2s^2\ 2p\ ^2P_{1/2}-2s\ 2p^2\ ^2S_{1/2}$	4
Fe XIV	274.203	274.204	$3s^2\ 3p\ ^2P_{1/2}-3s\ 3p^2\ ^2S_{1/2}$	4
Si VII	275.368	275.354	$2s^2\ 2p^4\ ^3P_2-2s\ 2p^5\ ^3P_2$	4
Si VIII	276.850	276.850	$2s^2\ 2p^3\ ^2D_{3/2}-2s\ 2p^4\ ^2D_{3/2}$	4
Si VIII		276.865	$2s^2\ 2p^3\ ^2D_{3/2}-2s\ 2p^4\ ^2D_{5/2}$	
Mg VII	277.042	276.993	$2s^2\ 2p^2\ ^3P_1-2s\ 2p^3\ ^3S_1$	4
Si VIII		277.058	$2s^2\ 2p^3\ ^2D_{5/2}-2s\ 2p^4\ ^2D_{3/2,5/2}$	32
Si X	277.265	277.278	$2s^2\ 2p\ ^2P_{3/2}-2s\ 2p^2\ ^2S_{1/2}$	4
Mg VII	278.395	278.393	$2s^2\ 2p^2\ ^3P_2-2s\ 2p^3\ ^3S_1$	4
Si VII		278.444	$2s^2\ 2p^4\ ^3P_1-2s\ 2p^5\ ^3P_2$	31
Fe XV	284.160	284.163	$3s^2\ ^1S_0-3s\ 3p\ ^1P_1$	4
S XI	285.600	285.587	$2s^2\ 2p^2\ ^3P_1-2s\ 2p^3\ ^3D_1$	4
S XI	285.828	285.823	$2s^2\ 2p^2\ ^3P_1-2s\ 2p^3\ ^3D_2$	4
S XII	288.401	288.421	$2s^2\ 2p\ ^2P_{1/2}-2s\ 2p^2\ ^2D_{3/2}$	30
Fe XIV	289.160	289.151	$3s^2\ 3p\ ^2P_{3/2}-3s\ 3p^2\ ^2S_{1/2}$	4
Fe XII	291.007	291.010	$3s^2\ 3p^3\ ^2D_{5/2}-3s\ 3p^4\ ^2P_{3/2}$	30
Ni XVIII	291.988	291.984	$3s\ ^2S_{1/2}-3p\ ^2P_{3/2}$	30
Si IX	292.756	292.759	$2s^2\ 2p^2\ ^3P_1-2s\ 2p^3\ ^3P_2$	6
Si IX	292.858	292.809	$2s^2\ 2p^2\ ^3P_1-2s\ 2p^3\ ^3P_0$	6
Si IX	296.12	296.113	$2s^2\ 2p^2\ ^3P_2-2s\ 2p^3\ ^3P_2$	10
Si IX	296.23	296.211	$2s^2\ 2p^2\ ^3P_2-2s\ 2p^3\ ^3P_1$	10
S XII	299.534	299.541	$2s^2\ 2p\ ^2P_{3/2}-2s\ 2p^2\ ^2D_{5/2}$	30
Si XI	303.324	303.325	$2s^2\ ^1S_0-2s\ 2p\ ^1P_1$	30
He II	303.784	303.781	$1s\ ^2S_{1/2}-2p\ ^2P_{3/2}$	30
He II		303.786	$1s\ ^2S_{1/2}-2p\ ^2P_{1/2}$	
Mn XIV	304.874	304.840	$3s^2\ ^1S_0-3s\ 3p\ ^1P_1$	30
Fe XV		304.894	$3s\ 3p\ ^3P_2-3p^2\ ^3P_2$	
Fe XIII	311.552	311.552	$3s^2\ 3p^2\ ^3P_1-3s\ 3p^3\ ^3P_2$	5
Mg VIII	311.769	311.773	$2s^2\ 2p\ ^2P_{1/2}-2s\ 2p^2\ ^2P_{3/2}$	7
Fe XIII	312.172	312.109	$3s^2\ 3p^2\ ^3P_1-3s\ 3p^3\ ^3P_1$	7
Fe XIII	312.897	312.872	$3s^2\ 3p^2\ ^3P_1-3s\ 3p^3\ ^3P_0$	7
Mg VIII	313.734	313.744	$2s^2\ 2p\ ^2P_{1/2}-2s\ 2p^2\ ^2P_{1/2}$	7
Si VIII	314.358	314.356	$2s^2\ 2p^3\ ^4S_{3/2}-2s\ 2p^4\ ^4P_{1/2}$	7
Mg VIII	315.022	315.016	$2s^2\ 2p\ ^2P_{3/2}-2s\ 2p^2\ ^2P_{3/2}$	7
Si VIII	316.208	316.218	$2s^2\ 2p^3\ ^4S_{3/2}-2s\ 2p^4\ ^4P_{3/2}$	7
Mg VIII	317.017	317.028	$2s^2\ 2p\ ^2P_{3/2}-2s\ 2p^2\ ^2P_{1/2}$	7
Mg VII	319.023	319.019	$2s^2\ 2p^2\ ^1D_2-2s\ 2p^3\ ^1D_2$	30
Si VIII	319.841	319.840	$2s^2\ 2p^3\ ^4S_{3/2}-2s\ 2p^4\ ^4P_{5/2}$	5
Ni XVIII	320.559	320.566	$3s\ ^2S_{1/2}-3p\ ^2P_{1/2}$	5
Fe XIII	320.807	320.809	$3s^2\ 3p^2\ ^3P_2-3s\ 3p^3\ ^3P_2$	5
Fe XV	327.032	327.033	$3s\ 3p\ ^3P_2-3p^2\ ^1D_2$	5
Al X	332.784	332.790	$2s^2\ ^1S_0-2s\ 2p\ ^1P_1$	5
Fe XIV	334.173	334.180	$3s^2\ 3p\ ^2P_{1/2}-3s\ 3p^2\ ^2D_{3/2}$	5
Mg VIII	335.233	335.231	$2s^2\ 2p\ ^2P_{1/2}-2s\ 2p^2\ ^2S_{1/2}$	5
Fe XVI	335.404	335.410	$3s\ ^2S_{1/2}-3p\ ^2P_{3/2}$	7
Fe XXI		335.693	$2s^2\ 2p^2\ ^3P_1-2s^2\ 2p^2\ ^1S_0$	
Mg VIII	338.997	338.984	$2s^2\ 2p\ ^2P_{3/2}-2s\ 2p^2\ ^2S_{1/2}$	7
Fe XI	341.129	341.114	$3s^2\ 3p^4\ ^3P_2-3s\ 3p^5\ ^3P_1$	7
Si IX	341.971	341.951	$2s^2\ 2p^2\ ^3P_0-2s\ 2p^3\ ^3D_1$	7
Si IX	344.987	344.954	$2s^2\ 2p^2\ ^3P_1-2s\ 2p^3\ ^3D_1$	7
Si IX	345.138	345.121	$2s^2\ 2p^2\ ^3P_1-2s\ 2p^3\ ^3D_2$	30
Fe X	345.735	345.738	$3s^2\ 3p^5\ ^2P_{3/2}-3s\ 3p^6\ ^2S_{1/2}$	7
Fe XII	346.849	346.852	$3s^2\ 3p^3\ ^4S_{3/2}-3s\ 3p^4\ ^4P_{1/2}$	7
Si X	347.401	347.409	$2s^2\ 2p\ ^2P_{1/2}-2s\ 2p^2\ ^2D_{3/2}$	4
Fe XIII	348.179	348.183	$3s^2\ 3p^2\ ^3P_0-3s\ 3p^3\ ^3D_1$	7
Fe XI	349.032	349.047	$3s^2\ 3p^4\ ^3P_1-3s\ 3p^5\ ^3P_0$	7
Mg VI	349.149	349.109	$2s^2\ 2p^3\ ^2D_{5/2}-2s\ 2p^4\ ^2D_{3/2}$	7
Mg VI		349.125	$2s^2\ 2p^3\ ^2D_{3/2}-2s\ 2p^4\ ^2D_{3/2}$	
Mg VI		349.164	$2s^2\ 2p^3\ ^2D_{5/2}-2s\ 2p^4\ ^2D_{5/2}$	
Mg VI		349.180	$2s^2\ 2p^3\ ^2D_{3/2}-2s\ 2p^4\ ^2D_{5/2}$	
Si IX	349.871	349.792	$2s^2\ 2p^2\ ^3P_2-2s\ 2p^3\ ^3D_2$	7

Si IX		349.860	$2s^2\,2p^2\,{}^3P_2–2s\,2p^3\,{}^3D_3$	
Fe XII	352.112	352.106	$3s^2\,3p^3\,{}^4S_{3/2}–3s\,3p^4\,{}^4P_{3/2}$	7
Fe XI	352.674	352.662	$3s^2\,3p^4\,{}^3P_2–3s\,3p^5\,{}^3P_2$	7
Fe XIV	353.833	353.837	$3s^2\,3p\,{}^2P_{3/2}–3s\,3p^2\,{}^2D_{5/2}$	30
Si X	356.027	356.030	$2s^2\,2p\,{}^2P_{3/2}–2s\,2p^2\,{}^2D_{5/2}$	5
Si X		356.055	$2s^2\,2p\,{}^2P_{3/2}–2s\,2p^2\,{}^2D_{3/2}$	
Ne V	359.378	359.375	$2s^2\,2p^2\,{}^3P_2–2s\,2p^3\,{}^3S_1$	30
Ca VIII		359.367	$3s^2\,3p\,{}^2P_{3/2}–3s^2\,3d\,{}^2D_{5/2}$	
Fe XIII	359.644	359.642	$3s^2\,3p^2\,{}^3P_1–3s\,3p^3\,{}^3D_2$	30
Fe XIII	359.830	359.842	$3s^2\,3p^2\,{}^3P_1–3s\,3p^3\,{}^3D_1$	30
Fe XVI	360.754	360.759	$3s\,{}^2S_{1/2}–3p\,{}^2P_{1/2}$	30
Mg VII	363.753	363.749	$2s^2\,2p^2\,{}^3P_0–2s\,2p^3\,{}^3P_1$	30
Fe XII	364.468	364.467	$3s^2\,3p^3\,{}^4S_{3/2}–3s\,3p^4\,{}^4P_{5/2}$	30
Mg VII	365.210	365.163	$2s^2\,2p^2\,{}^3P_1–2s\,2p^3\,{}^3P_0$	30
Mg VII		365.221	$2s^2\,2p^2\,{}^3P_1–2s\,2p^3\,{}^3P_1$	
Mg VII		365.235	$2s^2\,2p^2\,{}^3P_1–2s\,2p^3\,{}^3P_2$	
Mg VII	367.675	367.659	$2s^2\,2p^2\,{}^3P_2–2s\,2p^3\,{}^3P_2$	30
Mg VII		367.672	$2s^2\,2p^2\,{}^3P_2–2s\,2p^3\,{}^3P_1$	
Mg IX	368.063	368.070	$2s^2\,{}^1S_0–2s\,2p\,{}^1P_1$	30
Fe XIII	368.163	368.171	$3s^2\,3p^2\,{}^3P_2–3s\,3p^3\,{}^3D_3$	30
Fe XI	369.163	369.154	$3s^2\,3p^4\,{}^3P_1–3s\,3p^5\,{}^3P_2$	30
Fe XV	372.758	372.798	$3s\,3d\,{}^3D_3–3p\,3d\,{}^3F_4$	30
Mn XV	384.745	384.764	$3s\,{}^2S_{1/2}–3p\,{}^2P_{1/2}$	30
Cr XIV	389.854	389.864	$3s\,{}^2S_{1/2}–3p\,{}^2P_{3/2}$	30
Mg VI	399.275	399.383	$2s^2\,2p^3\,{}^4S_{3/2}–2s\,2p^4\,{}^4P_{1/2}$	30
Ne VI	399.837	399.826	$2s^2\,2p\,{}^2P_{1/2}–2s\,2p^2\,{}^2P_{3/2}$	30
Mg VI	400.668	400.663	$2s^2\,2p^3\,{}^4S_{3/2}–2s\,2p^4\,{}^4P_{3/2}$	30
Ne VI	401.139	401.154	$2s^2\,2p\,{}^2P_{1/2}–2s\,2p^2\,{}^2P_{1/2}$	30
Ne VI	401.936	401.928	$2s^2\,2p\,{}^2P_{3/2}–2s\,2p^2\,{}^2P_{3/2}$	30
Mg VI	403.296	403.308	$2s^2\,2p^3\,{}^4S_{3/2}–2s\,2p^4\,{}^4P_{5/2}$	30
Ne VI			$2s^2\,2p\,{}^2P_{3/2}–2s\,2p^2\,{}^2P_{1/2}$	
Na VIII	411.164	411.166	$2s^2\,{}^1S_0–2s\,2p\,{}^1P_1$	30
Cr XIV	412.039	412.052	$3s\,{}^2S_{1/2}–3p\,{}^2P_{1/2}$	30
Fe XIII	412.997	413.032	$3s^2\,3p^2\,{}^1D_2–3s\,3p^3\,{}^3D_3$	30
Ne V	416.208	416.210	$2s^2\,2p^2\,{}^1D_2–2s\,2p^3\,{}^1D_2$	30
Fe XV	417.245	417.258	$3s^2\,{}^1S_0–3s\,3p\,{}^3P_1$	30
S XIV	417.640	417.661	$1s^2\,2s\,{}^2S_{1/2}–1s^2\,2p\,{}^2P_{3/2}$	30
C IV	419.718	419.715	$1s^2\,2p\,{}^2P_{3/2}–1s^2\,3s\,{}^2S_{1/2}$	30
Mg VII	429.132	429.123	$2s^2\,2p^2\,{}^3P_0–2s\,2p^3\,{}^3D_1$	30
Mg VIII	430.445	430.455	$2s^2\,2p\,{}^2P_{1/2}–2s\,2p^2\,{}^2D_{3/2}$	30
Mg VII	431.141	431.191	$2s^2\,2p^2\,{}^3P_1–2s\,2p^3\,{}^3D_1$	30
Mg VII	431.288	431.305	$2s^2\,2p^2\,{}^3P_1–2s\,2p^3\,{}^3D_2$	30
Ne VI	433.161	433.173	$2s^2\,2p\,{}^2P_{1/2}–2s\,2p^2\,{}^2S_{1/2}$	30
Mg VII	434.71	434.708	$2s^2\,2p^2\,{}^3P_2–2s\,2p^3\,{}^3D_2$	10
Mg VII	434.917	434.906	$2s^2\,2p^2\,{}^3P_2–2s\,2p^3\,{}^3D_3$	30
Ne VI	435.632	435.641	$2s^2\,2p\,{}^2P_{3/2}–2s\,2p^2\,{}^2S_{1/2}$	30
Ca VIII	436.089	436.138	$3s^2\,3p\,{}^2P_{3/2}–3s\,3p^2\,{}^2P_{3/2}$	30
Mg VIII	436.726	436.734	$2s^2\,2p\,{}^2P_{3/2}–2s\,2p^2\,{}^2D_{5/2}$	30
Mg IX	439.173	439.176	$2s\,2p\,{}^3P_1–2p^2\,{}^3P_2$	30
Mg IX	441.221	441.199	$2s\,2p\,{}^3P_0–2p^2\,{}^3P_1$	30
Mg IX	443.371	443.403	$2s\,2p\,{}^3P_1–2p^2\,{}^3P_1$	30
Mg IX	443.956	443.973	$2s\,2p\,{}^3P_2–2p^2\,{}^3P_2$	30
S XIV	445.660	445.701	$1s^2\,2s\,{}^2S_{1/2}–1s^2\,2p\,{}^2P_{1/2}$	30
Mg IX	448.279	448.293	$2s\,2p\,{}^3P_2–2p^2\,{}^3P_1$	30
Ne VII	465.219	465.220	$2s^2\,{}^1S_0–2s\,2p\,{}^1P_1$	4
Ca IX	466.233	466.240	$3s^2\,{}^1S_0–3s\,3p\,{}^1P_1$	4
Ne V	480.40	480.408	$2s^2\,2p^2\,{}^3P_0–2s\,2p^3\,{}^3P_1$	10
Ne V	481.28	481.291	$2s^2\,2p^2\,{}^3P_1–2s\,2p^3\,{}^3P_0$	14
Ne V	481.36	481.363	$2s^2\,2p^2\,{}^3P_1–2s\,2p^3\,{}^3P_1$	14
Fe XV	481.46	481.489	$3s\,3p\,{}^1P_1–3p^2\,{}^1D_2$	10
Ne V	482.99	482.985	$2s^2\,2p^2\,{}^3P_2–2s\,2p^3\,{}^3P_1$	10
Ne V		482.997	$2s^2\,2p^2\,{}^3P_2–2s\,2p^3\,{}^3P_2$	

continued.

Ion	Wvl. (Å)	Wvl.^{L} (Å)	Transition	Ref.	Ion	Wvl. (Å)	Wvl.^{L} (Å)	Transition	Ref.
S XIII	491.45	491.464	$2s^2\ ^1S_0$–$2s\ 2p\ ^3P_1$	10	Ne VII	561.725	561.728	$2s\ 2p\ ^3P_2$–$2p^2\ ^3P_2$	10
Si XII	499.405	499.407	$1s^2\ 2s\ ^2S_{1/2}$–$1s^2\ 2p\ ^2P_{3/2}$	4	Ne VI	562.802	562.703	$2s^2\ 2p\ ^2P_{3/2}$–$2s\ 2p^2\ ^2D_{5/2}$	10
O III	507.36	507.388	$2s^2\ 2p^2\ ^3P_0$–$2s\ 2p^3\ ^3S_1$	10	Ne VI		562.798	$2s^2\ 2p\ ^2P_{3/2}$–$2s\ 2p^2\ ^2D_{3/2}$	
O III		507.680	$2s^2\ 2p^2\ ^3P_1$–$2s\ 2p^3\ ^3S_1$		Ne VII	564.568	564.528	$2s\ 2p\ ^3P_2$–$2p^2\ ^3P_1$	8
He I	508.180	507.718	$1s^2\ ^1S_0$–$1s\ 8p\ ^1P_1$	4	Fe XX	567.825	567.867	$2s^2\ 2p^3\ ^4S_{3/2}$–$2s^2\ 2p^3\ ^2D_{5/2}$	27
He I	508.63	508.613	$1s^2\ ^1S_0$–$1s\ 8p\ ^1P_1$	10	Al XI	568.359	568.122	$1s^2\ 2s\ ^2S_{1/2}$–$1s^2\ 2p\ ^2P_{1/2}$	8
He I	510.03	509.998	$1s^2\ ^1S_0$–$1s\ 7p\ ^1P_1$	10	Ne V		568.422	$2s^2\ 2p^2\ ^3P_0$–$2s\ 2p^3\ ^3D_1$	
He I	512.09	512.098	$1s^2\ ^1S_0$–$1s\ 6p\ ^1P_1$	10	Ne V	569.820	569.759	$2s^2\ 2p^2\ ^3P_1$–$2s\ 2p^3\ ^3D_1$	10
He I	515.627	515.617	$1s^2\ ^1S_0$–$1s\ 5p\ ^1P_1$	5	Ne V		569.837	$2s^2\ 2p^2\ ^3P_1$–$2s\ 2p^3\ ^3D_2$	
Si XII	520.685	520.666	$1s^2\ 2s\ ^2S_{1/2}$–$1s^2\ 2p\ ^2P_{1/2}$	5	Ne V	572.140	572.113	$2s^2\ 2p^2\ ^3P_2$–$2s\ 2p^3\ ^3D_2$	5
He I	522.203	522.214	$1s^2\ ^1S_0$–$1s\ 4p\ ^1P_1$	5	Ne V		572.034	$2s^2\ 2p^2\ ^3P_2$–$2s\ 2p^3\ ^3D_1$	
O III	525.801	525.794	$2s^2\ 2p^2\ ^1D_2$–$2s\ 2p^3\ ^1P_1$	5	Ne V	572.37	572.336	$2s^2\ 2p^2\ ^3P_2$–$2s\ 2p^3\ ^3D_3$	5
He I	537.024	537.031	$1s^2\ ^1S_0$–$1s\ 3p\ ^1P_1$	5	Ca X	574.052	574.011	$3s\ ^2S_{1/2}$–$3p\ ^2P_{1/2}$	5
C III	538.012	538.080	$2s\ 2p\ ^3P_0$–$2s\ 3s\ ^3S_1$	5	Si XI	580.905	580.908	$2s^2\ ^1S_0$–$2s\ 2p\ ^3P_1$	5
C III		538.149	$2s\ 2p\ ^3P_1$–$2s\ 3s\ ^3S_1$		O II		580.971	$2s^2\ 2p^3\ ^2P_{3/2}$–$2s\ 2p^4\ ^2P_{3/2}$	
O II		537.833	$2s^2\ 2p^3\ ^2D_{3/2}$–$2s\ 2p^4\ ^2P_{1/2}$		Ca VIII	582.872	582.832	$3s^2\ 3p\ ^2P_{1/2}$–$3s\ 3p^2\ ^2D_{3/2}$	5
C III	538.299	538.312	$2s\ 2p\ ^3P_2$–$2s\ 3s\ ^3S_1$	5	He I	584.340	584.335	$1s^2\ ^1S_0$–$1s\ 2p\ ^1P_1$	5
O II		538.263	$2s^2\ 2p^3\ ^2D_{3/2,5/2}$–$2s\ 2p^4\ ^2P_{3/2}$		Ar VII	585.694	585.756	$3s^2\ ^1S_0$–$3s\ 3p\ ^1P_1$	5
O II	539.091	539.086	$2s^2\ 2p^3\ ^4S_{3/2}$–$2s^2\ 2p^2\ 3s\ ^4P_{5/2}$	5	Fe XIX	592.234	592.236	$2s^2\ 2p^4\ ^3P_2$–$2s^2\ 2p^4\ ^1D_2$	10
O II	539.631	539.549	$2s^2\ 2p^3\ ^4S_{3/2}$–$2s^2\ 2p^2\ 3s\ ^4P_{3/2}$	5	Fe XII	592.501	592.600	$3s\ 3p^4\ ^4P_{5/2}$–$3s^2\ 3p^2\ 3d\ ^4F_{9/2}$	5
Ne IV	541.116	541.127	$2s^2\ 2p^3\ ^4S_{3/2}$–$2s\ 2p^4\ ^4P_{1/2}$	5	Ca VIII	596.885	596.926	$3s^2\ 3p\ ^2P_{3/2}$–$3s\ 3p^2\ ^2D_{5/2}$	5
Ne IV	542.072	542.071	$2s^2\ 2p^3\ ^4S_{3/2}$–$2s\ 2p^4\ ^4P_{3/2}$	5	O III	599.596	599.590	$2s^2\ 2p^2\ ^1D_2$–$2s\ 2p^3\ ^1D_2$	5
Ne IV	543.881	543.887	$2s^2\ 2p^3\ ^4S_{3/2}$–$2s\ 2p^4\ ^4P_{5/2}$	10	Si XI	604.187	604.149	$2s\ 2p\ ^1P_1$–$2p^2\ ^1D_2$	5
Al XI	550.050	550.032	$1s^2\ 2s\ ^2S_{1/2}$–$1s^2\ 2p\ ^2P_{3/2}$	10	O IV	608.312	608.397	$2s^2\ 2p\ ^2P_{1/2}$–$2s\ 2p^2\ ^2S_{1/2}$	5
O IV	553.343	553.329	$2s^2\ 2p\ ^2P_{1/2}$–$2s\ 2p^2\ ^2P_{3/2}$	10	O IV	609.786	609.829	$2s^2\ 2p\ ^2P_{3/2}$–$2s\ 2p^2\ ^2S_{1/2}$	5
O IV	554.076	554.076	$2s^2\ 2p\ ^2P_{1/2}$–$2s\ 2p^2\ ^2P_{1/2}$	10	Mg X		609.794	$1s^2\ 2s\ ^2S_{1/2}$–$1s^2\ 2p\ ^2P_{3/2}$	
O IV	554.512	554.513	$2s^2\ 2p\ ^2P_{3/2}$–$2s\ 2p^2\ ^2P_{3/2}$	10	Si X	624.70	624.779	$2s^2\ 2p\ ^2P_{3/2}$–$2s\ 2p^2\ ^4P_{5/2}$	8
O IV	555.270	555.263	$2s^2\ 2p\ ^2P_{3/2}$–$2s\ 2p^2\ ^2P_{1/2}$	10	Mg X	624.95	624.943	$1s^2\ 2s\ ^2S_{1/2}$–$1s^2\ 2p\ ^2P_{1/2}$	8
Ca X	557.764	557.766	$3s\ ^2S_{1/2}$–$3p\ ^2P_{3/2}$	10	O V	629.735	629.732	$2s^2\ ^1S_0$–$2s\ 2p\ ^1P_1$	5
Ne VI	558.606	558.685	$2s^2\ 2p\ ^2P_{1/2}$–$2s\ 2p^2\ ^2D_{3/2}$	10	Ar VII	637.475	637.472	$3s\ 3p\ ^3P_1$–$3p^2\ ^3P_1$	8
Ne VII	561.396	561.378	$2s\ 2p\ ^3P_1$–$2p^2\ ^3P_1$	8	Al X	637.85	637.850	$2s^2\ ^1S_0$–$2s\ 2p\ ^3P_1$	16

Si X	638.94	639.036	$2s^2\ 2p\ ^2P_{3/2}-2s\ 2p^2\ ^4P_{3/2}$	8
Ca VII	639.18	639.151	$3s^2\ 3p^2\ ^3P_2-3s\ 3p^3\ ^3D_3$	8
O II	644.135	644.155	$2s^2\ 2p^3\ ^2P_{3/2}-2s\ 2p^4\ ^2S_{1/2}$	8
O II		644.163	$2s^2\ 2p^3\ ^2P_{1/2}-2s\ 2p^4\ ^2S_{1/2}$	
Si X	649.30	649.305	$2s^2\ 2p\ ^2P_{3/2}-2s\ 2p^2\ ^4P_{1/2}$	16
S IV	657.335	657.319	$3s^2\ 3p\ ^2P_{1/2}-3s^2\ 3d\ ^2D_{3/2}$	8
S IV	661.442	661.396	$3s^2\ 3p\ ^2P_{3/2}-3s^2\ 3d\ ^2D_{5/2}$	8
S IV		661.455	$3s^2\ 3p\ ^2P_{3/2}-3s^2\ 3d\ ^2D_{3/2}$	
S V	663.20	663.128	$3s\ 3p\ ^3P_2-3s\ 3d\ ^3D_3$	8
S V		663.167	$3s\ 3p\ ^3P_2-3s\ 3d\ ^3D_2$	
Al IX	670.02	670.054	$2s\ 2p\ ^1P_1-2p^2\ ^1D_2$	8
N II	671.39	671.387	$2s^2\ 2p^2\ ^3P_2-2s^2\ 2p\ 3s\ ^3P_2$	8
N II		671.412	$2s^2\ 2p^2\ ^3P_0-2s^2\ 2p\ 3s\ ^3P_1$	
N II	671.77	671.774	$2s^2\ 2p^2\ ^3P_1-2s^2\ 2p\ 3s\ ^3P_0$	8
N II		671.631	$2s^2\ 2p^2\ ^3P_1-2s^2\ 2p\ 3s\ ^3P_1$	
N II	672.01	672.002	$2s^2\ 2p^2\ ^3P_2-2s^2\ 2p\ 3s\ ^3P_1$	8
Si IX	676.49	676.710	$2s^2\ 2p^2\ ^3P_1-2s\ 2p^3\ ^5S_2$	8
Al IX	680.39	680.283	$2s^2\ 2p\ ^2P_{1/2}-2s\ 2p^2\ ^4P_{3/2}$	8
S III	680.70	680.677	$3s^2\ 3p^2\ ^3P_2-3s^2\ 3p\ 3d\ ^3D_3$	8
S III	680.97	680.925	$3s^2\ 3p^2\ ^3P_2-3s^2\ 3p\ 3d\ ^3D_2$	8
S III		680.974	$3s^2\ 3p^2\ ^3P_1-3s^2\ 3p\ 4s\ ^3P_2$	
S III	681.51	681.489	$3s^2\ 3p^2\ ^3P_0-3s^2\ 3p\ 4s\ ^3P_1$	8
S III		681.578	$3s^2\ 3p^2\ ^3P_2-3s^2\ 3p\ 3d\ ^3D_1$	
Na IX	681.68	681.720	$1s^2\ 2s\ ^2S_{1/2}-1s^2\ 2p\ ^2P_{3/2}$	8
S III	683.53	683.590	$3s^2\ 3p^2\ ^1D_2-3s^2\ 3p\ 3d\ ^1F_3$	8
N III	684.99	684.998	$2s^2\ 2p\ ^2P_{1/2}-2s\ 2p^2\ ^2P_{3/2}$	8
N III	685.50	685.515	$2s^2\ 2p\ ^2P_{1/2}-2s\ 2p^2\ ^2P_{1/2}$	8
N III	685.79	685.817	$2s^2\ 2p\ ^2P_{3/2}-2s\ 2p^2\ ^2P_{3/2}$	8
N III	686.33	686.336	$2s^2\ 2p\ ^2P_{3/2}-2s\ 2p^2\ ^2P_{1/2}$	8
C II	687.05	687.053	$2s^2\ 2p\ ^2P_{1/2}-2s^2\ 3d\ ^2D_{3/2}$	8
C II	687.35	687.345	$2s^2\ 2p\ ^2P_{3/2}-2s^2\ 3d\ ^2D_{5/2}$	8
C III	690.53	690.521	$2s\ 2p\ ^1P_1-2s\ 3s\ ^1S_0$	8
N III	691.40	691.397	$2s\ 2p^2\ ^2D_{3/2}-2s^2\ 3p\ ^2P_{1/2}$	8
Ca IX	691.40	691.206	$3s^2\ ^1S_0-3s\ 3p\ ^3P_1$	8
Mg IX	693.98	694.005	$2s^2\ ^1S_0-2s\ 2p\ ^3P_2$	8
Na IX	694.13	694.147	$1s^2\ 2s\ ^2S_{1/2}-1s^2\ 2p\ ^2P_{1/2}$	8
Si IX	694.70	694.904	$2s^2\ 2p^2\ ^3P_2-2s\ 2p^3\ ^5S_2$	8
S V	696.61	696.624	$3s\ 3p\ ^1P_1-3s\ 3d\ ^1D_2$	8
Fe VIII	697.14	697.157	$3p^6\ 4p\ ^2P_{1/2}-3p^6\ 4d\ ^2D_{3/2}$	8
S III	700.29	700.287	$3s^2\ 3p^2\ ^3P_1-3s^2\ 3p\ 3d\ ^3P_0$	8
Ar VIII		700.246	$3s\ ^2S_{1/2}-3p\ ^2P_{3/2}$	
O III	702.33	702.337	$2s^2\ 2p^2\ ^3P_0-2s\ 2p^3\ ^3P_1$	8
O III	702.89	702.838	$2s^2\ 2p^2\ ^3P_1-2s\ 2p^3\ ^3P_0$	8
O III		702.896	$2s^2\ 2p^2\ ^3P_1-2s\ 2p^3\ ^3P_1$	13
O III		702.900	$2s^2\ 2p^2\ ^3P_1-2s\ 2p^3\ ^3P_2$	
O III	783.87	703.851	$2s^2\ 2p^2\ ^3P_2-2s\ 2p^3\ ^3P_1$	8
O III		703.854	$2s^2\ 2p^2\ ^3P_2-2s\ 2p^3\ ^3P_2$	
Mg IX	706.02	706.060	$2s^2\ ^1S_0-2s\ 2p\ ^3P_1$	8
S VI	706.50	706.471	$3p\ ^2P_{1/2}-3d\ ^2D_{3/2}$	8
S VI	712.68	712.672	$3p\ ^2P_{3/2}-3d\ ^2D_{5/2}$	8
Ar VIII	713.80	13.813	$3s\ ^2S_{1/2}-3p\ ^2P_{1/2}$	8
O II	718.49	718.506	$2s^2\ 2p^3\ ^2D_{5/2}-2s\ 2p^4\ ^2D_{5/2}$	8
O II		718.465	$2s^2\ 2p^3\ ^2D_{5/2}-2s\ 2p^4\ ^2D_{3/2}$	
O II		718.568	$2s^2\ 2p^3\ ^2D_{3/2}-2s\ 2p^4\ ^2D_{3/2}$	
O II		718.610	$2s^2\ 2p^3\ ^2D_{3/2}-2s\ 2p^4\ ^2D_{5/2}$	
Fe VIII	721.23	721.257	$3p^6\ 4p\ ^2P_{3/2}-3p^6\ 4d\ ^2D_{5/2}$	8
Ne I	735.86	735.896	$2s^2\ 2p^6\ ^1S_0-2s^2\ 2p^5\ 3s\ ^1P_1$	8
Ne I	743.71	743.719	$2s^2\ 2p^6\ ^1S_0-2s^2\ 2p^5\ 3s\ ^3P_1$	8
S IV	744.91	744.904	$3s^2\ 3p\ ^2P_{1/2}-3s\ 3p^2\ ^2P_{3/2}$	8
Fe VII	745.38	745.380	$3d\ 4p\ ^3F_2-3d\ 4d\ ^3F_2$	8
N II	745.84	745.842	$2s^2\ 2p^2\ ^1S_0-2s\ 2p^3\ ^1P_1$	8
N II	746.99	746.986	$2s^2\ 2p^2\ ^1D_2-2s^2\ 2p\ 3s\ ^1P_1$	8

continued.

Ion	Wvl. (Å)	Wvl.^{L} (Å)	Transition	Ref.
S IV	748.40	748.393	$3s^2\ 3p\ ^2P_{1/2}-3s\ 3p^2\ ^2P_{1/2}$	8
Mg IX	749.54	749.551	$2s\ 2p\ ^1P_1-2p^2\ ^1D_2$	8
S IV	750.22	750.221	$3s^2\ 3p\ ^2P_{3/2}-3s\ 3p^2\ ^2P_{3/2}$	8
S IV	753.74	753.760	$3s^2\ 3p\ ^2P_{3/2}-3s\ 3p^2\ ^2P_{1/2}$	8
O V	758.68	758.677	$2s\ 2p\ ^3P_1-2p^2\ ^3P_2$	8
O V	759.43	759.442	$2s\ 2p\ ^3P_0-2p^2\ ^3P_1$	8
O V	760.21	760.227	$2s\ 2p\ ^3P_1-2p^2\ ^3P_1$	8
O V	760.43	760.446	$2s\ 2p\ ^3P_2-2p^2\ ^3P_2$	8
O V	761.13	761.128	$2s\ 2p\ ^3P_1-2p^2\ ^3P_0$	8
O V	761.99	762.004	$2s\ 2p\ ^3P_2-2p^2\ ^3P_1$	8
Mg VIII	762.65	762.662	$2s^2\ 2p\ ^2P_{1/2}-2s\ 2p^2\ ^4P_{3/2}$	8
N III	763.33	763.334	$2s^2\ 2p\ ^2P_{1/2}-2s\ 2p^2\ ^2S_{1/2}$	8
N III	764.36	764.351	$2s^2\ 2p\ ^2P_{3/2}-2s\ 2p^2\ ^2S_{1/2}$	8
N IV	765.15	765.148	$2s^2\ ^1S_0-2s\ 2p\ ^1P_1$	8
Mg VIII	769.38	769.356	$2s^2\ 2p\ ^2P_{1/2}-2s\ 2p^2\ ^4P_{1/2}$	8
Ne VIII	770.42	770.410	$1s^2\ 2s\ ^2S_{1/2}-1s^2\ 2p\ ^2P_{3/2}$	8
Mg VIII	772.31	772.262	$2s^2\ 2p\ ^2P_{3/2}-2s\ 2p^2\ ^4P_{5/2}$	8
N III	772.89	772.889	$2s\ 2p^2\ ^2D_{5/2}-2p^3\ ^2P_{3/2}$	8
N III		772.955	$2s\ 2p^2\ ^2D_{3/2}-2p^3\ ^2P_{1/2}$	
O V	774.51	774.518	$2s\ 2p\ ^1P_1-2p^2\ ^1S_0$	8
N II	775.95	775.967	$2s^2\ 2p^2\ ^1D_2-2s\ 2p^3\ ^1D_2$	8
S X	776.23	776.375	$2s^2\ 2p^3\ ^4S_{3/2}-2s^2\ 2p^3\ ^2P_{3/2}$	8
O IV	779.82	779.820	$2s\ 2p^2\ ^2D_{3/2}-2p^3\ ^2D_{3/2}$	8
O IV		779.736	$2s\ 2p^2\ ^2D_{5/2}-2p^3\ ^2D_{3/2}$	
O IV	779.91	779.912	$2s\ 2p^2\ ^2D_{5/2}-2p^3\ ^2D_{5/2}$	8
O IV		779.997	$2s\ 2p^2\ ^2D_{3/2}-2p^3\ ^2D_{5/2}$	
Ne VIII	780.30	780.325	$1s^2\ 2s\ ^2S_{1/2}-1s^2\ 2p\ ^2P_{1/2}$	8
Mg VIII	782.34	782.364	$2s^2\ 2p\ ^2P_{3/2}-2s\ 2p^2\ ^4P_{3/2}$	8
S XI	783.01	782.958	$2s^2\ 2p^2\ ^3P_1-2s^2\ 2p^2\ ^1S_0$	8
S V	786.47	786.470	$3s^2\ ^1S_0-3s\ 3p\ ^1P_1$	8
S X	787.43	787.558	$2s^2\ 2p^3\ ^4S_{3/2}-2s^2\ 2p^3\ ^2P_{1/2}$	16

Ion	Wvl. (Å)	Wvl.^{L} (Å)	Transition	Ref.
O IV	787.72	787.710	$2s^2\ 2p\ ^2P_{1/2}-2s\ 2p^2\ ^2D_{3/2}$	8
Mg VIII	789.43	789.411	$2s^2\ 2p\ ^2P_{3/2}-2s\ 2p^2\ ^4P_{1/2}$	8
Na VIII	789.78	789.785	$2s^2\ ^1S_0-2s\ 2p\ ^3P_1$	8
O IV	790.19	790.199	$2s^2\ 2p\ ^2P_{3/2}-2s\ 2p^2\ ^2D_{5/2}$	8
O IV		790.112	$2s^2\ 2p\ ^2P_{3/2}-2s\ 2p^2\ ^2D_{3/2}$	
O II	796.66	796.684	$2s^2\ 2p^3\ ^2P_{3/2}-2s\ 2p^4\ ^2D_{5/2}$	8
S X	804.14	804.119	$2s^2\ 2p^2\ 3p\ ^4D_{7/2}-2s^2\ 2p^2\ 3d\ ^4F_{9/2}$	8
Fe VII		804.160	$4p\ ^3F_4-3d\ 4d\ ^3G_5$	
C II	806.57	806.533	$2s\ 2p^2\ ^4P_{1/2}-2s\ 2p\ 3s\ ^4P_{3/2}$	8
C II		806.568	$2s\ 2p^2\ ^4P_{5/2}-2s\ 2p\ 3s\ ^4P_{5/2}$	
C II		806.677	$2s\ 2p^2\ ^4P_{3/2}-2s\ 2p\ 3s\ ^4P_{3/2}$	
C II	806.84	806.830	$2s\ 2p^2\ ^4P_{3/2}-2s\ 2p\ 3s\ ^4P_{1/2}$	8
C II		806.861	$2s\ 2p^2\ ^4P_{5/2}-2s\ 2p\ 3s\ ^4P_{3/2}$	
S IV	809.66	809.656	$3s^2\ 3p\ ^2P_{1/2}-3s\ 3p^2\ ^2S_{1/2}$	8
C II		809.676	$2s\ 2p^5\ ^2D_{5/2}-2s\ 2p\ 3d\ ^2D_{5/2}$	
Si IV	815.05	815.055	$3p\ ^2P_{1/2}-4s\ ^2S_{1/2}$	8
S IV	815.95	815.941	$3s^2\ 3p\ ^2P_{3/2}-3s\ 3p^2\ ^2S_{1/2}$	8
Si IV	818.15	818.130	$3p\ ^2P_{3/2}-4s\ ^2S_{1/2}$	8
Ca IX	821.23	821.269	$3s\ 3p\ ^1P_1-3p^2\ ^1D_2$	8
O II	832.75	832.760	$2s^2\ 2p^3\ ^4S_{3/2}-2s\ 2p^4\ ^4P_{1/2}$	8
O III	832.94	832.929	$2s^2\ 2p^2\ ^3P_0-2s\ 2p^3\ ^3D_1$	8
O II	833.32	833.332	$2s^2\ 2p^3\ ^4S_{3/2}-2s\ 2p^4\ ^4P_{3/2}$	8
O III	833.74	833.749	$2s^2\ 2p^2\ ^3P_1-2s\ 2p^3\ ^3D_2$	8
O III		833.715	$2s^2\ 2p^2\ ^3P_1-2s\ 2p^3\ ^3D_1$	
O II	834.45	834.467	$2s^2\ 2p^3\ ^4S_{3/2}-2s\ 2p^4\ ^4P_{5/2}$	8
O III	835.09	835.092	$2s^2\ 2p^2\ ^3P_2-2s\ 2p^3\ ^3D_2$	8
O III		835.059	$2s^2\ 2p^2\ ^3P_2-2s\ 2p^3\ ^3D_1$	
O III	835.28	835.289	$2s^2\ 2p^2\ ^3P_2-2s\ 2p^3\ ^3D_3$	8
S V	849.29	849.240	$3s\ 3p\ ^3P_1-3p^2\ ^3P_2$	8
S V	852.17	852.178	$3s\ 3p\ ^3P_0-3p^2\ ^3P_1$	8
Mg VII	854.71	854.724	$2s^2\ 2p^2\ ^3P_1-2s\ 2p^3\ ^5S_2$	8

S V	854.844	854.870	$3s\,3p\,^3P_1$–$3p^2\,^3P_1$	25	O I		922.460	$2s^2\,2p^4\,^1D_2$–$2s^2 2p^3\,3d\,^1P_1$	
C II	858.04	858.093	$2s^2\,2p\,^2P_{1/2}$–$2s^2\,3s\,^2S_{1/2}$	8	H I	923.155	923.150	$1s\,^2S_{1/2}$–$9p\,^2P_{3/2}$	25
C II	858.53	858.561	$2s^2\,2p\,^2P_{3/2}$–$2s^2\,3s\,^2S_{1/2}$	8	N IV		923.221	$2s\,2p\,^3P_2$–$2p^2\,^3P_2$	13
Mg VII	868.13	868.193	$2s^2\,2p^2\,^3P_2$–$2s\,2p^3\,^5S_2$	8	O I		923.200	$2s^2\,2p^4\,^3P_1$–$2s^2\,2p^3\,9d\,^3D_2$	
S IX	871.605	871.726	$2s^2\,2p^4\,^3P_1$–$2s^2\,2p^4\,^1S_0$	25	N IV	923.700	923.677	$2s\,2p\,^3P_1$–$2p^2\,^3P_0$	25
C II	903.949	903.963	$2s^2\,2p\,^2P_{1/2}$–$2s\,2p^2\,^2P_{1/2}$	25	O I		923.790	$2s^2\,2p^4\,^3P_0$–$2s^2\,2p^3\,9d\,^3D_1$	
C II	904.135	904.143	$2s^2\,2p\,^2P_{3/2}$–$2s\,2p^2\,^2P_{3/2}$	25	N IV	924.266	924.286	$2s\,2p\,^3P_2$–$2p^2\,^3P_1$	25
C II	904.46	904.482	$2s^2\,2p\,^2P_{3/2}$–$2s\,2p^2\,^2P_{1/2}$	8	H I	926.23	926.230	$1s\,^2S_{1/2}$–$8p\,^2P_{3/2}$	8
Si VII	905.02	904.789	$2s^2\,2p^3\,3p\,^5P_2$–$2s^2\,2p^3\,3d\,^5D_2$	8	He II	927.883	927.850	$2p\,^2P_{3/2}$–$15d\,^2D_{5/2}$	25
Si VII		904.936	$2s^2\,2p^3\,3p\,^5P_2$–$2s^2\,2p^3\,3d\,^5D_3$		O I	929.515	929.520	$2s^2\,2p^4\,^3P_2$–$2s^2\,2p^3\,7d\,^3D_3$	25
H I	914.023	914.023	$1s\,^2S_{1/2}$–$20p\,^2P_{3/2}$	25	He II	930.314	930.020	$2p\,^2P_{3/2}$–$14d\,^2D_{5/2}$	25
H I	914.272	914.290	$1s\,^2S_{1/2}$–$19p\,^2P_{3/2}$	25	H I	930.768	930.750	$1s\,^2S_{1/2}$–$7p\,^2P_{3/2}$	25
H I	914.534	914.580	$1s\,^2S_{1/2}$–$18p\,^2P_{3/2}$	25	S VI	933.416	933.380	$3s\,^2S_{1/2}$–$3p\,^2P_{3/2}$	25
H I	914.869	914.920	$1s\,^2S_{1/2}$–$17p\,^2P_{3/2}$	25	H I	937.830	937.800	$1s\,^2S_{1/2}$–$6p\,^2P_{3/2}$	25
H I	915.262	915.330	$1s\,^2S_{1/2}$–$16p\,^2P_{3/2}$	25	S VI	944.541	944.524	$3s\,^2S_{1/2}$–$3p\,^2P_{1/2}$	25
N II	915.678	915.613	$2s^2\,2p^2\,^3P_0$–$2s\,2p^3\,^3P_1$	25	He II	949.342	949.360	$2p\,^2P_{3/2}$–$10d\,^2D_{5/2}$	25
H I	915.897	915.820	$1s\,^2S_{1/2}$–$15p\,^2P_{3/2}$	25	Si VIII		949.354	$2s^2\,2p^3\,^4S_{3/2}$–$2s^2\,2p^3\,^2P_{1/2}$	
N II	916.005	915.964	$2s^2\,2p^2\,^3P_1$–$2s\,2p^3\,^3P_0$	25	H I	949.754	949.745	$1s\,^2S_{1/2}$–$5p\,^2P_{3/2,1/2}$	25
N II		916.013	$2s^2\,2p^2\,^3P_1$–$2s\,2p^3\,^3P_2$		Si IX	950.110	950.157	$2s^2\,2p^2\,^3P_1$–$2s^2\,2p^2\,^1S_0$	25
N II		916.022	$2s^2\,2p^2\,^3P_1$–$2s\,2p^3\,^3P_1$		He II	958.705	958.700	$2p\,^2P_{3/2}$–$9d\,^2D_{5/2}$	25
H I	916.352	916.430	$1s\,^2S_{1/2}$–$14p\,^2P_{3/2}$	25	O I	971.775	971.720	$2s^2\,2p^4\,^3P_2$–$2s^2\,2p^3\,4d\,^3D_3$	25
N II	916.703	916.703	$2s^2\,2p^2\,^3P_2$–$2s\,2p^3\,^3P_2$	25	He II	972.144	972.120	$2p\,^2P_{3/2}$–$8d\,^2D_{5/2}$	25
N II		916.711	$2s^2\,2p^2\,^3P_2$–$2s\,2p^3\,^3P_1$		H I	972.450	972.538	$1s\,^2S_{1/2}$–$4p\,^2P_{1/2,3/2}$	25
H I	917.179	917.180	$1s\,^2S_{1/2}$–$13p\,^2P_{3/2}$	25	O I	973.254	973.240	$2s^2\,2p^4\,^3P_1$–$2s^2\,2p^3\,4d\,^3D_2$	25
H I	918.129	918.130	$1s\,^2S_{1/2}$–$12p\,^2P_{3/2}$	25	Ne VII	973.35	973.330	$2s\,2p\,^1P_1$–$2p^2\,^1D_2$	8
H I	919.351	919.350	$1s\,^2S_{1/2}$–$11p\,^2P_{3/2}$	25	O I	973.893	973.890	$2s^2\,2p^4\,^3P_0$–$2s^2\,2p^3\,4d\,^3D_1$	25
H I	920.954	920.970	$1s\,^2S_{1/2}$–$10p\,^2P_{3/2}$	25	O I	976.460	976.440	$2s^2\,2p^4\,^3P_2$–$2s^2\,2p^3\,5s\,^3S_1$	25
N IV	921.905	921.995	$2s\,2p\,^3P_1$–$2p^2\,^3P_2$	25	C III	977.042	977.020	$2s^2\,^1S_0$–$2s\,2p\,^1P_1$	25
O I		922.010	$2s^3\,2p^4\,^1D_2$–$2s^2 2p^3\,3d\,^1F_3$		O I	977.937	977.960	$2s^2\,2p^4\,^3P_1$–$2s^2\,2p^3\,5s\,^3S_1$	25
O I	922.010	922.070	$2s^2\,2p^4\,^1D_2$–$2s^2 2p^3\,3d\,^1D_2$	25	O I	978.585	978.620	$2s^2\,2p^4\,^3P_0$–$2s^2\,2p^3\,5s\,^3S_1$	25
N IV	922.520	922.520	$2s\,2p\,^3P_0$–$2p^2\,^3P_1$	25	Fe III	981.331	981.400	$3d^5\,4s\,^5P_2$–$3d^5\,5p\,^5P_2$	25

continued.

Ion	Wvl. (Å)	Wvl.^{L} (Å)	Transition	Ref.
O I	988.755	988.750	$2s^2\ 2p^4\ {}^3P_2-2s^2\ 2p^3\ 3s\ {}^3D_3$	25
N III	989.843	989.799	$2s^2\ 2p\ {}^2P_{1/2}-2s\ 2p^2\ {}^2D_{3/2}$	25
O I	990.191	990.190	$2s^2\ 2p^4\ {}^3P_1-2s^2\ 2p^3\ 3s\ {}^3D_2$	25
O I		990.190	$2s^2\ 2p^4\ {}^3P_1-2s^2\ 2p^3\ 3s\ {}^3D_1$	
O I	990.799	990.790	$2s^2\ 2p^4\ {}^3P_0-2s^2\ 2p^3\ 3s\ {}^3D_1$	25
Fe III	991.290	991.230	$3d^6\ {}^3F_4-3d^5\ 4p\ {}^3D_3$	25
N III	991.591	991.511	$2s^2\ 2p\ {}^2P_{3/2}-2s\ 2p^2\ {}^2D_{3/2}$	25
Fe III	991.83	991.830	$3d^6\ {}^1G_4-3d^5\ 4p\ {}^1H_5$	8
He II	992.330	992.360	$2p\ {}^2P_{3/2}-7d\ {}^2D_{5/2}$	25
Si II	992.647	992.680	$3s^2\ 3p\ {}^2P_{3/2}-3s^2\ 4d\ {}^2D_{5/2}$	25
Ne VI		992.636	$2s^2\ 2p\ {}^2P_{1/2}-2s\ 2p^2\ {}^4P_{3/2}$	
Fe III	994.753	994.770	$3d^6\ {}^3G_3-3d^5\ 4p\ {}^3F_2$	25
S II	995.974	996.009	$3s^2\ 3p^3\ {}^2D_{5/2}-3s^2\ 3p^2\ ({}^3P)\ 3d\ {}^2F_{7/2}$	25
Fe III	997.065	997.090	$3d^6\ {}^3P_2-3d^5\ 4p\ {}^3P_2$	25
Ne VI	997.18	997.031	$2s^2\ 2p\ {}^2P_{1/2}-2s\ 2p^2\ {}^4P_{1/2}$	8
Si III	997.363	997.389	$3s\ 3p\ {}^3P_2-3s\ 4s\ {}^3S_1$	25
Fe III	997.516	997.640	$3d^6\ {}^3F_3-3d^5\ 4p\ {}^3G_4$	25
Ne VI	999.238	999.183	$2s^2\ 2p\ {}^2P_{3/2}-2s\ 2p^2\ {}^4P_{5/2}$	25
Fe III	999.443	999.380	$3d^6\ {}^3F_2-3d^5\ 4p\ {}^3G_3$	25
O I		999.500	$2s^2\ 2p^4\ {}^1D_2-2s^2\ 2p^3\ 3s\ {}^1P_1$	
Ar VI	1000.149	1000.180	$3s^2\ 3p\ {}^2P_{3/2}-3s\ 3p^3\ {}^4P_{5/2}$	25
S II	1000.452	1000.488	$3s^2\ 3p^3\ {}^2D_{3/2}-3s^2\ 3p^2\ 3d\ {}^2F_{5/2}$	25
Ne VI	1005.79	1005.694	$2s^2\ 2p\ {}^2P_{3/2}-2s\ 2p^2\ {}^4P_{3/2}$	8
C II	1010.363	1010.373	$2s\ 2p^2\ {}^4P_{5/2}-2p^3\ {}^4S_{3/2}$	25
Ne VI		1010.205	$2s^2\ 2p\ {}^2P_{3/2}-2s\ 2p^2\ {}^4P_{1/2}$	
S III	1012.398	1012.494	$3s^2\ 3p^2\ {}^3P_0-3s\ 3p^3\ {}^3P_1$	25
Fe III		1012.410	$3d^6\ {}^3P_0-3d^5\ 4p\ {}^3P_1$	
S III	1015.520	1015.498	$3s^2\ 3p^2\ {}^3P_1-3s\ 3p^3\ {}^3P_0$	25
S III		1015.563	$3s^2\ 3p^2\ {}^3P_1-3s\ 3p^3\ {}^3P_1$	
S III	1015.771	1015.777	$3s^2\ 3p^2\ {}^3P_1-3s\ 3p^3\ {}^3P_2$	25
Fe III	1017.251	1017.240	$3d^6\ {}^3H_6-3d^5\ 4p\ {}^3H_6$	25

Ion	Wvl. (Å)	Wvl.^{L} (Å)	Transition	Ref.
Fe III	1017.744	1017.750	$3d^6\ {}^3H_5-3d^5\ 4p\ {}^3H_5$	25
S III	1021.133	1021.107	$3s^2\ 3p^2\ {}^3P_2-3s\ 3p^3\ {}^3P_1$	25
S III	1021.340	1021.323	$3s^2\ 3p^2\ {}^3P_2-3s\ 3p^3\ {}^3P_2$	25
Fe III	1024.20	1024.200	$3d^6\ {}^3D_1-3d^5\ 4p\ {}^3P_0$	8
He II	1025.200	1025.246	$2s\ {}^2S_{1/2}-6p\ {}^2P_{3/2}$	25
He II		1025.302	$2s\ {}^2P_{3/2}-6d\ {}^2D_{5/2}$	
H I	1025.745	1025.724	$1s\ {}^2S_{1/2}-3p\ {}^2P_{3/2,1/2}$	25
O I		1025.762	$2s^2\ 2p^4\ {}^3P_2-2s^2\ 2p^3\ 3d\ {}^3D_3$	
O I	1027.431	1027.431	$2s^2\ 2p^4\ {}^3P_1-2s^2\ 2p^3\ 3d\ {}^3D_2$	25
O I	1028.131	1028.150	$2s^2\ 2p^4\ {}^3P_0-2s^2\ 2p^3\ 3d\ {}^3D_1$	25
S II	1030.900	1030.888	$3s^2\ 3p^3\ {}^2P_{1/2}-3s^2\ 3p^2\ 4s\ {}^2D_{3/2}$	8
Fe III		1030.920	$3d^6\ {}^3G_4-3d^5\ 4p\ {}^3G_4$	
S II	1031.380	1031.371	$3s^2\ 3p^3\ {}^2P_{3/2}-3s^2\ 3p^2\ 4s\ {}^2D_{5/2}$	25
O VI	1031.959	1031.914	$1s^2\ 2s\ {}^2S_{1/2}-1s^2\ 2p\ {}^2P_{3/2}$	25
Fe III	1033.30	1033.300	$3d^6\ {}^3G_3-3d^5\ 4p\ {}^3G_3$	8
N I		1033.420	$2s^2\ 2p^3\ {}^2D_{5/2}-2s^2\ 2p^2\ 9d\ {}^2D_{5/2}$	
Fe III	1035.808	1035.780	$3d^6\ {}^3F_3-3d^5\ 4p\ {}^3F_3$	25
C II	1036.358	1036.339	$2s^2\ 2p\ {}^2P_{1/2}-2s\ 2p^2\ {}^2S_{1/2}$	25
C II	1037.029	1037.020	$2s^2\ 2p\ {}^2P_{3/2}-2s\ 2p^2\ {}^2S_{1/2}$	25
O VI	1037.641	1037.615	$1s^2\ 2s\ {}^2S_{1/2}-1s^2\ 2p\ {}^2P_{1/2}$	25
Fe III	1038.332	1038.360	$3d^6\ {}^3F_2-3d^5\ 4p\ {}^3F_2$	25
O I	1039.245	1039.230	$2s^2\ 2p^4\ {}^3P_2-2s^2\ 2p^3\ 4s\ {}^3S_1$	25
O I	1040.943	1040.942	$2s^2\ 2p^4\ {}^3P_1-2s^2\ 2p^3\ 4s\ {}^3S_1$	25
O I	1041.673	1041.690	$2s^2\ 2p^4\ {}^3P_0-2s^2\ 2p^3\ 4s\ {}^3S_1$	25
Si VII	1049.200	1049.199	$2s^2\ 2p^4\ {}^3P_1-2s^2\ 2p^4\ {}^1S_0$	25
Al VII	1053.938	1053.998	$2s^2\ 2p^3\ {}^4S_{3/2}-2s^2\ 2p^3\ {}^2P_{3/2}$	25
Ar XII	1054.720	1054.687	$2s^2\ 2p^3\ {}^4S_{3/2}-2s^2\ 2p^3\ {}^2D_{3/2}$	25
Al VII	1056.887	1056.917	$2s^2\ 2p^3\ {}^4S_{3/2}-2s^2\ 2p^3\ {}^2P_{1/2}$	25
Al VIII	1057.875	1057.890	$2p^2\ {}^3P_1-2p^2\ {}^1S_0$	25
S IV	1062.754	1062.664	$3s^2\ 3p\ {}^2P_{1/2}-3s\ 3p^2\ {}^2D_{3/2}$	25
Fe III	1063.984	1063.870	$3d^6\ {}^3D_3-3d^5\ 4p\ {}^3D_3$	25

Fe III	1064.643	1064.650	$3d^6\ {}^3F_2-3d^5\ 4p\ {}^3G_3$	25
Si IV	1066.647	1066.616	$3d\ {}^2D_{5/2}-4f\ {}^2F_{7/2}$	25
Si IV		1066.638	$3d\ {}^2D_{5/2}-4f\ {}^2F_{5/2}$	
Si IV		1066.652	$3d\ {}^2D_{3/2}-4f\ {}^2F_{5/2}$	
O IV	1067.83	1067.830	$2s^2\ 3d\ {}^2D_{5/2}-2s^2\ 4f\ {}^2F_{7/2}$	8
S IV	1072.99	1072.974	$3s^2\ 3p\ {}^2P_{3/2}-3s\ 3p^2\ {}^2D_{5/2}$	25
S IV	1073.500	1073.518	$3s^2\ 3p\ {}^2P_{3/2}-3s\ 3p^2\ {}^2D_{3/2}$	25
Fe III	1074.97	1074.970	$3d^6\ {}^3G_3-3d^5\ 4p\ {}^3F_2$	8
S III	1077.169	1077.173	$3s^2\ 3p^2\ {}^1D_2-3s^2\ 3p\ 3d\ {}^1D_2$	25
N II	1084.000	1083.992	$2s^2\ 2p^2\ {}^3P_0-2s\ 2p^3\ {}^3D_1$	25
N II	1084.585	1084.582	$2s^2\ 2p^2\ {}^3P_1-2s\ 2p^3\ {}^3D_2$	25
N II		1084.564	$2s^2\ 2p^2\ {}^3P_1-2s\ 2p^3\ {}^3D_1$	
He II	1084.949	1084.910	$2p\ {}^2P_{1/2}-5d\ {}^2D_{3/2}$	25
N II	1085.542	1085.548	$2s^2\ 2p^2\ {}^3P_2-2s\ 2p^3\ {}^3D_2$	25
N II		1085.531	$2s^2\ 2p^2\ {}^3P_2-2s\ 2p^3\ {}^3D_1$	
N II	1085.725	1085.703	$2s^2\ 2p^2\ {}^3P_2-2s\ 2p^3\ {}^3D_3$	25
C II	1092.726	1092.730	$2s\ 2p^2\ {}^2P_{3/2}-2s\ 2p\ 3d\ {}^2P_{3/2}$	25
C I	1104.186	1104.160	$2s^2\ 2p^2\ {}^3P_2-2s^2\ 2p\ 20d\ {}^1P_1$	25
C I	1104.968	1104.960	$2s^2\ 2p^2\ {}^3P_2-2s^2\ 2p\ 18d\ {}^1P_1$	25
C I	1106.071	1106.070	$2s^2\ 2p^2\ {}^3P_2-2s^2\ 2p\ 16d\ {}^1P_1$	25
C I	1106.504	1106.500	$2s^2\ 2p^2\ {}^3P_1-2s^2\ 2p\ 15d\ {}^1P_1$	25
C I	1106.809	1106.800	$2s^2\ 2p^2\ {}^3P_2-2s^2\ 2p\ 15d\ {}^1P_1$	25
Si III	1108.37	1108.361	$3s\ 3p\ {}^3P_0-3s\ 3d\ {}^3D_1$	8
S IV		1108.451	$3s\ 3p^2\ {}^2S_{1/2}-3s^2\ 4p\ {}^2P_{3/2}$	
C I	1108.430	1108.440	$2s^2\ 2p^2\ {}^3P_1-2s^2\ 2p\ 13d\ {}^1P_1$	25
C I	1109.041	1109.030	$2s^2\ 2p^2\ {}^3P_0-2s^2\ 2p\ 13d\ {}^3D_1$	25
C I	1109.234	1109.240	$2s^2\ 2p^2\ {}^3P_1-2s^2\ 2p\ 13d\ {}^3D_1$	25
Si III	1109.975	1109.972	$3s\ 3p\ {}^3P_1-3s\ 3d\ {}^3D_2$	25
Si III		1109.943	$3s\ 3p\ {}^3P_1-3s\ 3d\ {}^3D_1$	
C I	1111.992	1111.960	$2s^2\ 2p^2\ {}^3P_2-2s^2\ 2p\ 11d\ {}^1P_1$	25
C I		1111.990	$2s^2\ 2p^2\ {}^3P_2-2s^2\ 2p\ 11d\ {}^1F_3$	
Si III	1113.256	1113.232	$3s\ 3p\ {}^3P_2-3s\ 3d\ {}^3D_3$	25
Si III		1113.206	$3s\ 3p\ {}^3P_2-3s\ 3d\ {}^3D_2$	
C I	1114.383	1114.390	$2s^2\ 2p^2\ {}^3P_2-2s^2\ 2p\ 10d\ {}^1F_3$	25
C I	1117.203	1117.200	$2s^2\ 2p^2\ {}^3P_1-2s^2\ 2p\ 9d\ {}^1P_1$	25
C I	1117.649	1117.580	$2s^2\ 2p^2\ {}^3P_2-2s^2\ 2p\ 9d\ {}^3P_2$	25
C I		1117.720	$2s^2\ 2p^2\ {}^3P_2-2s^2\ 2p\ 9d\ {}^3D_3$	
H_2	1119.10	1119.100	1 - 3 Q3 (C-X)	8
C I	1122.335	1122.330	$2s^2\ 2p^2\ {}^3P_2-2s^2\ 2p\ 8d\ {}^3D_3$	25
Si IV	1122.537	1122.487	$3p\ {}^2P_{1/2}-3d\ {}^2D_{3/2}$	25
Fe III	1128.103	1128.060	$3d^6\ {}^5D_3-3d^5\ 4p\ {}^5P_3$	25
Si IV	1128.379	1128.342	$3p\ {}^2P_{3/2}-3d\ {}^2D_{5/2}$	25
Si IV		1128.327	$3p\ {}^2P_{3/2}-3d\ {}^2D_{3/2}$	
Fe III	1128.768	1128.720	$3d^6\ {}^5D_2-3d^5\ 4p\ {}^5P_2$	25
Fe III	1129.169	1129.190	$3d^6\ {}^5D_1-3d^5\ 4p\ {}^5P_1$	25
Ca XIII	1133.671	1133.758	$2s^2\ 2p^4\ {}^3P_2-2s^2\ 2p^4\ {}^1D_2$	8
C I	1138.942	1138.950	$2s^2\ 2p^2\ {}^3P_2-2s^2\ 2p\ 6d\ {}^3P_1$	25
C II		1138.936	$2s\ 2p^2\ {}^2P_{1/2}-2s\ 2p\ 3d\ {}^2D_{3/2}$	
C I	1139.807	1139.810	$2s^2\ 2p^2\ {}^3P_2-2s^2\ 2p\ 6d\ {}^3D_3$	25
C I		1139.770	$2s^2\ 2p^2\ {}^3P_1-2s^2\ 2p\ 7s\ {}^3P_2$	
Ne V	1145.582	1145.596	$2s^2\ 2p^2\ {}^3P_2-2s\ 2p^3\ {}^5S_2$	25
O I	1152.15	1152.152	$2s^2\ 2p^4\ {}^1D_2-2s^2\ 2p^3\ 3s\ {}^1D_2$	25
C I	1155.829	1155.809	$2s^2\ 2p^2\ {}^3P_0-2s^2\ 2p\ 5d\ {}^3P_1$	25
C I	1156.016	1155.979	$2s^2\ 2p^2\ {}^3P_1-2s^2\ 2p\ 5d\ {}^3P_1$	25
S I		1156.000	$3s^2\ 3p^4\ {}^3P_2-3s^2\ 3p^3\ 3d\ {}^3P_2$	
S I	1156.273	1156.260	$3s^2\ 3p^4\ {}^3P_2-3s^2\ 3p^3\ 3d\ {}^3P_1$	25
C I	1156.542	1156.560	$2s^2\ 2p^2\ {}^3P_2-2s^2\ 2p\ 5d\ {}^3P_2$	25
C I	1157.376	1157.400	$2s^2\ 2p^2\ {}^3P_1-2s^2\ 2p\ 6s\ {}^1P_1$	25
C I	1157.788	1157.790	$2s^2\ 2p^2\ {}^3P_1-2s^2\ 2p\ 5d\ {}^3D_2$	25
C I	1158.031	1157.910	$2s^2\ 2p^2\ {}^3P_0-2s^2\ 2p\ 5d\ {}^3D_1$	25
C I		1158.019	$2s^2\ 2p^2\ {}^3P_2-2s^2\ 2p\ 5d\ {}^3D_3$	
C I		1158.130	$2s^2\ 2p^2\ {}^3P_2-2s^2\ 2p\ 5d\ {}^3D_2$	

continued.

Ion	Wvl. (Å)	Wvl.^{L} (Å)	Transition	Ref.
C I	1158.415	1158.320	$2s^2\,2p^2\,{}^3P_0-2s^2\,2p\,6s\,{}^3P_1$	25
C I		1158.400	$2s^2\,2p^2\,{}^3P_2-2s^2\,2p\,6s\,{}^3P_2$	
C I	1158.709	1158.670	$2s^2\,2p^2\,{}^3P_1-2s^2\,2p\,6s\,{}^3P_0$	25
C I		1158.730	$2s^2\,2p^2\,{}^3P_1-2s^2\,2p\,5d\,{}^3F_2$	
C I	1158.933	1158.970	$2s^2\,2p^2\,{}^3P_2-2s^2\,2p\,5d\,{}^3F_3$	25
S I	1161.350	1161.340	$3s^2\,3p^4\,{}^3P_1-3s^2\,3p^3\,3d\,{}^3P_2$	25
S I	1161.583	1161.595	$3s^2\,3p^4\,{}^3P_1-3s^2\,3p^3\,3d\,{}^3P_1$	25
S I	1161.750	1161.749	$3s^2\,3p^4\,{}^3P_1-3s^2\,3p^3\,3d\,{}^3P_0$	25
H_2	1163.838	1163.850	1 - 4 Q3 (C-X)	25
N I	1164.004	1163.884	$2s^2\,2p^3\,{}^2D_{5/2}-2s^2\,2p^2\,3d\,{}^2D_{5/2}$	25
S I		1163.980	$3s^2\,3p^4\,{}^3P_0-3s^2\,3p^3\,3d\,{}^3P_1$	
N I		1164.000	$2s^2\,2p^3\,{}^2D_{3/2}-2s^2\,2p^2\,3d\,{}^2D_{5/2}$	
N I	1167.451	1167.448	$2s^2\,2p^3\,{}^2D_{5/2}-2s^2\,2p^2\,3d\,{}^2F_{7/2}$	25
Si VII	1167.763	1167.775	$2s^2\,2p^3\,3s\,{}^3S_1-2s^2\,2p^3\,3p\,{}^3P_2$	25
S I	1168.041	1168.030	$3s^2\,3p^4\,{}^3P_2-3s^2\,3p^3\,{}^1D_2$	25
N I	1168.307	1168.334	$2s^2\,2p^3\,{}^2D_{3/2}-2s^2\,2p^2\,3d\,{}^4P_{5/2}$	25
N I		1168.536	$2s^2\,2p^3\,{}^2D_{3/2}-2s^2\,2p^2\,3d\,{}^2F_{5/2}$	
Fe II		1168.550	$3d^6\,{}^4H_{11/2}-3d^5\,4s\,4p\,{}^2H_{9/2}$	
C III	1174.933	1174.933	$2s\,2p\,{}^3P_1-2p^2\,{}^3P_2$	28
C III	1175.268	1175.263	$2s\,2p\,{}^3P_0-2p^2\,{}^3P_1$	28
C III	1175.60	1175.590	$2s\,2p\,{}^3P_1-2p^2\,{}^3P_1$	28
C III	1175.716	1175.711	$2s\,2p\,{}^3P_2-2p^2\,{}^3P_2$	28
C III	1175.986	1175.987	$2s\,2p\,{}^3P_1-2p^2\,{}^3P_0$	28
C III	1176.375	1176.370	$2s\,2p\,{}^3P_2-2p^2\,{}^3P_1$	28
Fe II	1180.426	1180.430	$3d^6\,4s\,{}^4D_{3/2}-3d^6\,4p\,{}^4F_{5/2}$	28
Fe II	1183.826	1183.829	$3d^6\,4s\,{}^4D_{7/2}-3d^5\,4s\,4p\,{}^4F_{9/2}$	28
Si VIII	1184.040	1183.995	$2s^2\,2p^2\,3s\,{}^4P_{3/2}-2s^2\,2p^2\,3p\,{}^4D_{5/2}$	25
Fe II	1187.442	1187.420	$3d^6\,4s\,{}^4D_{7/2}-3d^5\,4s\,4p\,{}^2G_{9/2}$	25
C I	1188.835	1188.833	$2s^2\,2p^2\,{}^3P_0-2s^2\,2p\,4d\,{}^3P_1$	28
C I	1188.995	1188.992	$2s^2\,2p^2\,{}^3P_1-2s^2\,2p\,4d\,{}^3P_0$	28
C I	1189.060	1189.065	$2s^2\,2p^2\,{}^3P_1-2s^2\,2p\,4d\,{}^3P_1$	28
C I	1189.255	1189.249	$2s^2\,2p^2\,{}^3P_1-2s^2\,2p\,4d\,{}^3P_2$	25
C I	1189.630	1189.631	$2s^2\,2p^2\,{}^3P_2-2s^2\,2p\,4d\,{}^3P_2$	28
Mg VII	1189.83	1189.841	$2s^2\,2p^2\,{}^3P_1-2s^2\,2p^2\,{}^1S_0$	28
S III	1190.17	1190.199	$3s^2\,3p^2\,{}^3P_0-3s\,3p^3\,{}^3D_1$	28
Si II	1190.432	1190.417	$3s^2\,3p\,{}^2P_{1/2}-3s\,3p^2\,{}^2P_{3/2}$	25
Mg VI	1191.63	1191.670	$2s^2\,2p^3\,{}^4S_{3/2}-2s^2\,2p^3\,{}^2P_{1/2}$	28
C I	1191.834	1191.838	$2s^2\,2p^2\,{}^3P_2-2s^2\,2p\,4d\,{}^1F_3$	34
Fe II	1192.025	1192.030	$3d^6\,4s\,{}^4D_{5/2}-3d^6\,4p\,{}^4P_{5/2}$	28
C I	1193.02	1193.009	$2s^2\,2p^2\,{}^3P_1-2s^2\,2p\,4d\,{}^3D_2$	25
C I	1193.29	1193.264	$2s^2\,2p^2\,{}^3P_1-2s^2\,2p\,4d\,{}^3D_1$	28
Si II		1193.289	$3s^2\,3p\,{}^2P_{1/2}-3s\,3p^2\,{}^2P_{1/2}$	
S III	1194.026	1194.049	$3s^2\,3p^2\,{}^3P_1-3s\,3p^3\,{}^3D_2$	28
C I	1194.065	1194.064	$2s^2\,2p^2\,{}^3P_2-2s^2\,2p\,5s\,{}^3P_2$	28
S III	1194.39	1194.443	$3s^2\,3p^2\,{}^3P_1-3s\,3p^3\,{}^3D_1$	28
C I		1194.406	$2s^2\,2p^2\,{}^3P_1-2s^2\,2p\,5s\,{}^3P_0$	
Si II	1194.50	1194.501	$3s^2\,3p\,{}^2P_{3/2}-3s\,3p^2\,{}^2P_{3/2}$	28
Si II	1197.394	1197.395	$3s^2\,3p\,{}^2P_{3/2}-3s\,3p^2\,{}^2P_{1/2}$	28
S V	1199.197	1199.136	$3s^2\,{}^1S_0-3s\,3p\,{}^3P_1$	25
Mn II	1199.37	1199.380	$3d^5\,4s\,{}^7S_3-3d^4\,4s\,4p\,{}^7P_3$	28
N I	1199.549	1199.552	$2s^2\,2p^3\,{}^4S_{3/2}-2s^2\,2p^2\,3s\,{}^4P_{5/2}$	28
N I	1200.219	1200.226	$2s^2\,2p^3\,{}^4S_{3/2}-2s^2\,2p^2\,3s\,{}^4P_{3/2}$	28
N I	1200.712	1200.712	$2s^2\,2p^3\,{}^4S_{3/2}-2s^2\,2p^2\,3s\,{}^4P_{1/2}$	28
S III	1200.972	1200.961	$3s^2\,3p^2\,{}^3P_2-3s\,3p^3\,{}^3D_3$	25
Si III	1206.511	1206.502	$3s^2\,{}^1S_0-3s\,3p\,{}^1P_1$	28
S I	1207.00	1207.015	$3s^2\,3p^4\,{}^3P_2-3s^2\,3p^3\,13d\,{}^3D_3$	28
H_2	1208.93	1208.940	$1\ C^1\Pi_u-5Q3\ X^1\Sigma_g^+$	28
Si III	1210.454	1210.460	$3p^2\,{}^1D_2-3s\,4f\,{}^1F_3$	28
H I	1215.67	1215.670	$1s\,{}^2S_{1/2}-2p\,{}^2P_{3/2}$	28
H I		1215.676	$1s\,{}^2S_{1/2}-2p\,{}^2P_{1/2}$	
O I	1217.65	1217.648	$2s^2\,2p^4\,{}^1S_0-2s^2\,2p^3\,3s\,{}^1P_1$	28
O V	1218.351	1218.344	$2s^2\,{}^1S_0-2s\,2p\,{}^3P_1$	28

N I	1228.41	1228.414	$2s^2\ 2p^3\ {}^2P_{3/2}-2s^2\ 2p^2\ 4d\ {}^2P_{1/2}$	28
N V	1238.824	1238.823	$1s^2\ 2s\ {}^2S_{1/2}-1s^2\ 2p\ {}^2P_{3/2}$	28
Fe XII	1242.00	1242.005	$3s^2\ 3p^3\ {}^4S_{3/2}-3s^2\ 3p^3\ {}^2P_{3/2}$	28
N V	1242.803	1242.806	$1s^2\ 2s\ {}^2S_{1/2}-1s^2\ 2p\ {}^2P_{1/2}$	28
C I	1245.943	1245.940	$2s^2\ 2p^2\ {}^1D_2-2s^2\ 2p\ 16d\ {}^1F_3$	28
C I	1246.863	1246.860	$2s^2\ 2p^2\ {}^1D_2-2s^2\ 2p\ 15d\ {}^1F_3$	28
S I	1247.160	1247.160	$3s^2\ 3p^4\ {}^3P_2-3s^2\ 3p^3\ 8s\ {}^3D_3$	28
C III	1247.384	1247.383	$2s\ 2p\ {}^1P_1-2p^2\ {}^1S_0$	28
C I	1247.862	1247.870	$2s^2\ 2p^2\ {}^1D_2-2s^2\ 2p\ 15d\ {}^3F_3$	28
C I	1248.003	1248.000	$2s^2\ 2p^2\ {}^1D_2-2s^2\ 2p\ 14d\ {}^1F_3$	28
C I	1249.002	1249.000	$2s^2\ 2p^2\ {}^1D_2-2s^2\ 2p\ 14d\ {}^3F_3$	28
C I	1249.404	1249.400	$2s^2\ 2p^2\ {}^1D_2-2s^2\ 2p\ 13d\ {}^1F_3$	28
Si II	1250.433	1250.430	$3s\ 3p^2\ {}^2D_{5/2}-3p^3\ {}^2D_{5/2}$	28
C I		1250.420	$2s^2\ 2p^2\ {}^1D_2-2s^2\ 2p\ 13d\ {}^3F_3$	
S II	1250.50	1250.587	$3s^2\ 3p^3\ {}^4S_{3/2}-3s\ 3p^4\ {}^4P_{1/2}$	28
Si II	1251.162	1251.160	$3s\ 3p^2\ {}^4P_{5/2}-3p^3\ {}^4S_{3/2}$	28
C I	1252.209	1252.200	$2s^2\ 2p^2\ {}^1D_2-2s^2\ 2p\ 12d\ {}^3F_3$	28
C I	1253.467	1253.467	$2s^2\ 2p^2\ {}^1D_2-2s^2\ 2p\ 11d\ {}^1F_3$	28
C I	1254.510	1254.510	$2s^2\ 2p^2\ {}^1D_2-2s^2\ 2p\ 11d\ {}^3F_3$	28
Si I	1256.493	1256.490	$3s^2\ 3p^2\ {}^3P_1-3s\ 3p^3\ {}^3S_1$	28
C I	1257.565	1257.580	$2s^2\ 2p^2\ {}^1D_2-2s^2\ 2p\ 10d\ {}^3F_3$	28
Si I	1258.796	1258.795	$3s^2\ 3p^2\ {}^3P_0-3s\ 3p^3\ {}^3S_1$	28
S II	1259.529	1259.521	$3s^2\ 3p^3\ {}^4S_{3/2}-3s\ 3p^4\ {}^4P_{5/2}$	28
Si II	1260.421	1260.424	$3s^2\ 3p\ {}^2P_{1/2}-3s^2\ 3d\ {}^2D_{3/2}$	28
C I	1260.612	1260.613	$2s^2\ 2p^2\ {}^1D_2-2s^2\ 2p\ 9d\ {}^1F_3$	28
C I	1261.121	1261.122	$2s^2\ 2p^2\ {}^3P_1-2s^2\ 2p\ 3d\ {}^3P_2$	28
C I	1261.551	1261.552	$2s^2\ 2p^2\ {}^3P_2-2s^2\ 2p\ 3d\ {}^3P_2$	28
C I	1261.720	1261.740	$2s^2\ 2p^2\ {}^1D_2-2s^2\ 2p\ 9d\ {}^3F_3$	28
Si II	1264.738	1264.740	$3s^2\ 3p\ {}^2P_{3/2}-3s^2\ 3d\ {}^2D_{5/2}$	28
Si II	1265.003	1265.004	$3s^2\ 3p\ {}^2P_{3/2}-3s^2\ 3d\ {}^2D_{3/2}$	28
C I	1266.417	1266.419	$2s^2\ 2p^2\ {}^1D_2-2s^2\ 2p\ 8d\ {}^1F_3$	28
C I	1267.599	1267.596	$2s^2\ 2p^2\ {}^1D_2-2s^2\ 2p\ 8d\ {}^3F_3$	28
S I	1270.783	1270.782	$3s^2\ 3p^4\ {}^3P_2-3s^2\ 3p^3\ 5d\ {}^3D_3$	28
Mg II	1271.93	1271.940	$3p\ {}^2P_{1/2}-9s\ {}^2S_{1/2}$	28
C I	1274.759	1274.756	$2s^2\ 2p^2\ {}^1D_2-2s^2\ 2p\ 7d\ {}^1P_1$	28
C I	1274.983	1274.984	$2s^2\ 2p^2\ {}^1D_2-2s^2\ 2p\ 7d\ {}^1F_3$	28
C I	1276.221	1276.220	$2s^2\ 2p^2\ {}^1D_2-2s^2\ 2p\ 8s\ {}^3P_1$	28
S I	1277.21	1277.198	$3s^2\ 3p^4\ {}^3P_1-3s^2\ 3p^3\ 5d\ {}^3D_1$	28
S I		1277.212	$3s^2\ 3p^4\ {}^3P_1-3s^2\ 3p^3\ 5d\ {}^3D_2$	
C I	1277.285	1277.282	$2s^2\ 2p^2\ {}^3P_1-2s^2\ 2p\ 3d\ {}^3D_2$	28
C I	1277.556	1277.550	$2s^2\ 2p^2\ {}^3P_2-2s^2\ 2p\ 3d\ {}^3D_3$	28
C I	1277.728	1277.723	$2s^2\ 2p^2\ {}^3P_2-2s^2\ 2p\ 3d\ {}^3D_2$	28
C I	1277.951	1277.954	$2s^2\ 2p^2\ {}^3P_2-2s^2\ 2p\ 3d\ {}^3D_1$	28
C I	1279.228	1279.229	$2s^2\ 2p^2\ {}^3P_2-2s^2\ 2p\ 3d\ {}^3F_3$	28
C I	1279.890	1279.890	$2s^2\ 2p^2\ {}^3P_1-2s^2\ 2p\ 4s\ {}^3P_2$	28
C I	1280.33	1280.333	$2s^2\ 2p^2\ {}^3P_2-2s^2\ 2p\ 4s\ {}^3P_2$	28
C I	1288.038	1288.037	$2s^2\ 2p^2\ {}^1D_2-2s^2\ 2p\ 6d\ {}^1P_1$	28
C I	1288.422	1288.422	$2s^2\ 2p^2\ {}^1D_2-2s^2\ 2p\ 6d\ {}^1F_3$	28
C I	1289.975	1289.977	$2s^2\ 2p^2\ {}^1D_2-2s^2\ 2p\ 6d\ {}^3F_3$	28
C I	1291.304	1291.304	$2s^2\ 2p^2\ {}^1D_2-2s^2\ 2p\ 6d\ {}^1D_2$	28
SI III	1294.544	1294.548	$3s\ 3p\ {}^3P_1-3p^2\ {}^3P_2$	28
P II	1294.644	1294.640	$3s^2\ 3p^2\ {}^3P_2-3s\ 3p^3\ {}^1D_2$	28
S I	1295.653	1295.653	$3s^2\ 3p^4\ {}^3P_2-3s^2\ 3p^3\ 4s\ {}^3P_2$	28
Fe II	1296.090	1296.084	$3d^7\ {}^4P_{5/2}-3d^6\ 4p\ {}^4S_{3/2}$	28
Si III	1296.726	1296.728	$3s\ 3p\ {}^3P_0-3p^2\ {}^3P_1$	28
Si III	1298.963	1298.948	$3s\ 3p\ {}^3P_2-3p^2\ {}^3P_2$	28
S I	1300.909	1300.907	$3s^2\ 3p^4\ {}^1D_2-3s^2\ 3p^5\ 5s\ {}^1D_2$	28
Si III	1301.149	1301.151	$3s\ 3p\ {}^3P_1-3p^2\ {}^3P_0$	28
O I	1302.170	1302.168	$2s^2\ 2p^4\ {}^3P_2-2s^2\ 2p^3\ 3s\ {}^3S_1$	28
Si III	1303.32	1303.325	$3s\ 3p\ {}^3P_2-3p^2\ {}^3P_1$	28
Si II	1304.375	1304.372	$3s^2\ 3p\ {}^2P_{1/2}-3s\ 3p^2\ {}^2S_{1/2}$	28
O I	1304.860	1304.858	$2s^2\ 2p^4\ {}^3P_1-2s^2\ 2p^3\ 3s\ {}^3S_1$	28

continued.

Ion	Wvl. (Å)	Wvl.^{L} (Å)	Transition	Ref.
O I	1306.028	1306.029	$2s^2\ 2p^4\ {}^3P_0–2s^2\ 2p^3\ 3s\ {}^3S_1$	28
Si II	1309.279	1309.277	$3s^2\ 3p\ {}^2P_{3/2}–3s\ 3p^2\ {}^2S_{1/2}$	28
C I	1311.362	1311.363	$2s^2\ 2p^2\ {}^1D_2–2s^2\ 2p\ 5d\ {}^1F_3$	28
C I	1311.925	1311.924	$2s^2\ 2p^2\ {}^1D_2–2s^2\ 2p\ 6s\ {}^1P_1$	28
C I	1313.462	1313.464	$2s^2\ 2p^2\ {}^1D_2–2s^2\ 2p\ 5d\ {}^3F_3$	28
C I	1315.918	1315.919	$2s^2\ 2p^2\ {}^1D_2–2s^2\ 2p\ 5d\ {}^1D_2$	28
S I	1316.542	1316.542	$3s^2\ 3p^4\ {}^3P_2–3s^2\ 3p^3\ 4d\ {}^3D_3$	28
Ni II	1317.221	1317.220	$3d^9\ {}^2D_{5/2}–3d^8\ 4p\ {}^2F_{7/2}$	28
N I	1318.99	1318.998	$2s^2\ 2p^3\ {}^2P_{1/2}–2s^2\ 2p^2\ 3d\ {}^2P_{1/2}$	28
N I		1319.005	$2s^2\ 2p^3\ {}^2P_{3/2}–2s^2\ 2p^2\ 3d\ {}^2P_{1/2}$	
S I	1323.52	1323.515	$3s^2\ 3p^4\ {}^3P_1–3s^2\ 3p^3\ 4d\ {}^3D_2$	28
S I		1323.522	$3s^2\ 3p^4\ {}^3P_1–3s^2\ 3p^3\ 4d\ {}^3D_1$	
S I	1326.63	1326.643	$3s^2\ 3p^4\ {}^3P_0–3s^2\ 3p^3\ 4d\ {}^3D_1$	28
C I	1328.833	1328.833	$2s^2\ 2p^2\ {}^3P_0–2s\ 2p^3\ {}^3P_1$	28
C I	1329.10	1329.086	$2s^2\ 2p^2\ {}^3P_1–2s\ 2p^3\ {}^3P_0$	28
C I		1329.100	$2s^2\ 2p^2\ {}^3P_1–2s\ 2p^3\ {}^3P_2$	
C I		1329.123	$2s^2\ 2p^2\ {}^3P_1–2s\ 2p^3\ {}^3P_1$	
C II	1334.535	1334.535	$2s^2\ 2p\ {}^2P_{1/2}–2s\ 2p^2\ {}^2D_{3/2}$	28
C II	1335.708	1335.710	$2s^2\ 2p\ {}^2P_{3/2}–2s\ 2p^2\ {}^2D_{5/2}$	28
Fe XII	1349.39	1349.400	$3s^2\ 3p^3\ {}^4S_{3/2}–3s^2\ 3p^3\ {}^2P_{1/2}$	28
Cl I	1351.657	1351.657	$3s^2\ 3p^5\ {}^2P_{1/2}–3s^2\ 3p^4\ 4s\ {}^2P_{1/2}$	28
O I	1355.598	1355.598	$2s^2\ 2p^4\ {}^3P_2–2s^2\ 2p^3\ 3s\ {}^5S_2$	28
C I	1355.843	1355.844	$2s^2\ 2p^2\ {}^1D_2–2s^2\ 2p\ 4d\ {}^1F_3$	28
C I	1357.134	1357.134	$2s^2\ 2p^2\ {}^1D_2–2s^2\ 2p\ 5s\ {}^1P_1$	28
C I	1357.658	1357.658	$2s^2\ 2p^2\ {}^1D_2–2s^2\ 2p\ 4d\ {}^3D_3$	28
O I	1358.512	1358.512	$2s^2\ 2p^4\ {}^3P_1–2s^2\ 2p^3\ 3s\ {}^5S_2$	28
C I	1359.276	1359.275	$2s^2\ 2p^2\ {}^1D_2–2s^2\ 2p\ 4d\ {}^3F_3$	28
C I	1359.440	1359.439	$2s^2\ 2p^2\ {}^1D_2–2s^2\ 2p\ 5s\ {}^3P_1$	28
Fe II	1360.175	1360.170	$3d^6\ 4s\ {}^6D_{1/2}–3d^6\ 4p\ {}^2D_{3/2}$	28
Fe II	1361.371	1361.373	$3d^7\ {}^4P_{5/2}–3d^6\ 4p\ {}^4D_{7/2}$	8
Cl I	1363.44	1363.447	$3s^2\ 3p^5\ {}^2P_{1/2}–3s^2\ 3p^4\ 4s\ {}^2P_{3/2}$	28
C I	1364.164	1364.164	$2s^2\ 2p^2\ {}^1D_2–2s^2\ 2p\ 4d\ {}^1D_2$	28
Fe II	1368.099	1368.098	$3d^7\ {}^4P_{3/2}–3d^6\ 4p\ {}^4D_{5/2}$	28
Ca II	1368.425	1368.423	$3d\ {}^2D_{3/2}–7f\ {}^2F_{5/2}$	28
Fe II	1369.705	1369.707	$3d^7\ {}^4P_{5/2}–3d^6\ 4p\ {}^2P_{7/2}$	28
O V	1371.295	1371.296	$2s\ 2p\ {}^1P_1–2p^2\ {}^1D_2$	28
Fe II	1376.673	1376.672	$3d^7\ {}^4P_{1/2}–3d^6\ 4p\ {}^4P_{3/2}$	28
Ni II	1381.296	1381.295	$3d^9\ {}^2D_{3/2}–3d^8\ 4p\ {}^2P_{1/2}$	28
Fe II	1392.816	1392.817	$3d^7\ {}^2H_{9/2}–3d^6\ 4p\ {}^2G_{7/2}$	28
Si IV	1393.76	1393.757	$3s\ {}^2S_{1/2}–3p\ {}^2P_{3/2}$	28
O IV	1399.775	1399.780	$2s^2\ 2p\ {}^2P_{1/2}–2s\ 2p^2\ {}^4P_{1/2}$	28
O IV	1401.163	1401.157	$2s^2\ 2p\ {}^2P_{3/2}–2s\ 2p^2\ {}^4P_{5/2}$	28
Si IV	1402.773	1402.772	$3s\ {}^2S_{1/2}–3p\ {}^2P_{1/2}$	28
S IV	1404.773	1404.808	$3s^2\ 3p\ {}^2P_{1/2}–3s\ 3p^2\ {}^4P_{1/2}$	28
O IV	1404.82	1404.806	$2s^2\ 2p\ {}^2P_{3/2}–2s\ 2p^2\ {}^4P_{3/2}$	28
S IV	1406.059	1406.016	$3s^2\ 3p\ {}^2P_{3/2}–3s\ 3p^2\ {}^4P_{5/2}$	28
O IV	1407.386	1407.382	$2s^2\ 2p\ {}^2P_{3/2}–2s\ 2p^2\ {}^4P_{1/2}$	28
Ni II	1411.071	1411.071	$3d^9\ {}^2D_{3/2}–3d^8\ 4p\ {}^2D_{3/2}$	28
S IV	1416.928	1416.887	$3s^2\ 3p\ {}^2P_{3/2}–3s\ 3p^2\ {}^4P_{3/2}$	28
S I	1425.027	1425.030	$3s^2\ 3p^4\ {}^3P_2–3s^2\ 3p^3\ 4d\ {}^3D_3$	28
S I	1425.187	1425.188	$3s^2\ 3p^4\ {}^3P_2–3s^2\ 3p^3\ 4d\ {}^3D_2$	28
C I	1431.597	1431.597	$2s\ 2p^3\ {}^5S_2–2s\ 2p^2\ 3s\ {}^5P_3$	28
C I	1432.107	1432.105	$2s\ 2p^3\ {}^5S_2–2s\ 2p^2\ 3s\ {}^5P_2$	28
Ca II	1432.504	1432.503	$3d\ {}^2D_{3/2}–6f\ {}^2F_{5/2}$	28
C I	1432.54	1432.530	$2s\ 2p^3\ {}^5S_2–2s\ 2p^2\ 3s\ {}^5P_1$	28
S I	1433.279	1433.279	$3s^2\ 3p^4\ {}^3P_1–3s^2\ 3p^3\ 3d\ {}^3D_2$	28
S I	1436.959	1436.958	$3s^2\ 3p^4\ {}^3P_0–3s^2\ 3p^3\ 3d\ {}^3D_1$	28
C I	1450.11	1450.100	$2s^2\ 2p^2\ {}^1S_0–2s^2\ 2p\ 20d\ {}^1P_1$	28
Ni II	1454.853	1454.852	$3d^7\ 4s^2\ {}^4F_{5/2}–3d^8\ 5p\ {}^2D_{3/2}$	28
C I	1457.488	1457.480	$2s^2\ 2p^2\ {}^1S_0–2s^2\ 2p\ 14d\ {}^3D_1$	28
C I	1459.032	1459.032	$2s^2\ 2p^2\ {}^1S_0–2s^2\ 2p\ 3d\ {}^1P_1$	28
C I	1463.335	1463.336	$2s^2\ 2p^2\ {}^1D_2–2s^2\ 2p\ 3d\ {}^1F_3$	28

Fe X	1463.49	1463.486	$3s^2\,3p^4\,3d\,^4F_{9/2}-3s^2\,3p^4\,3d\,^2F_{7/2}$	28
C I	1464.99	1464.980	$2s^2\,2p^2\,^1S_0-2s^2\,2p\,11d\,^3D_1$	28
H_2		1464.020	$0\,X^1\Sigma_g^+-6R5\,B^1\Sigma_u^+$	
Fe II	1465.048	1465.043	$3d^5\,4s^2\,^6S_{5/2}-3d^6\,5p\,^6P_{5/2}$	28
Fe XI	1467.06	1467.423	$3s^2\,3p^4\,^3P_1-3s^2\,3p^4\,^1S_0$	28
Ni II	1467.265	1467.265	$3d^9\,^2D_{5/2}-3d^8\,4p\,^2D_{3/2}$	28
C I	1467.406	1467.402	$2s^2\,2p^2\,^1D_2-2s^2\,2p\,4s\,^1P_1$	28
Ni II	1467.765	1467.762	$3d^9\,^2D_{5/2}-3d^8\,4p\,^2F_{7/2}$	28
C I	1467.877	1467.877	$2s^2\,2p^2\,^1D_2-2s^2\,2p\,3d\,^3D_3$	28
C I	1468.410	1468.410	$2s^2\,2p^2\,^1D_2-2s^2\,2p\,3d\,^3D_1$	28
C I	1469.111	1469.110	$2s^2\,2p^2\,^1S_0-2s^2\,2p\,10d\,^3D_1$	28
C I	1470.094	1470.094	$2s^2\,2p^2\,^1D_2-2s^2\,2p\,3d\,^3F_3$	28
Fe II	1470.18	1470.182	$3d^6\,4s\,^2H_{11/2}-3d5\,4s\,4p\,^4G_{11/2}$	28
Fe II		1470.203	$3d^6\,4s\,^4G_{5/2}-3d5\,4s\,4p\,^4G_{7/2}$	
S I	1471.833	1471.832	$3s^2\,3p^4\,^1D_2-3s^2\,3p^3\,4s\,^3P_2$	28
C I	1472.231	1472.231	$2s^2\,2p^2\,^1D_2-2s^2\,2p\,4s\,^3P_1$	28
S I	1472.971	1471.972	$3s^2\,3p^4\,^3P_2-3s^2\,3p^3\,3d\,^5D_3$	28
C I	1473.242	1473.242	$2s^2\,2p^2\,^1S_0-2s^2\,2p\,9d\,^1P_1$	28
Fe II	1473.832	1473.834	$3d^5\,4s^2\,^6S_{5/2}-3d^6\,5p\,^6P_{7/2}$	28
S I	1473.994	1473.995	$3s^2\,3p^4\,^3P_2-3s^2\,4p^3\,4s\,^3D_3$	28
S I	1474.378	1474.380	$3s^2\,3p^4\,^3P_2-3s^2\,4p^3\,4s\,^3D_2$	28
S I	1474.575	1474.572	$3s^2\,3p^4\,^3P_2-3s^2\,4p^3\,4s\,^3D_1$	28
C I	1474.748	1474.740	$2s^2\,2p^2\,^1S_0-2s^2\,2p\,9d\,^3D_1$	28
Ni II	1477.226	1477.227	$3d^9\,^2D_{5/2}-3d^8\,4p\,^2F_{5/2}$	28
C I	1480.99	1480.990	$2s^2\,2p^2\,^1S_0-2s^2\,2p\,8d\,^3P_1$	28
C I	1481.114	1481.114	$2s^2\,2p^2\,^1S_0-2s^2\,2p\,8d\,^1P_1$	28
Fe II	1481.35	1481.349	$3d^6\,4s\,^4G_{7/2}-3d^6\,4p\,^4F_{7/2}$	28
S I	1481.68	1481.665	$3s^2\,3p^4\,^3P_1-3s^2\,4p^4\,3d\,^5D_2$	28
S I		1481.712	$3s^2\,3p^4\,^3P_1-3s^2\,4p^4\,3d\,^5D_1$	
C I	1481.768	1481.764	$2s^2\,2p^2\,^1D_2-2s^2\,2p\,3d\,^1D_2$	28
Ni II	1482.392	1482.393	$3d^9\,^2D_{3/2}-3d^8\,4p\,^2P_{3/2}$	28
C I	1482.729	1482.730	$2s^2\,2p^2\,^1S_0-2s^2\,2p\,8d\,^3D_1$	28
Mg II	1482.894	1482.890	$3p\,^2P_{3/2}-6s\,^2S_{1/2}$	28
S I	1483.037	1438.039	$3s^2\,3p^4\,^3P_1-3s^2\,3p^3\,4s\,^3D_2$	28
S I	1483.231	1483.233	$3s^2\,3p^4\,^3P_1-3s^2\,3p^3\,4s\,^5D_1$	28
Si II	1485.227	1485.224	$3s\,3p^2\,^2D_{5/2}-3s^2\,7f\,^2F_{7/2}$	28
S I	1485.622	1485.622	$3s^2\,3p^4\,^3P_0-3s^2\,3p^3\,3d\,^5D_1$	28
N IV	1486.498	1486.499	$2s^2\,^1S_0-2s\,2p\,^3P_1$	28
S I	1487.149	1487.150	$3s^2\,3p^4\,^3P_0-3s^2\,3p^3\,4s\,^3D_1$	28
Ni II	1487.455	1487.455	$3d^9\,^2D_{3/2}-3d^8\,4p\,^2D_{5/2}$	28
Ca II	1488.77	1488.770	$4p\,^2P_{1/2}-12d\,^2D_{3/2}$	28
C I	1492.574	1492.570	$2s^2\,2p^2\,^1S_0-2s^2\,2p\,7d\,^3P_1$	28
Fe II		1492.577	$3d^6\,4s\,^4G_{11/2}-3d^5\,4s\,4p\,^4F_{9/2}$	
N I	1492.62	1492.625	$2s^2\,2p^3\,^2D_{5/2}-2s^2\,2p^2\,3s\,^2P_{3/2}$	28
C I	1492.738	1492.738	$2s^2\,2p^2\,^1S_0-2s^2\,2p\,7d\,^1P_1$	28
N I	1492.82	1492.820	$2s^2\,2p^3\,^2D_{3/2}-2s^2\,2p^2\,3s\,^2P_{3/2}$	28
Ni II	1500.435	1500.437	$3d^9\,^2D_{3/2}-3d^8\,4p\,^2D_{3/2}$	28
S V	1501.80	1501.766	$3s\,3p\,^1P_1-3p^2\,^1D_2$	28
Si III	1501.87	1501.870	$3s\,3d\,^3D_1-3s\,4f\,^3F_2$	28
Ni II	1501.97	1501.962	$3d^9\,^2D_{5/2}-3d^8\,4p\,^4P_{3/2}$	28
Ni II	1502.150	1502.150	$3d^9\,^2D_{5/2}-3d^8\,4p\,^4P_{5/2}$	28
H_2	1504.755	1504.756	$1\,X^1\Sigma_g^+-7P5\,B^1\Sigma_u^+$	28
Fe II	1506.894	1506.898	$3d^6\,4s\,^4G_{9/2}-3d^5\,4s\,4p\,^4F_{11/2}$	28
C I	1510.671	1510.668	$2s^2\,2p^2\,^1S_0-2s^2\,2p\,6d\,^3P_1$	28
H_2		1510.693	$1\,X^1\Sigma_g^+-7P6\,B^1\Sigma_u^+$	
Ni II	1510.856	1510.859	$3d^9\,^2D_{3/2}-3d^8\,4p\,^2F_{5/2}$	12
C I	1510.98	1510.981	$2s^2\,2p^2\,^1S_0-2s^2\,2p\,6d\,^1P_1$	28
C I	1511.908	1511.907	$2s^2\,2p^2\,^1S_0-2s^2\,2p\,7s\,^1P_1$	28
C I	1513.150	1513.150	$2s^2\,2p^2\,^1S_0-2s^2\,2p\,6d\,^3D_1$	28
Fe II	1513.31	1513.316	$3d^6\,4s\,^2H_{11/2}-3d^6\,4p\,^4H_{13/2}$	28
Fe II	1515.115	1515.117	$3d^6\,4s\,^2H_{11/2}-3d^6\,4p\,^2G_{9/2}$	28
Fe II	1515.936	1515.937	$3d^7\,^2D_{5/2}-3d^6\,4p\,^2F_{7/2}$	28

continued.

Ion	Wvl. (Å)	Wvl.^{L} (Å)	Transition	Ref.
Si I	1517.30	1517.306	$3s^2\,3p^2\ ^3P_1–3s^2\,3p\ 46d\ ^3P_{0,1}$	28
Si I	1517.707	1517.703	$3s^2\,3p^2\ ^3P_1–3s^2\,3p\ 40d\ ^3P_{0,1}$	28
Si I	1518.31	1518.307	$3s^2\,3p^2\ ^3P_1–3s^2\,3p\ 34d\ ^3P_{0,1}$	28
Si I	1520.800	1520.796	$3s^2\,3p^2\ ^3P_2–3s^2\,3p\ 44d\ ^3P_1,{}^3D_3$	28
Si I	1521.26	1521.246	$3s^2\,3p^2\ ^3P_2–3s^2\,3p\ 38d\ ^3P_1,{}^3D_3$	28
Si I	1521.97	1521.970	$3s^2\,3p^2\ ^3P_2–3s^2\,3p\ 32d\ ^3P_1,{}^3D_3$	28
Si I	1522.13	1522.131	$3s^2\,3p^2\ ^3P_2–3s^2\,3p\ 31d\ ^3P_1,{}^3D_3$	28
Si I	1522.314	1522.315	$3s^2\,3p^2\ ^3P_2–3s^2\,3p\ 30d\ ^3P_1,{}^3D_3$	28
Si I	1522.44	1522.433	$3s^2\,3p^2\ ^3P_1–3p\ 20d\ ^3P_{0,1}$	28
Si I	1523.12	1523.120	$3s^2\,3p^2\ ^3P_1–3p\ 19d\ ^3P_{0,1}$	28
Si I	1523.22	1523.200	$3s^2\,3p^2\ ^3P_0–3p\ 17d\ ^3D_1$	28
Si I		1523.200	$3s^2\,3p^2\ ^3P_1–3p\ 17d\ ^3D_1$	
Fe II	1523.366	1523.374	$3d^7\ ^4F_{9/2}–3d^6\,4p\ ^2H_{11/2}$	28
Si I	1523.544	1523.543	$3s^2\,3p^2\ ^3P_2–3s^2\,3p\ 25d\ ^3P_1$	28
Si I	1523.58	1523.568	$3s^2\,3p^2\ ^3P_2–3s^2\,3p\ 25d\ ^3D_3$	28
Si I	1523.917	1523.915	$3s^2\,3p^2\ ^3P_1–3s^2\,3p\ 18d\ ^3P_{0,1}$	28
Si I		1523.915	$3s^2\,3p^2\ ^3P_2–3s^2\,3p\ 24d\ ^3D_3$	
Si I	1524.198	1524.192	$3s^2\,3p^2\ ^3P_0–3s^2\,3p\ 16d\ ^3P_1$	28
Si I	1524.85	1524.852	$3s^2\,3p^2\ ^3P_1–3s^2\,3p\ 17d\ ^3P_{0,1}$	28
Si I	1524.972	1524.970	$3s^2\,3p^2\ ^3P_1–3s^2\,3p\ 17d\ ^3D_1$	28
Si I	1525.28	1525.285	$3s^2\,3p^2\ ^3P_2–3s^2\,3p\ 21d\ ^3D_3$	28
Si I	1525.764	1525.758	$3s^2\,3p^2\ ^3P_0–3s^2\,3p\ 15d\ ^3D_1$	28
Si I	1525.838	1525.832	$3s^2\,3p^2\ ^3P_2–3s^2\,3p\ 20d\ ^3P_1$	28
Si I	1525.984	1525.983	$3s^2\,3p^2\ ^3P_1–3s^2\,3p\ 16d\ ^3P_{0,1}$	28
Si I	1526.137	1526.141	$3s^2\,3p^2\ ^3P_1–3s^2\,3p\ 16d\ ^3D_1$	28
Si II	1526.708	1526.720	$3s^2\,3p\ 2P_{1/2}–3s^2\,4s\ ^2S_{1/2}$	28
Si I	1527.183	1527.185	$3s^2\,3p^2\ ^3P_1–3s^2\,3p\ 24d\ ^3D_2$	28
Si I	1527.28	1527.252	$3s^2\,3p^2\ ^3P_1–3s^2\,3p\ 17s\ ^3P_2$	28
Si I	1527.35	1527.343	$3s^2\,3p^2\ ^3P_1–3s^2\,3p\ 15d\ ^3P_0$	28
Si I		1527.367	$3s^2\,3p^2\ ^3P_2–3s^2\,3p\ 18d\ ^3P_2$	
Si I		1527.367	$3s^2\,3p^2\ ^3P_2–3s^2\,3p\ 15d\ ^3D_1$	
Si I	1527.461	1527.460	$3s^2\,3p^2\ ^3P_1–3s^2\,3p\ 15d\ ^3P_2$	28
Si I	1527.56	1527.555	$3s^2\,3p^2\ ^3P_2–3s^2\,3p\ 15d\ ^3D_1$	28
Si I	1527.588	1527.588	$3s^2\,3p^2\ ^3P_1–3s^2\,3p\ 23d\ ^3D_2$	28
Si I	1528.185	1528.185	$3s^2\,3p^2\ ^3P_2–3s^2\,3p\ 19s\ ^3P_2$	28
Si I	1528.280	1528.277	$3s^2\,3p^2\ ^3P_2–3s^2\,3p\ 17d\ ^3P_1$	28
Si I	1528.36	1528.372	$3s^2\,3p^2\ ^3P_2–3s^2\,3p\ 17d\ ^3D_3$	28
Si I	1529.024	1529.022	$3s^2\,3p^2\ ^3P_1–2s^2\,3p\ 14d\ ^3P_1$	28
Si I	1529.202	1529.202	$3s^2\,3p^2\ ^3P_1–3s^2\,3p\ 14d\ ^3P_2$	28
Si I		1529.202	$3s^2\,3p^2\ ^3P_1–3s^2\,3p\ 20d\ ^3D_2$	
Si I	1529.267	1529.268	$3s^2\,3p^2\ ^3P_2–3s^2\,3p\ 14d\ ^3P_1$	28
Si I		1529.268	$3s^2\,3p^2\ ^3P_2–3s^2\,3p\ 18s\ ^1P_1$	
Si I	1529.395	1529.396	$3s^2\,3p^2\ ^3P_0–3s^2\ 16d\ ^3P_1$	28
Si I	1529.45	1529.464	$3s^2\,3p^2\ ^3P_2–3s^2\ 16d\ ^3P_2$	28
Si I	1529.54	1529.532	$3s^2\,3p^2\ ^3P_2–3s^2\ 16d\ ^3D_3$	28
Si I	1529.779	1529.778	$3s^2\,3p^2\ ^3P_1–3s^2\ 19d\ ^3D_2$	28
Si I	1530.60	1530.602	$3s^2\,3p^2\ ^3P_2–3s^2\,3p\ 24d\ ^3D_2$	28
Si I	1530.654	1530.657	$3s^2\,3p^2\ ^3P_2–3s^2\,3p\ 17s\ ^3P_2$	28
Si I	1530.689	1530.690	$3s^2\,3p^2\ ^3P_1–3s^2\,3p\ 18d\ ^3D_2$	28
Si I	1530.79	1530.787	$3s^2\,3p^2\ ^3P_2–3s^2\,3p\ 15d\ ^3P_1$	28
Si I	1530.883	1530.875	$3s^2\,3p^2\ ^3P_2–3s^2\,3p\ 15d\ ^3P_2$	28
Si I	1530.929	1530.929	$3s^2\,3p^2\ ^3P_2–3s^2\,3p\ 15d\ ^3D_3$	28
Si I		1530.929	$3s^2\,3p^2\ ^3P_1–3s^2\,3p\ 15s\ ^3P_2$	
Si I	1531.001	1531.002	$3s^2\,3p^2\ ^3P_0–3s^2\,3p\ 16d\ ^1P_1$	28
Si I		1531.002	$3s^2\,3p^2\ ^3P_2–3s^2\,3p\ 23d\ ^1F_3,{}^1D_2$	
Si I	1531.274	1531.276	$3s^2\,3p^2\ ^3P_1–3s^2\,3p\ 13d\ ^3P_2$	28
Si I		1531.276	$3s^2\,3p^2\ ^3P_2–3s^2\,3p\ 15d\ ^3F_3$	
Si I	1531.602	1531.602	$3s^2\,3p^2\ ^3P_0–3s^2\,3p\ 14s\ ^1P_1$	28
Si I		1531.602	$3s^2\,3p^2\ ^3P_1–3s^2\,3p\ 17d\ ^3D_2$	
Si I		1531.602	$3s^2\,3p^2\ ^3P_2–3s^2\,3p\ 22d\ ^1F_3$	
Si I	1532.077	1532.077	$3s^2\,3p^2\ ^3P_2–3s^2\,3p\ 21d\ ^1F_3$	28
Si I	1532.298	1532.295	$3s^2\,3p^2\ ^3P_0–3s^2\,3p\ 12d\ ^3D_1$	28

Si I	1532.445	1532.446	$3s^2\ 3p^2\ ^3P_2–3s^2\ 3p\ 14d\ ^3P_1$	28
Si I		1532.446	$3s^2\ 3p^2\ ^3P_2–3s^2\ 3p\ 15d\ ^1P_1$	
Si I		1532.446	$3s^2\ 3p^2\ ^3P_2–3s^2\ 3p\ 20d\ ^3D_2$	
Si I	1532.50	1532.490	$3s^2\ 3p^2\ ^3P_1–3s^2\ 3p\ 13d\ ^3F_2$	28
P II		1352.510	$3s^2\ 3p^2\ ^3P_0–3s\ 3p^3\ ^3D_1$	
Si I	1532.63	1532.624	$3s^2\ 3p^2\ ^3P_1–3s^2\ 3p\ 18s\ ^3P_0$	28
Si I		1532.624	$3s^2\ 3p^2\ ^3P_2–3s^2\ 3p\ 14d\ ^3P_2$	
Si I		1532.603	$3s^2\ 3p^2\ ^3P_2–3s^2\ 3p\ 14d\ ^3D_3$	
Si I		1532.646	$3s^2\ 3p^2\ ^3P_2–3s^2\ 3p\ 14d\ ^1F_3$	
Si I	1532.815	1532.815	$3s^2\ 3p^2\ ^3P_1–3s^2\ 3p\ 16d\ ^1P_1$	28
Si I	1532.927	1532.929	$3s^2\ 3p^2\ ^3P_1–3s^2\ 3p\ 16d\ ^3D_2$	28
Si II	1533.432	1533.450	$3s^2\ 3p\ ^2P_{3/2}–3s^2\ 4s\ ^2S_{1/2}$	28
Si I	1534.11	1534.112	$3s^2\ 3p^2\ ^3P_0–3s^2\ 3p\ 14d\ ^1P_1$	28
Si I		1534.112	$3s^2\ 3p^2\ ^3P_1–3s^2\ 3p\ 12d\ ^3D_1$	
Si I		1534.112	$3s^2\ 3p^2\ ^3P_2–3s^2\ 3p\ 18d\ ^3D_2$	
Si I	1534.184	1534.184	$3s^2\ 3p^2\ ^3P_1–3s^2\ 3p\ 15d\ ^3D_2$	28
Si I	1534.23	1534.236	$3s^2\ 3p^2\ ^3P_2–3s^2\ 3p\ 18d\ ^1F_3$	28
Si I	1534.37	1534.236	$3s^2\ 3p^2\ ^3P_2–3s^2\ 3p\ 15s\ ^3P_2$	28
Si I	1534.430	1534.431	$3s^2\ 3p^2\ ^3P_2–3s^2\ 3p\ 18d\ ^1D_2$	28
Si I	1534.550	1534.547	$3s^2\ 3p^2\ ^3P_2–3s^2\ 3p\ 13d\ ^3P_1$	28
Si I	1534.707	1534.707	$3s^2\ 3p^2\ ^3P_2–3s^2\ 3p\ 13d\ ^3P2$	28
Si I	1534.775	1534.778	$3s^2\ 3p^2\ ^3P_2–3s^2\ 3p\ 13d\ ^3D_3$	28
H_2		1534.772	$0\ X^1\Sigma_g^+–7P5\ B^1\Sigma_u^+$	
Si I	1534.888	1534.883	$3s^2\ 3p^2\ ^3P_0–3s^2\ 3p\ 13s\ ^1P_1$	28
Si I	1534.997	1534.996	$3s^2\ 3p^2\ ^3P_2–3s^2\ 3p\ 17d\ ^1F_3$	28
Si I	1535.90	1535.896	$3s^2\ 3p^2\ ^3P_0–3s^2\ 3p\ 15s\ ^3P_1$	28
P II		1535.900	$3s^2\ 3p^2\ ^3P_1–3s\ 3p^3\ ^3D_2$	
Si I	1536.083	1536.085	$3s^2\ 3p^2\ ^3P_1–3s^2\ 3p\ 14d\ ^3D_2$	28
Si I	1536.259	1536.257	$3s^2\ 3p^2\ ^3P_2–3s^2\ 3p\ 16d\ ^1P_1$	28
Si I	1536.38	1536.366	$3s^2\ 3p^2\ ^3P_2–3s^2\ 3p\ 16d\ ^3D_2$	28
P II		1536.390	$3s^2\ 3p^2\ ^3P_1–3s\ 3p^3\ ^3D_1$	
Si I	1536.47	1536.473	$3s^2\ 3p^2\ ^3P_2–3s^2\ 3p\ 16d\ ^1F_3$	28
Ni II	1536.74	1536.746	$3d^9\ ^2D_{3/2}–3d^8\ 4p\ ^4P_{3/2}$	28
Si I	1536.82	1536.830	$3s^2\ 3p^2\ ^3P_1–3s^2\ 3p\ 13s\ ^3P_2$	28
CO	1536.868	1536.870	$1\ A^1\Pi–0R62\ X^1\Sigma^+$	28
Ni II	1536.945	1536.944	$3d^9\ ^2D_{3/2}–3d^8\ 4p\ ^4P_{5/2}$	28
Si I	1537.01	1537.016	$3s^2\ 3p^2\ ^3P_1–3s^2\ 3p\ 11d\ ^3P_0$	28
Si I	1537.084	1537.086	$3s^2\ 3p^2\ ^3P_1–3s^2\ 3p\ 11d\ ^3P_1$	28
Si I	1537.156	1537.159	$3s^2\ 3p^2\ ^3P_2–3s^2\ 3p\ 12d\ ^3P_1$	28
Si I	1537.297	1537.297	$3s^2\ 3p^2\ ^3P_2–3s^2\ 3p\ 12d\ ^3P_2$	28
Si I	1537.476	1537.474	$3s^2\ 3p^2\ ^3P_2–3s^2\ 3p\ 12d\ ^3D_3$	28
Si I	1537.550	1537.555	$3s^2\ 3p^2\ ^3P_2–3s^2\ 3p\ 12d\ ^3D_1$	28
Si I	1537.617	1537.618	$3s^2\ 3p^2\ ^3P_2–3s^2\ 3p\ 15d\ ^1F_3,^3D_2$	28
Si I	1537.935	1537.935	$3s^2\ 3p^2\ ^3P_2–3s^2\ 3p\ 12d\ ^3F_3$	28
Si I		1537.935	$3s^2\ 3p^2\ ^3P_1–3s^2\ 3p\ 13d\ ^3D_2$	
Si I	1538.878	1538.876	$3s^2\ 3p^2\ ^3P_0–3s^2\ 3p\ 12d\ ^1P_1$	28
Si I	1539.706	1539.705	$3s^2\ 3p^2\ ^3P_2–3s^2\ 3p\ 14d\ ^1F_3$	28
Si I	1540.545	1540.544	$3s^2\ 3p^2\ ^3P_2–3s^2\ 3p\ 11d\ ^3P_1$	28
Si I	1540.707	1540.707	$3s^2\ 3p^2\ ^3P_1–3s^2\ 3p\ 12d\ ^1P_1,^3D_2$	28
Si I	1540.782	1540.783	$3s^2\ 3p^2\ ^3P_2–3s^2\ 3p\ 11d\ ^3P_2$	28
Si I	1540.963	1540.963	$3s^2\ 3p^2\ ^3P_2–3s^2\ 3p\ 11d\ ^3D_3$	28
Si I	1541.200	1541.198	$3s^2\ 3p^2\ ^3P_1–3s^2\ 3p\ 12s\ ^3P_2$	28
Si I	1541.326	1541.322	$3s^2\ 3p^2\ ^3P_2–3s^2\ 3p\ 13d\ ^1F_3$	28
C I	1541.52	1541.510	$2s^2\ 2p^2\ ^1S_0–2s^2\ 2p\ 5d\ ^3P_1$	28
Si I	1541.57	1541.569	$3s^2\ 3p^2\ ^3P_1–3s^2\ 3p\ 10d\ ^3P_1$	28
C I	1542.75	1542.177	$2s^2\ 2p^2\ ^1S_0–2s^2\ 2p\ 5d\ ^1P_1$	28
Si I	1542.428	1542.432	$3s^2\ 3p^2\ ^3P_2–3s^2\ 3p\ 13d\ ^1D_2$	28
Si I		1542.432	$3s^2\ 3p^2\ ^3P_0–3s^2\ 3p\ 11d\ ^1P_1$	
Si I	1543.724	1543.724	$3s^2\ 3p^2\ ^3P_1–3s^2\ 3p\ 11d\ ^3D_2$	28
C I	1543.959	1543.960	$2s^2\ 2p^2\ ^1S_0–2s^2\ 2p\ 6s\ ^1P_1$	28
Si I	1544.186	1544.184	$3s^2\ 3p^2\ ^3P_2–3s^2\ 3p\ 12d\ ^3D_2$	28
Si I	1544.594	1544.591	$3s^2\ 3p^2\ ^3P_2–3s^2\ 3p\ 12s\ ^3P_2$	28

continued.

Ion	Wvl. (Å)	Wvl.^{L} (Å)	Transition	Ref.
Si I	1545.06	1545.045	$3s^2\,3p^2\,{}^3P_2{-}3s^2\,3p\,10d\,{}^3P_1$	28
Si I	1545.10	1545.099	$3s^2\,3p^2\,{}^3P_2{-}3s^2\,3p\,12d\,{}^1D_2$	28
Si I	1545.163	1545.163	$3s^2\,3p^2\,{}^3P_1{-}3s^2\,3p\,11d\,{}^1D_2$	28
C I	1545.247	1545.249	$2s^2\,2p^2\,{}^1S_0{-}2s^2\,2p\,5d\,{}^3D_1$	28
Si I	1545.575	1545.575	$3s^2\,3p^2\,{}^3P_2{-}3s^2\,3p\,10d\,{}^3D_3$	28
Si I	1545.66	1545.664	$3s^2\,3p^2\,{}^3P_2{-}3s^2\,3p\,10d\,{}^3D_1$	28
Si I	1545.749	1545.749	$3s^2\,3p^2\,{}^3P_2{-}3s^2\,3p\,10d\,{}^3P_2$	28
Si I	1546.591	1546.590	$3s^2\,3p^2\,{}^3P_2{-}3s^2\,3p\,11s\,{}^3F_3$	28
Si I	1546.671	1546.674	$3s^2\,3p^2\,{}^3P_0{-}3s^2\,3p\,9d\,{}^3D_1$	28
H_2	1547.35	1547.337	$1\ X^1\Sigma_g^+{-}8R3\ B^1\Sigma_u^+$	28
Si I		1547.362	$3s^2\,3p^2\,{}^3P_1{-}3s^2\,3p\,9d\,{}^3P_0$	
Si I	1547.455	1547.452	$3s^2\,3p^2\,{}^3P_1{-}3s^2\,3p\,9d\,{}^3P_1$	28
C IV	1548.19	1548.189	$1s^2\,2s\,{}^2S_{1/2}{-}1s^2\,2p\,{}^2P_{3/2}$	28
Si I	1548.51	1548.518	$3s^2\,3p^2\,{}^3P_1{-}3s^2\,3p\,9d\,{}^3D_1$	28
Si I	1548.71	1548.715	$3s^2\,3p^2\,{}^3P_1{-}3s^2\,3p\,10d\,{}^3D_2$	28
Fe II	1550.263	1550.260	$3d^7\,{}^4F_{9/2}{-}3d^6\,4p\,{}^4F_{7/2}$	28
C IV	1550.77	1550.775	$1s^2\,2s\,{}^2S_{1/2}{-}1s^2\,2p\,{}^2P_{1/2}$	28
Si I	1551.239	1551.240	$3s^2\,3p^2\,{}^3P_2{-}3s^2\,3p\,9d\,{}^3P_2$	28
Si I	1551.86	1551.860	$3s^2\,3p^2\,{}^3P_2{-}3s^2\,3p\,9d\,{}^3D_3$	28
Si I	1552.209	1552.209	$3s^2\,3p^2\,{}^3P_2{-}3s^2\,3p\,10d\,{}^1F_3$	28
Si I		1552.209	$3s^2\,3p^2\,{}^3P_2{-}3s^2\,3p\,10d\,{}^3D_2$	
Fe II	1552.81	1552.813	$3d^6\,4s\,{}^4D_{1/2}{-}3d^5\,4s\,4p\,{}^4D_{1/2}$	28
Si I	1552.950	1552.950	$3s^2\,3p^2\,{}^3P_0{-}3s^2\,3p\,9d\,{}^1P_1$	28
H_2		1552.958	$1\ X^1\Sigma_g^+{-}8P3\ B^1\Sigma_u^+$	
Si I	1553.371	1553.370	$3s^2\,3p^2\,{}^3P_0{-}3s^2\,3p\,10s\,{}^1P_1$	28
CO		1553.390	$0\ A^1\Pi{-}0P29\ X^1\Sigma^+$	
Ca II	1554.643	1554.642	$3d\,{}^2D_{5/2}{-}5f\,{}^2F_{7/2}$	28
Si I	1554.704	1554.702	$3s^2\,3p^2\,{}^3P_1{-}3s^2\,3p\,9d\,{}^3D_2$	28
Si I	1555.511	1555.516	$3s^2\,3p^2\,{}^3P_0{-}3s^2\,3p\,8d\,{}^3D_1$	28
Si I	1555.661	1555.660	$3s^2\,3p^2\,{}^3P_1{-}3s^2\,3p\,8d\,{}^3P_0$	28
Si I	1556.04	1556.043	$3s^2\,3p^2\,{}^3P_2{-}3s^2\,3p\,9d\,{}^3F_2$	28
Si I	1556.161	1556.160	$3s^2\,3p^2\,{}^3P_1{-}3s^2\,3p\,8d\,{}^3P_1$	28
Si I	1556.54	1556.527	$3s^2\,3p^2\,{}^3P_1{-}3s^2\,3p\,8d\,{}^3P_2$	28
Si I	1557.383	1557.382	$3s^2\,3p^2\,{}^3P_1{-}3s^2\,3p\,8d\,{}^3D_1$	28
Si I	1558.244	1558.240	$3s^2\,3p^2\,{}^3P_2{-}3s^2\,3p\,9d\,{}^3D_2$	28
Si I	1558.45	1558.453	$3s^2\,3p^2\,{}^3P_1{-}3s^2\,3p\,9d\,{}^1D_2$	28
Fe II	1558.545	1558.542	$3d^7\,{}^4F_{5/2}{-}3d^6\,4p\,{}^2D_{5/2}$	28
Fe II	1558.68	1558.690	$3d^7\,{}^4F_{3/2}{-}3d^6\,4p\,{}^2D_{3/2}$	28
Fe II	1559.085	1559.084	$3d^7\,{}^4F_{9/2}{-}3d^6\,4p\,{}^4F_{9/2}$	28
Si I	1559.362	1559.364	$3s^2\,3p^2\,{}^3P_2{-}3s^2\,3p\,9d\,{}^1F_3$	28
Si I	1560.07	1560.072	$3s^2\,3p^2\,{}^3P_2{-}3s^2\,3p\,8d\,{}^3P_2$	28
C I	1560.308	1560.310	$2s^2\,2p^2\,{}^3P_0{-}2s\,2p^3\,{}^3D_1$	28
C I	1560.684	1560.683	$2s^2\,2p^2\,{}^3P_1{-}2s\,2p^3\,{}^3D_2$	28
Fe II	1561.06	1561.067	$3d^7\,{}^4F_{9/2}{-}3d^6\,4p\,{}^4G_{7/2}$	28
C I	1561.41	1561.341	$2s^2\,2p^2\,{}^3P_2{-}2s\,2p^3\,{}^3D_2$	28
C I		1561.367	$2s^2\,2p^2\,{}^3P_2{-}2s\,2p^3\,{}^3D_1$	
C I		1561.438	$2s^2\,2p^2\,{}^3P_2{-}2s\,2p^3\,{}^3D_3$	
Si I	1561.822	1561.822	$3s^2\,3p^2\,{}^3P_1{-}3s^2\,3p\,8d\,{}^3D_2$	28
Si I	1562.005	1562.006	$3s^2\,3p^2\,{}^3P_2{-}3s^2\,3p\,9d\,{}^1D_2$	28
Si I		1562.002	$3s^2\,3p^2\,{}^3P_0{-}3s^2\,3p\,8d\,{}^1P_1$	28
Si I	1562.28	1562.286	$3s^2\,3p^2\,{}^3P_1{-}3s^2\,3p\,10s\,{}^3P_1$	28
Si I	1563.365	1563.364	$3s^2\,3p^2\,{}^3P_2{-}3s^2\,3p\,8d\,{}^1F_3$	28
Fe II	1563.788	1563.788	$3d^7\,{}^4F_{7/2}{-}3d^6\,4p\,{}^4F_{7/2}$	28
Si I	1564.613	1564.614	$3s^2\,3p^2\,{}^3P_1{-}3s^2\,3p\,8d\,{}^3F_2$	28
Si I	1565.31	1565.322	$3s^2\,3p^2\,{}^3P_0{-}3s^2\,3p\,9s\,{}^1P_1$	28
Si I	1565.38	1565.395	$3s^2\,3p^2\,{}^3P_2{-}3s^2\,3p\,8d\,{}^3D_2$	28
Fe II	1566.817	1566.819	$3d^7\,{}^4F_{9/2}{-}3d^6\,4p\,{}^4G_{9/2}$	28
Si I	1567.728	1567.726	$3s^2\,3p^2\,{}^3P_1{-}3s^2\,3p\,7d\,{}^3P_0$	28
Si I		1567.726	$3s^2\,3p^2\,{}^3P_1{-}3s^2\,3p\,9s\,{}^3P_2$	
Fe II	1568.016	1568.016	$3d^7\,{}^4F_{5/2}{-}3d^6\,4p\,{}^4F_{3/2}$	28
Si I	1568.196	1568.196	$3s^2\,3p^2\,{}^3P_1{-}3s^2\,3p\,7d\,{}^3P_1$	28
Si I		1568.196	$3s^2\,3p^2\,{}^3P_2{-}3s^2\,3p\,8d\,{}^3F_2$	

Si I	1568.619	1568.618	$3s^2\,3p^2\,{}^3P_0$–$3s^2\,3p\,7d\,{}^3D_1$	28
Si I	1569.319	1569.319	$3s^2\,3p^2\,{}^3P_2$–$3s^2\,3p\,8d\,{}^3F_3$	28
Fe II	1569.673	1569.674	$3d^7\,{}^4F_{9/2}$–$3d^6\,4p\,{}^4G_{11/2}$	28
Si I	1570.026	1570.027	$3s^2\,3p^2\,{}^3P_1$–$3s^2\,3p\,7d\,{}^3P_1$	28
Fe II	1570.242	1570.242	$3d^7\,{}^4F_{5/2}$–$3d^6\,4p\,{}^4F_{5/2}$	28
Si I	1570.516	1570.517	$3s^2\,3p^2\,{}^3P_1$–$3s^2\,3p\,7d\,{}^3D_1$	28
Fe II	1571.05	1571.065	$3d^6\,4s\,{}^4D_{5/2}$–$3d^6\,4p\,{}^4P_{3/2}$	28
Fe II	1571.140	1571.137	$3d^7\,{}^4F_{7/2}$–$3d^6\,4p\,{}^4G_{5/2}$	28
Si I	1571.407	1571.406	$3s^2\,3p^2\,{}^3P_2$–$3s^2\,3p\,8d\,{}^1D_2$	28
Si I	1571.797	1571.796	$3s^2\,3p^2\,{}^3P_2$–$3s^2\,3p\,7d\,{}^3P_1$	28
Si I	1572.74	1572.717	$3s^2\,3p^2\,{}^3P_0$–$3s^2\,3p\,9s\,{}^3P_1$	28
Fe II		1572.750	$3d^7\,{}^4F_{7/2}$–$3d^6\,4p\,{}^4F_{9/2}$	
Fe II	1572.99	1573.000	$3d^6\,4s\,{}^4D_{5/2}$–$3d^6\,4p\,{}^4P_{5/2}$	28
Si I	1573.634	1573.635	$3s^2\,3p^2\,{}^3P_2$–$3s^2\,3p\,7d\,{}^3P_2$	28
Fe II	1573.83	1573.825	$3d^7\,{}^4F_{5/2}$–$3d^6\,4p\,{}^4F_{7/2}$	28
Si I	1573.883	1573.884	$3s^2\,3p^2\,{}^3P_2$–$3s^2\,3p\,7d\,{}^3D_3$	28
Fe II	1574.03	1574.038	$3d^7\,{}^2H_{11/2}$–$3d^6\,4p\,{}^2G_{9/2}$	28
Si I	1574.809	1574.810	$3s^2\,3p^2\,{}^3P_1$–$3s^2\,3p\,9s\,{}^3P_0$	28
Fe II	1574.91	1574.923	$3d^7\,{}^4F_{3/2}$–$3d^6\,4p\,{}^4F_{3/2}$	28
Si I	1575.11	1575.127	$3s^2\,3p^2\,{}^3P_0$–$3s^2\,3p\,7d\,{}^1P_1$	28
Si I	1576.825	1576.825	$3s^2\,3p^2\,{}^3P_2$–$3s^2\,3p\,7d\,{}^1F_3$	28
Si I	1577.039	1577.044	$3s^2\,3p^2\,{}^3P_1$–$3s^2\,3p\,7d\,{}^1P_1$	28
Fe II	1577.164	1577.166	$3d^7\,{}^4F_{3/2}$–$3d^6\,4p\,{}^4F_{5/2}$	28
Si I	1578.48	1578.478	$3s^2\,3p^2\,{}^3P_2$–$3s^2\,3p\,7d\,{}^3D_2$	28
Fe II	1578.501	1578.497	$3d^7\,{}^2D_{5/2}$–$3d^6\,4p\,{}^2D_{5/2}$	28
Fe II	1580.625	1580.625	$3d^7\,{}^4F_{7/2}$–$3d^6\,4p\,{}^4G_{9/2}$	28
H_2	1581.106	1581.103	$2\ X^1\Sigma_g^+$–$9P4\ B^1\Sigma_u^+$	28
Fe II	1581.273	1581.274	$3d^7\,{}^4F_{5/2}$–$3d^6\,4p\,{}^4G_{5/2}$	28
Fe II	1584.947	1584.949	$3d^7\,{}^4F_{5/2}$–$3d^6\,4p\,{}^4G_{7/2}$	28
Fe II	1585.990	1585.985	$3d^7\,{}^2D_{3/2}$–$3d^6\,4p\,{}^2G_{3/2}$	28
Si I	1586.132	1586.137	$3s^2\,3p^2\,{}^3P_1$–$3s^2\,3p\,6d\,{}^3P_0$	28
Si I	1586.792	1586.791	$3s^2\,3p^2\,{}^3P_1$–$3s^2\,3p\,6d\,{}^3P_1$	28
Si I	1586.892	1586.892	$3s^2\,3p^2\,{}^3P_1$–$3s^2\,3p\,8s\,{}^3P_2$	28
Si I	1587.764	1587.762	$3s^2\,3p^2\,{}^3P_2$–$3s^2\,3p\,7d\,{}^1D_2$	28
Fe II	1588.286	1588.286	$3d^7\,{}^4F_{3/2}$–$3d^6\,4p\,{}^4G_{5/2}$	28
Si I	1590.479	1590.477	$3s^2\,3p^2\,{}^3P_2$–$3s^2\,3p\,6d\,{}^3P_1$	28
Si I	1591.126	1591.123	$3s^2\,3p^2\,{}^3P_1$–$3s^2\,3p\,6d\,{}^3D_1$	28
Si I	1592.425	1592.423	$3s^2\,3p^2\,{}^3P_2$–$3s^2\,3p\,6d\,{}^3P_2$	28
Si I	1594.567	1594.566	$3s^2\,3p^2\,{}^3P_2$–$3s^2\,3p\,6d\,{}^3D_3$	28
Si I	1595.756	1595.755	$3s^2\,3p^2\,{}^3P_0$–$3s^2\,3p\,6d\,{}^1P_1$	28
Si I	1597.719	1597.720	$3s^2\,3p^2\,{}^3P_1$–$3s^2\,3p\,6d\,{}^1P_1$	28
Si I	1597.963	1597.962	$3s^2\,3p^2\,{}^3P_2$–$3s^2\,3p\,6d\,{}^1F_3$	28
Fe II	1598.45	1598.455	$3d^7\,{}^2D_{3/2}$–$3d^6\,4p\,{}^2D_{5/2}$	28
Fe II	1600.013	1600.013	$3d^7\,{}^2H_{9/2}$–$3d^6\,4p\,{}^2G_{7/2}$	28
Fe II	1602.208	1602.210	$3d^7\,{}^4F_{9/2}$–$3d^6\,4p\,{}^2F_{7/2}$	28
Fe II	1602.599	1602.596	$3d^6\,4p\,{}^6F_{11/2}$–$3d^6\,5d\,{}^6G_{13/2}$	28
C I	1602.972	1602.972	$2s^2\,2p^2\,{}^1S_0$–$2s^2\,2p\,4d\,{}^1P_1$	28
Fe II	1604.582	1604.583	$3d^7\,{}^2G_{7/2}$–$3d^6\,4p\,{}^2D_{5/2}$	28
Fe II	1605.319	1605.318	$3d^7\,{}^2G_{9/2}$–$3d^6\,4p\,{}^2F_{7/2}$	28
Si I	1605.837	1605.837	$3s^2\,3p^2\,{}^3P_1$–$3s^2\,3p\,6d\,{}^3F_2$	28
C I	1606.962	1606.960	$2s^2\,2p^2\,{}^1S_0$–$2s^2\,2p\,5s\,{}^1P_1$	28
C I	1608.45	1608.438	$2s^2\,2p^2\,{}^1S_0$–$2s^2\,2p\,4d\,{}^3D_1$	28
Fe II		1608.456	$3d^6\,4s\,{}^6D_{9/2}$–$3d^6\,4p\,{}^6P_{7/2}$	
Si I	1608.915	1608.916	$3s^2\,3p^2\,{}^3P_2$–$3s^2\,3p\,6d\,{}^3F_3$	28
Fe II	1610.93	1610.921	$3d^7\,{}^4F_{9/2}$–$3d^6\,4p\,{}^4G_{9/2}$	28
Fe II	1611.197	1611.201	$3d^6\,4s\,{}^6D_{9/2}$–$3d^6\,4p\,{}^4F_{7/2}$	28
Fe II	1612.800	1612.802	$3d^7\,{}^4F_{9/2}$–$3d^6\,4p\,{}^4G_{11/2}$	28
C I	1613.375	1613.376	$2s^2\,2p^2\,{}^3P_0$–$2s^2\,2p\,3s\,{}^1P_1$	28
C I	1613.809	1613.803	$2s^2\,2p^2\,{}^3P_1$–$2s^2\,2p\,3s\,{}^1P_1$	28
C I	1614.51	1614.507	$2s^2\,2p^2\,{}^3P_2$–$2s^2\,2p\,3s\,{}^1P_1$	28
S XI		1614.495	$2s^2\,2p^2\,{}^3P_1$–$2s^2\,2p^2\,{}^1D_2$	
Si I	1615.951	1615.949	$3s^2\,3p^2\,{}^3P_1$–$3s^2\,3p\,5d\,{}^3P_0$	28

continued.

Ion	Wvl. (Å)	$Wvl.^L$ (Å)	Transition	Ref.
Si I	1616.57	1616.579	$3s^2\,3p^2\,{}^3P_1$–$3s^2\,3p\,5d\,{}^3P_1$	28
Fe II	1616.649	1616.652	$3d^7\,{}^4F_{7/2}$–$3d^6\,4p\,{}^2F_{7/2}$	28
Fe II	1618.464	1618.470	$3d^6\,4s\,{}^6D_{7/2}$–$3d^6\,4p\,{}^6P_{7/2}$	28
Fe II	1620.05	1620.061	$3d^6\,4s\,{}^6D_{9/2}$–$3d^6\,4p\,{}^4D_{7/2}$	28
Fe II	1621.246	1621.252	$3d^6\,4s\,{}^6D_{7/2}$–$3d^6\,4p\,{}^4F_{7/2}$	28
Fe II	1621.69	1621.685	$3d^6\,4s\,{}^6D_{7/2}$–$3d^6\,4p\,{}^6P_{5/2}$	28
Fe II	1621.86	1621.867	$3d^7\,{}^4F_{7/2}$–$3d^6\,4p\,{}^4G_{5/2}$	28
Si I	1622.876	1622.881	$3s^2\,3p^2\,{}^3P_2$–$3s^2\,3p\,5d\,{}^3P_2$	28
Fe II	1623.090	1623.091	$3d^7\,{}^4F_{7/2}$–$3d^6\,4p\,{}^4G_{7/2}$	28
O VII	1623.61	1623.609	$1s\,2s\,{}^3S_1$–$1s\,2p\,{}^3P_2$	28
Fe II	1623.708	1623.705	$3d^7\,{}^4F_{5/2}$–$3d^6\,4p\,{}^2F_{5/2}$	28
Fe II	1625.518	1625.520	$3d^7\,{}^4F_{7/2}$–$3d^6\,4p\,{}^4G_{9/2}$	28
Si I	1625.699	1625.705	$3s^2\,3p^2\,{}^3P_0$–$3s^2\,3p\,5d\,{}^3D_1$	28
Fe II	1625.907	1625.909	$3d^6\,4s\,{}^6D_{5/2}$–$3d^6\,4p\,{}^6P_{7/2}$	28
Si I	1627.04	1627.050	$3s^2\,3p^2\,{}^3P_1$–$3s^2\,3p\,7s\,{}^3P_0$	28
Fe II	1627.12	1627.130	$3d^6\,4s\,{}^6D_{5/2}$–$3d^6\,4p\,{}^2D_{3/2}$	28
Fe II	1627.382	1627.382	$3d^7\,{}^4F_{5/2}$–$3d^6\,4p\,{}^2F_{7/2}$	28
Fe II	1628.719	1628.722	$3d^6\,4s\,{}^6D_{5/2}$–$3d^6\,4p\,{}^4F_{7/2}$	28
Fe II	1629.15	1629.154	$3d^6\,4s\,{}^6D_{5/2}$–$3d^6\,4p\,{}^6P_{5/2}$	28
Fe II	1629.378	1629.376	$3d^7\,{}^2G_{7/2}$–$3d^6\,4p\,{}^2F_{5/2}$	28
Si I	1629.43	1629.438	$3s^2\,3p^2\,{}^3P_1$–$3s^2\,3p\,5d\,{}^3D_2$	28
Si I		1629.404	$3s^2\,3p^2\,{}^3P_2$–$3s^2\,3p\,7s\,{}^3P_1$	
Si I	1629.946	1629.948	$3s^2\,3p^2\,{}^3P_2$–$3s^2\,3p\,5d\,{}^3D_3$	28
Fe II	1632.30	1632.307	$3d^6\,4s\,{}^6D_{3/2}$–$3d^6\,4p\,{}^2D_{3/2}$	28
Fe II	1632.666	1632.668	$3d^7\,{}^4F_{5/2}$–$3d^6\,4p\,{}^4G_{5/2}$	28
Si I	1633.23	1633.223	$3s^2\,3p^2\,{}^3P_1$–$3s^2\,3p\,6d\,{}^1P_1$	28
Fe II	1633.906	1633.908	$3d^7\,{}^4F_{5/2}$–$3d^6\,4p\,{}^4G_{7/2}$	28
Fe II	1634.343	1634.345	$3d^6\,4s\,{}^6D_{3/2}$–$3d^6\,4p\,{}^6P_{5/2}$	28
Ni II	1635.34	1635.340	$3d^8\,4p\,{}^4G_{5/2}$–$3d^8\,4d\,{}^4F_{3/2}$	28
Fe II	1636.319	1636.321	$3d^6\,4s\,{}^6D_{3/2}$–$3d^6\,4p\,{}^6P_{3/2}$	28
Fe II	1637.39	1637.397	$3d^7\,{}^4F_{9/2}$–$3d^6\,4p\,{}^2H_{11/2}$	28
Fe II	1639.40	1639.403	$3d^6\,4s\,{}^6D_{1/2}$–$3d^6\,4p\,{}^6P_{3/2}$	28
O VII	1639.78	1639.861	$1s\,2s\,{}^3S_1$–$1s\,2p\,{}^3P_0$	28
Fe II	1640.15	1640.152	$3d^7\,{}^4F_{3/2}$–$3d^6\,4p\,{}^4G_{5/2}$	28
He II	1640.4	1640.336	$2p\,{}^2P_{1/2}$–$3d\,{}^2D_{3/2}$	28
He II		1640.348	$2s\,{}^2S_{1/2}$–$3p\,{}^2P_{3/2}$	
He II		1640.378	$2p\,{}^2P_{1/2}$–$3s\,{}^2S_{1/2}$	
He II		1640.394	$2s\,{}^2S_{1/2}$–$3p\,{}^2P_{1/2}$	
He II		1640.477	$2p\,{}^2P_{3/2}$–$3d\,{}^2D_{5/2}$	
He II		1640.493	$2p\,{}^2P_{3/2}$–$3d\,{}^2D_{3/2}$	
He II		1640.536	$2p\,{}^2P_{3/2}$–$3s\,{}^2S_{1/2}$	
S I	1641.31	1641.296	$3s^2\,3p^4\,{}^1D_2$–$3s^2\,3p^3\,3d\,{}^3D_2$	28
Fe II	1642.48	1642.496	$3d^7\,{}^2P_{3/2}$–$3d^6\,4p\,{}^2P_{3/2}$	28
Fe II	1643.578	1643.576	$3d^7\,{}^4F_{7/2}$–$3d^6\,4p\,{}^4D_{5/2}$	28
Fe II	1645.008	1645.016	$3d^7\,{}^4F_{9/2}$–$3d^6\,4p\,{}^2H_{11/2}$	28
Fe II	1647.16	1647.159	$3d^6\,4s\,{}^4D_{5/2}$–$3d^5\,4s\,4p\,{}^4P_{5/2}$	28
Fe II	1647.546	1647.546	$3d^7\,{}^2F_{7/2}$–$3d^6\,4p\,{}^2D_{5/2}$	28
Fe II	1648.398	1648.403	$3d^6\,4p\,{}^4D_{5/2}$–$3d^6\,5d\,{}^4G_{7/2}$	28
Fe II	1649.423	1649.423	$3d^7\,{}^4F_{5/2}$–$3d^6\,4p\,{}^4D_{3/2}$	28
Si I	1653.39	1653.376	$3s^2\,3p^2\,{}^3P_2$–$3s^2\,3p\,5d\,{}^3F_3$	28
CO	1653.82	1653.800	$3\ A\,{}^1\Pi$–$4Q25\ X\,{}^1\Sigma^+$	28
CO		1653.850	$0\ A\,{}^1\Pi$–$2R15\ X\,{}^1\Sigma^+$	
Fe II	1654.26	1654.263	$3d^7\,{}^4F_{9/2}$–$3d^6\,4p\,{}^2G_{7/2}$	28
Fe II	1654.47	1654.476	$3d^7\,{}^4F_{3/2}$–$3d^6\,4p\,{}^4D_{1/2}$	28
Fe II	1654.65	1654.670	$3d^7\,{}^4F_{5/2}$–$3d^6\,4p\,{}^4D_{5/2}$	28
C I	1656.266	1656.266	$2s^2\,2p^2\,{}^3P_1$–$2s^2\,2p\,3s\,{}^3P_2$	28
Fe II	1656.74	1656.743	$3d^7\,{}^2P_{1/2}$–$3d^5\,4s\,4p\,{}^6P_{3/2}$	28
C I	1656.926	1656.928	$2s^2\,2p^2\,{}^3P_0$–$2s^2\,2p\,3s\,{}^3P_1$	28
C I	1657.01	1657.008	$2s^2\,2p^2\,{}^3P_2$–$2s^2\,2p\,3s\,{}^3P_2$	28
C I	1657.38	1657.380	$2s^2\,2p^2\,{}^3P_1$–$2s^2\,2p\,3s\,{}^3P_1$	28
C I	1657.91	1657.907	$2s^2\,2p^2\,{}^3P_1$–$2s^2\,2p\,3s\,{}^3P_0$	28
C I	1658.12	1658.122	$2s^2\,2p^2\,{}^3P_2$–$2s^2\,2p\,3s\,{}^3P_1$	28

Fe II	1658.404	1658.401	$3d^7\,^4F_{9/2}$–$3d^5\,4s\,4p\,^6P_{7/2}$	28	Fe II	1689.822	1689.828	$3d^7\,^4P_{5/2}$–$3d^6\,4p\,^4D_{7/2}$	28
Fe II	1658.769	1658.771	$3d^7\,^4F_{9/2}$–$3d^6\,4p\,^4F_{9/2}$	28	Fe II	1690.75	1690.759	$3d^7\,^4P_{5/2}$–$3d^6\,4p\,^4D_{5/2}$	28
CO	1659.40	1659.400	$0\ A^1\Pi$–$2R34\ X^1\Sigma^+$	28	Fe II	1691.270	1691.271	$3d^7\,^4F_{3/2}$–$3d^6\,4p\,^4F_{3/2}$	28
Fe II	1659.484	1659.483	$3d^7\,^4F_{7/2}$–$3d^6\,4p\,^4D_{5/2}$	28	Fe II	1692.52	1692.516	$3d^7\,^4F_{9/2}$–$3d^6\,4p\,^4G_{7/2}$	28
O III	1660.802	1660.809	$2s^2\,2p^2\,^3P_1$–$2s\,2p^3\,^5S_2$	28	Fe II	1693.931	1693.936	$3d^7\,^4F_{3/2}$–$3d^6\,4p\,^4F_{5/2}$	28
Fe II	1661.31	1661.324	$3d^7\,^4F_{9/2}$–$3d^6\,4p\,^4F_{7/2}$	28	Fe II	1694.677	1694.681	$3d^7\,^4F_{3/2}$–$3d^6\,4p\,^2D_{3/2}$	28
Fe II	1662.363	1662.369	$3d^7\,^4F_{3/2}$–$3d^6\,4p\,^4D_{5/2}$	28	Fe II	1696.79	1696.794	$3d^7\,^4F_{9/2}$–$3d^6\,4p\,^4G_{9/2}$	28
CO	1662.557	1662.600	$0\ A^1\Pi$–$2P29\ X^1\Sigma^+$	28	Fe II	1698.13	1698.135	$3d^7\,^4F_{5/2}$–$3d^6\,4p\,^4D_{7/2}$	28
Fe II	1663.223	1663.221	$3d^7\,^4F_{5/2}$–$3d^6\,4p\,^4D_{3/2}$	28	Fe II	1699.186	1699.193	$3d^7\,^4P_{3/2}$–$3d^6\,4p\,^4D_{3/2}$	28
Fe II	1663.687	1663.697	$3d^7\,^4F_{5/2}$–$3d^6\,4p\,^4D_{7/2}$	28	Fe II	1702.03	1702.043	$3d^7\,^4F_{9/2}$–$3d^6\,4p\,^4G_{11/2}$	28
O III	1666.157	1666.150	$2s^2\,2p^2\,^3P_2$–$2s\,2p^3\,^5S_2$	28	Ni II	1703.400	1703.408	$3d^9\,^2D_{5/2}$–$3d^8\,4p\,^2D_{3/2}$	28
S I	1666.68	1666.688	$3s^2\,3p^4\,^1D_2$–$3s^2\,3p^3\,4s\,^1D_2$	28	Fe II	1704.66	1704.652	$3d^7\,^4F_{7/2}$–$3d^6\,4p\,^2D_{5/2}$	28
Fe II	1667.904	1667.913	$3d^7\,^2P_{1/2}$–$3d^6\,4p\,^2P_{1/2}$	28	Fe II	1706.143	1706.142	$3d^7\,^4F_{7/2}$–$3d^6\,4p\,^4G_{5/2}$	28
Fe II	1669.65	1669.663	$3d^7\,^4F_{7/2}$–$3d^6\,4p\,^2G_{7/2}$	28	Fe II	1707.397	1707.399	$3d^7\,^4P_{5/2}$–$3d^6\,4p\,^4P_{3/2}$	28
Al II	1670.782	1670.787	$3s^2\,^1S_0$–$3s\,3p\,^1P_1$	12	Fe II	1707.669	1707.669	$3d^7\,^4F_{7/2}$–$3d^6\,4p\,^4H_{9/2}$	28
Fe II	1673.459	1673.462	$3d^7\,^2G_{9/2}$–$3d^6\,4p\,^2F_{7/2}$	28	Fe II	1708.24	1708.250	$3d^7\,^4P_{3/2}$–$3d^6\,4p\,^4P_{1/2}$	28
Ca II	1673.87	1673.860	$4p\,^2P_{1/2}$–$7d\,^2D_{3/2}$	28	Fe II	1708.618	1708.621	$3d^7\,^4F_{7/2}$–$3d^6\,4p\,^4G_{7/2}$	28
Fe II	1674.253	1674.254	$3d^7\,^4F_{7/2}$–$3d^6\,4p\,^4F_{9/2}$	28	Ni II	1709.58	1709.598	$3d^9\,^2D_{5/2}$–$3d^8\,4p\,^2F_{5/2}$	28
Fe II	1674.43	1674.440	$3d^7\,^4F_{7/2}$–$3d^6\,4p\,^4F_{5/2}$	28	Fe II	1713.00	1712.997	$3d^7\,^4F_{7/2}$–$3d^6\,4p\,^4G_{9/2}$	12
Fe II	1674.715	1674.716	$3d^7\,^4F_{3/2}$–$3d^6\,4p\,^4D_{1/2}$	28	Fe II	1715.50	1715.503	$3d^7\,^4P_{3/2}$–$3d^6\,4p\,^4P_{5/2}$	12
Si I	1675.201	1675.205	$3s^2\,3p^2\,^3P_2$–$3s^2\,3p\,4d\,^3P_2$	28	Fe II	1716.57	1716.577	$3d^7\,^4F_{5/2}$–$3d^6\,4p\,^2D_{5/2}$	12
Fe II	1676.359	1676.361	$3d^7\,^4F_{7/2}$–$3d^6\,4p\,^2G_{9/2}$	28	Fe II	1718.09	1718.123	$3d^7\,^4F_{5/2}$–$3d^6\,4p\,^4G_{5/2}$	12
Fe II	1676.85	1676.856	$3d^7\,^4F_{7/2}$–$3d^6\,4p\,^4F_{7/2}$	28	Fe II	1720.61	1720.616	$3d^7\,^4F_{5/2}$–$3d^6\,4p\,^4G_{7/2}$	12
Fe II	1677.84	1677.847	$3d^7\,^2P_{1/2}$–$3d^6\,4p\,^2D_{3/2}$	28	Fe II	1724.87	1724.854	$3d^7\,^4F_{3/2}$–$3d^6\,4p\,^2D_{5/2}$	12
Fe II	1679.38	1679.381	$3d^7\,^2G_{7/2}$–$3d^6\,4p\,^2F_{5/2}$	28	Fe II	1726.39	1726.391	$3d^7\,^4F_{3/2}$–$3d^6\,4p\,^4G_{5/2}$	12
Fe II	1679.49	1679.494	$3d^7\,^4P_{5/2}$–$3d^6\,4p\,^2{}_{7/2}$	28	Ni II	1741.55	1741.547	$3d^9\,^2D_{5/2}$–$3d^8\,4p\,^2D_{5/2}$	12
Fe II	1681.113	1681.111	$3d^7\,^4F_{5/2}$–$3d^6\,4p\,^2G_{7/2}$	28	N III	1746.82	1746.823	$2s^2\,2p\,^2P_{1/2}$–$2s\,2p^2\,^4P_{3/2}$	12
Fe II	1685.954	1685.954	$3d^7\,^4F_{5/2}$–$3d^6\,4p\,^4F_{5/2}$	28	Fe II		1746.818	$3d^7\,^2G_{9/2}$–$3d^6\,4p\,^2G_{9/2}$	
Fe II	1686.449	1686.455	$3d^7\,^4F_{7/2}$–$3d^6\,4p\,^4D_{7/2}$	28	Ni II	1748.28	1748.285	$3d^9\,^2D_{3/2}$–$3d^8\,4p\,^2D_{3/2}$	12
Fe II	1686.69	1686.692	$3d^7\,^4F_{5/2}$–$3d^6\,4p\,^2D_{3/2}$	28	N III	1748.63	1748.646	$2s^2\,2p\,^2P_{1/2}$–$2s\,2p^2\,^4P_{1/2}$	12
Fe II	1688.40	1688.401	$3d^7\,^4F_{5/2}$–$3d^6\,4p\,^4F_{7/2}$	28	N III	1749.67	1749.674	$2s^2\,2p\,^2P_{3/2}$–$2s\,2p^2\,^4P_{5/2}$	12

continued.

Ion	Wvl. (Å)	$Wvl.^L$ (Å)	Transition	Ref.	Ion	Wvl. (Å)	$Wvl.^L$ (Å)	Transition	Ref.
Ni II	1751.91	1751.911	$3d^9\,^2D_{5/2}$–$3d^8\,4p\,^2F_{7/2}$	12	Fe II	1835.86	1835.874	$3d^7\,^2G_{9/2}$–$3d^6\,4p\,^2G_{9/2}$	12
N III	1752.12	1752.160	$2s^2\,2p\,^2P_{3/2}$–$2s\,2p^2\,^4P_{3/2}$	12	Al III	1854.72	1854.720	$3s\,^2S_{1/2}$–$3p\,^2P_{3/2}$	12
N III	1753.98	1753.995	$2s^2\,2p\,^2P_{3/2}$–$2s\,2p^2\,^4P_{1/2}$	12	Fe II	1859.74	1859.741	$3d^6\,4s\,^4D_{7/2}$–$3d^6\,4p\,^4D_{7/2}$	12
Ni II	1754.80	1754.808	$3d^9\,^2D_{3/2}$–$3d^8\,4p\,^2F_{5/2}$	12	Fe II	1860.04	1860.055	$3d^7\,^2G_{9/2}$–$3d^6\,4p\,^2F_{7/2}$	12
Fe II	1761.36	1761.379	$3d^7\,^2G_{7/2}$–$3d^6\,4p\,^2G_{7/2}$	12	Al III	1862.79	1862.793	$3s\,^2S_{1/2}$–$3p\,^2P_{1/2}$	12
Al II	1763.95	1763.952	$3s\,3p\,^3P_2$–$3p^2\,^3P_2$	12	Fe II	1864.72	1864.743	$3d^7\,^2H_{11/2}$–$3d^6\,4p\,^2I_{13/2}$	12
Fe II	1772.50	1772.509	$3d^7\,^2G_{9/2}$–$3d^6\,4p\,^2H_{11/2}$	12	Fe II	1876.83	1876.838	$3d^7\,^2G_{7/2}$–$3d^6\,4p\,^2F_{5/2}$	12
Ni II	1773.94	1773.949	$3d^9\,^2D_{5/2}$–$3d^8\,4p\,^2G_{7/2}$	12	Fe II	1877.46	1877.467	$3d^7\,^2H_{11/2}$–$3d^6\,4p\,^2H_{11/2}$	12
Ni II	1788.48	1788.485	$3d^9\,^2D_{3/2}$–$3d^8\,4p\,^2D_{5/2}$	12	Si III	1892.04	1892.033	$3s^2\,^1S_0$–$3s\,3p\,^3P_1$	12
Fe II	1793.36	1793.367	$3d^7\,^2G_{7/2}$–$3d^6\,4p\,^2H_{9/2}$	12	Fe III	1895.48	1895.457	$3d^5\,4s\,^7S_3$–$3d^5\,4p\,^7P_4$	12
Fe II	1798.15	1798.156	$3d^7\,^2D_{5/2}$–$3d^6\,4p\,^2P_{3/2}$	12	C III	1908.74	1908.734	$2s^2\,^1S_0$–$2s\,2p\,^3P_1$	12
Si II	1808.01	1808.014	$3s^2\,3p\,^2P_{1/2}$–$3s\,3p^2\,^2D_{3/2}$	12	Fe III	1914.07	1914.071	$3d^5\,4s\,^7S_3$–$3d^5\,4p\,^7P_3$	12
Si II	1816.93	1816.928	$3s^2\,3p\,^2P_{3/2}$–$3s\,3p^2\,^2D_{5/2}$	12	Fe II	1925.99	1925.983	$3d^7\,^2H_{11/2}$–$3d^6\,4p\,^2H_{11/2}$	12
Si II	1817.44	1817.451	$3s^2\,3p\,^2P_{3/2}$–$3s\,3p^2\,^2D_{3/2}$	12	Fe III	1926.31	1926.323	$3d^5\,4s\,^7S_3$–$3d^5\,4p\,^7P_2$	12
Fe II	1818.51	1818.509	$3d^6\,4s\,^4D_{7/2}$–$3d^6\,4p\,^4D_{7/2}$	12	C I	1930.91	1930.905	$2s^2\,2p^2\,^1D_2$–$2s^2\,2p\,3s\,^1P_1$	12

Wvl. is solar wavelength.
$Wvl.^L$ is laboratory wavelength.

Principal spectral lines in solar off-limb spectrum

Ion	Wvl. (Å)	Wvl.^{L} (Å)	Transition	Ref.	Ion	Wvl. (Å)	Wvl.^{L} (Å)	Transition	Ref.
Si XII	520.66	520.666	$1s^2\,2s\,^2S_{1/2}$–$1s^2\,2p\,^2P_{1/2}$	9	Ar XII	670.31	670.355	$2s^2\,2p^3\,^4S_{3/2}$–$2s^2\,2p^3\,^2P_{1/2}$	9
Ca XIV	545.23	545.239	$2s^2\,2p^3\,^4S_{3/2}$–$2s^2\,2p^3\,^2P_{3/2}$	9	Si IX	676.51	676.710	$2s^2\,2p^2\,^3P_1$–$2s\,2p^3\,^5S_2$	9
Al XI	550.04	550.032	$1s^2\,2s\,^2S_{1/2}$–$1s^2\,2p\,^2P_{3/2}$	9	Fe XX	679.27	679.261	$2s^2\,2p^3\,^2D_{5/2}$–$2s^2\,2p^3\,^2P_{3/2}$	9
S XI	552.09	552.356	$2s^2\,2p^2\,^3P_1$–$2s\,2p^3\,^5S_2$	9	Mg VIII	679.79	679.779	$2s\,2p^2\,^2P_{1/2}$–$2p^3\,^2D_{3/2}$	9
Ca XV	555.38	555.270	$2s^2\,2p^2\,^3P_1$–$2s^2\,2p^2\,^1S_0$	9	Al IX	680.39	680.283	$2s^2\,2p\,^2P_{1/2}$–$2s\,2p^2\,^4P_{3/2}$	9
Ca X	557.76	557.766	$3s\,^2S_{1/2}$–$3p\,^2P_{3/2}$	9	Na IX	681.70	681.720	$1s^2\,2s\,^2S_{1/2}$–$1s^2\,2p\,^2P_{3/2}$	9
Si XI	564.00	563.958	$2s^2\,^1S_0$–$2s\,2p\,^3P_2$	9	Al IX	688.24	688.289	$2s^2\,2p\,^2P_{1/2}$–$2s\,2p^2\,^4P_{1/2}$	9
Fe XX	567.83	567.867	$2s^2\,2p^3\,^4S_{3/2}$–$2s^2\,2p^3\,^2D_{5/2}$	9	Mg VIII	689.63	689.642	$2s\,2p^2\,^2P_{3/2}$–$2p^3\,^2D_{5/2}$	9
Al XI	568.16	568.122	$1s^2\,2s\,^2S_{1/2}$–$1s^2\,2p\,^2P_{1/2}$	9	Ca IX	691.42	691.206	$3s^2\,^1S_0$–$3s\,3p\,^3P_1$	9
Ca X	574.01	574.011	$3s\,^2S_{1/2}$–$3p\,^2P_{1/2}$	9	Al IX		691.574	$2s^2\,2p\,^2P_{3/2}$–$2s\,2p^2\,^4P_{5/2}$	
S XI	574.88	579.859	$2s^2\,2p^3\,^4S_{3/2}$–$2s^2\,2p^3\,^2P_{1/2}$	9	Mg IX	693.96	694.005	$2s^2\,^1S_0$–$2s\,2p\,^3P_2$	9
Si XI	580.91	580.908	$2s^2\,^1S_0$–$2s\,2p\,^3P_1$	9	Na IX	694.13	694.147	$1s^2\,2s\,^2S_{1/2}$–$1s^2\,2p\,^2P_{1/2}$	9
Ca VIII	582.86	582.832	$3s^2\,3p\,^2P_{1/2}$–$3s\,3p^2\,^2D_{3/2}$	9	Si IX	694.69	694.904	$2s^2\,2p^2\,^3P_2$–$2s\,2p^3\,^5S_2$	9
Ar VII	585.75	585.756	$3s^2\,^1S_0$–$3s\,3p\,^1P_1$	9	Fe VIII	697.17	697.157	$3p^6\,4p\,^2P_{1/2}$–$3p^6\,4d\,^2D_{3/2}$	9
Fe XXI	585.77	585.767	$2s^2\,2p^2\,^3P_1$–$2s^2\,2p^2\,^1D_2$	9	Ar VIII	700.24	700.246	$3s\,^2S_{1/2}$–$3p\,^2P_{3/2}$	9
Fe XIX	592.23	592.236	$2s^2\,2p^4\,^3P_2$–$2s^2\,2p^4\,^1D_2$	9	Al IX	703.64	703.691	$2s^2\,2p\,^2P_{3/2}$–$2s\,2p^2\,^4P_{3/2}$	9
Ca VIII	596.99	596.926	$3s^2\,3p\,^2P_{3/2}$–$3s\,3p^2\,^2D_{5/2}$	9	Mg IX	706.04	706.060	$2s^2\,^1S_0$–$2s\,2p\,^3P_1$	9
Si XI	604.15	604.149	$2s\,2p\,^1P_1$–$2p^2\,^1D_2$	9	Ar VIII	713.82	713.813	$3s\,^2S_{1/2}$–$3p\,^2P_{1/2}$	9
Mg X	609.80	609.794	$1s^2\,2s\,^2S_{1/2}$–$1s^2\,2p\,^2P_{3/2}$	9	Fe VIII	721.26	721.257	$3p^6\,4p\,^2P_{3/2}$–$3p^6\,4d\,^2D_{5/2}$	9
Si X	611.61	611.712	$2s^2\,2p\,^2P_{1/2}$–$2s\,2p^2\,^4P_{3/2}$	9	Fe XX	721.55	721.559	$2s^2\,2p^3\,^4S_{3/2}$–$2s^2\,2p^3\,^2D_{3/2}$	9
Si X	621.11	621.115	$2s^2\,2p\,^2P_{1/2}$–$2s\,2p^2\,^4P_{1/2}$	9	Mg IX	749.56	749.551	$2s\,2p\,^1P_1$–$2p^2\,^1D_2$	9
K IX	621.405	621.455	$3s\,^2S_{1/2}$–$3p\,^2P_{3/2}$	16	Mg VIII	762.65	762.662	$2s^2\,2p\,^2P_{1/2}$–$2s\,2p^2\,^4P_{3/2}$	9
Si X	624.71	624.779	$2s^2\,2p\,^2P_{3/2}$–$2s\,2p^2\,^4P_{5/2}$	9	Mg VIII	769.38	769.356	$2s^2\,2p\,^2P_{1/2}$–$2s\,2p^2\,^4P_{1/2}$	9
Mg X	624.97	624.943	$1s^2\,2s\,^2S_{1/2}$–$1s^2\,2p\,^2P_{1/2}$	9	Ne VIII	770.42	770.410	$1s^2\,2s\,^2S_{1/2}$–$1s^2\,2p\,^2P_{3/2}$	9
Al X	637.75	637.764	$2s^2\,^1S_0$–$2s\,2p\,^3P_1$	9	Mg VIII	772.29	772.262	$2s^2\,2p\,^2P_{3/2}$–$2s\,2p^2\,^4P_{5/2}$	9
Si X	638.92	639.036	$2s^2\,2p\,^2P_{3/2}$–$2s\,2p^2\,^4P_{3/2}$	9	Al VIII	772.53	772.691	$2p^2\,^3P_2$–$2s\,2p^3\,^5S_2$	9
Ar VII	644.47	644.377	$3s\,3p\,^3P_2$–$3p^2\,^3P_1$	9	S X	776.25	776.375	$2s^2\,2p^3\,^4S_{3/2}$–$2s^2\,2p^3\,^2P_{3/2}$	9
Ca XIII	648.68	648.701	$2s^2\,2p^4\,^3P_1$–$2s^2\,2p^4\,^1S_0$	9	Ne VIII	780.34	780.325	$1s^2\,2s\,^2S_{1/2}$–$1s^2\,2p\,^2P_{1/2}$	9
Si X	649.21	649.305	$2s^2\,2p\,^2P_{3/2}$–$2s\,2p^2\,^4P_{1/2}$	9	Mg VIII	782.37	782.364	$2s^2\,2p\,^2P_{3/2}$–$2s\,2p^2\,^4P_{3/2}$	9
Ar XIII	656.69	656.604	$2s^2\,2p^2\,^3P_1$–$2s^2\,2p^2\,^1S_0$	9	S XI	782.98	782.958	$2s^2\,2p^2\,^3P_1$–$2s^2\,2p^2\,^1S_0$	9
Al X	670.02	670.054	$2s\,2p\,^1P_1$–$2p^2\,^1D_2$	9	Fe XXI	786.03	786.162	$2s^2\,2p^2\,^3P_2$–$2s^2\,2p^2\,^1D_2$	9

Continued.

Ion	Wvl. (Å)	Wvl.^{L} (Å)	Transition	Ref.
S X	787.43	787.558	$2s^2\,2p^3\,{}^4S_{3/2}-2s^2\,2p^3\,{}^2P_{1/2}$	9
Mg VIII	789.44	789.411	$2s^2\,2p\,{}^2P_{3/2}-2s\,2p^2\,{}^4P_{1/2}$	9
Cr XVIII	793.18	793.086	$2s^2\,2p^3\,{}^4S_{3/2}-2s^2\,2p^3\,{}^2D_{3/2}$	9
Si XIII	814.72	814.693	$1s\,2s\,{}^3S_1-1s\,2p\,{}^3P_2$	9
Ca IX	821.22	821.269	$3s\,3p\,{}^1P_1-3p^2\,{}^1D_2$	9
Fe XX	821.71	821.789	$2s^2\,2p^3\,{}^2D_{3/2}-2s^2\,2p^3\,{}^2P_{1/2}$	9
Fe XXII	845.57	845.571	$2s^2\,2p\,{}^2P_{1/2}-2s^2\,2p\,{}^2P_{3/2}$	9
Mg VII	854.64	854.724	$2s^2\,2p^2\,{}^3P_1-2s\,2p^3\,{}^5S_2$	9
Mg VII	868.08	868.193	$2s^2\,2p^2\,{}^3P_2-2s\,2p^3\,{}^5S_2$	9
S IX	871.71	871.726	$2s^2\,2p^4\,{}^3P_1-2s^2\,2p^4\,{}^1S_0$	9
S VIII		871.710	$2s^2\,2p^4\,3s\,{}^4P_{3/2}-2s^2\,2p^4\,3p\,{}^2D_{5/2}$	
Ne VII	895.16	895.174	$2s^2\,{}^1S_0-2s\,2p\,{}^3P_1$	9
S VI	933.41	933.380	$3s\,{}^2S_{1/2}-3p\,{}^2P_{3/2}$	9
Ca XIV	943.63	943.587	$2s^2\,2p^3\,{}^4S_{3/2}-2s^2\,2p^3\,{}^2D_{3/2}$	9
Si VIII	944.35	944.467	$2s^2\,2p^3\,{}^4S_{3/2}-2s^2\,2p^3\,{}^2P_{3/2}$	9
S VI		944.524	$3s\,{}^2S_{1/2}-3p\,{}^2P_{1/2}$	
Si VIII	949.24	949.354	$2s^2\,2p^3\,{}^4S_{3/2}-2s^2\,2p^3\,{}^2P_{1/2}$	9
Si IX	950.16	950.157	$2s^2\,2p^2\,{}^3P_1-2s^2\,2p^2\,{}^1S_0$	9
Ne VII	973.33	973.330	$2s\,2p\,{}^1P_1-2p^2\,{}^1D_2$	9
Fe XVIII	974.84	974.860	$2s^2\,2p^5\,{}^2P_{3/2}-2s^2\,2p^5\,{}^2P_{1/2}$	9
Si VIII	994.58	994.581	$2s^2\,2p^2\,3p\,{}^4D_{7/2}-2s^2\,2p^2\,3d\,{}^4F_{9/2}$	9
Mg XI	997.44	997.455	$1s\,2s\,{}^3S_1-1s\,2p\,{}^3P_2$	9
Ar XII	1018.75	1018.726	$2s^2\,2p^3\,{}^4S_{3/2}-2s^2\,2p^3\,{}^2D_{5/2}$	9
Fe X	1028.04	1028.024	$3s^2\,3p^4\,3d\,{}^4D_{7/2}-3s^2\,3p^4\,3d\,{}^2F_{7/2}$	9
Fe X		1028.082	$3s^2\,3p^4\,3d\,{}^4D_{5/2}-3s^2\,3p^4\,3d\,{}^2F_{7/2}$	
O VI	1031.93	1031.914	$1s^2\,2s\,{}^2S_{1/2}-1s^2\,2p\,{}^2P_{3/2}$	9
Ni XV	1033.01	1033.216	$3s^2\,3p^2\,{}^3P_1-3s^2\,3p^2\,{}^1S_0$	9
Ni XIV	1034.40	1034.880	$3s^2\,3p^3\,{}^4S_{3/2}-3s^2\,3p^3\,{}^2P_{3/2}$	9
O VI	1037.63	1037.615	$1s^2\,2s\,{}^2S_{1/2}-1s^2\,2p\,{}^2P_{1/2}$	9
Mg XI	1043.28	1043.299	$1s\,2s\,{}^3S_1-1s\,2p\,{}^3P_0$	9
Si VII	1049.26	1049.199	$2s^2\,2p^4\,{}^3P_1-2s^2\,2p^4\,{}^1S_0$	9
Al VII	1053.88	1053.998	$2s^2\,2p^3\,{}^4S_{3/2}-2s^2\,2p^3\,{}^2P_{3/2}$	9
Ar XII	1054.59	1054.687	$2s^2\,2p^3\,{}^4S_{3/2}-2s^2\,2p^3\,{}^2D_{3/2}$	9
Al VII	1056.79	1056.917	$2s^2\,2p^3\,{}^4S_{3/2}-2s^2\,2p^3\,{}^2P_{1/2}$	9
Al VIII	1057.86	1057.890	$2p^2\,{}^3P_1-2p^2\,{}^1S_0$	9
Fe XXIII	1079.42	1079.414	$2s\,2p\,{}^3P_1-2s\,2p\,{}^3P_2$	9
Ca XV	1098.48	1098.420	$2s^2\,2p^2\,{}^3P_1-2s^2\,2p^2\,{}^1D_2$	9
Fe XIX	1118.07	1118.057	$2s^2\,2p^4\,{}^3P_2-2s^2\,2p^4\,{}^3P_1$	9
Si VII	1132.80	1132.774	$2s^2\,2p^3\,3s\,{}^3D_3-2s^2\,2p^3\,3p\,{}^3F_4$	9
Ca XIII	1133.76	1133.758	$2s^2\,2p^4\,{}^3P_2-2s^2\,2p^4\,{}^1D_2$	9
Si VII	1135.39	1135.353	$2s^2\,2p^3\,3s\,{}^5S_2-2s^2\,2p^3\,3p\,{}^5P_3$	9
Si VII	1142.50	1142.441	$2s^2\,2p^3\,3s\,{}^5S_2-2s^2\,2p^3\,3p\,{}^5P_2$	9
Si VII	1143.58	1143.549	$2s^2\,2p^3\,3s\,{}^3P_2-2s^2\,2p^3\,3p\,{}^3D_3$	9
Si VI	1148.70	1148.700	$2s^2\,2p^4\,3s\,{}^4P_{3/2}-2s^2\,2p^4\,3p\,{}^4D_{5/2}$	9
Fe XVII	1153.16	1153.165	$2s^2\,2p^5\,3s\,{}^1P_1-2s^2\,2p^5\,3s\,{}^3P_0$	9
Ni XIV	1174.64	1174.720	$3s^2\,3p^3\,{}^4S_{3/2}-3s^2\,3p^3\,{}^2P_{1/2}$	9
Si VIII	1189.52	1189.487	$2s^2\,2p^2\,3s\,{}^4P_{5/2}-2s^2\,2p^2\,3p\,{}^4D_{7/2}$	9
Mg VII	1189.84	1189.841	$2s^2\,2p^2\,{}^3P_1-2s^2\,2p^2\,{}^1S_0$	9
Mg VI	1190.09	1190.124	$2s^2\,2p^3\,{}^4S_{3/2}-2s^2\,2p^3\,{}^2P_{3/2}$	9
Mg VI	1191.64	1191.670	$2s^2\,2p^3\,{}^4S_{3/2}-2s^2\,2p^3\,{}^2P_{1/2}$	9
S X	1196.21	1196.217	$2s^2\,2p^3\,{}^4S_{3/2}-2s^2\,2p^3\,{}^2D_{5/2}$	9
S X	1212.97	1212.932	$2s^2\,2p^3\,{}^4S_{3/2}-2s^2\,2p^3\,{}^2D_{3/2}$	9
Si VIII	1232.65	1232.650	$2s^2\,2p^2\,3s\,{}^4P_{5/2}-2s^2\,2p^2\,3p\,{}^4D_{5/2}$	9
N V	1238.82	1238.823	$1s^2\,2s\,{}^2S_{1/2}-1s^2\,2p\,{}^2P_{3/2}$	9
Fe XII	1242.00	1242.005	$3s^2\,3p^3\,{}^4S_{3/2}-3s^2\,3p^3\,{}^2P_{3/2}$	9
N V	1242.81	1242.806	$1s^2\,2s\,{}^2S_{1/2}-1s^2\,2p\,{}^2P_{1/2}$	9
Ne IX	1248.08	1248.284	$1s\,2s\,{}^3S_1-1s\,2p\,{}^3P_2$	9
Ni XIII	1277.22	1277.233	$3s^2\,3p^4\,{}^3P_1-3s^2\,3p^4\,{}^1S_0$	9
Ne IX	1277.79	1277.702	$1s\,2s\,{}^3S_1-1s\,2p\,{}^3P_0$	9
Ca XIV	1291.61	1291.538	$2s^2\,2p^3\,{}^2D_{3/2}-2s^2\,2p^3\,{}^2P_{3/2}$	9
Fe XIX	1328.79	1328.906	$2s^2\,2p^4\,{}^3P_2-2s^2\,2p^4\,{}^3P_0$	9
Ar XIII	1330.53	1330.374	$2s^2\,2p^2\,{}^3P_1-2s^2\,2p^2\,{}^1D_2$	9

Fe XII	1349.37	1349.400	$3s^2\,3p^3\,{}^4S_{3/2}$–$3s^2\,3p^3\,{}^2P_{1/2}$	9	Fe X	1463.50	1463.486	$3s^2\,3p^4\,3d\,{}^4F_{9/2}$–$3s^2\,3p^4\,3d\,{}^2F_{7/2}$	9
Fe XXI	1354.06	1354.067	$2s^2\,2p^2\,{}^3P_0$–$2s^2\,2p^2\,{}^3P_1$	9	Fe XI	1467.07	1467.423	$3s^2\,3p^4\,{}^3P_1$–$3s^2\,3p^4\,{}^1S_0$	9
Ca XV	1375.95	1375.935	$2s^2\,2p^2\,{}^3P_2$–$2s^2\,2p^2\,{}^1D_2$	9	K XIV	1477.16	1476.911	$2s^2\,2p^2\,{}^3P_2$–$2s^2\,2p^2\,{}^1D_2$	9
Ar XI	1392.10	1392.098	$2s^2\,2p^4\,{}^3P_2$–$2s^2\,2p^4\,{}^1D_2$	9	Cr X	1489.02	1489.040	$3s^2\,3p^3\,{}^4S_{3/2}$–$3s^2\,3p^3\,{}^2P_{3/2}$	9
Cr XVI	1410.60	1410.599	$2s^2\,2p^5\,{}^2P_{3/2}$–$2s^2\,2p^5\,{}^2P_{1/2}$	9	Ca XIV	1504.27	1504.278	$2s^2\,2p^3\,{}^2D_{3/2}$–$2s^2\,2p^3\,{}^2P_{1/2}$	9
Si VIII	1440.51	1440.510	$2s^2\,2p^3\,{}^4S_{3/2}$–$2s^2\,2p^3\,{}^2D_{5/2}$	9	Fe XX	1586.36	1586.247	$2s^2\,2p^3\,{}^2P_{1/2}$–$2s^2\,2p^3\,{}^2P_{3/2}$	9
Si VIII	1445.76	1445.737	$2s^2\,2p^3\,{}^4S_{3/2}$–$2s^2\,2p^3\,{}^2D_{3/2}$	9					

Wvl. is solar wavelength.
Wvl.^{L} is laboratory wavelength.

References

1 Acton *et al.* (1985)
2 Austin *et al.* (1966)
3 Behring *et al.* (1972)
4 Behring *et al.* (1976)
5 Brooks *et al.* (1999)
6 Brosius *et al.* (1998b)
7 Brosius *et al.* (2000)
8 Curdt *et al.* (2001)
9 Curdt *et al.* (2004) – includes Feldman *et al.* (1997b)
10 Dere (1978)
11 Doschek (1972)
12 Doschek *et al.* (1976b)
13 Dupree *et al.* (1973)
14 Feldman (1987)
15 Feldman *et al.* (1980a)
16 Feldman *et al.* (1997b)
17 Kastner *et al.* (1974)
18 Landi & Phillips (2005)
19 Malinovsky & Héroux (1973)
20 Manson (1972)
21 McKenzie & Landecker (1982)
22 McKenzie *et al.* (1980b)
23 McKenzie *et al.* (1985)
24 Neupert *et al.* (1973)
25 Parenti *et al.* (2005)
26 Phillips *et al.* (1982)
27 Sandlin *et al.* (1976)
28 Sandlin *et al.* (1986)
29 Sylwester *et al.* (2003)
30 Thomas & Neupert (1994)
31 Tousey *et al.* (1965)
32 Vernazza & Reeves (1978)
33 Walker *et al.* (1974)
34 Zhitnik *et al.* (2005)

Glossary

absorption coefficient (κ_ν) The amount of energy absorbed from a beam with specific intensity I_ν per unit solid angle per frequency interval per unit area per unit time. In Equation (2.13), this is taken to include scattering as well as absorbing processes.

absorption line A spectral line that has reduced flux compared with neighbouring continuum radiation, generally formed by cooler gas nearer to the observer absorbing radiation that is emitted by hotter gas further from the observer. The solar spectrum at wavelengths $\gtrsim$1600 Å is characterized by absorption lines from cooler regions in the upper photosphere or temperature minimum above the hotter, lower photosphere.

abundance The amount of a particular element present in the Sun or other astronomical object. The *absolute* abundance of an element is the number of atoms of the element (all ionization stages) as a fraction of the number of hydrogen atoms or ions (i.e. protons). The *relative* abundance is the number of atoms of the element expressed relative to the number of atoms of some other element.

active region A region of the solar atmosphere associated with relatively strong magnetic fields. Sunspots and faculae characterize an active region in the photosphere; bright plage and dark filaments characterize the chromospheric active region, including enhanced emission in the EUV. In the corona, active regions have loop structures visible in the EUV, soft X-rays, and (during eclipses and with white-light coronagraphs) visible-wavelength radiation. They may last from a few hours to several months, depending on their size.

Alfvén speed The speed of an Alfvén wave, given by $B/(4\pi\rho)^{1/2}$ where B is the magnetic field strength and ρ the mass density of the plasma.

Alfvén wave A magnetohydrodynamic wave propagating in a magnetized plasma in which the velocity and change in magnetic field are both perpendicular to the direction of the mean magnetic field.

allowed line A spectral line that is allowed according to the selection rules of dipole radiation. The corresponding transition normally has a relatively high Einstein coefficient or transition rate.

arch filament system A system of small loop-like filament structures often occurring in strongly developing active regions.

Balmer series A series of spectral lines due to hydrogen in the visible spectrum formed by transitions of the electron from total quantum number $n = 2$ to 3, 4, 5, etc., corresponding to spectral lines Hα (6563 Å), Hβ (4861 Å), Hγ (4340 Å). The Balmer limit (where the upper quantum number tends to infinity) is at 3646 Å.

Bent Crystal Spectrometer (BCS) Collimated bent crystal spectrometer on the *SMM* spacecraft, operating between 1980 and 1989, observing flare-excited X-ray lines of highly ionized Ca and Fe in the 1.8–3.2 Å wavelength range.

black body A hypothetical body that perfectly absorbs all radiation incident on it.

black-body radiation Radiation emitted by a black body, the flux of which has a wavelength dependence given by Equation (2.17), the Planck function. The total intensity of this radiation is $(\sigma/\pi)T^4$ (T is temperature in K, σ is the Stefan–Boltzmann constant).

Bohr atomic model Model for the hydrogen atom in which the electron can occupy certain circular orbits having quantized angular momenta equal to $nh/2\pi$. In this model, emission or absorption of radiation with frequency $h\nu$ occurs through electron jumps between orbits. The model can be extended to all hydrogen-like ions but it is unsatisfactory in that non-circular orbits are not considered and fine structure of the Balmer lines is not predicted.

Bragg Crystal Spectrometer (BCS) Uncollimated bent crystal spectrometer on *Yohkoh* spacecraft operating between 1991 and 2001, observing X-ray lines of highly ionized S, Ca, and Fe from flares and active regions in the 1.8–5.1 Å wavelength range.

Bragg's law The relation between the wavelength of radiation (particularly X-rays) incident on a crystal surface, the crystal plane separation, and the angle of incidence (measured from the crystal surface) (Equation (7.15)).

bremsstrahlung (also known as free–free or braking radiation) Radiation emitted by an electron passing by and decelerated by the Coulomb field of a positively charged ion. The difference in energy before and after the encounter appears as a photon, and as the electron follows an unbound orbit the initial and final states are unquantized; thus many such encounters result in a continuum. If the electrons are part of a Maxwell–Boltzmann distribution, *thermal bremsstrahlung* results; if, as is typical during the impulsive stages of flares, the electrons are part of a non-thermal (e.g. power-law in energy) distribution, *non-thermal bremsstrahlung* results.

chromosphere Region of solar atmosphere 500 km above the base of the Sun's photosphere (defined by the $\tau_{5000} = 1$ level) to approximately 9000 km above this level. Up to 1500 km above the $\tau_{5000} = 1$ level the chromosphere is fairly continuous, but above this level it is strongly influenced by the magnetic fields of the supergranulation, arranged in polygonal cells. At the edge of the supergranular cells, there is enhanced EUV emission and groups of Hα spicules. At the centres of the supergranular cells there are small bright features in Ca II H and K line images which may be associated with heating processes.

chromospheric network The network pattern visible in EUV and Ca II H and K line emission marking the outlines of supergranular cells, associated with enhanced magnetic fields. A corresponding *photospheric* network consists of small, bright dots generally along dark, intergranular lanes, the site of strong ($\sim$1000 G) magnetic fields.

configuration The arrangement of electrons in an atom, specified by the n and l quantum numbers of the individual electrons. Thus, for neutral carbon, the ground configuration is $1s^2 2s^2 2p^2$.

continuous spectrum A featureless distribution of radiation over a range of wavelengths.

continuum Region of spectrum between two emission or absorption lines. In the case of solar ultraviolet or X-ray emission, the continuum arises because of free–free or free–bound processes.

contribution function Function describing the electron temperature (T_e) dependence of a spectral line's emission for a unit emission measure (1 cm^{-3}), normally expressed by $G(T_e)$. For many solar ultraviolet and soft X-ray lines excited by electron collisions, G depends only on electron temperature T_e and atomic physics parameters, but there may be a dependence on electron density also. Equation (4.86) gives the definition.

corona A region of the solar atmosphere above the chromosphere characterized by very high temperatures. In the quiet Sun corona, the temperature is around 1.2×10^6 K, but is a little less ($\sim$800 000 K) in a coronal hole. Active regions have a coronal extension for which the temperature is enhanced, up to $\sim 5 \times 10^6$ K. In solar ultraviolet and soft X-ray images, the quiet Sun corona consists of material contained in magnetic loop structures, with footpoints located in the photosphere. In coronal holes, the ultraviolet and soft X-ray emission is much reduced, but X-ray bright points (normally unresolved small loops) are generally present. The plasma making up the corona is highly ionized, consisting of free protons, electrons, and helium nuclei, with small amounts of ions of heavier elements and a small ($<10^{-6}$) fraction of neutral hydrogen formed by recombinations of electrons and protons. The corona's ultraviolet and soft X-ray flux and complexity of form depend on the eleven-year solar activity cycle.

coronagraph A device in which the faint coronal emission, in white-light or ultraviolet, is made visible by the use of an occulting disk which blots out the intense photospheric light from the Sun's surface.

Coronal Diagnostic Spectrometer (CDS) EUV spectrometer on *SOHO* operating in the wavelength range 150–800 Å, with a grazing-incidence telescope feeding a normal-incidence spectrometer (NIS) and grazing-incidence spectrometer (GIS).

coronal hole A region of the solar corona in which the temperature and density are reduced compared with the quiet Sun corona. There is also reduced ultraviolet and soft X-ray emission. Such regions are common at each solar pole, particularly near times of solar minimum, from which open magnetic field lines extend out into interplanetary space, along which are carried particles making up fast solar wind streams.

coronal loops Loops within the quiet Sun or active region solar corona, the geometry of which is defined by magnetic flux tubes that extend high in the corona and with footpoints in the solar photosphere, as indicated by photospheric magnetograms.

coronal mass ejection (CME) Outward-moving large cloud of coronal material, with associated magnetic field, often but not invariably associated with solar flares. There is a leading edge visible with spacecraft coronagraphs, followed by a dark cavity and a core which appears to be an Hα prominence. The ejection speeds range from 10 km s^{-1} to 1000 km s^{-1} or more. Interplanetary spacecraft detect shocks and enhanced densities and solar wind speeds when a CME is directed at them. A 'halo' CME is one that is directed at the Earth or a spacecraft, giving the initial appearance of a halo round the Sun.

CORONAS Series of Russian spacecraft studying the solar atmosphere and solar–terrestrial relations. *CORONAS-I* was launched in 1994, *CORONAS-F* in 2001; a further spacecraft, *CORONAS-PHOTON* is due for launch in 2008. The instruments on board include those operating from gamma-rays to radio waves.

differential rotation Rotation of the Sun's interior and surface layers in which the rotation rate depends on latitude, surface rotation rates being higher at the equator than near the poles. A depth dependence of rotation rate occurs down as far as the tachoclyne, separating the convective from the radiative zones of the Sun's interior.

Doppler shift The shift of a spectral line towards longer or shorter wavelengths because of receding or approaching material. Doppler shifts of ultraviolet or soft X-ray spectral lines are frequently observed for many solar features, e.g. flares show transient Doppler shifts towards shorter wavelengths (blue shifts) at their impulsive stages, and there is a persistent Doppler shift towards longer wavelengths (red shifts) for quiet Sun transition region ultraviolet lines. The terms blue and red shifts are commonly used in solar physics for ultraviolet and soft X-ray lines despite the fact that these are strictly only applicable to visible-light spectral lines.

effective temperature The Sun's effective temperature is the temperature of a black body having the same radius as the Sun and emitting the same amount of energy per unit surface area. Current measurements of the mean solar irradiance give the solar effective temperature as 5778 K.

electron spin Angular momentum associated with a particle, in particular an electron, which by quantum mechanics is given by $\pm\hbar/2$.

electronvolt (eV) Amount of energy equal to the work done on an electron in moving it through a potential difference of 1 volt, equal to 1.602×10^{-12} erg. It is frequently used for X-ray photon energies and for particle energies in flares and other energetic phenomena in the Sun, for which kiloelectronvolt (keV) or megaelectronvolt (MeV) are used.

emerging flux region Bipolar region on the Sun associated with new magnetic flux rising from beneath the photosphere.

emission coefficient (j_ν) The amount of energy added per unit length to a beam of radiation per unit solid angle per frequency interval per unit area per unit time (Equation (2.12)).

emission line A bright spectral line seen against a background of low or zero flux, generally formed by an incandescent gas or plasma. The solar spectrum at ultraviolet ($\lesssim$1600 Å) and soft X-ray wavelengths is characterized by emission lines formed by the chromosphere, transition region, or corona, or specific hot regions such as flares.

emission measure (EM) The flux of a spectral line is related to an integral over volume (V) or height (h) of the contribution function $G(T_e)$ multiplied by electron density squared (N_e^2): $\int G(T_e) N_e^2 \, dV$. If the temperature dependence of $G(T_e)$ is sufficiently narrow, the $G(T_e)$ function can be taken out of the integral to leave a volume emission measure ($\int N_e^2 \, dV$) or surface (or column) emission measure ($\int N_e^2 \, dh$) which characterizes the emitting region, either a specific structure like a flare or a layer of the solar atmosphere respectively. *Differential* emission measure is generally used for a temperature-dependent function $\varphi(T_e)$ whereby the line flux is related to an integral over T_e, $\int \varphi(T_e)\, G(T_e)\, dT_e$.

emissivity Amount of emission per unit volume per unit time of a volume of emitting material.

Evershed effect Velocity flow ($\sim$2 km s^{-1}) away from a sunspot umbra along penumbral filaments seen in the photosphere.

excitation Process by which energy is imparted to an atom or ion, which is raised from a lower (often ground) quantum state to an upper quantum state (the energy difference is the *excitation energy*). In the corona, excitation is generally by collisions with free electrons, though for small excitation energies excitation by photons may be significant.

extreme ultraviolet (EUV) Region of spectrum in wavelength range 100 Å $> \lambda >$ 1000 Å. Such radiation is emitted by regions of the solar atmosphere from the chromosphere to the corona, consisting mostly of lines and a weak free–free continuum.

Extreme-ultraviolet Imaging Telescope (EIT) EUV imaging instrument on *SOHO* operating in four bandpasses (171 Å, 195 Å, 284 Å, and 304 Å), defined by multilayer coatings on quadrants of the telescope mirror, designed to study solar features in the chromosphere and corona.

Extreme-ultraviolet Imaging Spectrometer (EIS) EUV imaging spectrometer on *Hinode* operating in two bands, 170–211 Å and 246–291 Å, and with spectral resolution (with fine slits) capable of resolving velocities to a few km s^{-1}. Monochromatic images of solar features can be produced with broad slots.

facula Brighter and slightly hotter region of photosphere, associated with strong magnetic field and with chromospheric plage.

fibril Dark elongated features in Hα spectroheliograms near active regions, apparently following the local magnetic field.

filament See prominence.

FIP bias Defined (by Equation (11.1)) as $\mathrm{FIP_{bias}} = A_{\mathrm{El}}^{\mathrm{SUA}} / A_{\mathrm{El}}^{\mathrm{Phot}}$ where the relative abundances of an element El, A_{El}, are for the solar upper atmosphere (SUA) and photosphere (Phot).

FIP effect Effect whereby the ratio of abundances between the low- ($\lesssim$10 eV) and high-FIP ($\gtrsim$10 eV) groups in the quiet corona is a factor of 3–4 higher than the corresponding photospheric ratios. In Chapter 11, it is argued that this is due to the enrichment of low-FIP element abundances in the corona rather than the depletion of high-FIP elements.

flare A sudden release of energy in the corona arising from the relaxation of magnetic field configurations, resulting in radiation in wavelengths from gamma-rays to long-wavelength radio waves, with huge increases in both line and continuum radiation in the EUV and soft X-rays. An impulsive phase in hard X-rays occurs near the onset, often preceded by a slow rise in soft X-rays. The peak and decay of a flare is characterized by emission in loop structures that are bright in EUV and soft X-rays, with a pair of bright ribbons, visible in Hα and ultraviolet continuum at $\sim$1600 Å, marked by the loop footpoints. Large flares may have a total energy of $\sim 10^{32}$ erg.

flare star A type of dwarf star of late spectral type (K or M) with Hα and other Balmer lines in emission showing sudden, flare-like increases of visible, EUV, and soft X-ray emission, thought to be analogues of solar flares. The classical type of flare star is that typified by UV Ceti, but other categories are recognized, e.g. contact binary stars (W UMa) and binary systems containing evolved stars called RS CVn stars in which the magnetic fields of each star are thought to interact, giving rise to flares much more powerful than solar flares.

fluorescence Radiation from a material after it has been excited by an external source. In the context of fluorescent X-rays during solar flares, the radiation is from the photosphere which is excited by a soft X-ray-emitting flare source several thousand km above; the radiation has been observed in the form of Fe Kα and Kβ line emission. For crystal spectrometers viewing the solar X-ray spectrum, solar X-rays fluoresce the crystal material, and the fluorescence radiation may be emitted in all directions forming a continuous background.

flux For radiation received, *flux* is widely used in solar physics to denote irradiance (amount of radiant flux or power received per unit area). The c.g.s. units are erg cm^{-2} s^{-1} (SI units W m^{-2}). Spectral flux is flux per unit wavelength or frequency interval. For solar and stellar atmospheres, flux at frequency ν is the net rate of energy flow across a unit area in the atmosphere, i.e. $\mathcal{F}_\nu = \int_{4\pi} I_\nu(r) \cos\theta \, d\omega$ (Equation (2.2)).

forbidden line A spectral line that is forbidden, i.e. disobeys at least one of the selection rules applying to dipole radiation. The corresponding transition probability is normally very low compared with allowed lines, and so forbidden lines only occur in low-density gas or plasma such as the solar corona. Many coronal forbidden lines disobey the parity rule, i.e. the transition has no parity change.

Fraunhofer lines Absorption lines (including molecular bands) in the solar visible-wavelength spectrum first labelled by J. von Fraunhofer (1814), most of which are of solar origin (e.g. the D lines of Na I and H and K lines of Ca II) though some are due to constituents of the Earth's atmosphere.

free–bound radiation Radiation produced when a free electron is captured by an ion. If the recombination occurs to the ion's ground state, the minimum energy of the radiation is the ionization potential. The spectrum of the radiation thus has an edge at this energy with a continuum having a flux that decreases with increasing energy. (Also known as recombination radiation.)

free–free radiation See bremsstrahlung.

full width at half maximum (FWHM) A measure of the width of a profile, e.g. a spectral line profile, defined by the half-power points of the profile.

Geostationary Orbiting Environment Satellites (GOES) Geostationary Earth-orbiting satellites monitoring solar soft X-ray emission in spectral bands 0.5–4 Å and 1–8 Å.

granulation Cellular pattern seen in the photosphere, with individual granules that are polygonal, typical sizes $\sim$100–1100 km and lasting $\sim$18 minutes, separated by dark intergranular lanes.

hard X-rays Region of spectrum with wavelength range <1 Å (photon energies $\gtrsim$ 12 keV). Such radiation is emitted at the impulsive stage of solar flares, and consists of a continuum (mostly free–free radiation at photon energies $\gtrsim$ 20 keV) whose flux has a nearly power-law dependence on photon energy and is non-thermal in origin.

Hard X-ray Telescope (HXT) Instrument on the *Yohkoh* spacecraft for observing flares operating in the hard X-ray range ($\sim$15–100 keV).

helmet streamer Feature of white-light or EUV corona consisting of a loop-like base with linear streamer above extending away from the Sun. A prominence is often present within the cavity below the loop.

Hinotori Japanese solar spacecraft operating between 1981 and 1982 designed to investigate high-energy phenomena in solar flares.

Hinode Japanese solar spacecraft in a polar, Sun-synchronous orbit, launched in 2006, containing optical, EUV, and soft X-ray instruments designed to study the interaction of the solar magnetic field and the corona.

homologous flares Flares occurring repeatedly in the same location with similar development and other properties.

intensity For radiation received, *intensity* is generally used in solar physics to denote radiance (the amount of radiant intensity per unit area), with units erg cm^{-2} s^{-1} sr^{-1}. For solar or stellar atmospheres, *specific intensity* of radiation is the amount of energy passing through unit area (perpendicular to the beam of radiation) per unit time per unit frequency interval into unit solid angle.

intercombination line A spectral line in which the transition involves a change of total spin quantum number S.

ionization The removal of a bound electron from an ion or atom. *Collisional* ionization is when ionization occurs as a result of a collision with a particle, normally a free electron in the case of solar coronal plasmas. In autoionization, a doubly excited ion formed by dielectronic capture ionizes by removal of one of the excited electrons.

ionization potential The minimum energy required to remove an electron from an atom or an ion to infinity, normally expressed in eV. The *first ionization potential* (FIP) is the energy required to remove the least strongly bound of the electrons of the neutral atom.

isoelectronic sequence Sequence of ions in which the number and configuration of orbital electrons are identical. The 'carbon' isoelectronic sequence is one in which the ions all have the same electron configuration, $1s^2\,2s^2\,2p^2$. The spectral lines emitted by ions in an isoelectronic sequence thus have similar patterns.

***jj* coupling** A coupling scheme for multi-electron ions or atoms in which the magnetic interactions between the l and s momenta of individual electrons are large compared with the angle-dependent part of the electrostatic interactions, so that the total angular momentum J is the vector sum of the individual j momenta. jj coupling applies to some ions common in the solar atmosphere, generally those with heavier nuclei.

Large Angle Spectroscopic COronagraph (LASCO) An instrument on the *SOHO* spacecraft consisting of three coronagraphs (C1, C2, C3) with internal (C1) and external (C2, C3) occulters that image the corona from 1.1 to $30 R_\odot$.

level (atomic) Specification of the energy state of an atom by the quantum numbers L, S, and J, e.g. for neutral carbon the ground level is $1s^2 2s^2 2p^2\ ^3P_0$. Transitions between atomic levels give rise to spectral *lines*. An atomic *state* is further specified by magnetic quantum numbers (components of L, S, and J along a magnetic field direction); transitions between states give rise to spectral line *components*.

limb (solar) The edge of the Sun at a particular wavelength. For visible wavelengths there is a *limb darkening*, i.e. falling off of surface flux towards the limb owing to the $\tau = 1$ level occurring at higher and cooler layers of the solar photosphere. For most of the EUV and soft X-ray region, *limb brightening* occurs because of an increase in the path length through the optically thin material giving rise to the emission.

local thermodynamic equilibrium (LTE) Equilibrium state in which all thermodynamic properties of matter are equal to their thermodynamic equilibrium values at the local values of temperature and density. In LTE, which applies in the deep layers of the photosphere, some radiation escapes but sufficiently little that the source function may be assumed to be the Planck function ($S_\nu = B_\nu$).

***LS* (Russell–Saunders) coupling** A coupling scheme for multi-electron ions or atoms in which the orbital angular momenta of individual electrons add vectorially to give a total angular momentum L and spin angular momenta add vectorially to give a total spin S. In this coupling scheme, spin–orbit coupling is weak compared with the coupling of orbital momenta among themselves and coupling of spins among themselves. LS coupling applies to most atoms and ions with light nuclei represented in solar spectra.

Lyman series A series of spectral lines due to hydrogen occurring in the ultraviolet spectrum due to transitions of the electron in hydrogen from total quantum number $n = 1$ to 2, 3, 4, etc., corresponding to the spectral lines Ly-α (1216 Å), Ly-β (1025 Å), Ly-γ (972 Å). The Lyman limit (where the upper quantum number tends to infinity) is at 912 Å.

magnetic pressure In c.g.s. units, magnetic pressure is $B^2/8\pi$ where B is the magnetic field strength in gauss.

Maxwell–Boltzmann (Maxwellian) distribution Distribution of particle velocities in kinetic equilibrium: see Equation (4.9).

metastable level Excited state of an atom that has a relatively long lifetime. For the solar corona, a metastable state may connect to the ground level of an atom by a radiative transition, giving rise to a forbidden line, depending on the density of the coronal gas.

microflare Small flare-like outbursts with energies of 10^{-6} times that of a large flare. A *nanoflare* is a flare-like outburst that is 1000 times smaller than a microflare, hypothesized to account for coronal heating.

Moreton wave An extensive shock-wave-like disturbance, visible in Hα particularly, propagating horizontally away from the site of a large flare with velocities of ~1000 km s^{-1}. Apparently identifiable with similar EUV disturbances that are associated with coronal mass ejections.

near ultraviolet (NUV) Region of spectrum in wavelength range 2000 Å $< \lambda <$ 3900 Å. Such radiation is mostly emitted by the quiet Sun, and consists of a continuum whose flux strongly rises with wavelength and some absorption lines, notably the h and k lines of Mg II.

optical depth A measure of how far a beam of light will travel through a partially transparent medium such as the solar atmosphere. A light beam at frequency ν passing through a length element dl sees an optical depth element $d\tau_\nu = -\kappa_\nu\, ds$ where κ_ν is the total absorption coefficient (including scattering).

penumbra Region surrounding the umbra of a well-developed sunspot, consisting of bright and dark filaments radiating from the umbra. In the chromosphere, a system of radiating fibrils larger than the photospheric penumbra defines the superpenumbra.

photoionization Ionization of an atom or ion by electromagnetic radiation, with photon energy larger than the ionization potential of the atom or ion.

photosphere Lowest region of the solar atmosphere, visible in white light. The base of the photosphere is generally taken to be where the optical depth at 5000 Å, τ_{5000}, is unity, and where the temperature is approximately 6400 K.

photospheric network Network pattern formed by small, bright dots generally aligned along dark intergranular lanes, the site of strong (1000 G) magnetic fields.

plage Bright emission seen in Hα and Ca II H and K lines associated with active regions.

plasma beta (β) Ratio of gas (thermal) pressure Nk_BT (N = particle number density, T = temperature) to magnetic pressure ($B^2/8\pi$ in c.g.s. units, where B is the magnetic field strength in G).

polar plume High-density, linear structures occurring within polar coronal holes, visible in white light, EUV, and soft X-rays, associated with unipolar magnetic fields, and extending into interplanetary space. They last approximately a day but possibly recur at the same location.

polarity inversion line (PIL) Line where the observed photospheric line-of-sight magnetic field is zero (also known as magnetic inversion line, and inaccurately the neutral line).

pore Small, short-lived dark area in the photosphere, often the birthplace of a sunspot.

prominence Mass of relatively cool and dense material in the corona, appearing as bright features in chromospheric lines such as He I 584 Å, He II 304 Å, and Hα. On the solar disk, prominences appear at dark *filaments* in Hα, and also in some EUV lines.

quantum numbers Each electron of an atom is characterized by the four quantum numbers: the principal quantum number n, the orbital or azimuthal quantum number l (describing the angular momentum of the electron), the magnetic quantum number m_l (governing the electron energy in a magnetic field), and the spin quantum number m_s, equal to $\pm 1/2$. By Pauli's exclusion principle, no two electrons can be in the same *quantum state*, i.e. have the same set of four quantum numbers.

recombination The capture of a free electron by an ion or atom. In solar coronal plasmas, *radiative* recombination is when the excess energy of the recombination is carried away by a photon, and *dielectronic* recombination is when the recombined ion is in a doubly excited state which de-excites into its ground state.

REntgenovsky Spektrometr s Izognutymi Kristalami (RESIK) Crystal spectrometer on the Russian *CORONAS-F* spacecraft operating in the wavelength range 3 Å$< \lambda <$ 6 Å.

resolution For a spectrometer, spectral resolution is the smallest difference of wavelength, $\Delta\lambda$, for which two spectral lines with wavelength $\sim\lambda$ can just be separated. A spectrometer's *resolving power* is $\lambda/\Delta\lambda$. For a telescope, angular or spatial resolution is the smallest difference of angle between two points (e.g. features on the Sun) that can just be separated.

resonance line A spectral line with high transition probability, often applicable to the first members of a spectral sequence.

Reuven Ramaty High Energy Solar Spectroscopic Imager (RHESSI) Earth-orbiting spacecraft observing the Sun with cooled germanium detectors and modulation collimators, operating in the energy range 3 keV (4 Å) to 17 MeV, i.e. soft and hard X-rays and gamma-rays.

Schrödinger equation An equation in wave mechanics for the wave function ψ of a particle, in particular an electron: see Equation (3.15).

sidereal rotation period (solar) Rotation period of the Sun (normally latitude-dependent) with reference to the fixed stars (25.6 days at the solar equator). The *synodic period* is the rotation period with reference to the Earth (27.3 days at the solar equator).

Skylab Mission NASA space station, launched in 1973, including the Apollo Telescope Mount (ATM) that contained telescopes viewing the Sun in soft X-ray, ultraviolet, and visible wavelengths operated by astronauts during three manned missions. The spacecraft re-entered the Earth's atmosphere in 1979.

soft X-rays Region of spectrum in wavelength range 1 Å $< \lambda <$100 Å. Such radiation is emitted by hot coronal plasmas associated with active regions and flares, and consists of lines and continuum emitted by thermal plasmas.

Soft X-ray Telescope (SXT) Instrument on the *Yohkoh* spacecraft operating in the soft X-ray region (3–50 Å range).

solar coordinates Solar coordinates may be expressed as heliographic latitude and longitude or (as is commonly the case with spacecraft images) on a heliocentric scheme with (0,0) being the origin and x in the east–west direction (west being positive) and y in the south–north direction (north positive). The tilt of the solar axis (± 7 degrees approximately) is measured by B_0, the latitude of the apparent centre of the solar disk. Longitudes (actually central meridian distances, CMDs) are expressed as east and west longitudes from the Sun's central meridian, though more precisely longitudes are with respect to a zero longitude that rotates at the synodic rotation period of the Sun. As seen from the Earth, the north rotation pole of the Sun is different from celestial north by a position angle P (positive when the solar north pole is east of the celestial north point of the Sun).

solar cycle A nearly cyclical variation in solar activity, with period of ~11 years, the most familiar manifestation being sunspot number. The magnetic field disposition of spot pairs alternates with successive cycles, giving a magnetic cycle of ~22 years.

Solar and Heliospheric Observatory (SOHO) ESA/NASA solar spacecraft, launched in 2005, in halo orbit round the L1 Lagrangian point between the Earth and Sun (1.5×10^6 km from Earth), with twelve major instruments on board studying the Sun with continuous coverage from the solar interior to the solar wind.

solar energetic particles (SEP) Electrons and ions with energies from a few keV up to GeV, observed in interplanetary space in association with flares or other energetic events on the Sun.

Solar Maximum Mission (SMM) NASA solar spacecraft operating between 1980 and 1989 observing solar activity, particularly flares and coronal mass ejections, in wavelengths ranging from gamma-rays to white-light emission. The attitude control unit and some other systems were repaired during a Space Shuttle mission in 1984.

solar wind Outward flow of particles – mainly electrons and protons – and magnetic field from the Sun into interplanetary space. The *slow-speed* solar wind has velocities of ~400 km s^{-1}, the *fast-speed* solar wind has velocities of ~750 km s^{-1}.

Solar TErrestrial RElations Observatory (STEREO) Two nearly identical spacecraft launched in 2006, both along the orbit of the Earth, one behind, one ahead of the Earth, designed to provide stereoscopic images and measurements of the Sun and to study the nature of coronal mass ejections.

Solar Ultraviolet Measurements of Emitted Radiation (SUMER) Spectrometer on *SOHO* studying the solar atmosphere in the wavelength range 500 Å $< \lambda <$1610 Å. With a spectral resolving power of up to 38000, spatial resolution close to 1 arcsec, and time resolution of less than 1 s, SUMER has unique capabilities for observing dynamic phenomena in the solar atmosphere, with diagnostic information from pairs of spectral lines sensitive to densities, temperatures, etc.

source function Ratio of emission coefficient to absorption coefficient, $S_\nu = j_\nu/\kappa_\nu$.

spectrograph Instrument that produces a photograph of a spectrum (*spectrogram*). A *spectrometer* detects a spectrum with a photoelectric device such as a CCD or proportional counter.

spectroheliograph Instrument that produces photographs (*spectroheliograms*) of the Sun at specific wavelengths, particularly strong Fraunhofer lines in the visible spectrum.

spicule Spike-like jet features seen in the upper chromosphere on the limb (as in Hα or EUV images) or on the disk, distributed along supergranule cell boundaries.

sunspot Dark area in the photosphere, with reduced temperature and associated with strong (4000 G) magnetic fields. Spots may be single or appear in pairs or groups. There is a chromospheric counterpart that consists of radiating bright plumes seen in EUV images and dark filamentary structures in Hα.

sunspot number A measure of sunspot activity. *Relative sunspot number* is $k(10g + f)$ where k is an adjustable factor for a particular observatory, g the number of sunspot groups, and f the total number of spots.

supergranulation Convective pattern in the quiet Sun consisting of large (~30 000 km) cells lasting ~30 hours where there is a mostly horizontal, outward flow from cell centres and a downflow at the cell boundaries. Each supergranule has a chromospheric counterpart in the network.

temperature minimum region Region of solar atmosphere where the temperature reaches a minimum (~4400 K) with height above the photosphere.

term (atomic) Group of atomic levels specified by L and S quantum numbers of an atom, e.g. the ground term of neutral carbon is $1s^2 2s^2 2p^2\ {}^3P$. Terms with odd parity are written with superscript o (e.g. the ground level of boron is $1s^2 2s^2 2p\ {}^2P^{\mathrm{o}}_{1/2}$). A transition between two terms leads to a *multiplet*.

thermodynamic equilibrium Equilibrium state in a black body in which matter is in equilibrium with the radiation field. The source function S_ν equals B_ν, or the Planck function divided by π (Equation (2.17)).

Thomson scattering Scattering of electromagnetic radiation by charged particles, in particular free electrons in the corona, in which the photon energy is small compared with the energy of the electron.

transition region Region of solar atmosphere characterized by temperatures between 20 000 K and 10^6 K, formerly thought to be a narrow layer separating chromosphere and corona but now considered by some to be a mass of unresolved elements in the atmosphere subject to much dynamic activity.

Transition Region and Coronal Explorer (TRACE) A single-instrument spacecraft in a polar Sun-synchronous orbit launched in 1998. It has several spectral bands in the VUV, and three EUV bands (171 Å, 195 Å, and 284 Å). Its field of view observes about one-tenth of the solar disk at any time, and has 1 arcsec spatial resolution. It is designed to study the transport of energy by magnetic fields from the Sun's surface to the transition region and corona.

Ultraviolet Coronagraph Spectrometer (UVCS) Coronagraph on *SOHO* observing the solar corona with ultraviolet spectroscopy and polarimetry instruments. The spectrometer has channels observing the H I Ly-α line and the O VI coronal lines (plus neighbouring wavelength regions), and a white-light polarimeter.

Ultraviolet Spectrometer/Polarimeter (UVSP) Ultraviolet instrument on *SMM* operating in the range 1170–1800 Å (second order) with telescope and spectrometer capable of producing 'Dopplergrams' (images of the Sun in the red and blue parts of line profiles to give velocity information).

umbra The inner region of a fully developed sunspot, characterized by temperatures of ~4200 K and large (4000 G) magnetic fields.

unipolar region Large area on the Sun showing weak photospheric magnetic field of a single polarity.

vacuum ultraviolet (VUV) Region of spectrum in wavelength range 1000 Å $<\lambda<$ 2000 Å. Solar radiation in this range is emitted mostly by the chromosphere and transition region, and consists of emission lines up to wavelengths ~1600 Å, with continuum rising strongly with wavelength at >1600 Å.

wavefunction ψ Function in Schrödinger's equation involving the space coordinates of a particle, in particular an electron in an atom (the time-dependent wave function is written Ψ). The square of the absolute value of ψ, $|\psi|^2$, is the probability of finding the particle within a small volume at the (x, y, z) coordinates of the function.

X-ray bright point Small X-ray-emitting region in the corona associated with a bipolar magnetic field.

X-ray jet Transient X-ray phenomenon consisting of a collimated flow of material, associated with X-ray bright points, active regions, and flares.

X-ray Telescope (XRT) Soft X-ray telescope on *Hinode* with 2 arcsec spatial resolution forming solar images in various X-ray ranges defined by broad-band filters, designed to study the corona.

Yohkoh Japanese solar mission operating between 1991 and 2001 containing X-ray instruments designed to look at solar flares and other activity.

Further reading

Aschwanden, M.: *Physics of the Solar Corona: An Introduction.* New York: Springer-Verlag, and Chichester: Praxis Publishing, 2004.

Böhm-Vitense, E.: *Introduction to Stellar Astrophysics*, Vol. 2: *Stellar Atmospheres.* Cambridge University Press, 1989.

Bray, R. J., Cram, L. E., Durrant, C. J., & Loughhead, R. E.: *Plasma Loops in the Solar Corona.* Cambridge University Press, 2006.

Dwivedi, B. (ed.): *Dynamic Sun.* Cambridge University Press, 2003.

Foukal, P. V.: *Solar Astrophysics.* New York: John Wiley and Sons, 1990.

Golub, L., & Pasachoff, J. M.: *The Solar Corona.* Cambridge University Press, 1997.

Harra, L. K., & Mason, K. O. (eds.): *Space Science.* London: Imperial College Press, 2004.

Lang, K. R.: *The Cambridge Encyclopedia of the Sun.* Cambridge University Press, 2001.

Lang, K. R.: *Astrophysical Formulae*, Vol. 1: *Radiation, Gas Processes, and High Energy Astrophysics.* New York: Springer-Verlag, 1999.

Mariska, J. T.: *The Solar Transition Region.* Cambridge University Press, 1992.

Phillips, K. J. H.: *Guide to the Sun.* Cambridge University Press, 1992.

Schrijver, C. J., & Zwaan, C.: *Solar and Stellar Magnetic Activity.* Cambridge University Press, 2000.

Stix, M.: *The Sun.* New York: Springer-Verlag, 1989.

Tayler, R. J.: *The Sun as a Star.* Cambridge University Press, 1996.

Tennyson, J.: *Astronomical Spectroscopy: An Introduction to the Atomic and Molecular Physics of Astronomical Spectra.* London: Imperial College Press, 2005.

Zirker, J. B.: *Total Eclipses of the Sun.* Harvard: Princeton University Press, 1995.

References

Acton, L. W., *et al.* (1980). *Solar Phys.* **65**, 53.
Acton, L. W., Bruner, M. E., Brown, W. A., *et al.* (1985). *Astrophys. J.* **291**, 865.
Akiyama, S., Doschek, G. A., & Mariska, J. T. (2005). *Astrophys. J.* **623**, 540.
Anders, E., & Grevesse, N. (1989). *Geochim. Cosmochim. Acta* **53**, 197.
Anderson, L. S. (1989). *Astrophys. J.* **339**, 558.
Andretta, V., & Jones, H. P. (1997). *Astrophys. J.* **489**, 375.
Antonucci, E., Gabriel, A. H., Doyle, J. G., *et al.* (1984). *Astr. Astrophys.* **133**, 239.
Arnaud, M., & Raymond, J. (1992). *Astrophys. J.* **398**, 394.
Arnaud, M., & Rothenflug, R. (1985). *Astr. Astrophys.* **60**, 425.
Aschwanden, M. J. (2004). *Physics of the Solar Corona: An Introduction*, Chichester, UK: Springer-Verlag and Praxis Publishing Ltd.
Aschwanden, M. J., & Benz, A. O. (1997). *Astrophys. J.* **480**, 82.
Aschwanden, M. J., *et al.* (1999). *Astrophys. J.* **515**, 842.
Athay, R. G. (1976). *The Solar Chromosphere and Corona: Quiet Sun*, Boston: Reidel.
Athay, R. G., Gurman, J. B., Henze, W., & Shine, R. A. (1983). *Astrophys. J.* **265**, 519.
Austin, W. E., Purcell, J. D., Tousey, R., & Widing, K.G. (1966). *Astrophys. J.* **145**, 373.
Ayres, T. R. (1981). *Astrophys. J.* **244**, 1064.
Ayres, T. R. (1990). In *Solar Photosphere: Structure, Convection, and Magnetic Fields*, ed. J. O. Stenflo, Dordrecht: Kluwer Academic Publishers, p. 23.
Ayres, T. R., Testerman, L., & Brault, J. W. (1986). *Astrophys. J.* **304**, 542.
Bai, T. (1979). *Solar Phys.* **62**, 113.
Bambynek, W., Crasemann, B., Fink, R. W., *et al.* (1972). *Rev. Mod. Phys.* **44**, 716.
Banerjee, D., Teriaca, L., Doyle, J. G., & Wilhelm, K. (1998). *Astr. Astrophys.* **339**, 208.
Bar-Shalom, A., Klapisch, M., & Oreg, J. (1988). *Phys. Rev. A* **38**, 1773.
Bartoe, J.-D. F., Brueckner, G. E., Purcell, J. D., & Tousey, R. (1977). *Appl. Opt.* **16**, 879.
Bautista, M. A. (2000). In *Atomic Data Needs for X-ray Astronomy*, eds. M. A. Bautista, T. Kallman, & A. K. Pradhan, NASA/CP-2000-209968, p. 25.
Behring, W. E., Cohen, L., & Feldman, U. (1972). *Astrophys. J.* **175**, 493.
Behring, W. E., Cohen, L., Feldman, U., & Doschek, G. A. (1976). *Astrophys. J.* **203**, 521.
Beiersdorfer, P., Phillips, T., Jacobs, V. L., *et al.* (1993). *Astrophys. J.* **409**, 846.
Beiersdorfer, P., Scofield, J. H., & Osterheld, A. L. (2003). *Phys. Rev. Lett.* **90**, 235003.
Bely-Dubau, F., Gabriel, A. H., & Volonté, S. (1979). *Mon. Not. R. Astr. Soc.* **186**, 405.
Bely-Dubau, F., Dubau, J., Faucher, P., & Gabriel, A. H. (1982a). *Mon. Not. R. Astr. Soc.* **198**, 239.
Bely-Dubau, F., Dubau, J., Faucher, P., *et al.* (1982b). *Mon. Not. R. Astr. Soc.* **201**, 1155.
Berrington, K. A., Eissner, W., & Norrington, P. H. (1995). *Computer Physics Comm.* **92**, 290.
Bewsher, D., Parnell, C., & Harrison, R. A. (2002). *Solar Phys.* **206**, 21.
Bhatia, A., Fawcett, B. C., Lemen, J. R., Mason, H. E., & Phillips, K. J. H. (1989). *Mon. Not. R. Astr. Soc.* **240**, 421.
Bitter, M., *et al.* (1979). *Phys. Rev. Lett.* **43**, 129.
Bocchialini, K., Costa, A., Domenech, G., *et al.* (2001). *Astr. Astrophys.* **199**, 133.
Bohlin, J. D., & Sheeley Jr., N. R. (1978). *Solar Phys.* **56**, 125.

Bohlin, J. D., *et al.* (1975). *Astrophys. J.* **197**, 133.
Böhm-Vitense, E. (1989). *Introduction to Stellar Astrophysics*, Vol. 2, *Stellar Atmospheres*, Cambridge University Press.
Boland, B. C., Engstrom, S. F. T., Jones, B. B., & Wilson, R. (1973). *Astr. Astrophys.* **22**, 161.
Bommier, V., & Sahal-Brechot, S. (1982). *Solar Phys.* **78**, 157.
Bowyer, S., & Malina, R. F. (1991). In *Extreme Ultraviolet Astronomy*, eds. R. F. Malina & S. Bowyer, New York: Pergamon Press.
Bradshaw, S. J., & Mason, H. E. (2003). *Astr. Astrophys.* **401**, 699.
Braginskii, S. I. (1965). In *Reviews of Plasma Physics*, New York: M. A. Leontovich Consultants Bureau, p. 205.
Bray, R. J., Cram, L. E., Durrant, C. J., & Loughhead, R. E. (1991). *Plasma Loops in the Solar Corona*, Cambridge University Press.
Breeveld, E. R., Culhane, J. L., Normal, K., Parkinson, J. H., & Gabriel, A. H. (1988). *Astrophys. Lett. Comm.*, **27**, 155.
Brekke, P., Hassler, D. M., & Wilhelm, K. (1997). *Solar Phys.* **175**, 349.
Brković, A., Landi, E., Landini, M., Rüedi, I., & Solanki, S. K. (2002). *Astr. Astrophys.* **383**, 661.
Brooks, D. H., Fischbacher, G., Fludra, A., *et al.* (1999). *Astr. Astrophys.* **347**, 277.
Brosius, J. W., Davila, J. M., & Thomas, R. J. (1998a). *Astrophys. J. Lett.* **497**, L113.
Brosius, J. W., Davila, J. M., & Thomas, R. J. (1998b). *Astrophys. J. Supp.* **119**, 255.
Brosius, J. W., Thomas, R. J., Davila, J. M., & Landi, E. (2000). *Astrophys. J.* **543**, 1016.
Brosius, J. W., Landi, E., Cook, J. W., *et al.* (2002). *Astrophys. J.* **574**, 453.
Brosius, J. W., & Phillips, K. J. H. (2004). *Astrophys. J.* **613**, 580.
Brown, J. C., Dwivedi, B. N., Sweet, P. A., & Almleaky, Y. M. (1991). *Astr. Astrophys.* **249**, 277.
Brueckner, G. E., & Bartoe, J.-D. F. (1983). *Astrophys. J.* **272**, 329.
Brussaard, P. J., & van de Hulst, H. C. (1962). *Rev. Mod. Phys.* **34**, 507.
Bruzek, A. (1964). *Astrophys. J.* **140**, 746.
Bryans, P., Badnell, N. R., Gorczyca, T. W., *et al.* (2006). *Astrophys. J. Supp.* **167**, 343.
Burgess, A. (1965). *Astrophys. J.* **141**, 1588.
Burke, P. G., & Robb, W. D. (1975). *Adv. Atom. Molec. Phys.* **11**, 143.
Burton, W. M., & Wilson, R. (1961). *Proc. Phys. Soc.* **78**, 1416.
Cameron, A. G. W. (1970). *Space Sci. Rev.* **15**, 121.
Carlsson, M., & Stein, R. F. (1997). *Astrophys. J.* **481**, 500.
Carlsson, M., Judge, P. G., & Wilhelm, K. (1997). *Astrophys. J.* **486**, L63.
Chae, J., Schühle, U., & Lemaire, P. (1998a). *Astrophys. J.* **505**, 957.
Chae, J., Wang, H., Lee, C.-Y., Goode, P. R., & Schühle, U. (1998b). *Astrophys. J.* **497**, L109.
Chae, J., Wang, H., Lee, C.-Y., Goode, P. R., & Schühle, U. (1998c). *Astrophys. J.* **504**, L123.
Chae, J., Yun, H. S., & Poland, A. I. (1998d). *Astrophys. J. Supp.* **114**, 151.
Cheng, C.-C., Feldman, U., & Doschek, G. A. (1979). *Astrophys. J.* **233**, 736.
Cornille, M., Dubau, J., Faucher, P., Bely-Dubau, F., & Blancard, C. (1994). *Astr. Astrophys. Supp.* **105**, 77.
Cowan, R. D. (1981). *The Theory of Atomic Structure and Spectra*. Berkeley, Los Angeles, London: University of California Press.
Craig, I. J. D., & Brown, J. C. (1976). *Astr. Astrophys.* **49**, 239.
Cranmer, S. R., *et al.* (1999). *Astrophys. J.* **511**, 481.
Culhane, J. L., & Phillips, K. J. H. (1970). *Solar Phys.* **11**, 117.
Culhane, J. L., Sanford, P. W., Shaw, M. L., *et al.* (1969). *Mon. Not. R. Astr. Soc.* **145**, 435.
Culhane, J. L., Hiei, E., Doschek, G. A., *et al.* (1991). *Solar Phys.* **136**, 89.
Culhane, J. L., Harra, L. K., James, A. M., *et al.* (2007). *Solar Phys.* **243**, 19.
Cuntz, M., Rammacher, W., & Musielak, Z. E. (2007). *Astrophys. J.* **657**, L57.
Curdt, W., & Heinzel, P. (1998). *Astrophys. J.* **503**, L95.
Curdt, W., Brekke, P., Feldman, U., *et al.* (2001). *Astr. Astrophys.* **375**, 591.
Curdt, W., Landi, E., & Feldman, U. (2004). *Astr. Astrophys.* **427**, 1045.
DeForest, C. E., Hoeksema, J. T., Gurman, J. B., *et al.* (1997). *Solar Phys.* **175**, 393.
DeForest, C. E., & Gurman, J. B. (1998). *Astrophys. J.* **501**, L217.
Delaboudinière, J.-P., Artzner, G. E., Brunard, J., *et al.* (1995). *Solar Phys.* **162**, 291.
Del Zanna, G., & Mason, H. E. (2003). *Astr. Astrophys.* **406**, 1089.
Del Zanna, G., Bromage, B. J. I., Landi, E., & Landini, M. (2001) *Astr. Astrophys.* **379**, 708.
de Moortel, I., Ireland, J., Walsh, R. W., & Hood, A. W. (2002). *Solar Phys.* **209**, 61.
Dennis, B. R. (1985). *Solar Phys.* **100**, 465.

Dere, K. P. (1978). *Astrophys. J.* **221**, 1062.

Dere, K. P., & Mason, H. E. (1993). *Solar Phys.* **144**, 217.

Dere, K. P., Bartoe, J.-D. F., Brueckner, G. E., Cook, J. W., & Socker, D. G. (1987). *Solar Phys.* **114**, 223.

Dere, K. P., Bartoe, J.-D. F., & Brueckner, G. E. (1989). *Solar Phys.* **123**, 41.

Dere, K. P., Landi, E., Mason, H. E., Monsignori Fossi, B. C., & Young, P. R. (1997). *Astr. Astrophys. Supp.* **125**, 149.

Donnelly, R. F., & Unzicker, A. (1974). NOAA Tech. Memo. ELR SEL-72.

Doschek, G.A. (1972). *Space Sci. Rev.* **13**, 765.

Doschek, G. A. (1983). *Solar Phys.* **86**, 9.

Doschek, G. A., & Feldman, U. (1981). *Astrophys. J.* **251**, 792.

Doschek, G. A., & Feldman, U. (2000). *Astrophys. J.* **529**, 599.

Doschek, G. A., Meekins, J. F., Kreplin, R. W., Chubb, T. A., & Friedman, H. (1971). *Astrophys. J.* **170**, 573.

Doschek, G. A., Feldman, U., & Bohlin, J. D. (1976a). *Astrophys. J.* **205**, L177.

Doschek, G. A., Feldman, U., VanHoosier, M. E., & Bartoe, J.-D. F. (1976b). *Astrophys. J. Supp.* **31**, 417.

Doschek, G. A., Kreplin, R. W., & Feldman, U. (1979). *Astrophys. J. Lett.* **233**, L157.

Doschek, G. A., Feldman, U., Kreplin, R. W., & Cohen, L. (1980). *Astrophys. J.* **239**, 725.

Doschek, G. A., Feldman, U., Landecker H. R., & McKenzie, D. L. (1981). *Astrophys. J.* **249**, 372.

Doschek, G. A., Feldman, U., & Seely, J. F. (1985). *Mon. Not. R. Astr. Soc.* **217**, 317.

Doschek, G. A., Strong, K. T., & Tsuneta, S. (1995). *Astrophys. J.* **440**, 370.

Doyle, J. G., Madjarska, M. S., Roussev, I., Teriaca, L., & Giannikakis, J. (2002). *Astr. Astrophys.* **396**, 255.

Drake, J. F. (1971). *Solar Phys.* **16**, 152.

Dubau, J. (1994). *Atom. Data Nucl. Data Tab.* **57**, 21.

Dubau, J., Gabriel, A. H., Loulergue, M., Steenman-Clark, L., & Volonté, S. (1981). *Mon. Not. R. Astr. Soc.* **195**, 705.

Dubau, J., Cornille, M., Jacquemot, S., & Bely-Dubau, F. (1994). *X-ray Solar Physics from Yohkoh*, Tokyo: Universal Academy Press, Inc., p. 135.

Dupree, A. K., Huber, M. C. E., Noyes, R. W., *et al.* (1973). *Astrophys. J.* **182**, 321.

Dwivedi, B. N., Curdt, W., & Wilhelm, K. (1999). *Astrophys. J.* **517**, 516.

Dyall, K. G., Grant, I. P., Johnson, C. T., Parpia, F. A., & Plummer, E. P. (1989). *Computer Physics Comm.* **55**, 425.

Eissner, W. (1998). *Computer Physics Comm.* **114**, 295.

Eissner, W., Jones, M., & Nussbaumer, H. (1974). *Computer Physics Comm.* **8**, 274.

Emslie, A. G. (1985). *Solar Phys.* **98**, 281.

Emslie, A. G., Phillips, K. J. H., & Dennis, B. R. (1986). *Solar Phys.* **103**, 89.

Fano, U. (1961). *Phys. Rev.* **124**, 1866.

Favata, F., Flaccomio, E., Reale, F., *et al.* (2005). *Astrophys. J. Supp.*, **160**, 469.

Fawcett, B. C., Jordan, C., Lemen, J. R., & Phillips, K. J. H. (1987). *Mon. Not. R. Astr. Soc.* **225**, 1013.

Feldman, U. (1983). *Astrophys. J.* **275**, 367.

Feldman, U. (1987). *Astrophys. J.* **320**, 426.

Feldman, U. (1992). *Physica Scripta* **46**, 202.

Feldman, U. (1993). *Astrophys. J.* **411**, 896.

Feldman, U. (1996). *Phys. Plasmas*, **3**, 3203.

Feldman, U., & Doschek, G. A. (1977). *Astrophys. J.* **212**, 913.

Feldman, U., & Doschek, G. A. (1978a). *Astrophys. J. Supp.* **37**, 443.

Feldman, U., & Doschek, G. A. (1978b). *Astr. Astrophys.* **65**, 215.

Feldman, U., & Laming, J. M. (1994). *Astrophys. J.* **434**, 370.

Feldman, U., & Laming, J. M. (2000). *Phys. Scripta* **61**, 222.

Feldman, U., & Widing, K. G. (1990). *Astrophys. J.* **363**, 292.

Feldman, U., & Widing, K. G. (1993). *Astrophys. J.* **414**, 381.

Feldman, U., & Widing, K. G. (2003). *Space Sci. Rev.* **107**, 665.

Feldman, U., & Widing, K. G. (2007). *Space Sci. Rev.* **130**, 115.

Feldman, U., Doschek, G. A., Nagel, D. J., Cowan, R. D., & Whitlock, R. R. (1974). *Astrophys. J.* **192**, 213.

Feldman, U., Doschek, G. A., & Patterson, N. P. (1976a). *Astrophys. J.* **209**, 270.

Feldman, U., Doschek, G. A., Vanhoosier, M. E., & Purcell, J. D. (1976b). *Astrophys. J. Supp.* **31**, 445.

Feldman, U., Doschek, G. A., Prinz, D. K., & Nagel, D. J. (1976c). *Appl. Phys.* **47**, 1341.

Feldman, U., Doschek, G. A., & Mariska, J. T. (1979). *Astrophys. J.* **229**, 369.

Feldman, U., Doschek, G. A., & Kreplin, R. W. (1980a). *Astrophys. J.* **238**, 365.

Feldman, U., Doschek, G. A., Kreplin, R.W., & Mariska, J. T. (1980b). *Astrophys. J.* **241**, 1175.

Feldman, U., Doschek, G. A., & McKenzie, D. L. (1984). *Astrophys. J.* **276**, L53.
Feldman, U., Purcell, J. D., & Dohne, B. C. (1987). *Atlas of Extreme Ultraviolet Spectrograms from 170 to 625 Angstroms*, NRL 90-4100, Washington: NRL.
Feldman, U., Hiei, E., Phillips, K. J. H., Brown, C. M., & Lang, J. (1994). *Astrophys. J.* **421**, 843.
Feldman, U., Seely, J. F., Doschek, G. A., *et al.* (1995). *Astrophys. J.* **446**, 860.
Feldman, U., Doschek, G. A., Behring, W. E., & Phillips, K. J. H. (1996). *Astrophys. J.* **460**, 1034.
Feldman, U., Doschek, G. A., & Klimchuk, J. A. (1997a). *Astrophys. J.* **474**, 511.
Feldman, U., Behring, W. E., Curdt, W., *et al.* (1997b). *Astrophys. J. Supp.* **113**, 195.
Feldman, U., Schühle, U., Widing, K. G., & Laming, J. M. (1998). *Astrophys. J.* **505**, 999.
Feldman, U., Laming, J. M., Doschek, G. A., Warren, H. P., & Golub, L. (1999a)., *Astrophys. J. Letters* **511**, L61.
Feldman, U., Doschek, G. A., Schühle, U., & Wilhelm, K. (1999b). *Astrophys. J.* **518**, 500.
Feldman, U., Landi, E., Doschek, G. A., Dammasch, I., & Curdt, W. (2003a). *Astrophys. J.* **593**, 1226.
Feldman, U., Dammasch, I. E., Wilhelm, K., Lemaire, P., & Hassler, D. (2003b). *Images of the Solar Upper Atmosphere from SUMER on SOHO*, ESA SP-1274, Noordwijk.
Feldman, U., Dammasch, I., Landi, E., & Doschek, G.A. (2004). *Astrophys. J.* **609**, 439.
Feldman, U., Landi, E., & Laming, J. M. (2005a). *Astrophys. J.* **619**, 1142.
Feldman, U., Landi, E., Doschek, G. A., & Mariska, J. T. (2005b). *Astr. Astrophys.* **441**, 1211.
Feldman, U., Landi, E., & Doschek, G. A. (2007). *Astrophys. J.* **660**, 1674.
Fineschi, S., van Ballegooijen, A., & Kohl, J. L. (1999). ESA SP-446, p. 317.
Fink, R. W., Jopson, R. C., Mark, H., & Swift, C. D. (1966). *Rev. Mod. Phys.* **38**, 513.
Fisher, R., & Guhathakurta, M. (1995). *Astrophys. J.* **447**, L139.
Fludra, A., & Schmelz, J. T. (1998). *Astr. Astrophys.* **348**, 286.
Fludra, A., Bentley, R. D., Culhane, J. L., *et al.* (1990). In *Proceedings of the EPS 6th European Solar Meeting*, p. 266.
Fludra, A., Culhane, J. L., Bentley, R. D., *et al.* (1993). *Adv. Space Res.* **13**, 395.
Fludra, A., Saba, J. L. R., Hénoux, J.-C., *et al.* (1999). In *The Many Faces of the Sun*, eds. K. T. Strong, J. L. R. Saba, B. M. Haisch, & J. T. Schmelz, p. 89.
Fontenla, J., Reichmann, E. J., & Tandberg-Hanssen, E. (1988). *Astrophys. J.* **329**, 464.
Fontenla, J. M., Avrett, E. H., & Loesser, R. (1990). *Astrophys. J.* **355**, 700.
Forbes, T. G., & Acton, L. W. (1996). *Astrophys. J.* **459**, 330.
Fossum, A., & Carlsson, M. (2005). *Nature* **435**, 919.
Foukal, P. V., Huber, M. C. E., Noyes, R. W., *et al.* (1974). *Astrophys. J.* **193**, L143.
Froese Fischer, C. (2000). *Computer Physics Comm.* **128**, 635.
Gabriel, A. H. (1972). *Mon. Not. R. Astr. Soc.* **160**, 99.
Gabriel, A. H. (1976). *Phil. Trans. R. Soc. Lond. A* **281**, 339.
Gabriel, A. H., & Jordan, C. (1969). *Mon. Not. R. Astr. Soc.* **145**, 241.
Gabriel, A. H., & Jordan, C. (1973). *Mon. Not. R. Astr. Soc.* **186**, 327.
Gabriel, A. H., & Phillips, K. J. H. (1979). *Mon. Not. R. Astr. Soc.* **189**, 319.
Gabriel, A. H., *et al.* (1971). *Astrophys. J.* **169**, 595.
Gallagher, P. T., Phillips, K. J. H., Harra-Murnion, L. K., & Keenan, F. P. (1998). *Astr. Astrophys.* **335**, 733.
Gallagher, P. T., Phillips, K. J. H., Harra-Murnion, L. K., Baudin, F., & Keenan, F. P. (1999a). *Astr. Astrophys.* **348**, 251.
Gallagher, P. T., Mathioudakis, M., Keenan, F. P., Phillips, K. J. H., & Tsinganos, T. (1999b). *Astrophys. J.* **524**, L133.
Garcia, H. A., & McIntosh, P. S. (1992). *Solar Phys.* **141**, 109.
Gebbie, K. B., *et al.* (1981). *Astrophys. J.* **251**, L115.
Georgakilas, A. A., Muglach, K., & Christopoulou, E. B. (2002). *Astrophys. J.* **576**, 561.
Gingerich, O., Noyes, R. W., Kalkofen, W., & Cuny, Y. (1971). *Solar Phys.* **18**, 347.
Goldman, S. P., & Drake, G. W. F. (1981). *Phys. Rev. A* **24**, 183.
Goldschmidt, V. M., (1937). *Skrifter Norske Videnskapt-Akad. Oslo, Math. Naturew. Kl.* **4**, 99.
Golub, L., Krieger, A. S., Harvey, J. W., & Vaiana, G. S. (1977). *Solar Phys.* **53**, 111.
Grevesse, N., & Sauval, A. J. (1991). *The Infrared Spectral Region of Stars*, eds. C. Jaschek, & T. Andrillat, Cambridge University Press.
Grevesse, N., & Sauval, A. J. (1998). *Space Sci. Rev.* **85**, 161.
Grevesse, N., Asplund, M., & Sauval, A. J. (2005). *EAS Publications Series* **17**, 21.
Griffin, W. G., & McWhirter, R. W. P. (1962). *Proc. Conf. Optical Instruments and Techniques*, London: Chapman & Hall, p. 14.

Grineva, Yu. I., Karev, V. I., Korneyev, V. V., *et al.* (1973). *Solar Phys.* **29**, 441.
Gu, M. F. (2003). *Astrophys. J.* **582**, 1241.
Gudiksen, B. V., & Nordlund, A. (2005). *Astrophys. J.* **618**, 1020.
Haisch, B., & Schmitt, J. H. M. M. (1996). *Pub. Astr. Soc. Pac.* **108**, 113.
Handy, B. N., *et al.* (1999). *Solar Phys.* **187**, 229.
Harra-Murnion, L. K., *et al.* (1996). *Astr. Astrophys.* **306**, 670.
Harra, L. K., Gallagher, P. T., & Phillips, K. J. H. (2000). *Astr. Astrophys.* **362**, 371.
Harrison, R. A. (1997). *Solar Phys.* **175**, 467.
Harrison, R. A., Sawyer, E. C., Carter, M. K., *et al.* (1995). *Solar Phys.* **162**, 233.
Hassler, D., Wilhelm, K., Lemaire, P., & Schühle, U. (1997). *Solar Phys.* **175**, 375.
Héroux, L. (1964). *Proc. Phys. Soc.* **83**, 121.
Hibbert, A. (1975). *Computer Physics Comm.* **9**, 141.
Hill, K. W., Von Goeler, S., Bitter, M., *et al.* (1979). *Phys. Rev. A* **19**, 1770.
Holweger, H., & Müller, E. A. (1974). *Solar Phys.* **39**, 19.
Hudson, H. S. (1991). *Solar Phys.* **133**, 357.
Hunter, W. R. (1988). *Experimental Methods in the Physical Sciences*, New York: Academic Press, Vol. 31, p. 205.
Hurford, G. J., Schmahl, E. J., Schwartz, R. A., *et al.* (2002). *Solar Phys.* **210**, 61.
Innes, D. E., Brekke, P., Germerott, D., & Wilhelm, K. (1997a). *Solar Phys.* **175**, 341.
Innes, D. E., Inhester, B., Axford, W. I., & Wilhelm, K. (1997b). *Nature* **386**, 811.
Innes, D. E., Curdt, W., Schwenn, R., *et al.* (2001). *Astrophys. J. Lett.* **549**, L249.
Jakimiec, J., Tomczak, M., Falewicz, R., Phillips, K. J. H., & Fludra, A. (1997). *Astr. Astrophys.* **334**, 1112.
Jefferies, J. T., Orrall, F. Q., & Zirker, J. B. (1972). *Solar Phys.* **22**, 307.
Jordan, C. (1969). *Mon. Not. R. Astr. Soc.* **142**, 501.
Jordan, C., Brueckner, G. E., Bartoe, J.-D. F., Sandlin, G. D., & VanHoosier, M. E. (1978). *Astrophys. J.* **226**, 687.
Jordan, C., Ayres, T. R., Linsky, J. L., & Simon, T. (1987). *Mon. Not. R. Astr. Soc.*, **225**, 903.
Judge, P. G., Hubeny, V., & Brown, J. C. (1997). *Astrophys. J.* **475**, 275.
Kahler, S. W., Krieger, A. S., & Vaiana, G. S. (1975). *Astrophys. J. Lett.* **199**, L58.
Karzas, W. J., & Latter, R. (1961). *Astrophys. J. Supp.* **6**, 167.
Kashyap, V., & Drake, J. J. (2000). *Bull. Astr. Soc. India* **28**, 475.
Kastner, S. O., Neupert, W. M., & Swartz, M. (1974). *Astrophys. J.* **191**, 261.
Keenan, F. P., & McCann, S. M. (1987). *Solar Phys.* **109**, 31.
Keenan, F. P., Phillips, K. J. H., Harra, L. K., Conlon, E. S., & Kingston, A. E. (1992). *Astrophys. J.* **393**, 815.
Kelly, R. L. (1987). Atomic and ionic spectrum lines below 2000 Angstroms: Parts 1, 2, 3. *J. Phys. Chem. Ref. Data* **16**, Suppl. No. 1.
Kępa, A., Sylwester, J., Sylwester, B., & Siarkowski, M. (2004). *Proc. IAU Symp.*, **223**, 461.
Kingston, A. E., Doyle, J. G., Dufton, P. L., & Gurman, J. B. (1982). *Astrophys. J.* **81**, 47.
Kohl, J. L., Esser, R., Gardner, I. D., *et al.* (1995). *Solar Phys.* **162**, 313.
Kohl, J. L., Noci, G., Cranmer, S. R., & Raymond, J. C. (2006). *Astr. Astrophys. Rev.* **13**, 31.
Kosugi, T., Makishima, K., Murakami, T., *et al.* (1991). *Solar Phys.* **136**, 17.
Koutchmy, S. (1977). *Solar Phys.* **51**, 399.
Koutchmy, S., Hara, H., Suematsu, Y., & Reardon, K. (1997). *Astr. Astrophys.* **320**, L33.
Kramers, H. A. (1923). *Phil. Mag.* **46**, 836.
Krucker, S., Benz, A. O., Bastion, T. S., & Acton, L. W. (1997). *Astrophys. J.* **488**, 499.
Kurochka, L. N. (1970). *Astr. Zh.* **44**, 368.
Kurochka, L. N., & Maslennikova, L. B. (1970). *Solar Phys.* **11**, 33.
Laming, J. M., Drake, J. J., & Widing, K. G. (1995). *Astrophys. J.* **443**, 416.
Landi, E., & Feldman, U. (2004). *Astrophys. J.* **611**, 537.
Landi, E., & Landini M. (1997). *Astr. Astrophys.* **327**, 1230.
Landi, E., & Landini, M. (2004). *Astrophys. J.* **608**, 1133.
Landi, E., & Phillips, K. J. H. (2005). *Astrophys. J. Supp.* **160**, 286.
Landi, E., Landini, M., Pike, C. D., & Mason, H. E. (1997). *Solar Phys.* **175**, 553.
Landi, E., Mason, H. E., Lemaire, P., & Landini, M. (2000). *Astron. Astrophys.* **357**, 743.
Landi, E., Doron, R., Feldman, U., & Doschek, G. A. (2001). *Astrophys. J.* **556**, 912.
Landi, E., Feldman, U., & Dere, K. P. (2002). *Astrophys. J. Supp.* **139**, 281.
Landi, E., Feldman, U., Innes, D. E., & Curdt, W. (2003). *Astrophys. J.* **582**, 519.
Landi, E., Storey, P. J., & Zeippen, C. J. (2004). *Astrophys. J.* **607**, 640.

Landi, E., Del Zanna, G., Young, P. R., Dere, K. P., Mason, H. E., & Landini, M. (2006a) *Astrophys. J. Supp.* **162**, 261.

Landi, E., Feldman, U., & Doschek, G. A. (2006b). *Astrophys. J.* **643**, 1258.

Landini, M., & Monsignori Fossi, B. C. (1975). *Astr. Astrophys.* **42**, 213.

Landini, M., & Monsignori Fossi, B. C. (1981). *Astr. Astrophys.* **102**, 391.

Lee, T. N. (1974). *Astrophys. J.* **190**, 467.

Lemen, J. R., Phillips, K. J. H., Cowan, R. D., Hata, J., & Grant, I. P. (1984). *Astr. Astrophys.* **135**, 313.

Lenz, D. D, De Luca, E. E., Golub, L., *et al.* (1999). *Astrophys. J.* **517**, L155.

Levine, R. H. (1974). *Astrophys. J.*, **190**, 447.

Lima, J. J. G., & Priest, E. R. (1993). *Astr. Astrophys.* **268**, 641.

Lin, R. P., Schwartz, R. A., Pelling, R. M., & Hurley, K. C. (1981). *Astrophys. J.* **251**, L109.

Lin, R. P., Schwartz, R. A., Kane, S. R., Pelling, R. M., & Hurley, K. C. (1984). *Astrophys. J.* **283**, 421.

Lin, R. P., Dennis, B. R., Hurford, G. J., *et al.* (2002). *Solar Phys.* **210**, 3.

Linford G. A., & Wolfson, C. J. (1988). *Astrophys. J.* **331**, 1036.

Linhui, S., Holman, G. D., Dennis, B. R., *et al.* (2002). *Solar Phys.* **200**, 245.

Lites, B. W., Bruner, E. C., Chipman, E. G., *et al.* (1976). *Astrophys. J.* **210**, L111.

Lites, B. W., Thomas, J. H., Bogdan, T. J., & Cally, P. S. (1998). *Astrophys. J.* **497**, 464.

Lodders, K. (2003). *Astrophys. J.* **591**, 1220.

Madjarska, M. S., & Doyle, J. G. (2003). *Astr. Astrophys. Supp.* **403**, 731.

Makishima, K. (1974). In *Proc. Hinotori Symp. on Solar Flares*, eds. Y. Tanaka, *et al.*, Tokyo: ISAS, p. 120.

Malinovsky, M., & Héroux, L. (1973). *Astrophys. J.* **181**, 1009.

Maltby, P., Avrett, E. H., Carlsson, M., *et al.* (1986). *Astrophys. J.* **306**, 284.

Maltby, P., Brynildsen, N., Brekke, P., *et al.* (1998). *Astrophys. J.* **496**, L117.

Manson, J. E. (1972). *Solar Phys.* **27**, 107.

Mariska, J. T. (1992). *The Solar Transition Region*, Cambridge University Press.

Mariska, J. T. (1994). *Astrophys. J.* **434**, 756.

Mariska, J. T., Doschek, G. A., & Bentley, R. D. (1993). *Astrophys. J.* **419**, 418.

Mason, H. E., Doschek, G. A., Feldman, U., & Bhatia, A. K. (1979). *Astr. Astrophys.* **73**, 74.

Masuda, S., Kosugi, T., Hara, H., Tsuneta, S., & Ogawara, Y. (1994). *Nature* **371**, 495.

Matthews, S. A., & Harra-Murnion, L. K. (1997). *Solar Phys.* **175**, 541.

Mazzotta, P., Mazzitelli, G., Colafrancesco, S., & Vittorio, N. (1998). *Astr. Astrophys. Supp.* **133**, 403.

McIntosh, S. W. (2000). *Astrophys. J.* **533**, 1043.

McComas, D. J., Riley, P., Gosling, J. T., Balogh, A., & Forsyth, R. (1998). *J. Geophys. Res.* **103**, 1955.

McKenzie, D. L., & Feldman, U. (1992). *Astrophys. J.* **389**, 764.

McKenzie, D. I., & Feldman, U. (1994). *Astrophys. J.* **420**, 892.

McKenzie, D. L., & Landecker, P. B. (1982). *Astrophys. J.* **254**, 309.

McKenzie, D. L., Broussard, R. M., Landecker, H. R., *et al.* (1980a). *Astrophys. J. Lett.* **238**, L43.

McKenzie, D. L., Landecker, P. B., Broussard, R. M., *et al.* (1980b). *Astrophys. J.* **241**, 409.

McKenzie, D. L., Landecker, P. B., Feldman, U., & Doschek, G. A. (1985). *Astrophys. J.* **289**, 849.

McTiernan, J. M., Fisher, G. H., & Li, P. (1999). *Astrophys. J.*, **514**, 472.

Metropolis, N., Rosenbluth, M. N., Teller, A. H., & Teller, E. (1953). *J. Chem. Phys.* **21**, 1087.

Mewe, R., Kaastra, J. S., & Liedahl, D. A. (1995). *Legacy* **6**, 16.

Meyer, J.-P. (1985). *Astrophys. J. Supp.* **57**, 173.

Mihalas, D. (1970). *Stellar Atmospheres*, New York: W. H. Freeman and Company.

Minaeva, L. A. (1968). *Astr. Zh.* **45**, 578.

Minaeva, L. A., & Sobel'man, I. I. (1968). *J. Quant. Spectr. Rad. Transfer* **8**, 783.

Mitchell, W. E. (1981). *Solar Phys.* **69**, 391.

Monsignori Fossi, B. C., Landini, M., Thomas, R. J., & Neupert, W. M. (1994). *Adv. Space Res.* **14**, 163.

Moore, R. L., Tang, F., Bohlin, J. D., & Golub, L. (1977). *Astrophys. J.* **218**, 286.

Morita, S., & McIntosh, S. W. (2005). *Astr. Soc. Pac. Conf. Ser.* **346**, 317.

Nakariakov, V. M., & Ofman, L. (2001). *Astr. Astrophys.* **372**, L53.

Neal, R. M. (1993). Technical Report CRG-TR-93-1, Dept. Comput. Sci., University of Toronto.

Neugebauer, M. (1999). *Rev. Geophys.* **37**, 107.

Neupert, W. M., & Swartz, M. (1970). *Astrophys. J.* **160**, 189.

Neupert, W. M., Gates, W., Swartz, M., & Young, R. (1967). *Astrophys. J.* **149**, L79.

Neupert, W. M., Swartz, M., & Kastner, S. O. (1973). *Solar Phys.* **31**, 171.

Neupert, W. M., Epstein, G. L., Thomas, R. J., & Thompson, W. T. (1992), *Solar Phys.* **137**, 87.

Neupert, W. M., Newmark, J., Delaboudinière, J.-P., *et al.* (1998). *Solar Phys.* **183**, 305.
Noci, G., Kohl, J. L., & Withbroe, G. L. (1987). *Astrophys. J.* **315**, 706.
Ofman, L., Romoli, M., Poletto, G., Noci, G., & Kohl, J. L. (1997). *Astrophys. J.* **491**, L111.
Pallavicini, R., Serio, S., & Vaiana, G. S. (1977). *Astrophys. J.* **216**, 108.
Parenti, S., Vial, J.-C., & Lemaire, P. (2005). *Astr. Astrophys.* **443**, 679.
Parker, E. N. (1988). *Astrophys. J.* **330**, 474.
Parmar, A. N., Wolfson, C. J., Culhane, J. L., *et al.* (1984). *Astrophys. J.* **279**, 866.
Parnell, C. E., & Jupp, P. E. (2000). *Astrophys. J.* **529**, 554.
Peter, H., & Judge, P. G. (1999). *Astrophys. J.* **522**, 1148.
Petrasso, R. D., *et al.* (1975). *Astrophys. J. Lett.* **199**, L127.
Phillips, K. J. H. (1991). *Phil. Trans. R. Soc. Lond. A* **336**, 461.
Phillips, K. J. H. (2004). *Astrophys. J.* **605**, 921.
Phillips, K. J. H., & Feldman, U. (1991). *Astrophys. J.* **379**, 401.
Phillips, K. J. H., & Feldman, U. (1995). *Astr. Astrophys.* **304**, 563.
Phillips, K. J. H., & Feldman, U. (1997). *Astrophys. J.* **477**, 502.
Phillips, K. J. H., & Neupert, W. M. (1973). *Solar Phys.* **32**, 209.
Phillips, K. J. H., & Zirker, J. B. (1977). *Solar Phys.* **53**, 41.
Phillips, K. J. H., Neupert, W. M, & Thomas, R. J. (1974). *Solar Phys.* **36**, 383.
Phillips, K. J. H., Leibacher, J. W., Wolfson, C. J., *et al.* (1982). *Astrophys. J.* **256**, 774.
Phillips, K. J. H., Lemen, J. R., Cowan, R. D., Doschek, G. A., & Leibacher, J. W. (1983). *Astrophys. J.* **265**, 1120.
Phillips, K. J. H., Harra, L. K., Keenan, F. P., Zarro, D. M., & Wilson, M. (1993). *Astrophys. J.* **419**, 426.
Phillips, K. J. H., Pike, C. D., Lang, J., Watanabe, T., & Takahashi, M. (1994a). *Astrophys. J.* **435**, 888.
Phillips, K. J. H., Keenan, F. P., Harra, L. K., *et al.* (1994b). *J. Phys. B: At. Mol. Opt. Phys.* **27**, 1939.
Phillips, K. J. H., Bhatia, A. K., Mason, H. E., & Zarro, D. M. (1996). *Astrophys. J.* **446**, 519.
Phillips, K. J. H., Greer, C. J., Bhatia, A. K., *et al.* (1997). *Astr. Astrophys.* **324**, 381.
Phillips, K. J. H., Mathioudakis, M., Huenemoerder, D. P., Williams, D. R., Phillips, M. E., & Keenan, F. P. (2001). *Mon. Not. R. Astr. Soc.* **325**, 1500.
Phillips, K. J. H., Sylvwester, J., Sylwester, B., & Landi, E. (2003). *Astrophys. J.* **589**, L113.
Phillips, K. J. H., Feldman, U., & Harra L. K. (2005a). *Astrophys. J.* **634**, 641.
Phillips, K. J. H., Chifor, C., & Landi, E. (2005b). *Astrophys. J.* **626**, 1110.
Phillips, K. J. H., Rainnie, J. A., Harra, L. K., *et al.* (2005c). *Astr. Astrophys.* **416**, 765.
Phillips, K. J. H., Chifor, C., & Dennis, B. R. (2006a). *Astrophys. J.* **647**, 1480.
Phillips, K. J. H., Dubau, J., Sylwester, J., & Sylwester, B. (2006b). *Astrophys. J.* **638**, 1154.
Pike, C. D., Phillips, K. J. H., Lang, J., *et al.* (1996). *Astrophys. J.* **464**, 487.
Pneuman, G. W., & Kopp, R. A. (1978). *Solar Phys.*, **57**, 49.
Porter, J. G., Moore, R. L., Reichmann, E. J., Engvold, O., & Harvey, K. L. (1987). *Astrophys. J.* **323**, 380.
Pottasch, S. R. (1963). *Astrophys. J.* **137**, 945.
Pottasch, S. R. (1964). *Space Sci. Rev.* **3**, 816.
Pradhan, A. K., Norcross, D. W., & Hummer, D. G. (1981). *Astrophys. J.* **246**, 1031.
Preś, P., & Phillips, K. J. H. (1999). *Astrophys. J.* **510**, L73.
Priest, E. R., Foley, C., Heyvaerts, J., *et al.* (2000). *Astrophys. J.* **539**, 1002.
Rapley, C. G., Culhane, J. L., Acton, L. W., *et al.* (1977). *Rev. Sci. Instrum.*, **48**, 123.
Raymond, J. C. (2004). In *The Sun and the Heliosphere as an Integrated System*, eds. G. Poletto, & S. T. Suess, Boston/Dordrecht/London: Kluwer Academic Publishers, p. 353.
Raymond, J. C., & Doyle, J. G. (1981). *Astrophys. J.* **247**, 686.
Raymond, J. C., Kohl, J. L., Noci, G., *et al.* (1997). *Solar Phys.* **175**, 645.
Reale, F. (2002). *Astrophys. J.* **580**, 56.
Reale, F., Peres, G., & Serio, S. (1997). *Astr. Astrophys.* **318**, 506.
Reames, D. V. (1994). *Adv. Space Res.* **14**, 177.
Reames, D. V. (1998). *Space Sci. Rev.* **85**, 327.
Reeves, E. M. (1976). *Solar Phys.* **46**, 53.
Robbrecht, E., Verwichte, R., Berghmans, D., *et al.* (2001). *Astr. Astrophys.* **370**, 591.
Rosner, R., Tucker, W. H., & Vaiana, G. S. (1978). *Astrophys. J.* **247**, 686.
Russell, H. N. (1929). *Astrophys. J.* **70**, 11.
Rutherford, E. (1911). *Phil. Mag.* **21**, 669.
Saba, J. L. R. (1995). *Adv. Space Res.* **15**, 13.

Saba, J. L. R., & Strong, K. T. (1992). In *Proc. 1st SOHO Workshop*, ed. V. Domingo, Noordwijk: ESA, p. 347.
Sakurai, J. J. (1967). *Advanced Quantum Mechanics*, Reading, MA; Menlo Park, CA; London, Ontario: Addison-Wesley Publishing Co.
Sampson, D. H., & Clark, R. E. H. (1980). *Astrophys. J. Supp.* **44**, 169.
Samson, J. A. R. (1967). *Techniques of Vacuum Ultraviolet Spectroscopy*, New York, London, Sydney: John Wiley & Sons, Inc.
Sandlin, G. D., Brueckner, G. E., Scherrer, V. E., & Tousey, R. (1976). *Astrophys. J. Lett.* **205**, L47.
Sandlin, G. D., Bartoe, J.-D. F., Brueckner, G. E., Tousey, R., & Vanhoosier, M. E. (1986). *Astrophys. J. Supp.* **61**, 801.
Schmelz, J. T., & Fludra, A. (1993). *Adv. Space Res.* **13**, 325.
Schmelz, J. T., Nasraoui, K., Roames, J. K., Lippner, L. A., & Garst, J. W. (2005). *Astrophys. J.* **627**, L81.
Schrijver, C., Title, A. M., Harvey, K. L., *et al.* (1998). *Nature* **394**, 152.
Seaton, M. J. (1964). *Mon. Not. R. Astr. Soc.* **127**, 191.
Seely, J. F., Feldman, U., Schühle, U., Wilhelm, K., Curdt, W., & Lemaire, P. (1997). *Astrophys. J.* **484**, L87.
Seely J. F., Brown, C. M., Windt, D. L., Donguy, S., & Kjornrattanawanich, B. (2004). *Appl. Optics* **43**, 1463.
Serio, S., Peres, G., Vaiana, G. S., *et al.* (1981). *Astrophys. J.* **243**, 288.
Sheeley, N. R., Jr. (1969). *Solar Phys.* **9**, 347.
Sheeley, N. R., Jr. (1995). *Astrophys. J.* **440**, 884.
Sheeley, N. R., Jr. (1996). *Astrophys. J.* **469**, 423.
Shibata, K., *et al.* (1992). *Pub. Astr. Soc. Japan* **44**, L173.
Shibata, K., Nitta, N., Strong, K. T., *et al.* (1994). *Astrophys. J.* **431**, L51.
Shimizu, T. (1995). *Pub. Astr. Soc. Japan* **47**, 251.
Shimizu, T., Tsuneta, S., Acton, L. W., Lemen, J. R., & Uchida, Y. (1992). *Pub. Astr. Soc. Japan* **44**, 147.
Shimizu, T., Tsuneta, S., Acton, L. W., Lemen, J. R., Ogawara, Y., & Uchida, Y. (1994). *Astrophys. J.* **422**, 906.
Shull, J. M., & Van Steenberg, M. (1982). *Astrophys. J. Supp.* **48**, 95 and *Astrophys. J. Supp.* **49**, 351.
Siarkowski, M., Sylwester, J., Jakimiec, J., & Tomczak, M. (1996). *Acta Astr.* **46**, 15.
Smith, R. K., Brickhouse, N. S., Liedahl, D. A., & Raymond, J. C. (2001). *Astrophys. J.* **556**, 91.
Spadaro, D., Antiochos, S. K., & Mariska, J. T. (1991). *Solar Phys.* **382**, 338.
Spicer, D. S., Feldman, U., Widing, K. G., & Rilee, M. (1998). *Astrophys. J.* **494**, 450.
Spitzer, L. (1962). *Physics of Fully Ionized Gases* (2nd edition). New York: Interscience Publishers, John Wiley.
Sterling, A. C., Doschek, G. A., & Feldman, U. (1993). *Astrophys. J.* **404**, 394.
Sterling, A. C., Doschek, G. A., & Pike, C. D. (1994). *Astrophys. J.* **435**, 898.
Sterling, A. C., Hudson, H. S., Lemen, J. R., & Zarro, D. A. (1997). *Astrophys. J. Supp.* **110**, 115.
Strong, K. T., Clafin, E. S., Lemen, J. R., & Linford, G. A. (1988). *Adv. Space Res.* **8**, 167.
Strong, K. T., Lemen, J. R., & Linford, G. A. (1991). *Adv. Space Res.* **11**, 151.
Summers, H. P. (1972). *Mon. Not. R. Astr. Soc.* **158**, 255.
Summers, H. P. (1974). *Mon. Not. R. Astr. Soc.* **169**, 663.
Sylwester, B., Sylwester, J., Bentley, R. D., & Fludra, A. (1990). *Solar Phys.* **126**, 177.
Sylwester, J., *et al.* (2002). In *Proc. SOLSPA: The Second Solar Cycle and Space Weather Euroconference*, Vico Equense, Italy, Sept. 2001 (ESA-SP477), p. 597.
Sylwester, J., Sylwester, B., Culhane, J. L., *et al.* (2003). In *Proc. ISCS 2003 Symposium*, Noordwijk (ESA-SP535), p. 733.
Sylwester, J., Gaicki, I., Kordylewski, Z., *et al.* (2005). *Solar Phys.* **226**, 45.
Tanaka, K. (1986). *Pub. Astr. Soc. Japan* **38**, 225.
Tanaka, K., Watanabe, T., Nishi, K., & Akita, K. (1982). *Astrophys. J.* **254**, L59.
Thomas, R. J., & Neupert, W. M. (1994). *Astrophys. J. Supp.* **91**, 461.
Thompson, A. M. (1990). *Astr. Astrophys.* **240**, 209.
Tousey, R., Austin, W. E., Purcell, D., & Widing, K. G. (1965). *Annales Astroph.* **28**, 755.
Tousey, R., Bartoe, J.-D. F., Brueckner, G. E., & Purcell, J. D. (1977). *Appl. Optics* **16**, 870.
Tsuneta, S., *et al.* (1991). *Solar Phys.* **136**, 37.
Tsuneta, S., Masuda, S., Kosugi, T., & Sato, J. (1997). *Astrophys. J.* **478**, 787.
Tu, C.-Y., Marsch, E., Wilhelm, K., Curdt, W. (1998). *Astrophys. J.* **503**, 475.
Ugarte-Urra, I., Doyle, J. G., & Del Zanna, G. (2005). *Astr. Astrophys.* **435**, 1169.
Underwood, J. H., Milligan, J. E., deLoach, A. C., & Hoover, R. B. (1977). *Appl. Optics* **16**, 858.
Vaiana, G. S., & Rosner, R. (1978). *Ann. Rev. Astr. Astrophys.* **16**, 393.
Vaiana, G. S., Van Speybroek, L., Zombeck, M. V., *et al.* (1977). *Space Sci. Instr.* **3**, 19.
Vainshtein, L. A., & Safronova, U. I. (1978). *Atomic Data and Nuclear Data Tables*, **21**, 49.

Vainshtein, L. A., & Safronova, U. I. (1985). *Physica Scripta*, **31**, 519.
Veigele, W. J. (1970). *Astrophys. J.* **162**, 357.
Vernazza, J. E., & Reeves, E. M. (1978). *Astrophys. J. Supp.* **37**, 485.
Vernazza, J. E., Avrett, E. H., & Loeser, R. (1981). *Astrophys. J. Supp.* **45**, 635.
von Steiger, R., Geiss, J., & Gloeckler, G. (1997). In *Cosmic Winds and the Heliosphere*, eds. J. R. Jokipii, C. P. Sonett, & M. S. Giampapa, Tucson: University of Arizona Press.
Wadsworth, F. L. O. (1896) *Astrophys. J.* **3**, 54.
Walker, A. B. C., Rugge, H. R., & Weiss, K. (1974). *Astrophys. J.* **188**, 423.
Wang, T. J., Solanki, S. L., Curdt, W., Innes, D. E., & Dammasch, I. E. (2002). *Astrophys. J.* **574**, L101.
Wang, T. J., Solanki, S., Curdt, W., *et al.* (2003). *Astr. Astrophys.* **446**, 1105.
Warren, H. P. (1999). *Solar Phys.* **190**, 363.
Weber, M. A., Schmelz, J. T., DeLuca, E. E., & Roames, J. K. (2005). *Astrophys. J.* **635**, L101.
Widing, K. G. (1997). *Astrophys. J.* **480**, 400.
Widing, K. G., & Cheng, C.-C., (1974). *Astrophys. J. Lett.* **194**, L111.
Widing, K. G., & Dere, K. P. (1977). *Solar Phys.* **55**, 431.
Widing, K. G., & Feldman, U. (1992). In *Solar Wind Seven*, eds. E. Marsch, & R. Schwenn, Oxford: Pergamon Press, p. 405.
Widing, K. G., & Feldman, U. (1995). *Astrophys. J.* **442**, 446.
Widing, K. G., & Feldman U. (2001). *Astrophys. J.* **555**, 426.
Widing, K. G., Feldman, U., & Bhatia, A. K. (1986). *Astrophys. J.* **308**, 982.
Widing, K. G., Landi, E., & Feldman, U. (2005). *Astrophys. J.* **622**, 1211.
Wikstøl, O., Judge, P. G., & Hansteen, V. (1997). *Astrophys. J.* **483**, 972.
Wilhelm, K., Curdt, W., Marsch, E., *et al.* (1995). *Solar Phys.* **162**, 189.
Wilhelm, K., Lemaire, P., Curdt, W., *et al.* (1997). *Solar Phys.* **170**, 75.
Wilhelm, K., Dwivedi, B. N., Marsch, E., & Feldman, U. (2004). *Space Sci. Rev.* **111**, 415.
Winebarger, A., Warren, H., van Ballegooijen, A., DeLuca, E. E., & Golub, L. (2002). *Astrophys. J.* **567**, L89.
Withbroe, G. L. (1975). *Solar Phys.* **45**, 301.
Withbroe, G. L., & Noyes, R. W. (1977). *Ann. Rev. Astr. Astrophys.* **15**, 363.
Withbroe, G. L., Kohl, J. L., Weiser, H., & Munro, R. H. (1982). *Space Sci. Rev.* **33**, 17.
Young, P. R. (2005). *Astr. Astrophys.* **439**, 361.
Young, P. R., & Mason, H. E. (1997). *Solar Phys.* **175**, 523.
Young, P. R., & Mason, H. E. (1998). *Space Sci. Rev.* **85**, 315.
Young, P. R., Landi, E., & Thomas, R. J. (1998) *Astr. Astrophys.* **329**, 291.
Young, P. R., Del Zanna, G., Landi, E., *et al.* (2003). *Astrophys. J. Supp.* **144**, 135.
Zhang, H., & Sampson, D. H. (1987). *Astrophys. J. Supp.* **63**, 487.
Zhitnik, I. A., Kuzin, S. V., Urnov, A. M., *et al.* (2005). *Astr. Lett.*, **31**, 37.

Index